D0554872

World Atlas of
GREAT APES
and their Conservation

Photograph by Sebastião Salgado/Amazonas images

Published in association with UNEP-WCMC by the University of California Press
University of California Press
Berkeley and Los Angeles, California
University of California Press, Ltd.
London, England

UNEP-WCMC
219 Huntingdon Road
Cambridge, CB3 0DL, UK
Tel: +44 (0) 1223 277314
Fax: +44 (0) 1223 277136
E-mail: info@unep-wcmc.org
Website: www.unep-wcmc.org

Clothbound edition ISBN: 0-520-24633-0

Cataloging-in-Publication data is on file with the Library of Congress.

Citation: Caldecott, J., Miles, L., eds (2005) *World Atlas of Great Apes and their Conservation.* Prepared at the UNEP World Conservation Monitoring Centre. University of California Press, Berkeley, USA.

World Atlas of
GREAT APES
and their Conservation

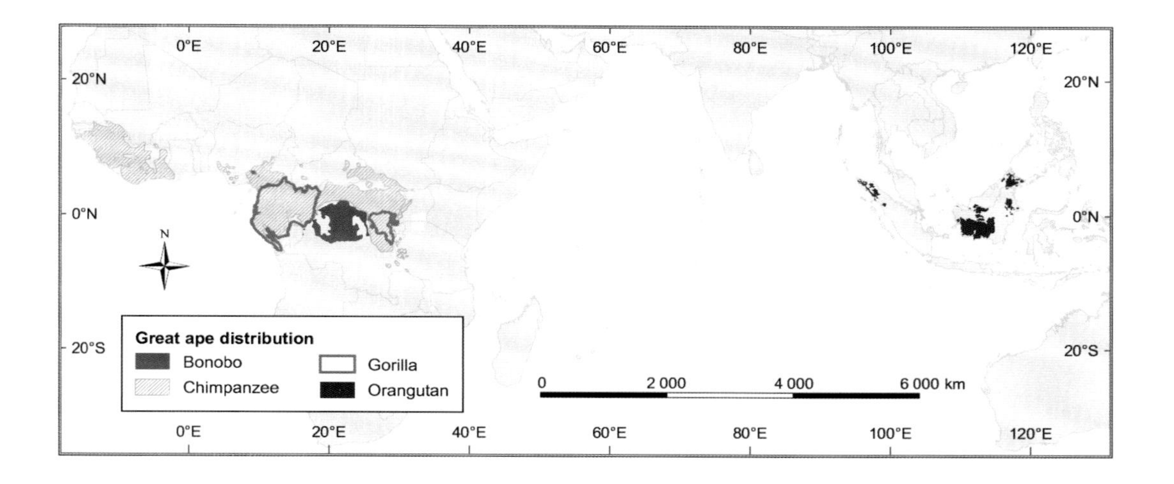

EDITED BY **JULIAN CALDECOTT** AND **LERA MILES**

Foreword by Kofi A. Annan

UNIVERSITY OF CALIFORNIA PRESS
Berkeley Los Angeles London

World Atlas of Great Apes and their Conservation

Prepared at
UNEP World Conservation
Monitoring Centre
219 Huntingdon Road
Cambridge CB3 0DL, UK
Website: www.unep-wcmc.org

Editors
Julian Caldecott
Lera Miles

Cartography
Lee Shan Khee
Matthew Doughty
Mary Edwards

Research assistant
Brigid Barry

Production editors
Helen de Mattos
Angela Jameson
Laura Kirby
Jane Lyons
Valerie Neal
Tim Osmond

Index
Jill Dormon

Layout
Raul López Cabello

Origination
Swaingrove Imaging

Printed and bound by Butler and Tanner, UK

A **Banson** Production
17e Sturton Street
Cambridge CB1 2QG, UK
banson@ourplanet.com

SUPPORTING ORGANIZATIONS

The United Nations Environment Programme is the principal United Nations body in the field of the environment. Its role is to be the leading global environmental authority that sets the global environmental agenda, that promotes the coherent implementation of the environmental dimension of sustainable development within the United Nations system, and that serves as an authoritative advocate for the global environment. Its objectives include analysis of the state of the global environment and assessment of global and regional environmental trends, provision of policy advice and early warning information on environmental threats, and to catalyze and promote international cooperation and action, based on the best scientific and technical capabilities available. Website: www.unep.org

The UK Department for Environment, Food and Rural Affairs is working for sustainable development: a better quality of life for everyone, now and for generations to come. This includes a better environment at home and internationally, and sustainable use of natural resources; economic prosperity through sustainable farming, fishing, food, water, and other industries that meet consumers' requirements; thriving economies and communities in rural areas and a countryside for all to enjoy. Website: www.defra.gov.uk

The Ernest Kleinwort Charitable Trust

Acknowledgments

The editors would like to record their immense gratitude to all those who committed the resources needed to make this atlas a reality. First, we must thank the organizations who lent their financial support: the United Nations Environment Programme Division of Environmental Conventions (UNEP DEC) and Division of Early Warning and Assessment (UNEP DEWA); the UK Department for Environment, Food and Rural Affairs (Defra); and the Ernest Kleinwort Charitable Trust.

Authors, peer reviewers, and providers of spatial and other data are named in the individual chapters, and we thank them again here. Their help has made a tremendous difference. Most of the images in this book were generously contributed by their creators, who are credited alongside each one. Many others contributed in diverse ways that may not be reflected in the form of names associated with particular sections – through strategic conversations, networking, providing introductions and anonymous inputs, by helping with mundane but essential tasks, or by providing moral support at critical times. Virtually everyone at UNEP-WCMC and the GRASP Secretariat in Nairobi, as well as many people in the nongovernmental conservation organizations within the GRASP network and the IUCN/SSC Primate Specialist Group Section on Great Apes, falls into this category.

With deep apologies for any omissions, we would like to offer particular thanks as follows. Jared Bakusa, Brian Groombridge, Florence Jean, Kim McConkey, and Adrian Newton were instrumental in getting the project started. Simon Blyth and Lucy Fish provided much support to the cartographers. Ian May set up the interactive map service that helped reviewers to audit the data. Pragati Tuladar helped to locate some of the places named in the text. Simon Burr and Maria Murphy helped us with the logistics of the peer review. Mary Cordiner helped us to obtain various vital pieces of literature. Brigid Barry and Lee Shan Khee each devoted months to the book, Brigid concentrating on the text and photos, and Shan Khee on the maps. Finally, our thanks to Phillip Fox, Jerry Harrison, David Jay, Tim Johnson, Rebecca Kormos, Mark Leighton, Kirsty Mackay, Daniel Malonza, Corinna Ravilious, Ian Redmond, Melanie Virtue, Matt Woods, and Kaveh Zahedi for their ongoing support.

We hope that this book does justice to the generosity of all involved, and that it will kindle an equal interest in great ape conservation among a new and larger audience.

Julian Caldecott and Lera Miles

Foreword

KOFI A. ANNAN

The great apes are our kin. Like us, they are self-aware and have cultures, tools, politics, and medicines. They can learn to use sign language, and have conversations with people and with each other. Sadly, however, we have not treated them with the respect they deserve, and their numbers are now declining, the victims of logging, disease, loss of habitat, capture, and hunting.

Nevertheless there are signs of hope. In some places, governments have taken the lead in conservation efforts, often cooperating across national frontiers. It has become increasingly clear that whoever initiates actions, be it central governments, local governments, international nongovernmental organizations, or individual citizens, local communities need to be involved. It is they who live with the great apes, and it is they who need to have the incentives – such as sharing in revenues from tourism – to conserve them.

This atlas tells the story of great ape conservation. It describes both the progress that has been achieved and what we must do if the great apes are to survive. Often, people treat great apes better when they treat each other better, as a result of education, good governance, and reduced poverty. But saving the great apes is also about saving people. By conserving the great apes, we can also protect the livelihoods of the many people who rely on forests for food, clean water, and much else. Indeed, the fate of the great apes has both practical and symbolic implications for the ability of human beings to move to a sustainable future.

Great apes cannot be conserved for free. The Great Apes Survival Project documented in this publication can help by mobilizing resources. But this is only part of the answer, and other good ideas on how to protect the great apes are also needed. We need ordinary people in their millions to love and protect them. We need governments and companies to 'adopt' them and the places where they live. We need to turn the tide of extinction that threatens our nearest living relatives.

Kofi A. Annan
Secretary-General of the United Nations

Contents

Using the maps

The maps in this book show forest cover, designated areas, and species distributions. Much information is condensed into each map, so some explanatory notes are given here.

For Africa, three types of species data are shown. The 'estimated ranges' are the maximum possible range of the species or subspecies. The 'confirmed ranges' are smaller areas within which the species is known to be present. The point data show species presence (squares or circles), alleged presence (?), or known local extinctions (X). With the exception of extinctions, these are maps of observations rather than of definitive presence or absence: presence should not be ruled out anywhere within the estimated ranges shown. The observations mapped include those of great ape nests, tracks, or other signs as well as of the animals themselves.

The dates shown in the legends indicate the date of last observation. If the date is an old one, there are no recent records, but this may be because the site has not recently been visited by a researcher rather than because of the absence of the species. Where a species has definitely been lost from a location, this is shown explicitly. Extinctions include only those that are recorded to have occurred since 1940. It is difficult to say definitively that a species is no longer present unless all suitable habitat has been destroyed, so local extinctions are inevitably under-represented.

The density of observation points is unlikely to indicate the actual density of apes; it is at least as likely to indicate density of survey effort. For example, Map 5.1 shows a large number of recent observations in the Salonga National Park, because a survey was carried out there in 2004. Please consult the species chapter or country profile text for information about the relative density of the species in different parts of the range.

The tree cover shown on the maps in shades of green is based on satellite imagery, and includes plantations and degraded forests as well as intact natural forest. Park and reserve data are shown in three categories: those areas known to have the highest level of official protection (a national designation assigned to IUCN Category I to IV or a designation associated with an international convention, such as a UNESCO World Heritage Site or Biosphere Reserve, or a Ramsar Site); those that have some form of designation not covered by the above; and those that are proposed for designation. Abbreviations have been used for the designation names, such as FR for Forest Reserve. These are listed opposite.

The African maps present a compilation of recorded observations and estimated range areas of great apes, put together by Thomas Butynski (Conservation International) and updated at UNEP-WCMC with help from conservationists and researchers worldwide. The Southeast Asian maps illustrate the forest blocks in which orangutans are present, together with their estimated density. These data were put together by researchers attending the Population and Habitat Viability Assessment Workshop held in Jakarta, Indonesia, in January 2004, and by the team at the Leuser Management Unit in Sumatra, Indonesia (see Chapter 11).

KEY TO ALL BACKGROUND DATA ON MAPS

Protected areas (boundary unknown)
- ▲ Nationally (IUCN Cat. I–IV) or internationally protected area
- ▲ Other designated area
- ▲ Proposed as a protected area

Protected areas (boundary delineated)
- ☐ Nationally (IUCN Cat. I–IV) or internationally protected area
- ☐ Other designated area
- ☐ Proposed as a protected area

Other features
- ■ National capital
- • Other city
- International boundary
- Primary or secondary road
- River
- Coastline
- Water body

Tree cover (percent)
- 0
- 1–10
- 11–40
- 41–60
- 61–100

ABBREVIATIONS USED ON THE MAPS

BR	Biosphere Reserve
CA	Conservation Area
CF	Classified Forest
CFR	Commercial Forest Reserve
FaR	Faunal Reserve
FFR	Forest and Floral Reserve
FNR	Forest Nature Reserve
FR	Forest Reserve
GR	Game Reserve
HA	Hunting Area
HP	Hunting Park
HR	Hunting Reserve
HZ	Hunting Zone
IER	Integral Ecological Reserve
NCU	Nature Conservation Unit
NF	National Forest
Non-HFR	Non-Hunting Forest Reserve
NP	National Park
NR	Nature Reserve
NHM	National Historic Monument
PF	Protection Forest
PFR	Protection Forest Reserve
PL	Protected Landscape
RC	Rehabilitation Center
RR	Resource Reserve
RS	Ramsar Site
RSR	Reserve for Scientific Research
ScR	Scientific Reserve
SNR	Strict Nature Reserve
SR	Special Reserve
WHS	World Heritage Site
WMA	Wildlife Management Area
WR	Wildlife Reserve
WS	Wildlife Sanctuary

MAP DATA SOURCES

Protected areas

IUCN (1994) *Guidelines for Protected Area Management Categories.* IUCN, Cambridge, UK and Gland, Switzerland. http://www.unep-wcmc.org/protected_areas/categories/eng/index.html.

World Commission on Protected Areas (2004) *World Database on Protected Areas.* UNEP-WCMC. http://sea.unep-wcmc.org/wdbpa/index.htm. Accessed September 2004.

Tree cover

Hansen, M., DeFries, R., Townshend, J.R., Carroll, M., Dimiceli, C., Sohlberg, R. (2003) *500m MODIS Vegetation Continuous Fields.* Global Land Cover Facility, College Park, Maryland. http://glcf.umiacs.umd.edu/data/modis/vcf/. Accessed September 13 2004.

Rivers

Petroconsultants (CES) Ltd (1990) *Mundocart/CD: Version 2.0.* Petroconsultants (CES) Ltd, London.

Roads, country boundaries, coasts, inland water bodies

DMA (1992) *Digital Chart of the World.* Defense Mapping Agency, Fairfax, Virginia.

Cities

ESRI (2003) *ESRI Data & Maps 2003.* ESRI, Redlands, California.

See the map data sources in the country profiles in Chapters 16 and 17 for great apes data sources.

Citations are numbered separately in each chapter and country profile. The corresponding numbered reference lists are not included in this volume, but can be accessed online at: http://www.unep-wcmc.org/resources/publications/WAGAC. A list of further reading is given at the end of each chapter and country profile. Data sources for the maps are also listed and usually represent additional further reading.

Introducing great apes

Richard Leakey

I became personally aware of apes when I was very young and long before they were as threatened as they are today. In 1949 my mother Mary Leakey discovered a fossil skull of a primitive ape, known as *Proconsul africanus*, in a then remote fossil site on Rusinga Island in Lake Victoria. The 17 million year old find had to be flown to England for study, and the press had a field day covering the skull and my mother's arrival in the country. As an inquisitive five year old, I was anxious to know what all the fuss was about and from this I learned about apes, especially fossil ones! The name *Proconsul* was inspired by Consul, a famous captive chimpanzee in the London Zoo. And so it was that chimpanzees as well as fossil apes became imprinted on my young brain.

Much later, I was further 'ape conditioned' by my father, Louis Leakey, who was very involved in getting Jane Goodall established on her wild chimpanzee studies at Gombe in the United Republic of Tanzania. Jane's work and the publicity that was generated through the National Geographic Society and other media probably did more than anything else to alert the general public to the existence of chimpanzees and, by extension, the other great apes. In time, other people and other studies continued to enthral the public and so build an awareness of our closest relatives living in the shrinking forests of Africa and Southeast Asia.

Molecular biologists and geneticists have demonstrated how biologically close we are to chimpanzees and bonobos, gorillas and orangutans. It must be stressed that in calling for measures to protect the remaining populations and to improve on the care and husbandry of those in captivity, we are speaking of our own relatives, not simply some hairy abstract beasts. The great apes and ourselves are so close that it is obvious that a fundamental error was made when classifying ourselves as something separate. We talk of six great ape species remaining, all of which are threatened, while there are in fact seven. That six should be at the mercy of one is a sad testimony and a poor reflection upon our claim to being the most intelligent of the set.

This atlas of great apes is timely and sets out a great deal of information that many people are ignorant of. The threats to orangutans, gorillas, bonobos, and chimpanzees are many but the loss of habitat as remaining forests are plundered is surely a major concern. Disease too is a worry, particularly where there are increasing contacts through tourism.

I believe we have an obligation to our descendants as well as to our ancestors: the remaining wild great ape populations must be protected for all time. As humans, we need to advance a new moral imperative to ensure the survival of these wonderful relatives, and getting to know them better is surely a good start.

Richard Leakey

CHAPTER 1

Evolution, dispersal, and discovery of the great apes

MARTIN JENKINS

The family Hominidae is one of the smaller families of mammals, with seven living species. Six species are confined to various forested or wooded habitats in the Old World tropics (i.e. the tropical parts of Africa and Eurasia); all of these are considered under threat of extinction, some critically so. The seventh species is ubiquitous and enormously abundant, probably the most numerous large animal that has ever lived: *Homo sapiens*, our own species. It is entirely thanks to the activities of humans that each of the other species is currently in such a precarious state. We are, however, now in the unique position where, if we have the collective will, we can reverse this trend and ensure that our closest relatives have a viable future on the planet. This volume attempts to chart a course by which this might be achieved.

Apart from human beings (*Homo*) there are three genera of great apes alive today: gorillas (*Gorilla*), chimpanzees (*Pan*), and orangutans (*Pongo*). The first two of these are confined to Africa, while the third occurs in Southeast Asia. Until recently, there were generally accepted to be one species of gorilla (*Gorilla gorilla*), two species of chimpanzee (the robust or common chimpanzee, *Pan troglodytes* and the pygmy chimpanzee or bonobo, *Pan paniscus*), and one species of orangutan (*Pongo pygmaeus*). Recent studies and changes in approach to taxonomy have led to the populations of gorilla and orangutan each being classified as two separate species.[31] It is this taxonomy, as endorsed by the Primate Specialist Group of IUCN–The World Conservation Union, that is used in this book, although it has not been universally accepted.[38] Hence the gorillas are considered to comprise the eastern gorilla (*Gorilla beringei*) and the western gorilla (*Gorilla gorilla*), while the orangutans are separated into the Sumatran orangutan (*Pongo abelii*) and the Bornean orangutan (*Pongo pygmaeus*).

There is a second group of living apes: the gibbons, which are briefly covered in Chapter 12. These are generally placed in a separate family, the Hylobatidae, although some taxonomists regard them too as members of the family Hominidae. Currently about 12 species in four genera are recognized, ranging through Southeast Asia from Assam to Java, south China, Borneo, and Sumatra. The gibbons share a lineage with the other apes and the forests of the orangutans, but show their own unique features of lifetime monogamy, duet singing, territoriality within small home ranges, and a specialization for rapid movement by swinging from their hands like a pendulum beneath tree branches. Many gibbon species are also highly endangered.

DISCOVERY OF THE GREAT APES

A long history

Apes and humans are no strangers to each other. Cave deposits in Viet Nam that date from around half a million years ago (mya) contain the remains of orangutans mingled with those of the early human *Homo erectus*,[21] while cooked orangutan bones dating to around 35 000 years ago have been found in the Great Caves of Niah in Sarawak, Borneo. In Africa, although direct fossil evidence is lacking, humans and apes have undoubtedly shared the same forests for millennia. To the western world, however, these creatures remained half known and little understood until comparatively recently. Indeed it was not until the early 20th century that the last species was described scientifically and a reasonably clear picture of this previously enigmatic group of animals emerged.

First recorded contacts

The Carthaginian general Hanno made a voyage down the west coast of Africa in the 5th century BC, and the *Periplus* (account) of this has survived. In it, there is an intriguing reference to wild, hairy people, called 'gorillae' by local interpreters, living on an island in a lake. Apart from this, the first convincing written record of man-like apes, at least according to Thomas Huxley,[37] who reviewed historical accounts and then current knowledge in 1863 (and from whom most of the following is drawn), is in Philip Pigafetta's *Description of the Kingdom of Congo* published in 1598.[60] This was based on the notes of Eduardo Lopez, a Portuguese sailor. In it, there is a passage that states (Huxley's translation): "in the Songan country, on the banks of the Zaire, there are multitudes of apes, which afford great delight to the nobles by imitating human gestures".

This 'group of young primates' appeared in *The Childhood of Animals* by E. Yarrow Jones in 1912.

Mary Evans Picture Library

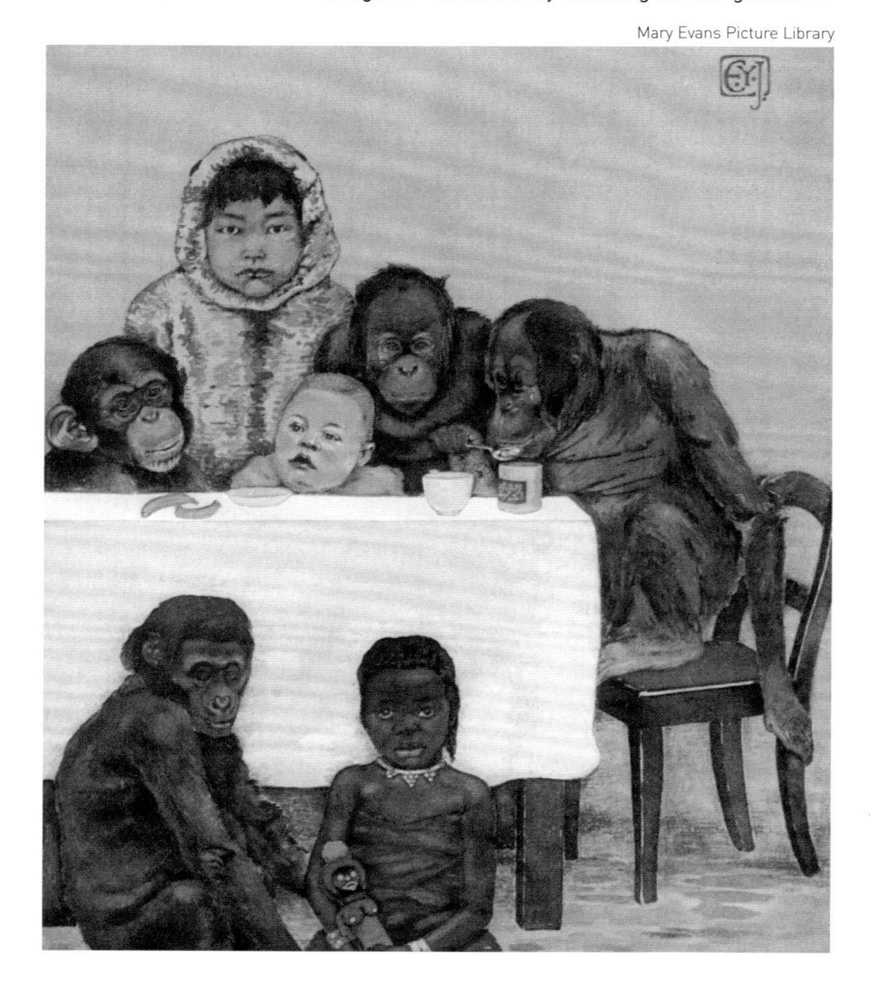

The word 'ape' was applied in generalized fashion to a number of Old World primates, particularly macaques (*Macaca* spp.), until at least the 19th century and there is no reason to assume from this verbal account that the creatures referred to were apes in the modern sense. However, in a subsequent chapter of the volume is a plate by the brothers De Bry of two of these creatures, which look very much as if they are based on a faithful description of a gorilla or chimpanzee.

Much more detailed and convincing descriptions are found in two books published in the following decades by the English clergyman Samuel Purchas. He records the accounts given to him by Andrew Battell, a soldier who had lived in equatorial Africa for many years. The longest is in the second volume,[62] in which Purchas recounts Battell's description of forests along a river:

The woods are so covered with baboones, monkies, apes and parrots, that it will feare any man to travaile in them alone. Here are also two kinds of monsters, which are common in these woods, and very dangerous. The greatest of these two monsters is called Pongo in their language, and the lesser is called Engeco. This Pongo is in all proportion like a man; but that he is more like a giant in stature than a man; for he is very tall, and hath a man's face, hollow-eyed, with long haire upon his browes. His face and eares are without haire, and his hands also. His bodie is full of haire, but not very thicke; and it is of a dunnish colour. He differeth not from a man but in his legs; for they have no calfe. Hee goeth alwaies upon his legs, and carrieth his hands clasped in the nape of his necke when he goeth upon the ground. They sleepe in the trees, and build shelters for the raine. They feed upon fruit that they find in the woods, and upon nuts, for they eate no kind of flesh. They cannot speake, and have no understanding more than a beast. The people of the countrie, when they travaile in the woods make fires where they sleepe in the night; and in the morning when they are gone, the Pongoes will come and sit about the fire till it goeth out; for they have no understanding to lay the wood together.

There is an added marginal note from Purchas:

The Pongo a giant ape. He told me in conference with him, that one of these pongoes tooke a negro boy of his which lived a moneth with them. For they hurt not those which they surprise at unawares, except they look on them; which he avoyded. He said their highth was like a man's, but their bignesse twice as great. I saw the negro boy. What the other monster should be he hath forgotten to relate; and these papers came to my hand since his death, which, otherwise, in my often conferences, I might have learned. Perhaps he meaneth the Pigmy Pongo killers mentioned.

The description of the 'pongo', down to its vegetarian habits and building of nests in trees, is clearly that of a gorilla, while the name 'enche-eko' (a phonetic version of Battell's 'engeco') was still in use in Gabon at least until the early 19th century for the chimpanzee.[69]

Making sense of ape descriptions

A generation later can be found the first account of an ape in Europe. The *Observationes Medicae* by Nicholas Tulpius, published in 1641, contains a description of what may have been a young chimpanzee, brought back from the region of Angola and presented to Frederick Henry, Prince of Orange. Tulpius states that the animal concerned is referred to by the Indians as 'Orange-autang, or Man-of-the-Woods' and by the Africans as 'Quoias Morrou'. It is possible, given Dutch contact with both the East Indies and southern Africa at the time, that this ape was not a chimpanzee from Angola, but an orangutan from Angkola in western Sumatra.[66]

By this time, evidently, there was already an understanding that similar animals occurred in both Asia and Africa, although confusion between the various kinds persisted for well over a century afterwards. This is in part explained by the fact that, during the late 17th century, accounts of apes in Asia were garbled and often verged on the ridiculous. The first apparent illustration of an Asian ape[7] is evidently of a rather hairy female human. This picture was added after the author's death, but the description, which was indeed penned by Bontius in 1658, may really have described an orangutan.

The very end of the 17th century saw the publication of the first scientific account of one of the apes – a treatise by Tyson published by the Royal Society in London in 1699 and entitled, *Orang-outang, sive 'Homo Sylvestris', or the Anatomy of a Pygmie compared with that of a 'Monkey' an 'Ape' and a 'Man'*. The description is of a young chimpanzee, also brought back from Angola. Tyson reviewed existing literature of the time and concluded that the animal was not one of those previously described by Battell, Tulpius, or Bontius, but was probably identical with the so-called pygmies of the ancients. He enumerated a large number of characteristics in which his animal "more resembled a Man than Apes and Monkeys do" and then a slightly shorter list of those in which it "differ'd from a Man and resembled more the Ape and Monkey kind." He concluded that though it "does so much resemble a 'Man' in many of its parts, more than any of the ape kind, or any other 'animal' in the world, that I know of: yet by no means do I look upon it as the product of a 'mixt' generation – 'tis a 'Brute-Animal *sui generis*' and a particular 'species of Ape'."

An anonymous description, accompanied by a drawing by Scotin, of an undoubted chimpanzee appeared in 1739. It seems to have been the same animal later described and depicted by the great French naturalist Buffon, who not only examined a live young chimpanzee, but also came into possession of an adult specimen of a gibbon from Asia (now known to be a lar gibbon *Hylobates lar*). This, described in detail by Buffon and Daubenton under the name 'jocko' (erroneously derived from Battell's engeco), was the first adult ape recorded in Europe and the last to be seen there for many years. On the basis of these specimens and existing accounts, Buffon finally concluded that there were two species of 'orang', or manlike ape: a large one, the pongo of Battell, from Africa; and a small one, the jocko, in the East Indies.[11] He thought that the small apes recorded by himself and Tulpius were young pongoes.

Meanwhile, in 1779 the Dutch anatomist Peter Camper published a detailed treatise on the orangutan, based on dissections of several young females and a young male.[12] He stated: "The true Orang, that is to say, that of Asia, that of Borneo, is consequently ... neither the Pongo nor the Jocko, nor the Orang of Tulpius, nor the Pigmy of Tyson – it is an animal of a peculiar species."

Shortly after this, the first adult orangutan specimen reached Europe; its skeleton was displayed in the Museum of the Prince of Orange and first seen by Camper in 1784. Camper was evidently unsure of the relationship between this large animal, which stood over 1.2 m tall and which he called a pongo after Battell, and the (juvenile) orangutans he had described so meticulously five years earlier. For some years it was assumed that it was a separate species from the chimpanzee, gibbon, and orangutan. In 1810, Blumenbach, a German, suggested the animal was in fact an adult orangutan.[4] Richard Owen finally demonstrated this in persuasive fashion.[57, 82] His monograph also contained the first recorded description of the skeleton of an adult chimpanzee – clearly the adult version of the African animals considered by Buffon and Tulpius to be young pongoes.

By this time, then, the orangutan and the chimpanzee had finally become well characterized as distinct entities, the former living in Asia and the latter in Africa, and both known in their juvenile and adult states. It was also established that the only other apes in eastern Asia were various species of gibbon. Understanding of the last of the apes to be scientifically described – Battell's true 'pongo' – lagged far behind.

In 1819 a traveler, Thomas Bowdich, noted that local people in the region of the Gaboon (Gabon) River reported the existence of a second great ape in addition to the engeco, called the 'ingena' and described as "five feet high and four across the shoulders."[8] However, there was little further evidence for the existence of this animal until 1847, when an American missionary, Dr Thomas Savage,

Mary Evans Picture Library

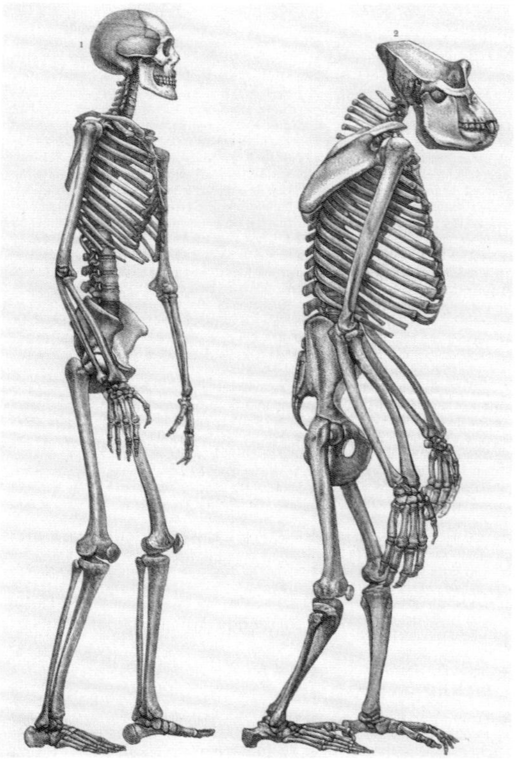

A late 19th century print of the skeleton of a human compared with that of a gorilla.

came across the skull of an unknown ape in the house of the Rev. Mr Wilson, a missionary resident on the Gaboon River.

From the skull and the descriptions of the animal provided by local people, Savage concluded that the animal in question was a new species of ape.[69] Savage and Wilson obtained a good account of the habits of this creature in the wild and enough physical material to allow Jeffries Wyman, an American anatomist, to publish a detailed description.[70] It was Savage who applied the name 'gorilla', taken from the *Periplus* of Hanno, to this species, although making no claim to its being actually the animal described in the *Periplus*.

It had taken over 200 years but, finally, the remarkable accuracy of Battell's original account of the African apes was confirmed. As Huxley put it, the gorilla had of all the apes "the singular fortune of being the first to be made known to the general world and the last to be scientifically investigated."

By the middle of the 19th century, therefore, it was known that there were four distinct 'kinds' of apes: in eastern Asia, the gibbons and the orangutans; in western Africa, the chimpanzees and the gorilla. Understanding of the natural history of the gibbons and orangutans was quite well advanced, thanks to the observations of a number of naturalists including Müller, Duvaucel, Bennett, Wallace, and Brooke. There were known to be several species of gibbon, but the question of whether there were several, two, or only one species of orangutan was regarded by Huxley at least as unresolved. Given the wide geographic area over which chimpanzees occurred, he thought it possible that there might be more than one species. The recognition of the bonobo as a separate species in the early 20th century confirmed this. In contrast, he assumed there to be only one species of gorilla. He also noted that, despite the accounts of Savage and Wilson, knowledge of both the chimpanzee and the gorilla in the wild was much less complete than that of the Asiatic apes.

Apes as human relatives

Although they were undoubtedly regarded as fascinating in their own right, the great question regarding apes in western scientific and wider intellectual circles at this time was, of course, where they stood in relation to humans. Darwin had published his revolutionary *On the Origin of Species* in 1859, spurred on by Wallace, who had independently developed similar ideas largely as a result of his observations in Asia. As evolutionary concepts took hold, it became ever clearer that similarities between species might be a sign of evolutionary affinities: that is the more similar different species were to each other, the more likely they were to have shared a recent common ancestor. The notion of the transmutability of species implicit in this, although by no means new, was difficult for many to accept, as was the idea of selection acting on random inheritable variation – the cornerstone of Darwinian theory.

There were enough problems in accepting these concepts when applied to other organisms, but such difficulties paled in comparison to those encountered in considering the place of humans in the scheme of things. It was incontrovertible by now that the great apes had startlingly similar anatomical features to humans. The idea that this might mean that they and humans had shared a recent ancestor and were thus closely related to each other was – and still is – anathema to many. Ironically, one of those who found this idea most repugnant was Richard Owen, the anatomist who had produced by far the most accurate and detailed descriptions to date of the morphology of the chimpanzee (now considered the closest relative to humans). So determined was he to demonstrate the separateness of humans that he erected an entire mammalian subclass – the Archencephala – to contain them alone, based very largely on the presence of a small structure in the human brain, the hippocampus minor, supposedly absent in all other apes.[58]

To Darwin, Huxley, and the increasing number of others who embraced evolutionary theory in all its ramifications, the question was not whether the great apes and humans were closely related – that was taken as read – but which of the species was the closest living relative to humans. On anatomical grounds, Huxley concluded that it was either the gorilla or the chimpanzee, but believed there was insufficient evidence to determine which of the two was actually the closest. This question was not satisfactorily resolved until over a century later, with the development of new techniques in molecular biology.

EVOLUTION OF THE GREAT APES

Reconstructing phylogenies: fossils and genes

There are two major sources of evidence that can be used to establish the relatedness of organisms: the living organisms themselves and fossil remains. In both cases, two basic premises are used: that species sharing a large number of characters are likely to be related, and that species sharing some highly complex and specialized feature are likely to be more closely related than species not possessing it. Until recently, the features that could be analyzed were essentially anatomical or morphological ones, although occasionally animal behavior has also been used. There are, however, difficulties with this approach. These arise because characters can evidently appear and disappear through evolution and, most importantly, because the same characters can arise independently in different lineages. Wings for powered flight, for example, are present in birds, mammals (bats), and insects, and have therefore arisen independently at least three times in the course of evolution. Moreover, among birds and insects there are species or groups of species that can no longer fly and which may, in the case of some insect groups, no longer possess recognizable wings at all.

Advances in molecular biology since the 1970s have revolutionized approaches to taxonomy and systematics. The opportunity now exists to compare directly the genetic material of different individuals, populations, and species, and often to gain a far clearer insight into the degree of relatedness between them. These techniques allow us to determine which living organisms are most closely related to each other with increasing confidence. However, describing the evolutionary route by which these organisms arrived at their current state still

William H. Calvin (www.williamcalvin.com)

A century after evolutionary theory took hold, the discovery of tool use among great apes further revolutionized the way that humans perceive themselves. Here, a bonobo at San Diego Zoo soaks up juice with the mashed end of a stick.

Box 1.1 WHAT MAKES A PRIMATE?

There is no unique feature that is characteristic of all primates, fossil and living, by which they can be distinguished from all other mammals. Rather, there are a number of features that are not common in other mammals, each of which is found in most primates; taken together, these allow primates to be distinguished from other groups.[63] The most important of these are:

- hands and feet that can grasp, usually with big toes and thumbs that can be opposed to the other digits (although humans, for example, have lost the ability to oppose their big toes);
- claws that have been modified to form a flat nail (although some species have modified nails, called 'toilet-claws', for grooming on one or two toes, and the aye-aye from Madagascar and the New World marmosets and tamarins have re-evolved claws from nails on all digits except the big toe);
- eyes that are at the front of the face and look forward, with overlapping visual fields, allowing binocular vision and accurate judging of distances;
- relatively large brains compared to those normally found in other mammals of comparable size;
- small litter sizes – usually of only one young – and young that mature slowly compared to most mammals of equivalent size;
- a distinctive origin within the skull of the auditory bulla (the bony case that protects the underside of the inner and middle ears).

Together, the above features allow us to assign a range of living animals to the order Primates, including tarsiers, lemurs, lorises, monkeys, and apes (including humans). There is now widespread agreement that the monkeys and apes all share a more recent common ancestor with each other than they do with the other primates. They therefore form a monophyletic group, the simians. The precise relationships between simians and other primates, however, as well as the origin of the primates as a whole, remain much less settled.[23]

Martin Jenkins

A bonobo's foot, with opposing big toe and precision grasp.

William H. Calvin (www.williamcalvin.com)

requires evidence from the past, usually obtained from fossil remains. The problem is that the fossil record is an extremely incomplete and biased sample of life in past times. The vast majority of individual organisms leave no lasting physical trace when they die, largely because decomposing and scavenging organisms are so efficient at their work. Even if some or all of an organism is preserved, the chance that the remains will survive in recognizable form through geological time to be recovered today is extremely small. Where they do survive, the fossil materials are usually very incomplete. Although trace fossils may sometimes give exquisitely detailed indications of the soft tissues of organisms, fossils are generally formed only from those parts of the organism that were hard in life – for example, shells of mollusks, bony skeletons, and, in the case of many vertebrates, teeth. Only in exceptional cases of recent preservation in the form of subfossil remains is there any likelihood of genetic material being recovered.

Overall, it is estimated that known fossils may represent perhaps 1 percent of the species that have ever existed.[64] For the primates, the fossil record has been estimated to be rather more complete, at up to 7 percent representation.[81] The geographical clustering of fossil finds can yield even more complete series than this overall figure would suggest.[75] Still, in any one lineage, many more species have existed than we currently know about, even among the primates, which are particularly intensively searched for and studied. One corollary of this is that it is very unlikely that any known fossil organism is the direct ancestor of any living one. This simple observation is very often overlooked in debates about genealogy and phylogenetics, perhaps nowhere more so than in attempts to recreate the human ancestral line.

Most recent attempts to recreate the calendar of evolution use molecular-clock methods, which bring together molecular studies and the fossil record. They are based on the assumption that random mutations accumulate in various kinds of DNA at constant rates, with no reverse mutations, and that code differences between lineages can be calibrated with the fossil record and made equivalent to time since lineages became separate. Some DNA (e.g. mitochondrial DNA or mtDNA) appears to accumulate mutations more quickly than others (e.g. the globin genes), so their molecular clocks can be said to run at different rates. These different rates can be exploited to infer evolutionary relationships at different taxonomic levels, i.e. more or less far back in time. For example, mtDNA can be used to establish relationships between modern human populations, globin genes for relationships between modern mammals, and cytochrome c for relationships in 'deep time' between eukaryote lineages. The reason why the cytochrome c 'clock' runs so slowly is that the structure and function of this molecule is vital to metabolism and cannot be changed through significant mutation without lethal effect. These molecules therefore remain essentially the same across very many species, but nevertheless accumulate minor differences from harmless mutations. Relating this to an actual measure of years requires the fossil record to offer one or more reliable estimates of divergence time in the lineages under study. Both the assumption of a regular rate of change in DNA through time and the setting of calibration points are problematic, but this approach has nevertheless produced results that are gaining wide general acceptance, as in the case of the apes discussed below.

Primate origins

The earliest fossils that are unequivocally identified as primates date from the early part of the Eocene epoch (Table 1.1), some 54–55 mya.[34, 81] They represent a diverse collection of lineages that have radiated from common ancestors to form two distinct groups of species, the superfamilies Omomyoidea and Adapoidea, recorded from a range of fossil sites in the northern hemisphere. These groups were prosimians (i.e. not simians, see Box 1.1), and some 200 species in over 70 genera have so far been described.[22] The vast majority of them had disappeared by the end of the Eocene (at least from the fossil record), apparently falling prey to deteriorating global climates. Of the species known from the fossil record, only nine are known from post-Eocene deposits.[75] The body sizes of these early primates are estimated to have ranged from around 50 g (the weight of the smallest living primates: the mouse lemurs, *Microcebus* spp.) to up to 7–8 kg (the weight of the larger guenons, *Cercopithecus* spp., or the smaller mangabeys, *Cercocebus* spp.).

By the early Eocene, the primates were already evidently well established as a group, indicating that their origin lay earlier, at a time for which no relevant fossils have yet been found. Exactly how much earlier is a subject of debate. Until relatively recently it was widely argued that the temporal origin of primates, along with most other major mammalian groups, lay relatively close to the Cretaceous/Tertiary boundary. At around 65 mya, this is the point at which the dinosaurs, plesiosaurs, pterosaurs, and many other groups finally disappeared. It was assumed that the primates and others diverged from early mammalian stock at this time and then underwent a relatively rapid radiation, evolving into a wide range of forms over what is, geologically speaking, a fairly short time.

More recent analysis, however, using molecular clocks[26, 54] and applying statistical analyses to the fossil record,[81] suggests that divergence of the major mammalian groups, including primates, can

Table 1.1 Epochs of the Cenozoic era, the 'age of mammals'[45]

	Start (mya)	End (mya)
Paleocene Epoch	65	54
Eocene Epoch	54	38
Oligocene Epoch	38	26
Miocene Epoch	26	5
Pliocene Epoch	5	1.6
Pleistocene Epoch	1.6	0.01
Holocene Epoch	0.01	now

be traced back much earlier, to the mid-Cretaceous era some 90 mya, with the last common ancestor of all the living primates believed to have lived somewhat over 80 mya.

Reconstructing the ecology and behavior of the Eocene primates is highly problematic, and ideas about their hypothetical ancestors are even more speculative. For many years it was argued that the major impetus for primate evolution was the adoption of an arboreal lifestyle,[41, 74] involving grasping hands and feet for holding onto branches, binocular vision for judging distances, an increased brain size for processing complex spatial information, and a decreased dependence on smell (of lesser importance in a windy canopy above the ground). There is, however, one major problem with the notion that an arboreal lifestyle *per se* explains primate adaptations: the fact that many other mammals are at least as arboreal as many primates and yet do not possess these attributes, or at least not all of them together.[16]

More recently it has been argued that diet was likely to have been a principal impetus, with some primatologists[17] arguing that the ancestral primate was an already arboreal animal that became adapted to hunting by sight in order to capture insects in the fine branches of the forest canopy or in bushy undergrowth. Others[80] note that the rise of primate species coincides with the rise of the flowering plants (angiosperms). In particular, the apparent radiation of primate species in the Eocene coincides with a marked radiation in flowering plants and particularly the development of complex tropical forests, with a range of fruiting trees. They argue that primates evolved specifically to take advantage of this new range of nutritious fruits and flowers growing in the fine, terminal branches of trees and bushes. Studies of modern lemurs such as *Microcebus* and the dwarf lemurs, *Cheirogaleus*, believed to be quite similar in many ways to these early primates, suggest that the spur to developing fullblown primate features may have been the need to adapt to a combination of these two food sources. The fruits and flowers on the fine terminal branches would themselves have attracted a range of insects and other invertebrates, and the early primates would have fed on both the plant matter and, perhaps more opportunistically, on the congregating invertebrates.[49] As will be seen, diet and, particularly, changes in diet are widely held to have played a crucial role in the evolution of primates from the origins of the group up to the appearance of modern humans.

By the early 19th century, the chimpanzees of Africa were well characterized and known to be quite distinct from the orangutans of Southeast Asia.

Michael Huffman

The origin of the simians

The earliest evolutionary history of the primates is still a mystery, as is the exact path that led to the emergence of the simian line. Possible fossil simians are now known from as far back as the early Eocene, some 50 mya, from a range of sites in North Africa, although most Eocene simians are from the late Eocene, and the earliest known from outside Africa (from the Arabian Peninsula) date from the early Oligocene. The relationship of these early simians to the Omomyoids and Adapoids is unclear – it is far from certain that either of these two groups actually gave rise to the simians. Nor is it clear where or when the simian line arose. As the initial discoveries of early simian fossils were made in North Africa (specifically the Fayum deposits in Egypt), it was widely assumed that simians arose in Africa; it is, however, perfectly possible that they arose in Asia and spread subsequently to Africa.

Old World Simians and the Miocene radiation
Fossils are known from the late Eocene and early Oligocene, 30–40 mya, that are indisputably early 'catarrhine' primates, of the lineage that includes both the apes (the hominoids) and the Old World monkeys (the cercopithecoids). Most of these have been found in the Fayum deposits in Egypt and the Taqah sediments in Oman, but there is also a single tooth from Angola.[63] These fossils, of which the best known is *Aegyptopithecus,* provide a link between the early Eocene simians and the living monkeys and apes. The nature of the fossils, and molecular-clock analysis, indicate that these so-called dawn apes pre-date the period at which the apes diverged from the Old World monkeys.

The Miocene – the age of the apes
For the Old World primates, as indeed for many other groups of animals, the Oligocene represents an important gap in the fossil record. During the early Miocene, around 22 mya, however, fossils begin to reappear in much greater numbers. By this time, the apes were evidently firmly established as a separate evolutionary lineage, indicating a split with the Old World monkeys some time between 22 and 30 mya. The Miocene, which ended around 5 mya, can be seen as the era of the apes, during which this lineage became remarkably widespread and diverse in the Old World. Up to 100 species in perhaps 40 genera have so far been identified at numerous locations in Africa, Europe, and Asia. Research in this field continues, and the rate of discovery of new fossils indicates that the many species named so far constitute only a fraction of the likely true diversity of Miocene apes. A new genus and species, *Pierolapithecus catalaunicus,* was reported in 2004 from mid-Miocene deposits in Spain.[55] Many of the finds, however, are known from only very partial remains – chiefly teeth and associated fragments of jawbone – making it very difficult to determine phylogenetic relationships or reconstruct the adaptations and ecologies of the species concerned.

Known ape fossils from the first part of the Miocene, until 15–17 mya, are confined to Africa. From what can be determined, it seems that these early Miocene apes, of which the best known are in the genus *Proconsul,* were a variable group, ranging in probable body size from around 3 kg to well over 80 kg. They appear to have been largely frugivorous, although diet undoubtedly varied from species to species. *Proconsul* at least lacked a tail – a lack characteristic of all modern apes – and appears to have been adapted to an arboreal way of life. However, it had not developed the highly flexible, powerful forelimbs of modern apes, that enable them to travel by swinging along beneath branches, and was thus more suited to traveling along the tops of branches on all fours.[1]

KOCP

Orangutans are believed to have diverged from the common ancestral line of the great apes some 11 million years ago.

It seems that some time between 15 and 17 mya, the apes invaded Eurasia from Africa along with a range of other mammals, taking advantage of a new land bridge between the two continents. Such land bridges had existed before this migration, however, so it is likely that something had changed in the biology of the apes (or the ecology of Eurasia) to allow them to exploit and occupy Eurasian ecosystems: perhaps a dietary factor which enabled them to subsist on coarser herbage and/or harder fruits. Fossil evidence indicates rapid dispersal and diversification, with apes being widespread and diverse in the Eurasian fossil record from around 14 mya to around 8 mya.[85] In contrast, the fossil record for large apes in Africa for this period is very sparse. African

Martha M. Robbins

The gorillas, while the first to be recorded, were the last of the great apes to be scientifically described.

fossil remains from around 5 mya onwards, which consist almost entirely of apes believed to be more or less closely associated with the hominid line, are much better represented in collections; this is undoubtedly in part because of the large amount of effort spent in searching for them.

From the fossil record, it is clear that the living apes represent a small vestige of former ape diversity, but it is hard to make sense of the descent of living species since the fossil record is so scanty. It is also very hard to demonstrate close anatomical relationships between any of the Miocene ape fossils and living species (including humans). These factors have led to much speculation as to the exact relationships between the living species, and regarding the nature of their immediate progenitors and where they may be found.

Some of this speculation – at least that concerning the relatedness of the living apes – has now been largely resolved by molecular analysis. It is now widely accepted that the gibbons were the first to split off from the branch that has led to the other living apes. The orangutans are the next to have diverged. Gorillas are believed to have split next from the line leading to chimpanzees, bonobos, and humans, with chimpanzees and bonobos being the most recent to diverge from each other.

Although the sequence of events is no longer disputed, the timing of each split is much more open to debate. This is because it depends both on reliable calibration from the fossil record, and on assumptions of a relatively uniform rate of mutation in the DNA sequences that are analyzed. One study that makes use of a range of sequences, and uses a date of divergence between Old World monkeys and apes set at 23.3 mya, gives divergence times as follows (with 95 percent confidence limits): gibbons 14.9 ± 2 mya; orangutans 11.3 ± 1.3 mya; gorillas 6.4 ± 1.5 mya; and chimpanzees and humans 5.4 ± 1.1 mya.[76] These generally agree well with a number of earlier studies, although the divergence times for the gibbons and the gorillas are somewhat more recent than those given by, for example, Pilbeam, which are 17–16 mya and around 9 mya respectively.[61] Changing the calibration point by shifting the assumed divergence time of the Old World monkeys and the apes back into the late Oligocene would shift the calculated divergence times of the various apes back, but the order of divergence and degree of relatedness between the species would remain the same.[39]

Possible orangutan relatives

The difficulty of linking fossil apes to living ones has not stopped paleoanthropologists from trying to do just that, most frequently in the case of the orangutan. Several genera of fossil apes from Asia have been associated with this lineage, most importantly *Gigantopithecus* from late Miocene deposits in the Siwalik region of India and Pakistan, and Pleistocene deposits in southern China and Viet Nam; *Lufengpithecus* from the late Miocene and possibly early Pliocene in southern China; *Sivapithecus* from the late Miocene (12.5–8.5 mya) in the Siwalik region; and *Khoratpithecus* from the middle and late Miocene in Thailand.

The latter three have been proposed as close relatives to ancestral orangutans, although strong arguments have also been made against this in the case of *Sivapithecus* and *Lufengpithecus*.[19, 43] Currently there seems to be agreement that, while *Sivapithecus* is not particularly closely related to the ancestral orangutan, it is at least closer to orangutans than to any other living primates. There is, however, much less consensus on whether this is the case for *Lufengpithecus*.

The new genus *Khoratpithecus* (which includes two species, of which one, *K. chiangmuanensis*, was originally placed in *Lufengpithecus* when described) seems at present to be the best candidate for a near relative to the ancestral orangutan.[19] Not only does it share some highly distinctive features with living and Pleistocene fossil orangutans, it also comes from an area where orangutans are known to have occurred during the Pleistocene, and is associated with a tropical flora. This is in contrast to both *Sivapithecus* and *Lufengpithecus*, both of which appear to have lived in areas with temperate or seasonal and relatively open rather than forested habitats. Interestingly, the earlier of the two *Khoratpithecus* species, dated to the middle Miocene (just over 11 mya), is found in association with a flora that shows strong African affinities, indicating a temporary floral and faunal dispersal corridor between Southeast Asia and Africa at this time. This fits well with the estimated time of divergence (11.3 mya) between the orangutan lineage and that leading to the gorillas, chimpanzees, and humans, as derived from molecular-clock analysis.

It might speculatively be argued from this that the gibbons represent the survivors of the first wave of apes to invade Eurasia during the early Miocene, while the orangutans represent the survivors of a subsequent, mid-Miocene, invasion from Africa. Alternatively, it has been argued that the surviving African apes, including hominids, arose from a secondary invasion of that continent by Eurasian apes, some time during the middle or late Miocene. Modern proponents of this hypothesis regard one of two Eurasian genera, *Dryopithecus*, known from Spain to eastern Europe, or *Ouranopithecus* (or *Graecopithecus*) from Greece, as likely to be close to the ancestor of African apes and humans.[1] There is currently no consensus on this, and it is unclear how such a consensus might be reached, although an improvement in the fossil record, particularly of mid- and late-Miocene African apes, would help.

Gigantopithecus, the other probable orangutan relative, is unique in being the only extinct nonhominid ape genus so far known to have survived later than the early Pliocene. Pleistocene remains from southern China and Viet Nam, ascribed to the species *Gigantopithecus blackii*, and consisting of extremely large teeth and lower jaws (mandibles), indicate an ape larger than any other known, living or dead. Although at one time thought to be perhaps close to the ancestral human line, the genus appears to be most similar to *Sivapithecus* and has been tentatively grouped with the latter in the same taxonomic tribe (the Sivapithecini).[43] In surviving into the Pleistocene, this genus was contemporary with at least early humans; indeed, remains ascribed to this species have been found mixed with those of *Homo erectus* and orangutans in deposits in a Vietnamese cave dated to around 475 000 years ago. The timing and cause of its extinction remain, tantalizingly, a mystery.

Chimpanzees and humans

While much in the evolution of the great apes remains uncertain or strenuously debated, it is now almost universally accepted among scientists that chimpanzees are our nearest relatives, that the split between the lineage leading to them and that leading to modern humans took place between 4.3 and 6.6 mya, and that this phase of human and chimpanzee evolution took place in Africa. What the last common ancestor of the two groups and the earliest distinct hominid (and indeed the ancestral chimpanzee) looked like, how they behaved, and why the two lineages went their separate ways are all the subject of much speculation. Characteristics that are widely considered to be very important in distinguishing the human lineage from that of other apes, and particularly the chimpanzee lineage, are:

- bipedalism, modern humans being the only living hominids that habitually engage in an erect, bipedal striding gait;
- brain size relative to body size, which is far larger in humans than in any other primates; and
- tooth structure and wear, which is strongly correlated with diet.

Great importance is attached to teeth because these are the most abundant, and sometimes the only available, fossil remains. Interpreting tooth structure from often worn and broken fossils to construct evolutionary arguments is highly contentious. Similarly, although it is widely accepted that bipedalism is a fundamental feature distinguishing the human lineage from that of other living apes, there is no consensus as to when it arose. There are currently too few relevant fossil remains to construct convincing arguments. Brain size is of limited use in constructing early phylogenies as the major increase in relative brain size in the human line began only around 2 mya, long after the split from the chimpanzee lineage.

Currently there are three major candidates for the earliest fossil of a distinctly human lineage.[2, 29] These are:

- *Ardipithecus kadabba* from Ethiopian deposits dated at 5.6–5.8 mya;[32, 33]
- *Orrorin tugenensis* from deposits estimated to date from slightly under 6 mya in the Tugen Hills in Kenya;[73] and
- *Sahelanthropus tchadensis* from the Durab Desert in northern Chad, dated to between 6 and 7 mya.[10]

Each of these (very incomplete) remains appears to have a mosaic of so-called primitive characteristics, that is those shared with earlier fossil apes and to some extent with living apes, and derived characteristics, that is those more closely aligned with the hominid lineage and not shared with other living apes. Each has its own fervent supporters as the earliest known representative of the human lineage, and each has its own equally fervent detractors. If the most recent molecular-clock analysis for the split between chimpanzees and humans is accepted (5.4 ± 1.1 mya),[76] and the dates for *Orrorin* and *Sahelanthropus* are reliable, then these two at least may pre-date the time when the chimpanzees and humans diverged. However, those who advocate either of these as early human ancestors maintain that the molecular clock is wrongly calibrated and that divergence times were earlier than it indicates.

While the fossil record provides scant evidence for the changes that took place leading to the divergence of the chimpanzee and human lines, some light might be shed on this by comparing the genetic material or DNA code that comprises the genome of each species. This is one major impetus for the current undertaking to sequence fully the chimpanzee genome. The genomes of humans and chimpanzees are around 98.8 percent the same, and it is hoped that examining the remaining 1.2 percent will give some insight into what, at least at the genetic level, makes us distinctively human.[59] One interesting preliminary finding is that enzymes for breaking down amino acids (the building blocks of proteins) have been positively selected for in the human lineage, compared with the chimpanzee lineage. These enzymes are associated with a meat-eating diet and indicate that increasing carnivory may have played an important part in human evolution, related perhaps to increasing brain size.

A final point is that not all authors agree that sufficient genetic divergence has yet occurred between chimpanzees and humans to warrant their being placed in separate genera.[86] If this argument were accepted, then for taxonomic purposes chimpanzees, bonobos, and humans would all be assigned to the genus *Homo*, further emphasizing the sibling nature of our evolutionary relationship with these great apes.

RESEARCH ON WILD GREAT APES

The first 'primatologist' might have been Charles Darwin, who spent weeks at London Zoo watching the monkeys before he wrote *Expression of the Emotions in Man and Animals* (1872). Field primatology began in the early years of the 20th century, with the study of chacma baboons (*Papio ursinus*) in South Africa by Eugène Marais, though this was not published until long afterwards.[48] Primatology continued with attempts by Henry Nissen[56] and Harold Bingham[3] to observe chimpanzees and gorillas in the wild – described by Alison Jolly as "difficult quarry in impossible terrain, for people who had no idea what primate research would mean."[40] Clarence Ray Carpenter accomplished field studies of howler monkeys (*Alouatta palliata*) and lar gibbons (*Hylobates lar*) in the 1930s,[13, 14] and of rhesus macaques (*Macaca mulatta*) in the 1940s.[15] Field primatology began to flower in the 1950s and 1960s, when an interest in the study of the great apes in the wild emerged. Japanese researchers were among the first in the field, with an exploratory trip by Imanishi and Mitani in 1958.[50] Dutch researchers followed soon after with Kortlandt's field trips to the then Belgian Congo (now Democratic Republic of the Congo) from 1960 onwards.[44]

Early research focused on eastern chimpanzees, mountain gorillas, and orangutans. George Schaller pioneered studies of mountain gorillas in the 1950s and 1960s;[71, 72] in 1967, these were followed up by Dian Fossey in Rwanda, leading to the establishment of the Karisoke Research Center where she worked until her death in 1985.[27] Jane Goodall's study on eastern chimpanzees began in 1960, and led to the establishment of the Gombe Stream Research Center in the United Republic of Tanzania.[30] Both women began their studies under the direction of Louis Leakey, a renowned paleo-anthropologist, and their work helped to create the first significant international awareness of great apes, particularly through the pages of their principal sponsor's magazine, *National Geographic*. The killing of Digit, one of the Karisoke study animals, in 1977 shattered the Eden-like quality of those early studies. The British Broadcasting Corporation TV series *Life on Earth*, first screened in 1979 and featuring David Attenborough face to face with mountain gorillas, and the sensational murder of Dian Fossey (followed by the film of her life, *Gorillas in the Mist*), also helped to alert the world to mountain gorillas and the perils facing them. Research at both Karisoke and Gombe Stream continues to date.

The first significant field study of orangutans began in the 1960s, when John MacKinnon started out in Sabah, Malaysia and Renun, Sumatra.[46, 47] Biruté Galdikas began to work at the Tanjung Puting Reserve in central Borneo, Indonesia in 1971,[28] again with the involvement of Louis Leakey (her field site is called Camp Leakey); this project has continued to date. Detailed studies of the Sumatran orangutan continued in the early 1970s,[67] at about the same time that research on the western chimpanzee began in earnest,[5, 77, 78] followed by the western gorilla and the central chimpanzee in the 1980s.[83, 84] A systematic distribution survey of the northern range of the bonobo was first carried out in 1973,[42] and the threats from logging and bushmeat hunting were well understood,[79] but it was not until the 1990s that further significant research on bonobos in the wild was carried out.

Field research developed in parallel with captive studies, and the disciplines asked and answered questions of each other regarding great ape behavior, learning ability, ecology, evolution, cognition, and communication.[35] Chimpanzees, which had been used for decades in biomedical laboratories, were the focus of most early behavioral research, though they became less prominent as studies of other great ape species multiplied. The discovery in the early 1970s that great apes could be taught human sign languages, and the birth of sociobiology in the late 1970s, both contributed to intense interest in great ape behavior and evolution.[51] The discovery of culture,[87] tool use,[6] and even recognizable politics among chimpanzees[25] further revolutionized the way we humans thought about great apes and defined ourselves in relation to them.

By the 1980s, a movement had begun among primate biologists, philosophers, and others to recognize great apes and humans as belonging to a single 'community of equals', with common rights to freedom from torture and arbitrary imprisonment.[18] Public interest in and concern for the great apes has continued to deepen, and research on these animals is still increasing at universities and scientific institutions throughout the world. Some research centers have also played an important direct role in conservation of the great apes, including Ketambe in Sumatra, Camp Leakey in Borneo, Karisoke in Rwanda, and Gombe in Tanzania. By their simple presence, they have attracted political attention and deterred poachers and loggers on the ground. In some cases, the areas around the field stations have been declared national parks.

Martha M. Robbins

Mountain gorillas being observed in the wild, Bwindi National Park, Uganda.

RESEARCH ON CAPTIVE GREAT APES

A population of captive animals allows noninvasive research to be conducted that is difficult or impossible in the wild, including studies of social interactions, animal health, and reproductive biology. Great apes and humans share many basic features at the levels of whole-body physiology and metabolism, organ function, cell structure, and even gene organization; this means that great apes are excellent models for studies of human health and disease. These range from field-based studies of disease in the wild, to laboratory studies of captive animals, and even the involvement of apes in space research.

A fruitful avenue for research with apes is

Box 1.2 CRYPTIC APES

Cryptic species are those that are reported anecdotally, or from physical signs (such as footprints), or from photographic evidence, but that have not been described unambiguously from a live or dead specimen by a professional biologist. The great ape community has attracted its own share of cryptic species, of which three deserve a mention in this volume because some primate biologists believe that there is a significant chance that they may one day give rise to new validated species or subspecies. These are the 'yeti' of the Himalayas,[52] 'orang pendek' of Sumatra,[20] and 'Bili-Bondo ape' of the northern Democratic Republic of the Congo (DRC).[88]

Yeti

There have been many reports by western, Chinese, Tibetan, Nepali, and other local observers of large bipedal ape-like beings in the Himalayan Mountains of Sikkim, Bhutan, and Nepal, and nearby ranges such as the Pamirs of Tajikistan. The earliest published report was that of B.H. Hodson, the British representative in Nepal, in 1832. Many reports since 1921 have been by mountaineers participating in expeditions to the Mount Everest area, variously including observations of the animals themselves (e.g. by Alan Cameron in 1923, Tenzing Norgay in 1949, Don Whillans in 1970, and Craig Calonica in 1998) or their footprints (e.g. by L.A. Waddell in 1889, H.W. Tilman in 1936, Edmund Hillary and Tenzing Norgay in 1953, and Lord Hunt in 1978), sometimes supported by findings of food scraps and dung (e.g. by the London *Daily Mail* expedition in 1954, and Norman Dyrenfurth in 1958). The animals are sometimes said to injure people (e.g. by Hodson in 1832, Waddell in 1889, Jan Frostis in 1948, and the Thomas Slick expedition in 1957), giving rise to the yeti's other name of the 'abominable snowman', but there are also reports of injured climbers being helped by them (e.g. by d'Auvergne in 1938).

All of this anecdotal and fragmentary evidence is suggestive, but also shows how myth and imagination guided by expectation based on legend can interact with observation to confuse the issue in the absence of hard evidence. On the other hand, the presumed yeti habitat of high mountains and dense *Rhododendron* forests in remote areas is not conducive to observation, camera trapping, or the hunting and capturing of shy and elusive animals. Following an expedition to Nepal, the mountaineer Reinhold Messner concluded that the yeti is a Himalayan black bear (*Ursus thibetanus*).[52] The bear rears onto its hind legs at times, and when moving often puts the back foot in the footprint of the front foot, which gives the tracks the appearance of belonging to a bipedal animal. Messner also indicates that when villagers took him to see a yeti, it was a bear that was encountered.

Orang pendek

An undescribed bipedal ground-dwelling ape appears to exist in the Kerinci-Seblat National Park and surrounding areas in west-central Sumatra, known locally as 'orang pendek' (little person). Is it a gibbon that has escaped competition with siamang and agile gibbon in the canopy? Or is it a descendant of the ancestral orangutan that used to inhabit this part of the Sunda shelf? The modern orangutan has spread southwards around the faunal barrier of Lake Toba to the west,[66] but there is some evidence to suggest that the southern populations of Kerinci-Seblat are different, as was claimed by the Dutch around 1920.

A team from Fauna and Flora International, led by Debbie Martyr, has analyzed completed questionnaires from about 200 Kerinci people; these support the original Dutch claim of an ape that is not a gibbon or orangutan as we know them. Apart from numerous casts of a unique footprint, each team member has seen an orang pendek, and is totally convinced of its existence, despite the current lack of proof from the camera traps, managed so successfully otherwise by Jeremy Holden. Achmad Yanuar has surveyed orangutans in Borneo, and was very skeptical about the orang

the investigation of the aging processes. Because apes are so much like humans, they go through many of the same aging processes and suffer from many of the same age-related disorders as do humans. Thus studies of aging in great apes have explored menopause and the behavioral changes that accompany it, changes in sexual activity, and behavioral change associated with brain aging. The Great Ape Aging Project involves noninvasive monitoring of health, cognition, and behavior of the oldest great apes in research facilities and zoological gardens. Further, it promotes the study

pendek, until he saw one. It remains a challenge to validate the existence of this species sharing human lineage; meanwhile, there is the real fear that it will soon become extinct because the fires and illegal poaching and felling that are devastating Kerinci-Seblat National Park must be pushing this ape to the brink of extinction.

The context of speculation about the orang pendek was dramatically changed in October 2004 by the publication of descriptions of *Homo floresiensis*, a 1 m tall hominin (fossil of human lineage) that existed on the Indonesian island of Flores as recently as 18 000 years ago, with anecdotal evidence of its survival into modern times.[9, 24, 53] This extraordinary discovery encourages the thought that a confirmed, small, recent hominin in Flores might not be entirely unconnected with a reported, small, current, bipedal, ground-dwelling 'ape' in Sumatra. One obvious possibility is that both are related to *Homo erectus* whose fossils dating to 1.6 mya have been found on the island of Java, which lies between Sumatra and Flores.

The unknown ape of northern DRC

North of the Congo River, there is a gap of about 1 000 km between the known distributions of the eastern gorilla (*Gorilla beringei*) and the western (*G. gorilla*). This gap is inhabited by chimpanzees (*Pan troglodytes*). In its midst is an area surrounding the towns of Bili and Bondo, in which European hunters had killed gorillas and officials had obtained gorilla skulls in the 19th century. The area was visited in 1996 by Karl Ammann, who sighted a gorilla skull there but also collected anecdotal reports of anomalous behavior by the local great apes. These 'gorillas' were said to slip away from people and never to charge them in the intimidation display characteristic of male gorillas. The 1999–2002 war in DRC inhibited further surveys, but a team of wildlife scientists including George Schaller and Richard Wrangham visited the area in 2001 and found nests that were built on the ground. This is unexceptional for gorillas, but the nests were being used repeatedly and were often in swampy locations, which is unusual for gorillas. The animals themselves were not seen.

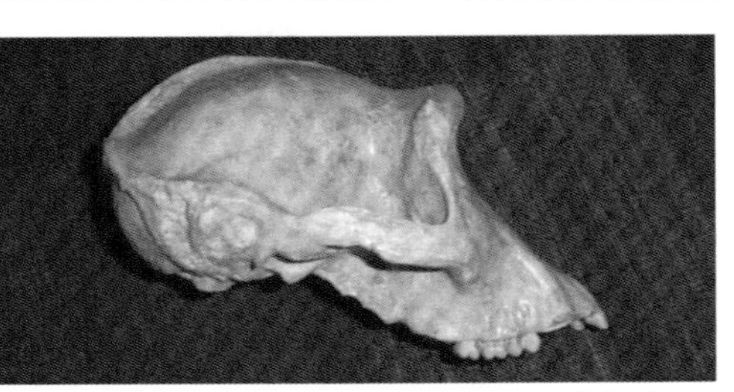

Colin Groves

This cast of the skull of the 'unknown ape' of Bili-Bondo is chimpanzee-like, but with a pronounced sagittal crest.

The following year, however, a team working with Shelly Williams succeeded in videotaping them, and Ammann in photographing them; they seemed to resemble gorillas in cranial anatomy but chimpanzees in postcranial anatomy, and to be anomalous in their fur.[88] Williams revisited the area in 2003 with Groves, who measured a number of skulls and pronounced them to be those of exceptionally large chimpanzees, one with an unusual sagittal crest. Casts of footprints were also longer than the longest recorded for either gorillas or chimpanzees, and the body weight estimated from evaluating a photograph of a dead individual exceeded that of the heaviest recorded chimpanzee. During the 2003 survey, Williams reported four of the apes being attracted to an imitation of the cry of a wounded duiker; they approached fast and apparently with deadly intent before fleeing silently on encountering humans. This behavior is suggestive of an active, if only opportunistic, hunting animal, more like a chimpanzee than a gorilla. Mitochondrial DNA analysis of hair and fecal samples had meanwhile established a chimpanzee identity for the animals, at least on the maternal side. Researchers from the University of Amsterdam launched a field study of these unusual chimpanzees in 2004.

Julian Caldecott

of the brains of apes that have aged and died in captivity, with a view to discovering more about the development of conditions related to Alzheimer's and Parkinson's syndromes, and other forms of neurological degeneration with age.

Chimpanzees are the only great apes currently used for laboratory-based biomedical research. They are uniquely susceptible to human hepatitis infections and serve as a model for this global health problem. Hepatitis research using chimpanzees has led to vaccines to protect people from hepatitis B and has also played an important role in the development of assays to reduce the risk of transmission of the hepatitis C virus through blood transfusions. Other areas of biomedical research on chimpanzees include the human immunodeficiency virus (HIV), although chimpanzees have proven to be poor models for HIV research; cognition; genetics; neurology; drug testing; and respiratory viruses. The value of genomic analyses of chimpanzees has also become established.[68] Until the late 1970s, the demand from biomedical labs for chimpanzees was largely met by imports from the wild, with infants being captured by killing the mother and any other defensive family members. Chimpanzees were listed on Appendix 1 of the Convention on International Trade in Endangered Species of Wild Fauna and Flora (CITES) in 1977, thereby banning international trade for primarily commercial purposes. The last such case in Europe was a controversial decision by Austria to issue an import permit for 20 chimpanzees from Sierra Leone in 1984, on export permits that had been issued prior to the ban.[65]

There is increasing resistance on ethical grounds to the use of captive apes in medical and other research. It was banned in the United Kingdom in 1997 and is forbidden in several other European Union countries, and there is a drive for a Europe-wide ban. The Humane Society of the United States is calling for an official ban on the use of great apes in biomedical research and testing in the United States, and for the permanent relocation of apes from research institutions to suitable sanctuaries. In the USA, approximately 1 300 chimpanzees now remain in research labs.[36] The National Institutes of Health maintain eight National Primate Research Centers for studying nonhuman primates, of which two use chimpanzees. The US federal government spent US$25–30 million on chimpanzee research at 23 institutions in 2001. Although a large total, this is less than 10 percent of that spent on research using monkeys. Fewer invasive procedures are used in research involving chimpanzees than is the case for other primates, probably because of the apes' cost.[36]

FURTHER READING

Begun, D.R. (2003) Planet of the apes. *Scientific American* **289**: 74–83.

Brown, P., Sutikna, T., Morwood, M.J., Soejono, R.P., Jatmiko, W.S.E., Rokus, A.D. (2004) A new small-bodied hominin from the Late Pleistocene of Flores, Indonesia. *Nature* **431**: 1055–1061.

Groves, C.P. (2001) *Primate Taxonomy.* Smithsonian Institution Press, Washington, DC.

Hartwig, W.C. (2002) *The Primate Fossil Record.* Cambridge University Press, Cambridge, UK.

Janke, A., Arnason, U. (2001) Primate divergence times. In: Galdikas, B.M.F., Briggs, N.E., Sheeran, L.K., *et al.*, eds, *All Apes Great and Small, vol. 2: African Apes.* Kluwer Academic/Plenum Publishers, Boston, Dordrecht, London, Moscow, New York. pp. 19–33.

Kavanagh, M. (1983). *A Complete Guide to Monkeys, Apes and Other Primates.* Jonathan Cape, London.

Tavaré, S., Marshall, C.R., Will, O., Soligo, C., Martin, R.D. (2002) Using the fossil record to estimate the age of the last common ancestor of extant primates. *Nature* **416**: 726–729.

Wildman, D.E., Uddin, M., Liu, G., Goodman, M. (2003) Implications of natural selection in shaping 99.4% nonsynonymous DNA identity between humans and chimpanzees: enlarging genus *Homo. Proceedings of the National Academy of Science (PNAS)* **100** (12): 7181–7188. http://www.pnas.org/cgi/content/abstract/1232172100v1.

Young, E. (2004) The beast with no name. *New Scientist* **184** (2468): 33–35.

ACKNOWLEDGMENTS

Many thanks to Peter Andrews (Natural History Museum), Colin Groves (Australian National University), Alexander Harcourt (University of California, Davis), David Pilbeam (Harvard University), Christophe Soligo (Natural History Museum, London), Michael Wilson (Gombe Stream Research Center), and David Woodruff (University of California, San Diego) for their valuable comments on the draft of this chapter. Thanks also to Valerie Kapos (UNEP-WCMC) for research and advice.

AUTHORS

Martin Jenkins, UNEP World Conservation Monitoring Centre

Box 1.1 Martin Jenkins, UNEP World Conservation Monitoring Centre

Box 1.2 Julian Caldecott, UNEP World Conservation Monitoring Centre

Great ape biology

JANE GOODALL

Since the first methodological studies of the apes in their natural habitat suddenly proliferated in the 1960s, after the Second World War, a vast amount of data has been collected from across Asia and Africa. Behavioral ecologists have made great strides in explaining the species differences in ape social structure and behavior on the basis of differences in diet, distribution of food, and risks from predators and conspecifics. And ongoing studies of diverse ape populations have been greatly aided by videotaping, and by new technologies such as genetic profiling of DNA (obtained noninvasively from fecal samples), and satellite imagery.

The apes of Asia and Africa exhibit a wide range of social systems and behaviors. They are found in group sizes ranging from over 100 (chimpanzees) to lone individuals (orangutans). By and large, the Asian apes live in smaller groups than those of Africa. The most solitary apes are the orangutans, and the most social are bonobos and chimpanzees. Gibbons live in pairs and are monogamous, maintaining their relationships and territories through daily bouts of loud and haunting duet singing and 'dancing' in the trees. The parents raise their offspring together, then drive each out when he or she reaches maturity. Siamangs are more social and can be found in larger groups.

Orangutans, the most arboreal of the great apes, live in semisolitude. Fully mature dominant males with fully developed cheek pads, known as 'flanged' males, live alone in forests that contain the home ranges of several females. These males, more than twice the size of females, advertise their whereabouts with loud long calls, and the females visit them around the time of ovulation. When two flanged males meet in the presence of a receptive female lethal competition can ensue. Younger or less dominant males sometimes succeed in mating with an ovulating female, but typically without her consent. When large quantities of food are available, immature or low-dominance males often join together with females to exploit it, but they will be chased away if a flanged male arrives. Although orangutans have been studied for many years there is still debate about whether there are permanent social bonds between individuals in dispersed local communities – although it would be surprising (to me) if there are not.

The African apes – gorillas, chimpanzees, and bonobos – are all highly social, but their groups have rather different structures. Gorillas live in groups of three to 50 individuals who are always together. A typical group contains one or two silverback males, a few younger, blackback males, and a number of adult females and young. Young females generally leave their natal groups, and mate with males of other groups in which they stay to raise their families. Most males also leave their natal groups as late adolescents, sometimes associating with other males until they are mature enough to lead a group of their own. Then they may capture females from other groups, sometimes committing infanticide.

Chimpanzees and bonobos live in multimale, multifemale groups or communities. These are fission-fusion social units, with individuals associating in smaller temporary subgroups within the community range. In some populations, chimpanzees, particularly females, often travel alone. When seasonal fruits ripen, large noisy gatherings congregate to feast together. Chimpanzee and bonobo males generally remain for life in the community in which they are born (this is known as philopatry) while many young females (like gorilla females), leave and join other communities before giving birth. This pattern is unusual among mammals, but is typical of many human societies.

Male chimpanzees and gorillas are clearly dominant to females in their group, and can show high levels of aggression. Male bonobos are generally less aggressive, and there is less sexual dimorphism. Indeed, females sometimes form alliances in order to dominate males. Conflicts in bonobo society are often resolved through sexual behavior.

Silverback gorillas are valiant in defense of group members, fighting those enemies (including human hunters) who are not intimidated by their impressive chest beating, roaring, and fast charge. Male chimpanzees cooperate to defend their community range, patrolling the boundaries, proclaiming their presence with their loud 'pant-hoot' distance call, and sometimes conducting lethal raids into adjoining territories, attacking and killing members of other communities – behavior with many similarities to primitive human warfare.

One behavior that has attracted much attention is tool use, the use of an object to attain a goal out of reach of hand, claw, mouth, or beak. Most apes are capable of tool using and tool making, when an object is modified in order to make it suitable for use as a tool. At one time this was thought to be the behavior which differentiated *Homo*, more than any other, from the rest of the animal kingdom. Tool use has now been seen in the wild in birds, monkeys, and three of the six species and subspecies of great apes. All the great apes make use of natural materials to construct their nests, but this is not strict tool use (any more than bird, mammal, or fish nests). Bornean orangutans, however, have been seen holding leaves over their heads for shelter from the rain. Sumatran orangutans, chimpanzees, and bonobos are the most prolific tool users and tool makers. All the great ape species, however, are able to acquire tool-using behaviors in captivity. Most fascinating is the fact that in all areas where they have been studied, chimpanzees use different objects for different purposes, and all available evidence suggests that these traditions can be described as primitive cultural behaviors, passed from one generation to the next through observation, imitation, and practice. This is likely to hold good for all ape tool-using skills.

We humans, of course, are also great apes. According to some studies, we share more of our DNA with chimpanzees than chimpanzees do with gorillas, placing us clearly within the ape family tree. The structure of the immune system, the composition of the blood, and the anatomy of the brain and nervous system, are strikingly similar in humans and other apes. This close biological, evolutionary relationship makes the study of the behavior of the other great apes particularly fascinating and important for us, providing insights into the evolution of much of our own behavior.

There can be no question that the apes resemble us in many aspects of social behavior. They have distinctive personalities, and show emotions similar (perhaps identical) to those we call joy, sadness, fear, and so on. Chimpanzees show political behavior (alliance forming and social manipulation), hunt mammals and share the kill, use various leaves for medicinal purposes, and communicate with kissing, embracing, patting on the back, swaggering, and so on. In captivity they show clear comprehension of human-type language and evidence of language capabilities. They have long periods of childhood dependence on the mother, and, certainly in chimpanzees, long-lasting and supportive affectionate bonds between mothers and their offspring, and between siblings. Like us, the apes have a dark side to their nature, but they are also capable of compassion and altruism. We seem to have inherited both a capacity for violence and a capacity for loving from our shared primate heritage.

Perhaps the greatest difference between *Homo* and our ape relatives is the fact that we have developed a sophisticated spoken – and now written and electronic – language that enables us to plan far into the future, learn from the distant past, teach about objects and events not present – even purely imaginary – and share and discuss ideas. Our highly evolved intellect gives us the ability to make decisions regarding the life and death of entire species. Only we can make the decision to preserve the apes. Let us hope we work harder to do so, both because they are worth it in their own right, and also so that we may continue to learn from them about their world, and about our own.

Jane Goodall
Founder, the Jane Goodall Institute
UN Messenger of Peace
www.janegoodall.org

CHAPTER 2

Great ape habitats: tropical moist forests of the Old World

JULIAN CALDECOTT AND VALERIE KAPOS

All great apes are associated at least to some degree with tropical forests, but not with all forest types. Whether a given tropical ecosystem supports great ape populations is determined by a mixture of biogeographical and ecological factors, combined with patterns of habitat conversion and disturbance resulting from human activity.

The most important ecological factor is the availability of an adequate supply of suitable food. Here, it is relevant that the great apes have simple, globular stomachs and lack any special adaptations to allow fermentative digestion. Other primate groups, such as the leaf monkeys (Colobinae), have sacculated, fermentative stomachs (with small 'bag-like' compartments where bacteria break down cellulose). This has profound ecological consequences; these primates are able to obtain nutrients from coarse materials such as mature leaves, which are otherwise difficult to digest and may be defended by toxic secondary metabolites. The diet of the great apes, on the other hand, is largely restricted to ripe, sugary fruit and to other easily digestible plant parts – shoots, palm hearts, flowerbuds, herbaceous foliage, ginger stems, seeds not defended by dangerous chemicals – and to the tissues of vertebrate or invertebrate animals.

Body size further determines the food supplies needed by primates. The bigger any mammal is, the less vulnerable it is to poisoning and the lower its energy demand per unit weight. The proportion of lower-quality foods (such as leaves) that a mammal with a nonfermentative digestive system can eat increases with its size. Smaller mammals of the same general design require foods that are richer in energy. These simple rules map well onto the great apes: the largest, the gorillas (around 90–220 kg), eat considerable amounts of herbaceous foliage; smaller (35–100 kg) orangutans favor ripe fruits but tolerate

Iroko Foundation

Mist rising off the rain forest near the village of Abo Obisu, Cross River National Park, Nigeria.

Map 2.1 Great ape habitats of Sumatra and Borneo

Data sources are provided at the end of this chapter

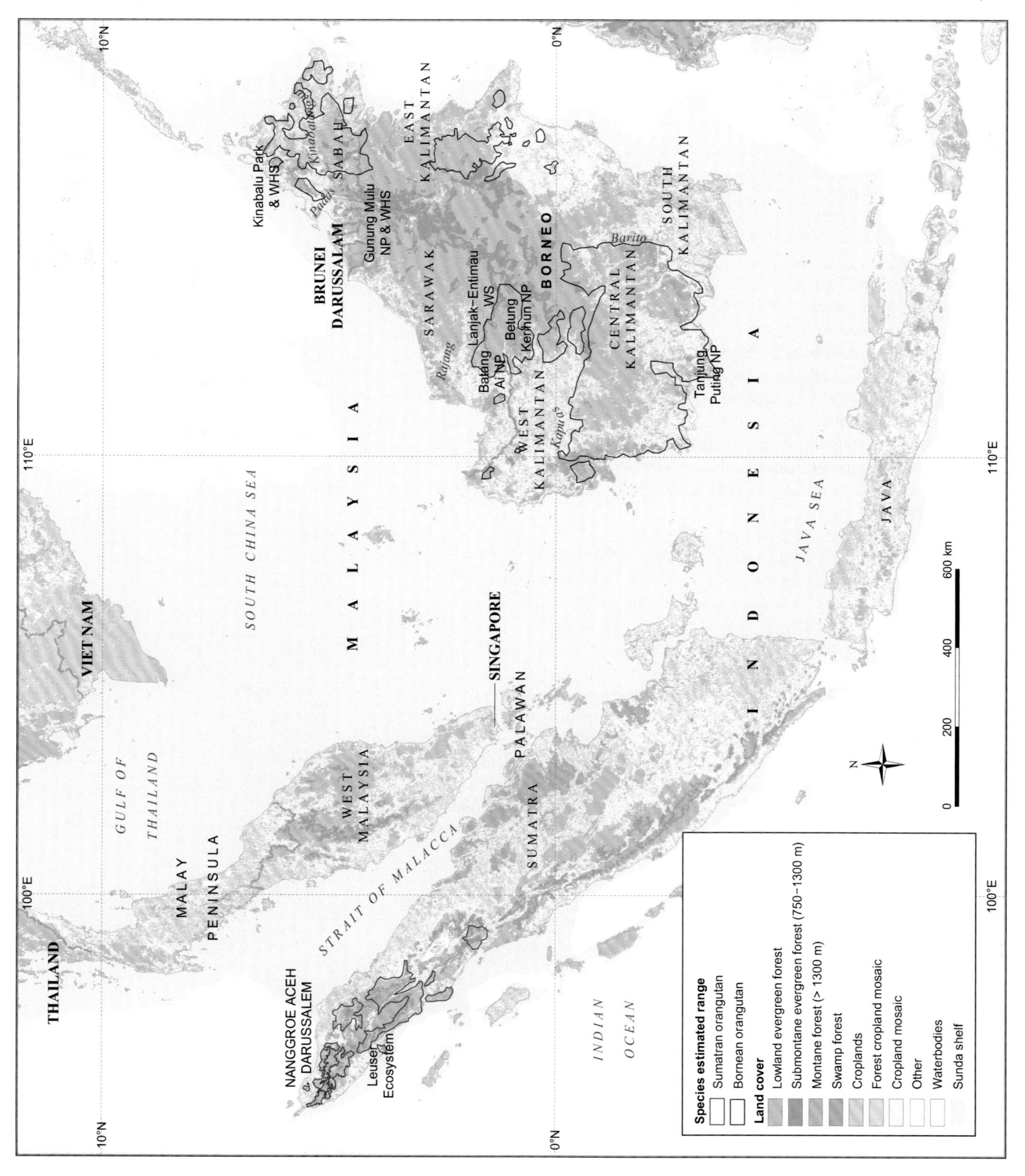

poorer fare, such as unripe fruits and tree bark; and the smallest, chimpanzees and bonobos (30–60 kg), also favor ripe fruit but are agile enough to catch vertebrate meat, and are diligent harvesters of insects. All of the species may eat young, poorly defended foliage.

To understand the ecology of a great ape, therefore, the key step is to consider how the ecosystems in which they live offer the kinds of foods that they are able to eat, and how this varies in space and time. The balance and timing of different seasonal phases (flowering, fruiting, flushing of new leaves) of forest plant species are thought to have strong influences on ape distribution. Particularly, where fruiting and flushing phases occur synchronously, seasonal food scarcity may be a serious problem for apes; this can be alleviated only if it is possible to move between habitats within a mosaic of forests with different seasonal conditions. Foraging and ranging behavior in great apes is therefore also strongly influenced by such patterns, and aspects of social behavior may well be indirectly affected by the same influences.

SOUTHEAST ASIA

Biogeography

Orangutans inhabit lowland rain forest on two large islands (Map 2.1): Sumatra (475 000 km^2) and Borneo (740 000 km^2). These lie on the Asian or Sunda continental shelf, which also supports the islands of Palawan and Java, and the Malay Peninsula. The continental shelf is covered by seas that are often less than 200 m deep, including the Java Sea, part of the South China Sea, the Gulf of Thailand, and the Strait of Malacca, but is bounded by much deeper water. The shallow modern seas over the Sunda shelf have come and gone, as the sea level has changed over the last 2 million years in response to successive global ice ages and warming events. This exerted a strong influence on the region's biogeography, as at times it promoted the dispersion of terrestrial species through the Sundaic land masses while limiting dispersion between them, and at other times it tended to promote isolation and local speciation within the various islands.[1, 15]

Most animal species native to Borneo and Sumatra have close relatives in Asia, where ancestral orangutans originated. Among plants, groups centered on the Sunda shelf include the tree families Dipterocarpaceae (dipterocarps) and Magnoliaceae (magnolias), the breadfruit (*Artocarpus*) and its relatives, and the climbing palms or rattans (Calamoideae).[45] Sumatra and Borneo are both part of the West Malesian botanical subregion.[34, 44] The ecologies of these islands have been reviewed by Whitten and colleagues for Sumatra,[46] and by MacKinnon and colleagues for Indonesian Borneo (Kalimantan).[18] These two islands both have moist equatorial climates with a mean annual rainfall of 2 500–5 000 mm. Both are extremely rich in species. Borneo has more species as a result of its larger size, and more of these are endemic (occurring nowhere else) because of its greater isolation (Table 2.1).[1, 15, 17, 40, 41, 47]

Ecology of Sundaic dipterocarp forests

The natural vegetation type that dominates the interiors of Sumatra and Borneo is tropical evergreen rain forest;[44] this changes with altitude from lowland mixed dipterocarp forest (below about 700 m), to hill dipterocarp forest (from about 700 to 1 200 m), to lower montane rain forest (from about 1 200 to 1 500 m), and eventually to upper montane rain forest (above about 1 500 m). Each forest type spreads over a wider altitudinal range on larger mountains, and is more compressed on

Table 2.1 Species richness and endemism in Sumatra and Borneo

Island	Birds	Mammals	Reptiles	Freshwater fish	Selected plant taxa
Number of native species					
Sumatra	465	194	217	272	820
Borneo	420	210	254	368	900
Percent endemic species					
Sumatra	2	10	11	11	11
Borneo	6	48	24	38	33

smaller ones. Significant areas of low-stature heath forest are found throughout the interior of Borneo in areas of white sand or other very nutrient-poor soils. Both Sumatra and Borneo have very large areas of freshwater swamp forests that often grow over deep peat deposits.

The lowland dipterocarp forests are the most species-rich of these forest types; up to 2 300 species of tree have been recorded in such ecosystems in Borneo, compared to about 850 in heath forests and fewer than 250 in peat-swamp forests.[44] Plant species richness declines with elevation, although endemism may increase; several hills in the northern Sarawak area have been identified as centers of plant diversity.[40] The same patterns hold true for birds, for example on Gunung Mulu in northern Sarawak, where there are 171 species in lowland forest but only 12 at 1 300 m.[44] The mountains of northern Borneo comprise an endemic bird area to which 26 restricted-range bird species are confined.[12]

Root system of a strangling fig, Malaysia.

KOCP

The lowland and hill 'dipterocarp' forests are so called because many of their large trees belong to the family Dipterocarpaceae. The abundance of these trees is a common feature of lowland and hill forests in Borneo, Java, Sumatra, and the Malay Peninsula, though Bornean forests typically have the greatest degree of dipterocarp dominance. Their fruiting patterns, which tend to be synchronized within and among species, add to the ecological influence exerted by their abundance. Fruiting is irregular, resulting in massive fruit crops (mastings) at unpredictable intervals of two to five years.[13] This is thought to reduce seed predation by overwhelming with food during mast years the populations of seed-eating animals that are limited in abundance by food scarcity at other times, reducing the level of predation inflicted on any one seed crop.

Water stress during occasional droughts is believed to provide the main stimulus for masting by dipterocarps. As many other tree taxa in Southeast Asia use the same environmental cue to prompt flowering, there is a tendency in dipterocarp forests for the foods available to fruit-eating and seed-eating animals to be either superabundant or almost absent at any given time. This helps to explain why such animals are collectively rare in dipterocarp forests compared with other rain forests, and also why the biomass of fruit-eating primates is inversely related to dipterocarp abundance within forests that are otherwise similar.[2, 4, 5, 19] The pattern of fruit and seed availability in dipterocarp forests favors high mobility and/or rapid reproduction among the animals that depend on such foods. The first adaptation allows them to track fruiting activity over wide areas, while the second allows their populations to respond swiftly to unpredictable food supply.

Orangutans are fruit-eating animals adapted to an environment in which fruit is fundamentally in rather poor supply. They manage this by being strongly adapted to arboreality, spending most of their time in the trees, and being deeply familiar with conditions in the canopy within a large home range. Orangutans can therefore track the seasonal changes in a patchy rain forest, where the timing of fruiting peaks can vary with elevation and aspect. They forage within the range in a typically zigzag way, unless they happen to know where large sources of fruit are to be found.

Tropical rain-forest structure changes markedly with altitude, as the trees become smaller in girth and lower in stature, more densely packed and with fewer large branches, and the canopy lowers from 25–40 m in lowland forest to 15–25 m in lower montane. This structural change alone would be expected to impose energy costs on a very arboreal animal like an orangutan. The total availability of fruit likely to be preferred by orangutans also declines with altitude, as does floristic diversity. Tree genera that typically comprise major components of orangutan diets progressively drop out of the canopy composition with increasing elevation: *Nephelium* (Sapindaceae), *Baccaurea* (Euphorbiaceae), *Artocarpus* (Moraceae), and *Aglaia* (Meliaceae) disappear between hill and upper dipterocarp forest (at 700–900 m); *Xanthophyllum* (Polygalaceae), *Mangifera* (Anacardiaceae), and *Garcinia* (Clusiaceae) disappear between upper dipterocarp and oak–laurel forest (at 1 200–1 400 m).[4] These factors combine to explain why orangutans are generally restricted to altitudes of less than about 750 m, except where there are exceptional concentrations of favored fruit trees.[27]

Trees that provide fruit suitable for orangutans are typically found at higher densities and bear fruit more continuously in Sumatran forest than in Bornean forest, although there is much patchiness and dynamism in the forests of both islands. Significantly, the lesser dominance of dipterocarp trees in Sumatra makes space for other tree species that collectively fruit more steadily. Some differences between Sumatran and Bornean orangutan behavior have been attributed to the different patterns of food supply.[33] Figs, in particular, occur at such high concentrations in some parts of Sumatra that these fruit alone are thought to have enabled greater density and sociability among orangutans in Sumatra than in Borneo.[9] Average home-range size, day-range length, and population density all respond to differences in the abundance and continuity of fruit availability between locations, seasons, and ecosystem types, both within and between Sumatra and Borneo.

Before deforestation, the distribution of orangutans was discontinuous both in Borneo and in Sumatra. Orangutans are absent from very large areas of apparently suitable habitat, for example in most of the southern two thirds of Sumatra, and between the Rajang River in central Sarawak and the Padas River in western Sabah. One explanation for this is that prehistoric hunting may have extirpated orangutans; ancient cave sites in areas now missing orangutans often contain their bones along with those of other prey. Moreover, some areas with abundant orangutans occur where local people have a strong cultural reluctance to eat orangutans. Examples are in the strongly Muslim Aceh Province of Sumatra and in the Batang Ai catchment in Sarawak, where a long-standing hunting taboo exists. An alternative explanation is that orangutans live close to the edge of an ecological niche that can become unviable with a slight shift in forest composition, for example

Ian Singleton/SOCP

Primary lowland rain forest, habitat of the Sumatran orangutan, Indonesia.

Map 2.2 Great ape habitats of Africa

Data sources are provided at the end of this chapter

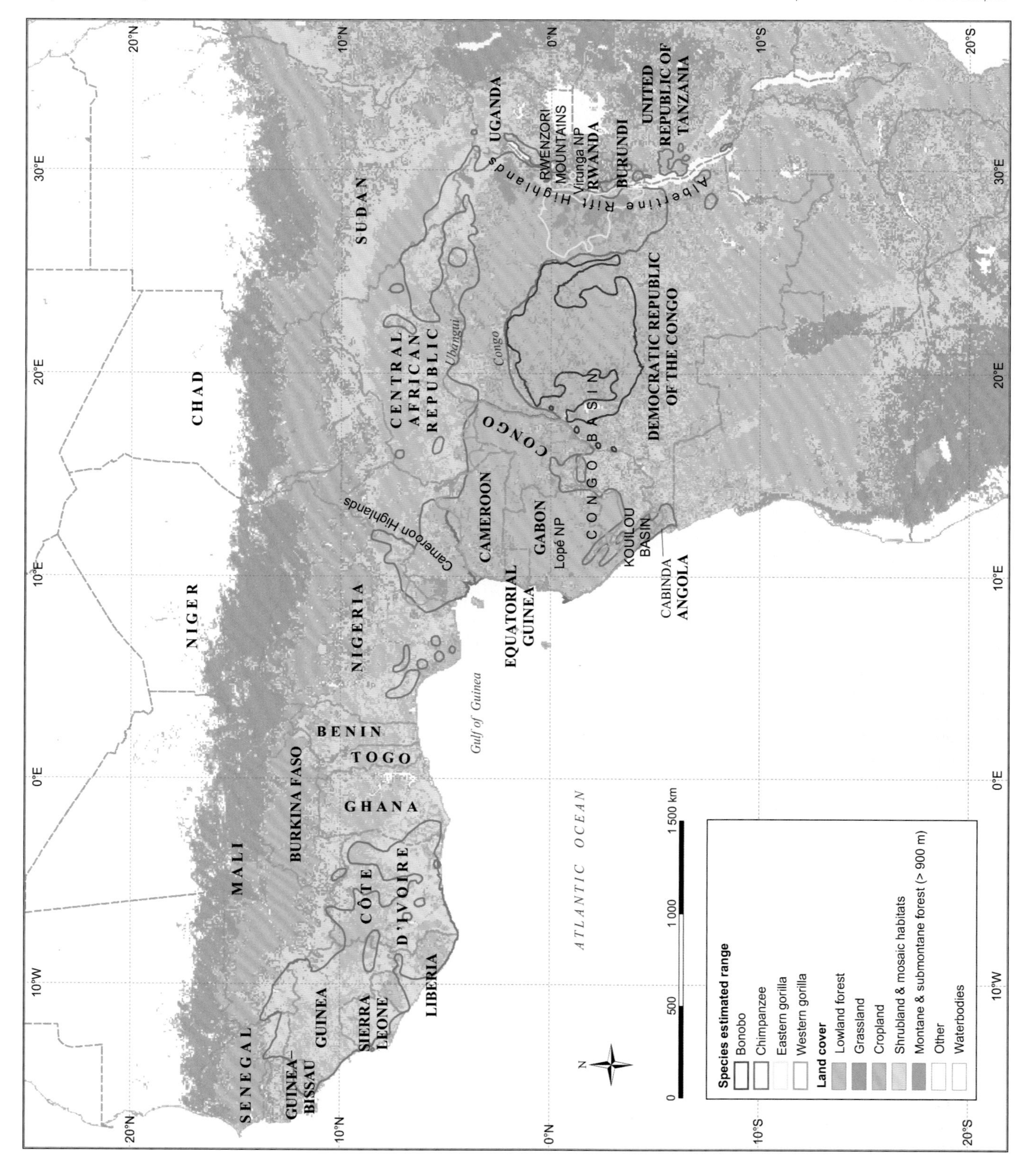

in favor of dipterocarps. According to this argument, patchiness in the distribution of breeding populations of orangutans reflects patchiness in the forest ecosystem. Each theory may explain the absence of orangutans from different places.

Impacts of human disturbance

It is no longer realistic to describe Sumatran or Bornean ecology without reference to human impacts. This is because the land cover of both islands seems to be in the process of rapid conversion, possibly in its entirety, from moist forest to plantations, farms, settlements, and fire-maintained grassland. This pessimistic scenario is informed by the recent history of land use in both islands. The latter has involved official and unofficial resettlement programs, widespread logging (both legal and illegal), rapid expansion of oil palm (*Elaeis guineensis*) and pulpwood plantations (e.g. of *Acacia mangium*, to produce the cellulose used in paper), and the widespread and poorly planned development of road infrastructure. The unrestrained use of fire as a means to clear land in an environment that is ever more loaded with fuel, and increasingly dry as a result of local, regional, and global climate change, has also been a contributory factor.

Given that Borneo and Sumatra are two of the most biologically rich islands on Earth, the implications of this change for global biodiversity are profound, and neither orangutan species could possibly survive for long in these circumstances. The only plausible strategy for safeguarding significant components of Sumatran and Bornean biodiversity, including viable populations of orangutans, depends on preserving intact areas of lowland forest large enough to retain resistance to fire under drought conditions. This is still just feasible in Sumatra within the 26 000 km^2 forests of the Gunung Leuser area (the Leuser Ecosystem, see Box 11.2), and in Borneo within the 11 000 km^2 transfrontier forests of the Lanjak-Entimau Wildlife Sanctuary in Sarawak and the Betung Kerihun National Park in West Kalimantan. The rich peat swamps of West Kalimantan and Central Kalimantan also have a strong potential for biodiversity conservation.

AFRICA

Biogeography

Together, the African apes occupy a wider range of ecosystem types than orangutans. All of the African ape species are closely associated with tropical moist forests and rain forests, but their habitats span wide altitudinal ranges both within and between species, and chimpanzees also use dry forest and savanna habitats.

The African rain forests are sustained by the flow of wet air off the Atlantic Ocean, with winds that shed their moisture as they rise and cool over land. The forests stretch from the mouth of the Congo River across the Congo Basin to the mountainous borders of Rwanda and Uganda, from the southern Democratic Republic of the Congo (DRC) to southern Nigeria (Map 2.2). In Benin and Togo, drier forests and savannas intervene, due to the configuration of the coast and prevailing winds – an ancient interruption known as the Dahomey Gap. Beyond this, a further expanse of rain forest runs from Ghana to Guinea, blending further up the coast and inland into bush savanna, grassland, and ultimately the Sahara Desert.

These lowland moist forests are known as the Guineo-Congolian formations.[42] They are most species-rich in Central Africa and relatively impoverished in West Africa. Overall, they are much less rich in species than the rain forests of Southeast Asia and South America. Their relatively low diversity is thought to result from past climatic fluctuations that greatly reduced their extent. There is good palynological (pollen) and fossil evidence of major vegetation changes in what is now the rain-forest zone of Africa. The climate has dried out repeatedly over geological time, most recently during the last northern hemisphere ice age, which ended approximately 14 000 years ago. These changes caused the restriction of moist forests and their species to a few remaining moist enclaves, and the expansion of the drier vegetation types. Indeed, the forests in some southern parts of the Congo Basin are now growing on what were once sand dunes of the Kalahari Desert. Many moist forest species disappeared entirely under these climatological stresses.

As the climate changed following the last glacial maximum, however, the remaining moist forest species spread out from the few centers where they had persisted. Those species that were less efficient dispersers tended to remain localized in the refugia, giving rise to centers of diversity and endemism. The lesser species richness of the West African rain forests is consistent with both the much greater reduction in rain-forest area of this zone during the Pleistocene, and the greater prevalence of drier and more seasonal current

Tamar Ron

Lowland forest chimpanzee habitat, Maiombe, Cabinda, Angola.

climates, which make these forests naturally more fragmented than in Central Africa.

Main forest types occupied by African apes

The Guineo-Congolian lowland rain forests grow on well drained soils throughout the region and are largely evergreen, despite some seasonal variation in their climates, especially in West Africa. Although a number of tree species shed their leaves at particular seasons, they are not synchronized in doing so. They normally remain leafless for only a few days, long enough to purge themselves of leaf-eating insects. These forests are characteristically tall, with canopies usually over 30 m. Emergents and sometimes even the main canopy can reach 50–60 m, especially in the wettest coastal zones.

Generally these forests are rich in tree species and relatively poor in herbaceous and understory species, except where canopies are disturbed or otherwise broken. In these areas, the giant ginger, *Aframomum giganteum*, other gingers (Zingiberaceae), and members of the Marantaceae family are notable features of the understory and crucial food plants for apes. A few areas are nearly as rich in woody plant species as parts of South America but, generally, plant species richness is a little lower. Most of the tree species are relatively widespread, but some areas are noted for their high concentrations of species with restricted distributions.

In some areas of the central Congo Basin, lowland rain forests are dominated by one or a few species belonging to the legume family Caesalpiniaceae. Such dominance has major implications for the distribution and foraging ecology of forest animals. *Gilbertiodendron dewevrei* and *Julbernardia seretii* are two of the species that most commonly succeed in dominating the canopy. These and other dominating Caesalpiniaceae are noted for producing dry fruits in a mast fruiting pattern[11] and for having relatively unpalatable leaves. Although these fruits are less attractive to primates than fleshy fruits, and their irregular production poses problems of food scarcity at certain times of year, there is evidence that some apes, especially lowland gorillas, move into these forests to take advantage of the plentiful and nutritious fruits when they are in season.

Inland swamp forests and mosaics of swamp forests with lowland rain forests also provide key habitat for apes in Africa, especially in Central Africa.[24] They are often as tall as the well drained (*terra firme*) forests, but with much more broken and uneven canopies. This unevenness allows for penetration of more sunlight than the continuous canopy of *terra firme* forest does and, as a result, there is much more growth of herbaceous understory plants. The Central African swamp forests are particularly important for bonobos, which are thought to have survived the last glaciation in enclaves of swamp forest. Elsewhere, local patches of swampy forest are also key habitat for gorillas, which like to eat herbaceous aquatic plants. The differing phenology of the plant species growing in swamp forests means that these habitat patches can be very important on a seasonal basis, in maintaining continuous food supplies for apes within their habitats.

Throughout the region, forests change in character with increasing elevation. As in Southeast Asia, they tend to decrease in stature and species richness, but increase in endemism due to their isolation from other similar areas. They also tend to have greater numbers of epiphytes and a more substantial herb layer than lowland forests. The two main areas of highland forest of relevance to African apes are the Cameroon Highlands and the Albertine Rift Highlands.

The Cameroon Highlands are volcanic uplands covering about 14 000 km^2 in western Cameroon and eastern Nigeria. They are of international importance for their endemic birds and amphibians (Table 2.2) and also have a high degree of floristic endemism. Due to the moister climate near the coast, montane forests occur at lower elevations (above 500–800 m) at the southwestern end of the region than inland (above 2 000 m), where the surrounding lowland vegetation is savanna. The montane forest grades into a zone of mixed *Podocarpus* (Podocarpaceae) and bamboo at higher elevations (>2 600 m) on the inland mountains.[14]

The Albertine Rift Highlands of eastern DRC, Rwanda, Burundi, and western Uganda cover around 56 000 km^2 and include the Itombwe, Virunga, and Rwenzori Mountains. They are very rich floristically and have large numbers of endemic bird and amphibian species (Table 2.2), many of which are at risk of extinction due to high rates of deforestation. The lowland rain forests and transitional or submontane forests that occur up to about 1 600 m in altitude are closely related floristically to the Guineo-Congolian lowland forests. They have canopies averaging 30–40 m tall. Above 1 500–1 600 m, montane forest is found, which has less than half as many woody species as the lowland forests and forms a canopy about 15–25 m tall. These forests are notable for the presence of coniferous *Podocarpus* species. Above 2 100 m, bamboo appears in the forests, and the frequency of bamboo stands increases with altitude; bamboo is interspersed with dwarf or 'elfin' forest and subalpine scrub above 2 500 m and may form continuous bamboo forests at elevations above 3 000 m. The high-altitude open forests and shrublands, which may also be mixed with grasslands, are characterized by *Hagenia abyssinica* (Rosaceae), *Hypericum revolutum* (Clusiaceae), ericaceous shrubs, and a dense herbaceous understory including *Galium ruwenzoriense* (Rubiaceae), thistles, wild celery, and other herbs that are important in the diet of mountain gorillas.[38]

The bamboo zones of these highland systems provide crucial resources to a number of animal species, including both mountain gorillas and eastern lowland gorillas. The bamboo, *Yushania alpina* (synonym: *Arundinaria alpina*), spreads via rhizomes and forms a dense canopy, especially at altitudes of 2 300–2 600 m. Gorillas move from mixed forest into the bamboo forest during the 'short rainy season', the season of bamboo sprouting (September–November),[49] when bamboo shoots may make up 70–90 percent of their diet. The persistence of the bamboo determines both forest structure and the abundance of other gorilla food plants; the canopy of the bamboo is so dense that it impedes the establishment of other plant species. Thus gorillas must use several other habitats besides the bamboo to obtain adequate food supplies throughout the year.[35]

Table 2.2 Endemic species in the highlands of Central Africa[3, 14, 23]

	Mammals	Birds	Amphibians
Cameroon Highlands	16	20	60
Albertine Rift Highlands	36	30–36	34

The origin of and factors determining the distribution of bamboo stands are the subjects of some debate. It has been suggested that the bamboo stands establish only in response to forest disturbance, which may include fire. Bamboos are monocarpic, i.e. they have only one fruiting season. This life cycle of mass flowering at infrequent intervals of 15+ years, followed by death, may help to maintain these pure stands as dead bamboo promotes fire that helps to suppress other species. Conversely, the density of the bamboo stands means that other trees can become established within them only during episodes of bamboo dieback. These episodes may be periods of significant food scarcity for local gorillas.

Chimpanzees are notable among the African apes for using drier tropical ecosystems as habitat. Particularly in the far west of their range (Senegal and Mali) and in East Africa (United Republic of

Tanzania) they make use of dry forests and woodland mosaics as well as savanna woodlands. The seasonally dry forests of West Africa's Sudanian belt[42] are now much fragmented by dense human settlements and agriculture. Their character is determined by a strongly seasonal distribution of rainfall, with the 900–1 200 mm of annual rainfall confined almost entirely to six or eight months of the year. The dry season is made more intense by the influence of the hot, dusty harmattan wind which blows from the Sahara, and periodic extreme droughts are characteristic of this zone. The forests that persist in parts of this area that are protected from fire are characteristically deciduous, and dominated by two species of caesalpiniaceous tree, *Gilletiodendron glandulosum* and *Guibourtia copallifera*. Along watercourses there is denser gallery forest, which can be evergreen in places.

More important as chimpanzee habitat, if only because it is more extensive, is the woodland mosaic and wooded savanna that has been created by extensive forest clearance and grazing in both the moister Guinean forest zone and the Sudanian *Isoberlinia* (Caesalpiniaceae) woodland. In these systems, dense grass cover and the seasonal fire regime impede the regeneration of forest trees. Similarly in East Africa, some chimpanzees occupy areas of savanna that result from the conversion and degradation of both wetter and drier miombo woodland formations, including areas with annual rainfall as low as 850 mm.[16]

Where chimpanzees inhabit open habitats, they are heavily dependent on any available trees and woodland for food and shelter. Their use of habitat patches within a mosaic varies seasonally, and it appears that the food sources of the more characteristically dry zone vegetation, such as the pods of *Isoberlinia* in West Africa and *Julbernardia* and *Brachystegia* (Caesalpiniaceae) in East Africa, may be most important for chimpanzees at the driest times of year.

Impacts of human disturbance

Like the Asian forests supporting orangutans, African ape habitat is increasingly subject to disturbance, exploitation, and conversion to other land uses. These pose significant threats to all of the species (see Chapter 13).

However, all persist to some extent in disturbed habitats and indeed depend on low-level ecosystem disturbance for a proportion of their food supplies. All of the African apes are to a greater or lesser degree folivorous, but cannot digest the mature leaves of canopy tree species. Disturbance, at least by the natural treefall dynamics of the forests, creates gap environments where herbaceous species thrive and provide palatable foliage. Similarly, the fruits of some species that are characteristic of regenerating forest, such as *Musanga* (Moraceae), can make up a major part of ape diets at some times of year.[10] For these reasons, lowland gorillas, chimpanzees, and bonobos make good use of disturbed habitats and forest edges as habitat for foraging.[8, 21, 28, 29]

It is also clear, however, that disturbance of forests by humans has significant negative impacts on apes and that all species avoid areas where people are active. There are data now showing that even low-intensity selective logging in Gabon can cause a significant long-term depression of chimpanzee numbers.[43] Given that most African moist forests where great apes live are now allocated to logging concessions, this adds poignancy to calls for ways to be found to collaborate with and regulate more closely the logging companies concerned, to make the whole process as 'great ape friendly' as possible. The general outlook is that the ecosystems that can support apes are shrinking at an alarming rate, with potentially disastrous consequences for ape populations.[7]

THE ROLE OF APES IN THEIR ECOSYSTEMS

While depending on their supporting ecosystems, apes play critical roles in determining the nature and persistence of those same ecosystems. Through their consumption of fruit they act as important dispersers of many forest tree species. In one Ugandan forest, chimpanzees were responsible for a disproportionately large fraction (about 45 percent) of the seeds defecated by frugivorous primates.[48] Meanwhile, studies in Gabon have shown that gorillas are the sole dispersers for the dominant tree species at Lopé National Park, and high-quality dispersers for others.[31, 36, 37] Both chimpanzees and gorillas disperse large numbers of viable seeds over far greater distances than other forest primates,[25] and therefore play a major role in maintaining those species that regenerate better farther from the parent tree. This in turn contributes to the diversity and heterogeneity of the forest.

The great apes also have impacts on forest structure and composition through their use of

Elizabeth A. Williamson

The edge of the Volcanoes National Park in Rwanda, showing cultivation right up to the park boundary, and pyrethrum growing in the foreground.

leaves and branches as both food and nesting material. Gorillas in particular may help to perpetuate the occurrence of the herbaceous plants they favor by causing significant structural disruption to the forest as they feed.[26, 39]

Apes are also ecologically important as competitors and, in some cases, predators. Chimpanzees and gorillas occurring in the same forest area use very similar resources, but this competition is modified by the greater quantity of herbaceous material in the gorillas' diet and because they avoid direct contact with each other.[30] Chimpanzees compete more directly with other frugivorous primates, especially cercopithecine monkeys,[20, 22, 32] and may drive them away from food sources. There is also considerable evidence that chimpanzees are important predators of some monkey species, especially of the red colobus (see Box 4.1),[6] and of other small mammals. Both gorillas and chimpanzees also prey on invertebrates, and may break open termite mounds as they forage.

CONCLUSIONS

Great apes belong to tropical ecosystems; they shape and are shaped by them. Through evolutionary time, however, these ecosystems have changed as forests and seas have flowed back and forth over deserts and grasslands, great swamps have filled and dried and filled again, and isolated populations of plants, vertebrates, and invertebrates have regained contact with one another only to be divided again later. Through all of this, within the constraints of their particular lineage, the apes have survived as well as possible under prevailing circumstances in each place, sometimes having to move, sometimes dying out before a river or mountain barrier, and steadily changing down the generations. The present distribution and success or otherwise of the apes in the forests where they live is the outcome of this long history of adaptation and movement. Their ecologies are defined by their abilities – to move in the trees or on the ground, to find and process their preferred foods and to tolerate others, to invent and use tools – interacting with their social systems and the ways in which foods are distributed in their environment. All of this complexity is rapidly being understood, but at a time when the long history of the tropical moist forests may be coming to an end at the hands of humans.

FURTHER READING

Chapman, C.A., Lambert, J.E. (2000) Habitat alteration and the conservation of African primates: case study of Kibale National Park, Uganda. *American Journal of Primatology* **50**: 169–185.

Furuichi, T., Hashimoto, C., Tashiro, Y. (2001) Fruit availability and habitat use by chimpanzees in the Kalinzu Forest, Uganda: examination of fallback foods. *International Journal of Primatology* **22**: 929–945.

Hart, T.B. (1995) Seed, seedling and subcanopy survival in monodominant and mixed forests of the Ituri Forest, Africa. *Journal of Tropical Ecology* **11**: 443–459.

Jenkins, M. (1992) Biological diversity. In: Sayer, J.A., Harcourt, C.S, Collins, N.M., eds, *The Conservation Atlas of Tropical Forests: Africa.* IUCN and Simon & Schuster, Cambridge, UK. pp. 26–32.

Lieth, H., Werger, M.J.A., eds (1989) *Ecosystems of the World, 14B. Tropical Rain Forest Ecosystems: Biogeographical and Ecological Studies.* Elsevier, Amsterdam.

MacKinnon, K.S., Hatta, G., Halim, H., Mangalik, A. (1996) *The Ecology of Kalimantan.* Periplus, Singapore.

Poulsen, J.R., Clark, C.J., Smith, T.B. (2001) Seed dispersal by a diurnal primate community in the Dja Reserve, Cameroon. *Journal of Tropical Ecology* **17**: 787–808.

Sayer, J.A., Harcourt, C.S., Collins, N.M., eds (1992) *The Conservation Atlas of Tropical Forests: Africa.* IUCN and Simon & Schuster, Cambridge, UK.

Tutin, C.E.G., Oslisly, R. (1995) *Homo, Pan* and *Gorilla* – coexistence over 60 000 years at Lopé in central Gabon. *Journal of Human Evolution* **28** (6): 597–602.

WCMC (1992) *Global Biodiversity: Status of the Earth's Living Resources.* Chapman & Hall, London.

White, F. (1983) *The Vegetation of Africa: A Descriptive Memoir to Accompany the UNESCO/AETFAT/UNSO Vegetation Map of Africa.* UNESCO, Paris.

White, L.J.T., Tutin, C.E.G. (2001) Why chimpanzees and gorillas respond differently to logging: a cautionary tale from Gabon. In: Weber, W., White, L.J.T., Vedder, A., Naughton-Treves, L., eds, *African Rain Forest Ecology and Conservation.* Yale University Press, New Haven. pp. 449–462.

Whitmore, T.C. (1984) *Tropical Rain Forests of the Far East,* 2nd edn. Clarendon, Oxford.

Whitten, A.J., Damanik, S.J., Anwar, J., Hisyam, N. (1984) *The Ecology of Sumatra.* Gadjah Mada University Press, Yogyakarta, Indonesia. Reprinted (2000). Periplus, Singapore.

Whitten, A.J., Whitten, J. (1992) *Wild Indonesia.* New Holland, London.

Wrangham, R.W., Chapman, C.A., Chapman, L.J. (1994) Seed dispersal by forest chimpanzees in Uganda. *Journal of Tropical Ecology* **10**: 355–368.

MAP DATA SOURCES

Maps 2.1 and 2.2 Data are based on the following sources:

Global Land Cover 2000 database. European Commission, Joint Research Centre, 2003. http://www.gvm.jrc.it/glc2000. Data accessed October 2004.

See species chapters for great apes data sources.

ACKNOWLEDGMENTS

Many thanks to Colin Groves (Australian National University), David Chivers (Wildlife Research Group, University of Cambridge), and Elizabeth A. Williamson (University of Stirling) for their valuable comments on the draft of this chapter, and also to Marc Languy (WWF Eastern Africa Regional Programme Office) for information on the Albertine Rift ecoregion.

AUTHORS

Julian Caldecott, UNEP World Conservation Monitoring Centre
Valerie Kapos, UNEP World Conservation Monitoring Centre

CHAPTER 3

Chimpanzee and bonobo overview

JULIAN CALDECOTT

There are two species of the genus *Pan*, the chimpanzee (*P. troglodytes*), with four subspecies, and the bonobo (*P. paniscus*). They are of a similar size, with adult males being 30–61 kg in weight and 82–91 cm in head and body length, and females being about 35 percent lighter and 4 percent shorter.[11, 37] Males are therefore rather more robustly built than females. Both species have black faces as adults, black fur, arms as long as their legs, and no tail. Diagnostic differences between the two species include:[11, 12]

- bonobos are born with black faces, chimpanzees with pink ones;
- bonobos have red lips, chimpanzees have brown or black ones;
- bonobos have hardly any beard on the chin, adult chimpanzees have white beards;
- bonobos are born with prominent side-whiskers, chimpanzees have none;
- bonobo adults retain a prominent tail tuft that is apparent only in juvenile chimpanzees;
- bonobos have short and very rounded skulls, chimpanzees have longer ones with a lower forehead and prominent brow ridges;
- chimpanzee eyes are comparatively deepset and close together; and
- the bonobo clitoris appears large compared to that of any other ape, and is shifted ventrally compared to that of chimpanzees.[17]

Based on differences in their mitochondrial DNA, it is thought that the bonobo and chimpanzee lineages diverged 1.3–3.0 million years ago (mya), with the median of the range reported as 1.5 mya and the mean 2.1 mya.[7, 21, 29, 32, 38] The common ancestor of the two species may have evolved in an open-country habitat,[23] and is thought to have colonized the Congo Basin in drier periods of the Pliocene (5.0–1.6 mya) or Pleistocene (1.6–0.01 mya; see Table 1.1), during high-latitude glaciations.

The chimpanzee ancestors spread through the drier forests and woodlands in a great arc from East

A female bonobo with her infant (Columbus Zoo and Aquarium).

David W. Liggett (www.daveliggett.com)

Michael Huffman

Adult female chimpanzee and her offspring, Mahale Mountains National Park, United Republic of Tanzania.

Africa, through North-Central Africa to West Africa, north of the Congo River. Meanwhile, the bonobo ancestors became isolated to the south of this river, in the heartland of the Congo Basin. This area became wetter after the glacial period, and the bonobo adapted its biology and behavior to survive. Many other species (e.g. the okapi, *Okapia johnstoni*, and the four-toed elephant shrew, *Petrodromus tetradactylus*) also seem to be descended from lineages that penetrated the same area at drier times, and then became effectively trapped among the rivers, swamps, and forests that grew up around them as the rains returned.[19]

ECOLOGY AND DISTRIBUTION

Chimpanzees live in a wide variety of habitats, from humid evergreen forests, through mosaic woodlands and deciduous forest, to dry savanna woodlands; they occupy a range of elevations from sea level in West Africa to 2 600 m in East Africa. Hence they have by far the most widespread distribution and the most cosmopolitan ecology of any great ape, and show many signs of adaptability and opportunism. They eat a great variety of foods within the broad constraints of their digestive system. They cannot cope with a large quantity of mature leaves, which contain both abundant fiber and secondary metabolites such as tannins and alkaloids. Hence their diets are dominated by fruits (especially ripe, sugar-rich ones when available), flowers, and seeds, but include some young leaves, algae, mushrooms, honey, and a variety of small mammals and invertebrates. As many as 330 food types can be eaten in a year.

Over 12 study sites, chimpanzees have been observed to hunt at least 32 species of mammals, of which the most important is the red colobus monkey (*Procolobus* spp.) (see Box 4.1). They have also been seen to eat 17 other species of primates, particularly forest monkeys (*Cercopithecus* and *Colobus* spp.), but also baboons (*Papio* spp.). Flying squirrels (Anomaluridae), tree pangolins (*Manis tricuspis*), elephant shrews (*Rhynchocyon cernei*), and various duikers (*Cephalophus* spp.) are all also reported as prey. It is almost always adult males that hunt, and the meat is often shared between community members, particularly in response to begging. It is notable that chimpanzees seem to hunt in 'binges' during which hunting is an almost daily occurrence, with a much lower frequency at other times.[10, 33] It is hard to explain this pattern solely in ecological or physiological terms on current evidence. One speculation is that it may be rooted in social psychology, in which case it would have more in common with a 'craze' or a 'fad' during which the animals reinforce each other's memory of recent hunts and excitement about hunting through further hunting behavior. Perhaps this shared enthusiasm ebbs away when all the easily killed prey has been caught and the chimpanzees start to forget, until the next time.

Dietary flexibility, coupled with ecological variation over a huge geographical range within which seasonality is important, can only result in very variable foraging and ranging behavior. A community of chimpanzees, typically of about 35 animals, nevertheless occupies an area with a rather limited range of 6–15 km^2, and not all parts of the larger ranges may actually be used. Males use an area 150–200 percent greater than that used by females, and are more likely to be seen near boundaries, supporting the scenario of females having small core areas within the defended home range of the males.[4] Chimpanzees are very mobile, and travel an average of about 4 km each day.[9] Each community spends much of its time divided into foraging parties. When resources are scarce, chimpanzees reduce their daily range and party size, spend more time feeding, and more frequently eat lower-quality food items.[6]

The main habitat where bonobos have been studied is primary lowland forest, but they also make use of open woodland savanna, dry forest, swamp forest, marsh grassland, and disturbed, secondary forests if possible; they apparently prefer habitats in which a variety of ecosystem types, and edges between them, are available.[34] Bonobos are mainly frugivorous, although their diet also includes leaves, pith, flowers, seeds, nuts, sprouts, mushrooms, and algae. Additional food sources such as high-quality terrestrial herbaceous vegetation, earthworms, larvae, termites, ants, honey, truffles, aquatic plants, invertebrates, and fish have also been reported.[1, 15, 20] Bonobos also consume small mammals occasionally, including flying squirrels, infant duikers, and bats, but there is little evidence that hunting is as important an activity for bonobos as it is for chimpanzees. Bonobos interact with monkeys at times, and may kill but have not been seen to eat them.[16, 28] Bonobos eat less animal protein than chimpanzees, which may be related to their greater use of the protein contained in nonreproductive plant parts,[23] especially the stems of aquatic or amphibious herbs and grasses in marshy grasslands.[35]

Bonobos live in communities that are slightly larger than those of chimpanzees (50–120 rather than 20–106 individuals is the range reported)[8] but, like chimpanzees, they often forage in smaller parties. They do spend longer than chimpanzees, however, in large groups and, since relations with neighboring communities are far more relaxed than among chimpanzees, these groups sometimes include parties from different communities. Both species use a mobile, flexible foraging strategy designed to obtain a fruit-rich but otherwise generalist diet of easily digestible and nontoxic food of great taxonomic diversity, exploiting as many available ecosystem types as they find useful. The minor differences in diet, ranging, and foraging behavior that have been observed cannot yet be assessed for significance. This conclusion is somewhat unsatisfactory in view of the different primary environments of the two species (the inner Congo Basin versus semideciduous woodlands across Africa), and implies a need for further research on the ecosystems and biogeography of the inner Congo Basin, and the feeding and foraging strategies of bonobos in many locations and circumstances.

SOCIETY AND PSYCHOLOGY

The social lives of chimpanzees and bonobos are similar in many ways, but deeply different in others.[10, 17] In both species, young females leave their natal community and migrate from one community

The Jane Goodall Institute (JGI)

Ian Redmond

Chimpanzee habitat ranges from humid forest (here in Gombe, United Republic of Tanzania) to dry savanna woodland (such as Bafing Reserve, Mali).

Box 3.1 USE OF HUMAN LANGUAGES BY CAPTIVE GREAT APES

Some believe that the extinction of nonhuman great apes is preferable to preserving them forever in captivity, on the grounds that their nobility is diminished in artificial habitats. Others hold that great apes in captivity can lead happy lives, that the value of the preserved genetic material will prove to be very great, and that the human psyche would be significantly damaged by the loss of these species. This view embraces preservation strategies that create a diversity of niches for great apes that include the wild, zoos, reserves, refuges, sanctuaries, and even laboratories.

Chimpanzees and bonobos have lived in a captive research facility at Georgia State University in the USA since 1971, most notably sponsored by the work of Duane Rumbaugh and Sue Savage-Rumbaugh. This research has explored the mental abilities and cognitive character of great apes, in the process significantly changing our view of *Pan* and how these nonhumans might exist in human-modified landscapes. Two methods have been used to teach human languages to great apes: one uses sign language; the other, explored here, uses graphical symbols that represent words (lexigrams). The following is a brief account of the research initiatives of the Rumbaughs, the great apes that have participated in the research at the Language Research Center of Georgia State University, and the future plans for their lives in coexistence with humans.

Lana project, 1971–1976[26]

Lana is a female chimpanzee born in 1970 at the Yerkes National Primate Research Center. Her name derives from the LANguage Analogue (LANA) project, which sought to develop a computer-based language training system in an effort to investigate the ability of chimpanzees to acquire language. Lana joined the research as a subject when she was two and a half years old. The research was the first to interface a keyboard with a chimpanzee. At that time, it was believed that only humans could use symbols.

Lana demonstrated that she could discriminate between lexigrams and associate them with ideas. As she progressed, she would sequence words and use them grammatically, later starting to create novel utterances in response to unplanned events that affected her life. For example, Lana would request that the research technician refill her computer vending device when it was empty of treats, or request an item she had seen outside her room that the computer had no facility to provide to her. Lana exhibited language learning, and her experimental accomplishments were extraordinary. Equally important to her legacy is the lexigram keyboard, developed by Duane Rumbaugh, which has served as the primary communicative interface for ape language research at Decatur, Georgia for the last several decades. This keyboard is composed of three panels with approximately 384 noniconic arbitrary symbols. When the apes depress a key, the word represented there is spoken by a digital voice and the lexigram is displayed on a video screen.[27]

Sherman and Austin research, 1975–1980[30]

Sue Savage-Rumbaugh argued that the essence of language does not exist outside sociality and began working with two young male chimpanzees, Sherman, born 1973, and Austin, born 1974, using the LANA keyboard. The issue of human cuing was overcome experimentally by focusing on peer communication rather than that between experimenter and subject. The receptive component of language was featured in chimp-to-chimp communication, in which they structured their interactions around statements of planned intent. Unlike Lana, Sherman and Austin could categorize,

to another before remaining in one to breed. If this were the only mixing between groups, communities would consist of unrelated females and closely related males, but there is evidence that male chimpanzees also sometimes transfer. Differences between the two species are most obvious in the relations between males and females.

Among chimpanzees, males associate closely with one another, grooming one another frequently and cooperating in hunting, in patrolling borders, in stalking and sometimes killing chimpanzees from neighboring communities,[36] and in guarding and mating with swollen females (see below). Among bonobos, grooming between individuals of the opposite sex is more frequent and occurs for longer periods of time than grooming between females or

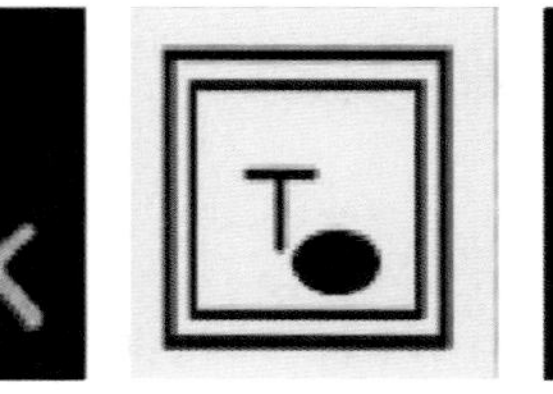

Lexigrams from the 384-word keyboard designed by Duane Rumbaugh for the Lana project.

pretend, plan, comprehend, and respond to each other. Attending Sherman's and Austin's more complex use of language features was an increase in sociability and cooperation. Despite these achievements, Sherman and Austin did not comprehend spoken English. Austin died in 1998, but the other apes at the Language Research Center have not forgotten him; they still make reference to him using his lexigram, and they enjoy seeing videotapes of him.

Kanzi research, 1980–1993[31]

This was the first research initiative to use bonobos in language investigations. It began with a wild-caught female named Matata and her adopted son Kanzi. Kanzi was a nine month old baby playing in the lab while Savage-Rumbaugh tried to teach his mother language. Kanzi was not a focus of the research because scientists thought him too young to learn these skills. When baby Kanzi was briefly separated from his mother, he began spontaneously to demonstrate productive competence for lexigrams and receptive competence for spoken English (something Matata had not achieved through direct training). Kanzi's acquisition of productive and receptive competence emerged following passive observational exposure. Later, as his language complex matured, Savage-Rumbaugh demonstrated that Kanzi's utterances included grammar, syntax, and semanticity. It also seemed that his language skill enhanced his ability to learn other skills, such as the manufacture of Oldowan-type rock tools. Kanzi's receptive competence for spoken English contrasted dramatically with the failure of the chimpanzees Sherman and Austin to do likewise. What was the basis for the difference between the language skills displayed by bonobo Kanzi and the chimpanzees? Savage-Rumbaugh had clearly demonstrated in Kanzi that language could be acquired spontaneously and observationally without planned training; that comprehension precedes production and drives language acquisition; and that early exposure to language can greatly improve the level of competency attained.

Panpanzee and Panbanisha research, 1986–1990[2]

Considering the question of receptive competence for spoken English, Savage-Rumbaugh proceeded to investigate the questions of species variables in a co-rearing study of a chimpanzee and a bonobo. In this study, Savage-Rumbaugh had hoped to have Kanzi's mother Matata raise chimpanzee Panpanzee and bonobo Panbanisha in identical environments. They were born within three weeks of each other. While Matata took good care of both babies, she would only allow Panbanisha to nurse. At that point, Savage-Rumbaugh and her human colleagues assumed the rearing of both babies until they were four years old. Based upon this study, Savage-Rumbaugh determined that the failure of the chimpanzees Sherman and Austin to comprehend spoken English is not a species-specific

continued overleaf

males only. Bonobo males are much more peaceful than chimpanzee males, interact less, compete less for copulation opportunities, are not as territorial, are less aggressive with males of other groups, and do not hunt other large mammals. Female bonobos maintain strong bonds with their sons, which in itself increases the frequency of grooming between males and females.[22]

Female chimpanzees show only infrequent social interactions but bonobo social structure is, by contrast, dominated by female coalitions that influence mating strategies and food allocation. Female bonobos use a number of techniques to establish and maintain their bonds, including sharing food and forming alliances between themselves against males.[24, 25] Female coalitions help to

variable, as both Panpanzee and Panbanisha developed receptive competence for English.

Panbanisha and Kanzi, 1990–present[13]

Kanzi lives in a bonobo community at a facility that includes a 20 ha wooded forest within a 125 ha woodland preserve. The bonobos spend as much time outdoors traveling and communicating as they do indoors with computers and joysticks. Locations in the forest are named with lexigrams, and the bonobos know this forest as well as humans might know their own village. The bonobos are able to plan where they will go and what they will do when they get there, and they talk about these plans on the communication boards.

Kanzi and Panbanisha continue to expand their linguistic world with music, art, writing, tool making, and tool using. Savage-Rumbaugh documented on film Kanzi's ability to 'rock knapp', breaking off flakes of stone to produce functional cutting tools as taught by archeologist Nicholas Toth; Panbanisha's ability to write lexigrams on the floor with chalk; and the ability of both to participate in musical performances with musicians.

Kanzi communicating with Sue Savage-Rumbaugh through the lexigram keyboard. Kanzi's presence at the project since infancy has greatly advanced understanding of apes' capacity for language.

Great Ape Trust of Iowa

Summary

An overview of language research with apes during the last 50 years provides strong evidence for their use of words (manual gestures or graphic patterns) as meaningful symbols that refer to things and their qualities (temperature, color, etc.), persons or peers, activities, or as places for foods, rest, chasing, and so on.[13] Apes can also comprehend new sentences with fairly complex structures. They can use language to achieve outcomes that they would otherwise not be able to achieve, for example to formulate names for new items based on novel word combinations. They can use manual signs and graphic symbols to communicate about things that are not present; they can learn to communicate their needs and to fulfill one another's requests for specific tools, foods, and games; they can integrate their language skills and apply them creatively even several years later in new contexts. If reared in a manner that approximates child rearing, apes can come to understand complex human speech and its syntax.

Language acquisition using lexigrams is optimized if it occurs in the course of social rearing in an environment that is language structured. Ideally, this provides a running vocal narrative to the apes as infants, describing what things are, what is about to happen, and so on; this narrative should be integrated with the use of graphic symbols that are to function as words. Results show that apes can enter the language domain as a result of human rearing and instruction, although their capacity for language is much more limited than that of humans. A great deal remains to be learned. Future research promises to continue to blur the boundary between the basic principles of human and animal learning, language, symbolic function, and complex behaviors.

Duane Rumbaugh and Bill Fields

offset the greater muscularity of males, and are maintained partly by a frequent behavior among bonobo females: collaborative genital rubbing. This is more often initiated by low-ranking females than by high-ranking ones,[14] and may regulate and reconcile social tension that sometimes arises following bouts of aggression or when food is monopolized.

All of this is correlated with marked differences in sexuality between chimpanzees and bonobos, even though the two species use the same 'building blocks' of sexual physiology.[5] In summary, at an age of about seven (in bonobos) or 10 (in chimpanzees), a female begins her first menstrual cycle. In chimpanzees, during the first (follicular) phase of such a cycle, estrogen levels rise; this causes the perineal skin to swell in a very visible way that greatly increases her attractiveness to males. This swelling reaches a maximum about nine days before ovulation and is sustained until about three days afterwards. This is the time of maximum female attractiveness as well as receptivity (willingness to accept mounting) and proceptivity (tendency to solicit mounting). The swelling collapses abruptly after ovulation, as estrogen levels drop and progesterone levels rise, thereafter remaining quiescent throughout the luteal phase of the cycle and the menstruation that follows it.

The same process occurs in bonobos, but with the important difference that their swellings last much longer and are in fact semipermanent. Bonobo swellings vary somewhat during the menstrual cycle in firmness (turgidity) and therefore attractiveness to males; the swellings peak around mid-cycle. Female bonobos are continuously receptive and there is little evidence that their proceptivity varies much.[5] A consequence is that there are far more females available for sex at any given time in a bonobo community than in a chimpanzee one. Since the status of bonobo females is not automatically lower than that of males, as is the situation for chimpanzees, and since so many females are willing and available, male bonobos are not able to sequester females and rarely dispute each other's access to them. In both species, a young female can be sexually active for some time before she conceives her first infant, and this may be an important time for her to build social relationships in several communities before settling down.[18] For a young female bonobo, this phase can last for up to six years while, in chimpanzees, it lasts for only a few months. This implies that a bonobo female is more likely to encounter familiar adults whenever her community meets another throughout her subsequent life.

CONSERVATION CONCERNS

The most recent estimated total population of wild chimpanzees is 173 000–300 000,[3] with downward trends in many but not all areas in which comparable surveys have been undertaken. The two West African subspecies are least abundant, which is consistent with the greater deforestation in their area of distribution. The species is threatened by a combination of factors that tend to multiply each other's impacts. Light selective logging causes only temporary disturbance and may not greatly reduce the forest's carrying capacity for chimpanzees, but more intense and/or repeated logging causes mounting disruption to the forest ecosystem, degrading its integrity, opening it up to drying winds and sunshine, and increasing its vulnerability to fire. Increased access to the logged-over area along logging roads encourages hunting, especially where there is a commercial trade in bushmeat, so the effects of hunting almost inevitably multiply with those of logging.

Volker Sommer

Researchers and field assistants at Gashaka-Gumti National Park, Nigeria. Field workers wear yellow hats that are visible from afar so that apes and monkeys are not scared by the sudden appearance of humans.

Mining and oil extraction can have similar effects on access, as well as destroying ecosystems locally. Increased access also leaves the forest open to invasion by land-hungry farmers, leading to further hunting and also the fragmentation of the forest by an expanding and eventually coalescing mosaic of farms and villages. Chimpanzees are bound to find it increasingly hard to survive in such a landscape, the more so as the reduced and fragmented populations come into more frequent contact with humans and become increasingly vulnerable to human diseases. Hence, the survival of chimpanzees is ultimately threatened by the whole process of advancing human use of tropical moist forests.

Bonobos are distributed patchily over a large geographical area of around 340 000 km^2, but are nowhere common. There are estimated to be fewer than 100 000 bonobos in the wild, perhaps many fewer. They are hunted for food in most places where they occur, although taboos provide partial protection in some places. Hungry soldiers, militiamen, and refugees during the civil war certainly killed many, but most bonobo populations escaped this direct impact. More serious was the increased hunting brought about by disruption of agriculture and trade, and the resulting food shortages in the late 1990s, with some more accessible areas reported to have lost 25–75 percent of their bonobos. Increasing trade with the end of warfare in most areas is likely to allow more traffic in bushmeat, however, which may shift the purpose of hunting from subsistence to profit, and will keep bonobo populations under hunting pressure. Peace is likely to bring other dangers too, including an expansion of industrial-scale logging, mining, and forest clearance for farming, all with associated improvements in access and the spread of hunting into new areas (see the Democratic Republic of the Congo (DRC) country profile, Chapter 16).

WHAT WE DO NOT KNOW

Hundreds of researcher-years have been dedicated to the study of wild chimpanzees at study sites across their range in the wild, for example from Gombe and Mahale in the United Republic of Tanzania, to Kibale and Budongo in Uganda, Lopé in Gabon, Taï in Côte d'Ivoire, and Bossou in Guinea. This fieldwork supplements thousands of person-years of captive research on every aspect of chimpanzee biology (admittedly much of it motivated by the use of chimpanzees as physiological proxies for humans), up to and including the imminent publication of the chimpanzee's entire genome. No comparable research effort has been directed to bonobos, although our knowledge of this species is catching up fast.

For bonobos, much more needs to be learned about communication in the wild, including both vocal and symbolic aspects; tool uses and culture; and the species' behavioral ecology in mosaic woodland and grassland habitats as well as in forest habitats. As noted above, further research is needed on the ecosystems and biogeography of the inner Congo Basin, and the feeding and foraging strategies of bonobos in many locations and circumstances. Research using existing captive populations, preferably based in DRC and combined with public education, is needed if we are to learn

Jo Thompson/Lukuru Wildlife Research Project

Forest and savanna mosaic habitat in the southern region of bonobo range, Lukuru, Democratic Republic of the Congo.

more about language development and all other aspects of cognition and communication, as well as the neuroendocrine control of sexual behavior and interindividual relationships among males and females.

As a surer understanding of bonobo biology is obtained, a raft of additional behavioral and ecological questions about differences and similarities with chimpanzees will surely be raised; this will then prompt a new generation of field studies on chimpanzees. There are already many issues in chimpanzee biology that require further exploration, notably the way in which "chimpanzee traditions ebb and flow, from community to community, across the continent of Africa";[36] their use of medically or psychically active plants; the origins, role, and significance of hunting; and the psychological dynamics of group existence. For both species, of course, research oriented to encouraging and enabling their survival in the wild is important and urgent. This would include obtaining a better understanding of how the management of forest and the forest farming of mosaic landscapes can be changed in partnership with human stakeholders to improve chimpanzee and bonobo survival. Researchers should also be on the lookout for ways to improve dissemination of their findings in local languages and other appropriate media. Local acceptance of the inherent value and interest of the two species of *Pan* is crucial to ensuring their survival.

FURTHER READING

Boesch, C., Hohmann, G., Marchant, L., eds (2002) *Behavioural Diversity in Chimpanzees and Bonobos.* Cambridge University Press, Cambridge, UK.

Dixson, A.F. (1998) *Primate Sexuality: Comparative Studies of the Prosimians, Monkeys, Apes, and Human Beings.* Oxford University Press, Oxford.

Fouts, R.S., Fouts, D.H. (1993) Chimpanzees' use of sign language. In: Cavalieri, P., Singer, P., eds, *The Great Ape Project.* St. Martin's Griffin, New York. pp. 28–41.

Gagneux, P. (2002) The genus *Pan*: population genetics of an endangered outgroup. *Trends in Genetics* **18** (7): 327–330.

Goodall, J. (1986). *The Chimpanzees of Gombe: Patterns of Behavior.* Harvard University Press, Cambridge, Massachusetts.

Hillix, W.A., Rumbaugh, D.M. (2004) *Animal Bodies, Human Minds: Ape, Dolphin, and Parrot Language Skills.* Plenum, New York.

Kormos, R., Boesch, C., Bakarr, M.I., Butynski, T.M., eds (2003) *West African Chimpanzees: Status Survey and Conservation Action Plan.* IUCN/SSC Primate Specialist Group. IUCN, Gland, Switzerland.

Parish, A.R., de Waal, F.B.M. (2000) The other "closest living relative" – how bonobos (*Pan paniscus*) challenge traditional assumptions about females, dominance, intra- and intersexual interactions, and hominid evolution. *Annals of the New York Academy of Science* **907**: 97–113.

Rumbaugh, D.M., ed. (1977) *Language Learning by a Chimpanzee: The LANA Project.* Academic Press, New York.

Savage-Rumbaugh, E.S. (1986) *Ape Language: From Conditioned Response to Symbol.* Columbia University Press, New York.

Stanford, C.B. (1998) The social behavior of chimpanzees and bonobos: empirical evidence and shifting assumptions. *Current Anthropology* **39**: 399–420.

Wrangham, R., Peterson, D. (1997) *Demonic Males: Apes and the Origins of Human Violence.* Bloomsbury, London.

ACKNOWLEDGMENTS

Many thanks to Colin Groves (Australian National University), Takeshi Furuichi (Meiji-Gakuin University), and to Hilde Vervaecke and Jeroen Stevens (both University of Antwerp) for their valuable comments on the draft of this chapter. Thanks also to Tim Inskipp and Carmen Lacambra (both UNEP-WCMC) for research into the literature.

AUTHORS

Julian Caldecott, UNEP World Conservation Monitoring Centre
Box 3.1 Duane Rumbaugh and Bill Fields, Georgia State University

Mike Powles/Still Pictures

CHAPTER 4

Chimpanzee (*Pan troglodytes*)

TIM INSKIPP

The chimpanzee (*Pan troglodytes* Blumenbach, 1775) is also known as the 'robust' or 'common' chimpanzee to distinguish it from the bonobo (*Pan paniscus* Schwarz, 1929), sometimes called the 'gracile' chimpanzee, which has a much more limited distribution. The chimpanzee has a thickset body, with a short neck and broad shoulders, arms longer than its legs, and no tail. It has a low forehead with prominent brow ridges and eyes that are deepset and close together. The nose is broad and flat and the hands and fingers are long, with the outer skin of the middle fingers thickened. The skin of the face is pink at birth, becoming pinkish brown to black by maturity. The fur is long and sparse and mainly black; adults have a white beard on the chin and juveniles have tufts of white hair above the buttocks.[87, 92] Chimpanzees remain on all fours most of the time, but occasionally adopt bipedal postures.

Four subspecies of chimpanzee are generally recognized: the central (*P. t. troglodytes* Blumenbach, 1799), the western (*P. t. verus* Schwarz, 1934), the eastern (*P. t. schweinfurthii* Giglioli, 1872), and the Nigeria-Cameroon (*P. t. vellerosus* Gray, 1862). The last has also been called the Nigeria chimpanzee,[90] though it is now thought to be more numerous in Cameroon than in Nigeria. The more neutral common name Nigeria-Cameroon is used here.

The central chimpanzee is larger and heavier than the other subspecies, with size varying between populations. The mean length of the head plus body in two sample areas of Cameroon was 819 mm and 914 mm for males, and 796 mm and 871 mm for females.[87] The mean weights of males were 60 kg in Cameroon and 52 kg in Gabon; two females from Cameroon both weighed 50 kg, whereas the mean weight of 19 females from Gabon was 44 kg.

The western chimpanzee is smaller, with a less broad head; it has a steeply descending occiput (back of the head), raised brow ridges, and a thicker, more rounded, white beard. The weights of two males were 46.3 and 48.5 kg, whereas one female was only 21.2 kg.[87, 167]

The eastern chimpanzee is smaller and shorter-limbed than the central chimpanzee, with a more rounded head, an elongated occiput and straight brow ridges, and a full but straggly white beard; weights from the United Republic of Tanzania ranged from 30.3 to 52 kg for males and from 22.7 to 45.5 kg for females,[167] whereas three males from the Democratic Republic of the Congo (DRC) weighed 52.5–61 kg.[87]

The external characters of the Nigeria-Cameroon chimpanzee are less well known because the subspecies has been recognized mainly on genetic characters;[81] however, photographs[125] and drawings[201] indicate that it has a more prominent brow ridge and much smaller ears. The taxonomist Colin Groves has noted that the skull of this subspecies has a closer similarity to those of the central and eastern chimpanzees than to the western chimpanzee.[87]

It has been reported that *P. t. marungensis* could be differentiated from *P. t. schweinfurthii* as a fifth subspecies.[89] A mitochondrial DNA (mtDNA) analysis suggested that the lineage of *P. t. verus* was so distinct that the taxon warranted specific rank.[75, 170] It has been argued that as *P. t. schweinfurthii* is also diagnosably different, either all three should be recognized as distinct species or the single species concept should be maintained.[87]

Map 4.1 Chimpanzee distribution (see country profiles for further detail)

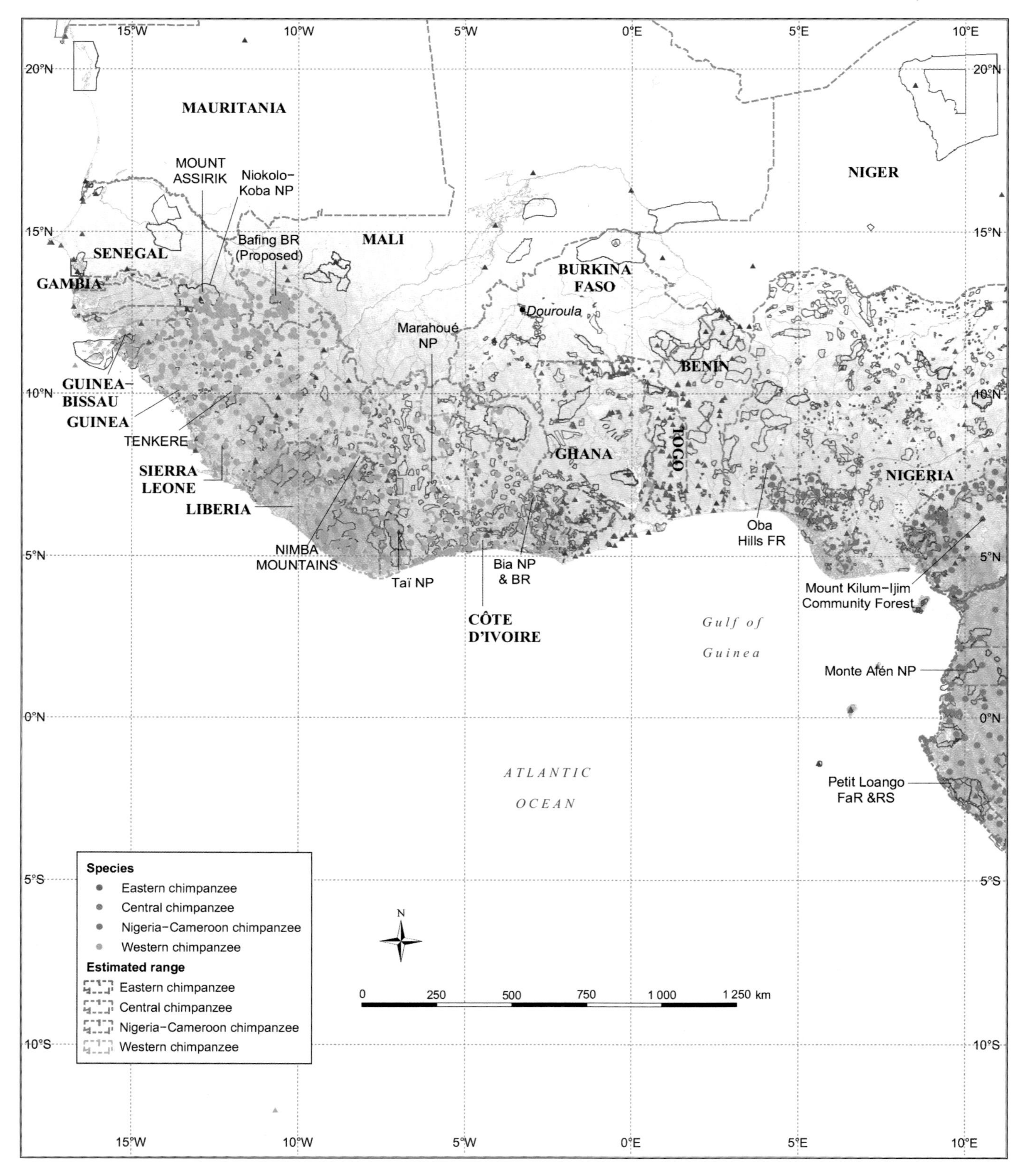

Data sources are provided at the end of this chapter

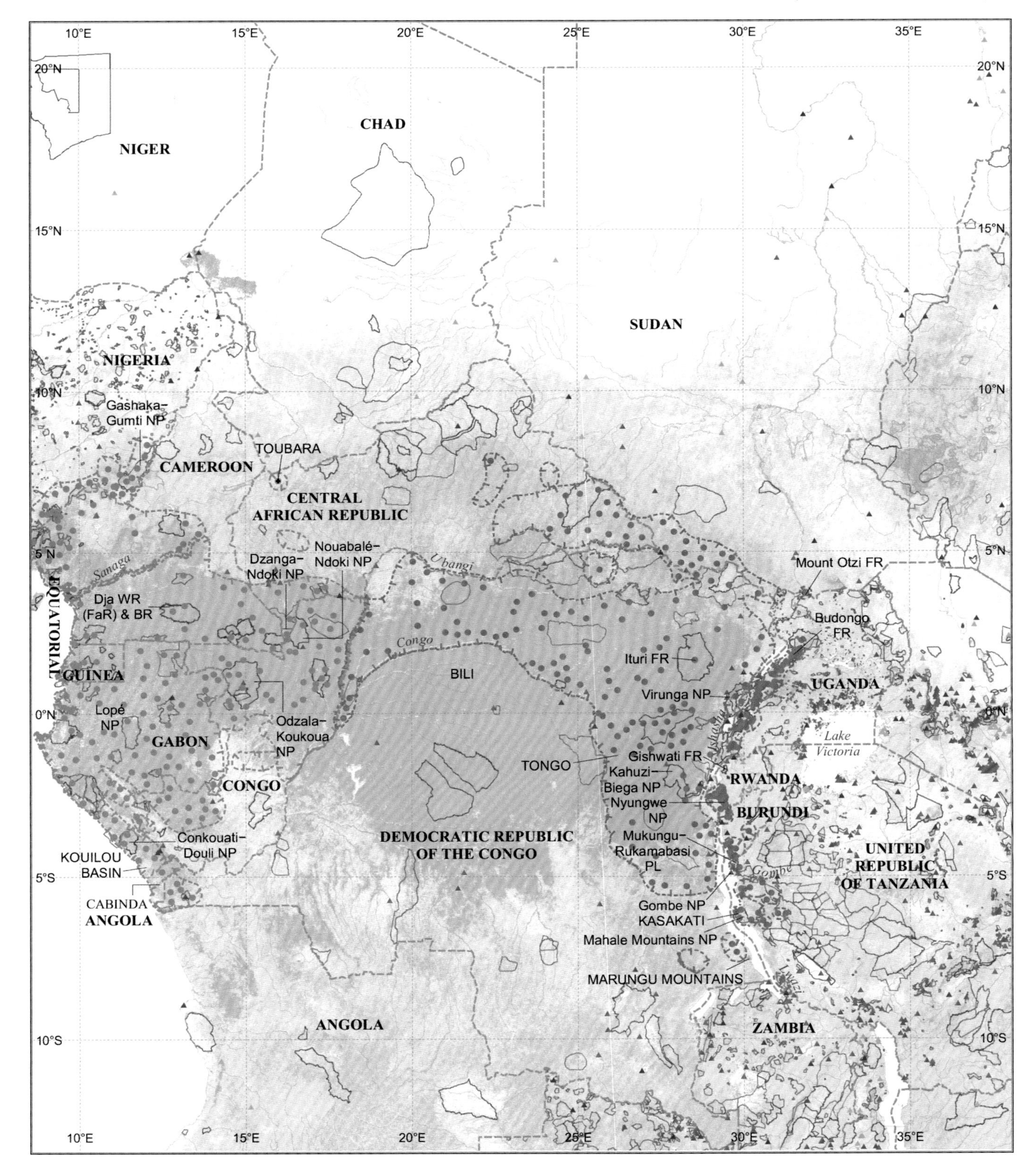

Some studies based on mtDNA sequences have, conversely, questioned the genetic distinction between the subspecies *P. t. schweinfurthii* and *P. t. troglodytes*.[74, 75] Indeed, comparisons of mtDNA sequences are increasingly leading towards the conclusion that there is a very close relationship between the two chimpanzees of Central and East Africa (*P. t. schweinfurthii* and *P. t. troglodytes*), and also between the two chimpanzees of West Africa (*P. t. verus* and *P. t. vellerosus*), such that it may one day be appropriate to recognize only two subspecies, the central/eastern (*P. t. troglodytes*) and the western (with *P. t. vellerosus* as the prior name).[202] The central, western, eastern and Nigeria-Cameroon subspecies are distinguished in the map presented here (Map 4.1).

DISTRIBUTION

The central chimpanzee (*P. t. troglodytes*) occurs fairly widely in southern Cameroon south of the Sanaga River.[165, 214] It extends east into the western part of the Central African Republic (CAR), where it is largely confined to the extreme southwest, in Dzanga-Ndoki National Park[17] and several locations at about 4°N. In the CAR it has also been found at an isolated locality at Toubara (7°15'N 15°55'E), where it was originally reported in the early 1960s,[128] and was apparently still present after 1983.[37] Southwards, the subspecies occurs in two areas of Equatorial Guinea,[134] including Monte Alén National Park.[76] It is widespread throughout Gabon[18] and the northern part of Congo north of the Equator.[13] It also occurs fairly widely in the Kouilou basin in southern Congo[63] and an isolated locality at 3°S 16°E near the border with DRC.[37] The southernmost localities are in the Cabinda province of Angola and in the extreme west of DRC, just north of the Congo River.[37] The geographic range of the subspecies is about 695 000 km² in area.[37]

Juichi Yamagiwa

Eastern chimpanzee, Kahuzi-Biega National Park, Democratic Republic of the Congo.

The western chimpanzee (*P. t. verus*) occurs over a large area southwards and eastwards from Mount Assirik (12°58'N 12°46'W) in southeast Senegal,[41] into southwest Mali, north to Djibashin water source (13°03'N 10°36'W)[64] and southern Guinea-Bissau.[79] It occurs more or less throughout Guinea,[126] Sierra Leone,[94] Liberia,[184] and much of Côte d'Ivoire,[103] and extends into southwest Ghana, east to about 0°30'W.[139] In Burkina Faso, unconfirmed reports have suggested that chimpanzees may migrate into the southwestern part of the country,[33] while Butynski[37] referred to strong anecdotal information that a few chimpanzees were still present along the Volta River near 'the bend' at the village of Douroula. The western chimpanzee occurred previously in Gambia, where it was apparently extirpated around the end of the 19th century;[40] in Togo, where it was last recorded as recently as 1971;[31] and in Benin, where it is believed to have disappeared in recent decades.[32] The geographic range of the subspecies is about 631 000 km².[37]

The eastern chimpanzee (*P. t. schweinfurthii*) has a fragmented range in the north, with few records since 1983: only two localities in the eastern CAR, only one locality in extreme southwest Sudan, and scattered localities east of the Ubangi River and south to the Equator in DRC. Between the Equator and 5°S in DRC, there are many localities with records since 1983[37, 91, 205] and there is an isolated record in the Marungu Mountains to the south. It extends east into western Uganda,[211] where a small population was discovered in the north in Otzi Forest Reserve, at the extreme northeast of the range of the species.[56] Further south it occurs in Rwanda, where it is known from the Nyungwe forest and possibly from the Gishwati forest;[166] Burundi,[273] where it is known from Kibira National Park, the Mabanda/Nyanza Lake and Mukungu-Rukamabasi

Protected Landscapes, and Rumonge Forest Reserve;[43] and in the extreme west of Tanzania[134] south to the Lwazi River, Rukwa region (8°12'S 31°08'E).[204] The geographic range comprises about 874 000 km^2.[37]

The Nigeria-Cameroon chimpanzee (*P. t. vellerosus*) occurs in southern Nigeria in small, highly fragmented populations from the Oba Hills Forest Reserve south and east to the southeastern Niger Delta, and also along the border with Cameroon, from Gashaka Gumti National Park southwest to both the Okwangwo and the Oban divisions of Cross River National Park.[201] The affinities of the western populations are unknown and it is possible that they belong to the western subspecies.[37] The subspecies also occurs in western Cameroon, mainly near the border with Nigeria, particularly in the Takamanda Forest Reserve and Korup National Park.[125, 214, 263] This population extends south to the Sanaga River, which is probably the distribution limit for the Nigeria-Cameroon chimpanzee. The subspecies also occurs in three areas farther inland and north of the Sanaga River.[37] The geographic range of the Nigeria-Cameroon chimpanzee encompasses about 142 000 km^2.[37]

Chimpanzee studies have been focused on a limited number of locations within this broad distribution. The main field-study sites for chimpanzees are listed in Table 4.1. Early research focused on eastern chimpanzees, with Jane Goodall's study beginning in 1967 and leading to the establishment of the Gombe Stream study area in Tanzania,[83] where research continues to date. Studies on western chimpanzees began in earnest in the mid-1970s, at Bossou, Guinea[236, 237] and Taï National Park, Côte d'Ivoire,[19] and were followed by work on the central chimpanzee starting in the 1980s, for example at Lopé National Park, Gabon.[250] Research has continued for many years at a number of field sites, allowing detailed demographic understanding of the chimpanzee populations to be reached. The Nigeria-Cameroon chimpanzee has not been the subject of a similarly long-term treatment.

Table 4.1 Main field-study sites and other locations mentioned in this chapter

Site	Country	Chimpanzee subspecies
Bafing (proposed Biosphere Reserve)[128]	Mali	western
Bossou, near the Nimba Mountains[235]	Guinea	western
Budongo Forest Reserve[218]	Uganda	eastern
Bwindi Impenetrable NP[36]	Uganda	eastern
Dzanga-Ndoki NP[17]	CAR	central
Gashaka Gumti NP	Nigeria	Nigeria-Cameroon
Gombe NP[83, 243]	United Rep. of Tanzania	eastern
Goualougo Triangle, Nouabalé-Ndoki NP[169]	Congo	central
Ishasha River[223]	DRC	eastern
Ituri Forest Reserve[95]	DRC	eastern
Kahuzi-Biega NP[91]	DRC	eastern
Kalinzu Forest Reserve[97, 98]	Uganda	eastern
Kasakati[119]	United Rep. of Tanzania	eastern
Kibale NP[77]	Uganda	eastern
Lopé NP[252]	Gabon	central
Mahale Mountains NP[186, 187]	United Rep. of Tanzania	eastern
Minkébé NP[112]	Gabon	central
Monte Alén NP[76, 121]	Equatorial Guinea	central
Mount Assirik, Niokolo-Koba NP[7, 8]	Senegal	western
Nimba Mountains[151]	Guinea	western
Nouabalé-Ndoki NP[130]	Congo	central
Odzala NP[13]	Congo	central
Semliki[115]	Uganda	eastern
Taï NP[3, 4, 27]	Côte d'Ivoire	western
Tenkere[1, 94]	Sierra Leone	western
Tongo, Virunga NP[132]	DRC	eastern
Ugalla[185]	United Rep. of Tanzania	eastern

NP: National Park

Adapted from Moore, J., Collier, M. (1999) *African Ape Study Sites*. http://weber.ucsd.edu/~jmoore/apesites/ApeSite.html. Updated January 28 1999; accessed October 26 2004.

BEHAVIOR AND ECOLOGY

Habitat and diet

Of all the great apes, chimpanzees are the least strongly associated with tropical lowland moist forests. They live in a wide variety of habitats, from humid evergreen forests, through mosaic woodlands and deciduous forest, to dry savanna woodlands.[128] Their habitats range in altitude from sea level in West Africa to 2 600 m in East Africa.[10] The availability of year-round surface water is usually important in limiting chimpanzee distribution but, in some areas, they have developed techniques for accessing water during dry periods. At Tongo, for example, chimpanzees live in forest on well drained volcanic soil and, when water is scarce, dig up tubers containing water.[132]

Chimpanzees are also very adaptable in the face of habitat disturbance. In the Kalinzu Forest Reserve, for instance, one group occupies logged

Box 4.1 CHIMPANZEES AS PREDATORS

Until the 1960s, it was widely believed that chimpanzees were entirely herbivorous, and they are indeed largely fruit eaters. Meat is consumed for only about 3 percent of the time they spend eating, which is less than in nearly all human societies.[228] Jane Goodall's pioneering work at Gombe first documented that wild chimpanzees relish meat and hunt a variety of species of other mammals. Today, hunting by chimpanzees at Gombe has been well documented,[83, 228, 242] and hunting patterns have been reported from most other sites in Africa where chimpanzees have been studied: these include Mahale Mountains National Park in Tanzania,[256] Kibale National Park in Uganda,[163] and Taï National Park in Côte d'Ivoire.[22, 25]

After four decades of research on eastern chimpanzees at Gombe, a great deal is known about their predatory patterns. Chimpanzees live in communities comprising 20 to over 100 animals that split into smaller parties for short periods of time. Such a community of chimpanzees may kill and eat more than 100 small- and medium-sized animals such as monkeys, wild pigs, and small antelopes each year. The most important vertebrate prey species in their diet is, however, the red colobus monkey. At Gombe, red colobus account for more than 80 percent of mammalian prey. An individual infant or juvenile colobus stands a greater chance of being caught than does an adult;[230] 75 percent of all colobus killed are immature. Adult and adolescent male chimpanzees do most of the hunting, making about 90 percent of the kills. Females also hunt, though more often they receive a share of meat from a male who either captured the meat or stole it from the captor. Lone chimpanzees, either male or female, sometimes hunt but hunts are most often social activities. In other hunting species such as lions and wolves, cooperation among hunters yields greater success rates than hunting alone; in both Gombe and the Taï forests, likewise, there is a strong positive relationship between the number of hunting chimpanzees and the odds of a successful hunt.[25, 230] Although most successful hunts result in a kill of a single colobus monkey, at times up to seven may be killed.

In her early years of research, Jane Goodall[83] noted that the Gombe chimpanzees tend to hunt in 'binges', during which they would hunt almost daily and kill large numbers of monkeys and other prey. The explanation for such binges has always been unclear. For example, the most intense hunting binge seen between 1960 and 1995 occurred in the dry season of 1990.[228] From late June through early September, a period of 68 days, the chimpanzees were observed to kill 71 colobus monkeys in 47 hunts. The total number of kills, including those resulting from hunts at which no human observer was present, may have been one third greater. During this time, the chimpanzees may have killed more than 10 percent of the entire colobus population within their hunting range, a predation rate that would certainly not have been sustainable in the long term. The sudden changes in hunting frequency observed at Gombe seem to be related to ecological, social, and demographic factors.

Chimpanzees are omnivores, eating a diet that is high in plant foods. Decisions about when to eat meat are based on the nutritional costs and benefits of obtaining prey, compared to the

forest dominated by *Musanga* spp. (Moraceae) and with many large figs (*Ficus* spp., Moraceae), while another group occupies an unlogged area, including forest dominated by *Parinari* spp. (Chrysobalanaceae), mature mixed forest, and hill forest.[97] Some groups survive in areas that have been logged and then almost totally converted to agriculture, where they travel among the few small remaining forest patches and raid crops.[46] In the Tomboronkoto region of southeastern Senegal, chimpanzees have been found resting and eating in caves during the dry season, perhaps to escape the high daytime temperatures in their savanna habitat.[215] In Bossou, chimpanzees spend more time in trees in the rainy season; a study concluded that this was not a response to the vertical distribution of the food but rather helped them avoid being cold and wet, as they would otherwise be on the damp ground away from the breezes of the canopy.[241]

Chimpanzees eat a wide range of foods, with an emphasis on fruits, flowers, and seeds, but including some young leaves and a variety of small mammals and invertebrates. As many as 330 food types (taxa and plant parts) can be eaten in a year. Diets can vary from area to area, mainly as a function of what is available, but may also reflect essential nutrients that the food provides relative to those available from plants. However, social influences such as party size and composition also seem to affect hunting behavior. A major goal of research on predatory behavior in chimpanzees is to understand when and why they decide to hunt colobus monkeys rather than forage for fruits, even though the hunt risks both injury from colobus canine teeth and failure to catch anything.

Early studies of this behavior suggested that meat eating and meat sharing had a strong social basis.[242] Hunting was seen as a form of social display, in which a male chimp tries to show his prowess to other members of the community.[127] In the 1970s, the first systematic study of chimpanzee behavioral ecology at Gombe concluded that although predation by chimps was nutritionally based, some aspects of hunting behavior were not well explained by nutritional needs alone.[274] More recently, researchers in the Mahale Mountains chimpanzee research project reported that the alpha male there, Ntilogi, used colobus meat for political gain, withholding it from rivals and doling it out to allies.[192] At Gombe, female chimpanzees that consistently receive generous shares of meat after a kill have more surviving offspring, indicating a reproductive benefit tied to meat eating.[153] Other researchers argue that male bonding is promoted by meat sharing, and is then useful in enhancing individual male reproductive success.[164]

There are many reasons why chimpanzees hunt; season, group composition, and individual personalities all play a role.[228] Future research in this area should be able to establish further the underlying motivations and strategies of hunting and sharing. Although most researchers have drawn comparisons between hunting behavior in chimpanzees and that of social carnivores such as wolves and lions, much more apt comparisons are to be found with human hunter-gatherers. In both humans and chimpanzees, meat is only a part of the diet and decisions must be made on a continual basis whether to hunt. People forage for meat and also gather plant foods though, as in chimpanzees, there are strong gender biases with males in most societies doing most of the organized hunting. Chimpanzees forage mainly for ripe fruit and hunt opportunistically when they happen to encounter prey. Their meat-sharing patterns are more systematic and more nepotistic than behavior seen in wild baboons, capuchin monkeys, or any other nonhuman primate.

Craig Stanford

Craig Stanford

A male chimpanzee at Gombe National Park eating red colobus meat.

local tradition and cultural variation.[113] There are also differences in the techniques used to process food[198] and in the medicinal use of plants.[111] Movement between foraging sites often takes place on the ground, but at least some tree cover is needed to provide food and nesting sites.

The diet is usually dominated by ripe fruits from forest trees; the fruits chosen tend to be those with a high calorie content in the form of sugars.[148] It appears that chimpanzees consume herbs mainly as a fallback source of carbohydrates when fruit is not freely available.[142] In the Budongo Forest Reserve, one community spent 65 percent of their feeding time eating fruit, particularly figs, 20 percent eating tree leaves, and a small amount of time feeding on herbaceous vegetation.[178] A further study in this reserve found that at least 15 species of figs were used, with the fruits and young leaves being eaten.[254] The foods selected tend to be low in tannins, although chimpanzees seem to be able to tolerate higher tannin levels than do monkeys living in the same forest, such as guerezas (*Colobus guereza*) or blue and redtail monkeys (*Cercopithecus* spp.).[219]

Some foods require specific and complex forms of processing. To access the edible part of a

The Jane Goodall Institute (JGI)

A young eastern chimpanzee feeding on leaves, United Republic of Tanzania.

Saba florida (Apocynaceae) fruit, for example, requires a chimpanzee to remove a thick outer layer. A study of how the skills involved are learned concluded that information gained by infants through observation of their mothers probably involves affordance learning (i.e. learning how to use an object by observing how it is used) or imitation (i.e. learning by copying the motor behavior involved) without intentional teaching. No evidence was found to suggest that teaching by the mother or imitation by gestural copying were involved.[38] Infants were able to process whole fruits when they were two years old, but mastery of the complex adult technique was not gained for a further two years.[53]

At Bossou, over 200 plant species, 30 percent of all those recorded there, have been seen to be used by chimpanzees.[238] A variety of other items may be eaten including algae, mushrooms, honey, termites, ants, and mammals such as tree pangolins (*Manis tricuspis*).[113] Fig trees are among the most important species but, in times of food shortage, the parasol tree (*Musanga cecropioides*, Moraceae) and the oil palm (*Elaeis guineensis*, Arecaceae) are commonly used.[280] *Musanga* fruits have been found to be an important fallback food in other areas, e.g. the Kalinzu Forest Reserve, Uganda,[72] presumably because they are available all year round. Chimpanzees tend to avoid being immersed in water but one individual at Mahale Mountains National Park was noted entering a stream several times to feed on algae,[221] and some ex-captive chimpanzees in Congo have also been photographed wading.[66]

In the Lopé National Park, insects are an important item in chimpanzee diet, and 31 percent of fecal samples contained insect remains.[251] The species most often eaten was the weaver ant (*Oecophylla longinoda*); others included two large ant species and bees (*Apis*). In the same area, chimpanzees feeding in forest fragments spent more time eating leaf petioles, bark, and pith, and less time feeding on flowers.[248] However, as in all study areas, fruit is the commonest item, making up 62 percent of the food consumed by the Lopé chimpanzees. They have been recorded eating the fruit of 114 species of plants. When preferred fruits are scarce, they maintain a relatively high intake of fruit by exploiting small arillate (fleshy, often brightly colored) fruits and palm nuts.[252]

In the montane forests of Kahuzi-Biega National Park, the highest location where chimpanzees live, they were recorded to eat 171 different food items. These were mainly plant materials belonging to 114 species, including 66 species of fruits. Figs were the most frequently eaten, being found in 92 percent of fecal samples.[10]

In Mahale Mountains National Park, chimpanzees have also been reported eating soil. It is postulated that they do this to obtain mineral supplements, medicinal chemicals with antacid and antidiarrheal properties, and to adsorb and detoxify alkaloids.[140] The hypothesis is that chimpanzees have learned, by personal experience or by observing others, that eating soil relieves an unconscious craving for micronutrients, or else makes them feel better when they are unwell.

In most study populations, chimpanzees have been observed hunting various prey species, and meat eating accounts for about 3 percent of time spent feeding. There is considerable variation among populations, however, and the percentage of fecal samples to contain animal remains ranges tenfold, from 0.6 percent at Kasakati to nearly 6 percent at Gombe National Park. Identified remains include crowned guenon (*Cercopithecus pogonias*), scaly tailed flying squirrel (*Idiurus* spp., Anomaluridae), and a duiker (*Cephalophus* sp.).[11, 130] Across 12 study sites, at least 32 species of mammals have been recorded as prey, but the most common were primates, particularly forest monkeys.[255]

Animals were killed relatively infrequently by

chimpanzees in Kahuzi-Biega National Park. In a 16 month study there, one group killed 18–30 mammals per year, mainly juvenile or subadult *Cercopithecus* monkeys.[11] Some other primates, for example vervet monkeys (*Cercopithecus aethiops*), are often hunted by chimpanzees,[256] but baboons (*Papio* spp.) are rarely preyed upon. The killing of baboons has been seen only at Gombe, where olive baboons (*P. anubis*) are occasionally hunted, and in the Mahale Mountains National Park, where yellow baboons (*P. cynocephalus*) are preyed upon.[172] In the Budongo Forest Reserve, the commonest mammalian prey is the guereza (*Colobus guereza*), but other prey included blue duiker (*Cephalophus monticola*) and an elephant shrew (*Rhynchocyon cernei*) captured opportunistically.[182] Chimpanzees are not carrion eaters, and the scavenging of meat from animals killed by other predators is rare. In one instance in Gombe, a freshly killed bushbuck (*Tragelaphus scriptus*) was largely ignored by a chimpanzee group and there was virtually no interest shown in consuming the carcass.[171]

The predation of red colobus (*Procolobus* spp.) by chimpanzees is discussed in detail in Box 4.1. It offers a useful example of ecological variation between sites. The chimpanzees in the smaller trees at Gombe National Park have to cope with more aggressive colobus than those in the tall trees at Taï National Park. Perhaps in response, Gombe chimpanzees only hunt opportunistically, whereas Taï chimpanzees adopt a planned and collaborative strategy[21] and tend to share meat more actively and more frequently.[25]

Males often share meat with other members of their group and this has been a fertile area for speculation about motives and strategies. It has been suggested that the energetic cost is less for sharing than for defending a carcass,[78] and that sharing promotes alliances that yield benefits in the form of grooming or support in dominance struggles,[57] or in the form of sexual favors.[231] Chimpanzees may also share plant food. These transfers are usually made from mothers to infants,[196] or between captive chimpanzees where food has been supplied by keepers,[152] but some cases of sharing vegetable matter among mature chimpanzees have also been described in the wild.[174] Sharing of termites has also been documented between mothers and offspring,[153] and between adult males.[179]

Primatologists believe that species (e.g. macaques) that consume high-energy foods, such as ripe fruits, tend to be more intelligent and interact more – all else being equal – than those (e.g. langurs) that eat mature leaves. Part of the rationale is that high-energy foods are clumped in the environment, so that whoever finds them first can often choose who gains access. This creates opportunities for reciprocal altruism and the maintenance of relationships between individuals, whether relatedness-driven or relationship-driven, or both. It is further argued that large brains are energetically expensive, so fresh meat, rich in fats and proteins, is a particularly desirable food for an intelligent animal such as a chimpanzee. By extension to the hominid lineage, an implication is that large brain size, meat eating, and political behavior all go together, though as in all such reasoning the causality soon becomes difficult to untangle.

Ranging behavior

In the Budongo Forest Reserve, the home range of the Sonso community was found to be about 6.8 km^2, among the smallest reported for habituated chimpanzees.[181] In contrast, the minimum-area polygon enclosing the home range of a community in the Kibale National Park was found to cover 14.9 km^2, though only 7.8 km^2 of this land may actually have been used. Males used an area 150–200 percent greater than that used by females, and were more likely to be seen near the boundaries. This supports the hypothesis that females use small core areas within the defended

David Watts

Chimpanzees on patrol, Kibale National Park, Uganda.

home range of the males.[47] There is some evidence from Budongo Forest Reserve, however, that males also spend most of their time within a restricted core area.[180]

In comparison with western gorillas, which generally cover less than 2 km per day, chimpanzees are more mobile, traveling an average of about 4 km per day.[80] When resources are scarce, e.g. during the dry season in Côte d'Ivoire, chimpanzees may reduce their daily range and party size, spending more time feeding, and feeding more frequently on lower-quality food items.[62]

Social behavior

It is hard to describe any aspect of chimpanzees' behavior and ecology in isolation from any other. Few of their day-to-day decisions seem to be made taking only one factor into account; individual personality, history, and relationships are all crucial influences in the expression of behavior in different contexts. The options for behavior are also very wide, since more complex behavior has been reported for this species than for any other non-human animal. A list of the behavioral patterns displayed by the Mahale chimpanzees included over 500 descriptive terms, of which more than 200 were patterns also commonly seen in humans and bonobos, and about 50 were common to several study populations of chimpanzees but not observed in bonobos.[194] For more on behavioral differences between chimpanzee populations, see Box 4.3.

A chimpanzee community has an average size of about 35 members, with a range of 20 to over 100 (occasionally as many as 130). There seem to be social differences between eastern and western populations, with western chimpanzees being generally less violent and more likely to form stable groups and female–female alliances.[276, 281] There is always a flexible, fission–fusion dynamic, with parties forming for short periods and with different members. This system may make it easier to exploit resources of various sizes, seasons, and locations within the community's home range.[113] The size and composition of parties is influenced by the threat of predators, including people; the presence of other mammal species; and the availability and distribution of water and nesting sites,[113, 222] as well as food abundance. One study of eastern chimpanzees at Budongo Forest Reserve suggested that both dispersal and abundance of food should be considered when assessing the impact of food supply on grouping patterns.[183] The importance of food as a factor in determining such patterns declined with increasing food abundance.

The presence of 'swollen' females, with prominent perineal swellings around the anus and vulva,[59] has more effect on party size than does food availability.[3, 4, 48, 100, 275] Large parties of chimpanzees can build up when a swollen adult female is attracted to a party consisting of a top-ranking male, often accompanied by large numbers of adult and adolescent males, with other males joining the group to be with her.[146] Meanwhile, in small communities, single high-ranking males may try to sequester and guard swollen females. In the unusually large community at Ngogo, Kibale National Park, where the number of males is very high, pairs or trios of top-ranking males sometimes cooperated to prevent swollen females from mating with other males and tolerated each other's mating activities.[265] From these observations, it seems that such females are apparently so attractive that the male response to them can easily overwhelm all other considerations in their decision making; this requires some explanation.

At the age of about 10 years, a female chimpanzee begins her menstrual cycle, with a periodicity of about 35 days.[12, 65, 102, 283] During the first (follicular) phase of the cycle, estrogen is secreted by the developing Graafian follicle – the capsule that protects a developing egg. This causes the perineal ('sexual') skin to swell in a very visible way. As estrogen output peaks just before ovulation, so does the swelling; this is accompanied by changes in

Ilka Herbinger/Wild Chimpanzee Foundation (WCF)

Western chimpanzees resting, Taï National Park, Côte d'Ivoire.

behavior and in relations with adult males, as well as in her attractiveness to males. Under the further influence of the ovarian and adrenal androgens, the female becomes proceptive and actively seeks sexual contact with males, behavior that reaches a peak at or just before ovulation when the swelling is greatest. She also becomes receptive to intromission and the maintenance of the copulatory posture until intravaginal ejaculation is achieved. This mid-cycle surge in sexual motivation and preparedness is conventionally called 'estrus' in mammals, and in most it is driven principally by the ovarian estrogens. In apes (and Old World monkeys), however, it is under more complex neuroendocrine and social control, so the term estrus is not wholly appropriate.[124]

After ovulation, estrogen levels drop sharply and the corpus luteum develops from the follicle and starts to secrete progesterone. In response, the swelling collapses, and remains quiescent throughout the luteal phase of the cycle and the menstruation that follows it. A young female can have these regular bouts of swelling and sexual proceptivity for months before she conceives her first infant, and this may be an important time for her to try out life in several communities before settling down.[123]

All this leads to the expression of a variety of social forms.[113] Males are usually found with females in mixed parties, whereas females often range alone or in small parties with other females. Males associate closely with each other, including mutual grooming, but social interactions between females are infrequent (in marked contrast with the bonobo).[73] More specific research suggests that adult males show no particular preference for associating with each other, whereas nonswollen females prefer each other as party members.[210] It seems that males choose to associate with partners on a tactical basis, rather than randomly grouping or independently selecting the same locations in which to forage.[176] There is little evidence that kinship strongly influences male relationships.[160]

Social interactions are frequently complicated, with varying degrees of cooperation, coalition, and alliance formation. Juveniles and adolescents tend to associate with their mothers. As they get older, males begin to associate more with adult males, while females often continue to associate with their mothers in early adolescence, before transferring to other communities in later adolescence.[101] In over 80 person-years of field observation of well known individuals in Tanzania, no male chimpanzee has been seen to transfer out of his natal group.[133] There are, however, records from 1976 to 1997 of disappearances of immature males from their natal groups at Bossou, which may represent emigration rather than mortality.[234]

The Jane Goodall Institute (JGI)

Grooming is particularly important between male chimpanzees.

In Kibale National Park, 35 males in the large Ngogo community were seen to associate in two separate subgroups, the members of which tended to stay close together and participate in boundary patrols.[267] Despite this clustering, there were low levels of aggression between individuals of different subgroups.[157]

Patrols along the boundaries of the range of a community are carried out by groups of males, usually adults, and may lead to lethal attacks on neighboring communities, often targeting their males or young.[264, 279] These combative patrols may benefit the community by extending their range, protecting other community members, and incorporating more females into the community; these advantages apparently outweigh the risks involved in carrying out the patrols and attacking conspecifics.

Within a chimpanzee community, it is rare to see aggressive behavior that is intense enough to cause the death of a community member. There are records from Tanzania, where an alpha male was killed,[188] and from Uganda, where the victim was a young adult male.[67] In Tanzania, a young adult male was violently attacked and ostracized by other males in the same community, apparently because he behaved in ways that the alpha male and his allies found provocative.[193]

Box 4.2 CHIMPANZEE VISION

During primate evolution, vision has become the most highly developed sense. It is an integral part of chimpanzee life and culture, allowing such activities as the sophisticated use of tools, feeding, recognition of individual conspecifics, and communication.

Vision and primate evolution

The first primates probably evolved from insectivorous mammals more than 65 and perhaps as long as 80 million years ago (mya). The theories surrounding the subsequent adaptive radiation of primates include the development of enhanced color vision as a key step, with various explanations for it. The presumed arboreal lifestyle of early primates led early researchers to suggest that this environment promoted an increased reliance on visual and tactile senses.[122, 227] Furthermore, orbital convergence and stereoscopic vision, together with grasping hands with nails rather than claws, could be particular adaptations for a visual predator.[42] In the forest canopy at night, increased visual acuity would allow discrimination between potential food items. Precise eye–hand coordination would also aid animals operating in the dark.[239] Locomotion may also have been an important force selecting for stereoscopic vision in arboreal animals that would need to locate branches or trunks precisely, in order to leap between them.[55] Today, this enhanced visual acuity is used by chimpanzees when manipulating tools.

Color vision

Primates have excellent color vision compared with that of most mammals. Chimpanzees, like other apes and Old World monkeys, have color vision based on three different types of cone photopigment in the retina of the eye.[120] Each of these photopigments absorbs a different wavelength of light, and this type of color vision is known as trichromacy. Trichromatic primates have particularly good discrimination in the red–green part of the spectrum when compared to primates with only two types of photopigment (dichromacy). This has prompted the suggestion that trichromacy evolved to allow identification of ripe fruits against a dappled background of forest leaves.[217] Trichromatic species are indeed better at this task than dichromatic ones.[39, 226] Trichromacy also benefits folivory,[60] and its evolution would have promoted the selective exploitation of young leaves (which are often red in tropical forests) as a food resource.[61]

Visual cues and social interactions

Frugivorous primates have relatively large brains with more neurons in the parvocellular system, one of the pathways for processing visual information.[9] This system primarily analyzes fine detail and color, supporting selection for the ability to detect fruits based on specific visual cues. The same trend in the parvocellular system is also observed in primates with larger group sizes.[9] This visual system is therefore likely to be used by chimpanzees in critical social interactions, such as interpreting facial expressions and gaze direction. The ability to recognize other individuals is crucial in the development and evolution of mammalian social systems. Chimpanzees also have the remarkable ability to identify mother–son relationships in unknown individuals, using only visual cues. Vision is therefore an important mechanism of kin recognition based on outward appearance, independent of previous experience.[209] This ability is very important to chimpanzees, where related individuals may spend some time apart and where political alliances may be formed on the basis of familial relationships.

Vision has an important role in chimpanzee communication, both in determining facial expressions and in the gestural basis of chimpanzee 'language'. The importance of vision in chimpanzee communication is supported by the observation that chimpanzees are readily able to learn and communicate hundreds of human 'signs'. Some researchers even believe that human language has evolved from a gestural origin, which would have relied heavily on vision.[52]

Alison Surridge

Infanticide has been reported in chimpanzee populations across Africa, but is most frequent in eastern chimpanzees. It has been recorded in the Budongo Forest Reserve,[6, 177, 240] at Gombe National Park,[83] at Mahale Mountains National Park,[186, 189, 191] and at Kibale National Park.[264, 264] Infanticide is usually carried out by males, and has been explained as a strategy to bring the mother back into a condition of fertility and sexual receptivity earlier than would otherwise be the case, potentially increasing the killer's chance of fathering her next offspring.[109] The dead infants are often subsequently eaten, with the meat being shared as usual.[93, 266] This suggests that nutrition (and the enhanced status that goes with control of a meat resource) may be another motivation for infanticide.[93] Observations of infanticide occurring within a Mahale community showed that victims were always unweaned males whose mothers had mainly mated with older adolescent or immature males.[93] Kidnapping of infants by males has also been observed, and may result in the deaths by starvation of unweaned individuals. It has been suggested that this is sometimes motivated by an interest in the infant as a 'possession' or plaything rather than an interest in its death.[200] Although it can be an important cause of death in a statistical sense,[264] infanticide is by no means the norm, and rarely affects weaned young.

Mutual grooming is important in chimpanzee societies, especially among males, where it is thought to play a major role in servicing relationships and coalitions that can improve status.[5] Chimpanzees groom in clusters varying from two to 23 individuals, with adult males grooming for longest in small clusters, and adult females grooming in clusters of five or more.[173] A custom of social scratching has been noted in association with grooming in the Mahale Mountains National Park[175] and at Ngogo, Kibale National Park,[191] but not at Gombe National Park, Kanyawara (Uganda), or anywhere in West Africa.

Chimpanzees from communities that have had frequent contact with humans behave differently from those that encounter humans for the first time. Encounters in the Goualougo Triangle (Congo) with chimpanzees that lacked prior experience with humans were characterized by a high frequency of curious responses.[169] Where chimpanzees have had more contact with people, the survivors tend to be very much more wary of or aggressive towards humans.

Development and reproduction

Wild chimpanzees have a very low reproductive rate. Females reach sexual maturity at 10–13 years of age and typically give birth every six years thereafter. Infants are very dependent on their mothers for the first five years. They generally have a life expectancy at birth of less than 15 years; the mean adult lifespan, after reaching sexual maturity, is about 15 years.[104] Females are therefore likely to have only three or four offspring during their lifetime. In the Taï National Park, dominant females invested about two years more than the average in caring for sons, whereas subordinate females invested about 11 months more in caring for daughters.[24] In Gombe National Park, high-ranking females have significantly higher rates of infant survival, faster-maturing daughters, and more rapid production of young.[216] Twins are rare: a single twin birth was recorded among 135 births over 29 years of study in Mahale Mountains National Park.[145]

The demography of the Mahale chimpanzees has been studied in detail.[191] The major identified cause of death was disease (48 percent), followed by senescence (24 percent), and within-species aggression (16 percent). Half of all chimpanzees died before weaning. Landmarks in a typical female life history include the first maximal swelling at an average of about 10 years, emigration to other groups at 11 years, and giving birth for the first time at 13 years. The fecundity of females was highest between 20 and 35 years old, with a birth rate of 0.2 per female per year. A similar study at Bossou from 1975 to 2001 found that the average age of giving birth for the first time (at about 11 years) was lower than in all other wild chimpanzee populations, and that the infant and juvenile survival rate was the highest. This suggested that the lifetime reproductive success was likely to be higher at Bossou than at any other long-term study site.[235]

Members of each community breed mostly with one another, with extragroup paternity being a minority event. A study in the Taï National Park,

Andrew Fowler

This Nigeria-Cameroon infant will remain dependent on its mother until it is five years old.

Box 4.3 CHIMPANZEE CULTURES

Midway through the 20th century, we knew next to nothing about wild chimpanzee behavior. Subsequent decades of field study at multiple long-term sites have made it possible to put together a comprehensive assessment of behavioral variation across Africa, much as anthropologists have done for human societies. The resulting picture shows chimpanzees to have a rich cultural complexity that was unsuspected before the fieldwork began.

Early steps in this pathway of discovery included attempts to identify differences in feeding habits between the Gombe and Mahale sites in Tanzania[198] and the identification of a social custom, the 'grooming handclasp', which is present at Mahale but not Gombe.[155] As more long-term field studies progressively yielded greater knowledge of local habits, researchers began to draw up comparative tables that suggested a growing list of behavioral differences right across Africa.[29, 83, 153] However, as these were based only on the data that happened to have been published from each site, the conclusions were inevitably limited.

In the late 1990s, the directors of all the sites where long-term field studies on chimpanzees were being undertaken agreed to pool their data for the first time, to create a more comprehensive understanding of local traditions.[268, 269, 270] In the first phase, site directors suggested patterns of behavior they suspected might represent local traditions, generating a list of 65 'candidate behaviors'. In the second phase, each behavior was coded by the core researchers at each site as either *customary* there (performed by all able-bodied individuals of at least one age-sex category), *habitual* (occurring repeatedly among several individuals, consistent with some degree of social transmission), or *absent*, either *with an apparent ecological explanation* (such as the absence of appropriate raw materials) or *with no such explanation*. Based on this last piece of information, putative traditions or 'cultural variants' were defined as those behavioral patterns that were either customary or habitual in at least one chimpanzee community, yet absent without ecological explanation at another. Genetic explanations were also excluded using various criteria, notably geographic proximity of behavioral variants.

The 39 cultural variants identified included forms of tool use, techniques for dealing with ectoparasites, social customs, and courtship gambits. This complexity contrasted strikingly with earlier reports of animal traditions, which typically recognized just a single behavioral variation. Moreover, each community was found to exhibit around 10 or more of these traditions so that, if we know enough about an individual wild chimpanzee, we can now assign it to its community on the basis of its cultural profile alone (see table), as we expect to do for people. Indeed, the fact that each community expresses these multiple traditions is one of the reasons that it makes sense to talk of 'culture'. It is precisely this multiplicity that we identify with the phenomenon of human culture, our inevitable reference point for such comparisons.[271]

This systematic two-phase procedure is now being applied in the second Collaborative Chimpanzee Cultures Project, which is investigating additional behavioral variants. Areas explicitly omitted from the first study, such as styles of hunting and carnivory, are being included. The procedure has also been applied to orangutans in an equivalent international collaborative effort that identified 19 cultural variants, including both tool use and social customs.[258] For both orangutans and chimpanzees, there is evidence that the variants are socially transmitted. First, the similarity of overall behavioral profile is correlated with geographic proximity, indicating that behaviors are transmitted from the location at which they first appear. Second, communities with higher indices of social proximity have larger cultural repertoires. Where apes spend more time together, they have greater opportunity to learn new behaviors from one another, so behaviors are more likely to spread throughout a group.

These discoveries are significant for both anthropology and conservation. From an anthropological perspective, the new picture helps us understand the roots of our own extraordinary cultural capacity. The common human/great ape ancestor of about 15 mya was likely to have exhibited cultures something like the simpler forms seen in chimpanzees and orangutans today. From the conservation perspective, the tragedy is that not only do we risk losing several subspecies of chimpanzee, but also and more imminently we risk losing their unique subcultures. As this very richness becomes more widely known, perhaps greater conservation effort will be mobilized.

Andrew Whiten

Chimpanzee cultural variation across long-term study sites

Distribution across the six most long-term study sites of 38[a] behavior patterns that meet the criterion of being customary or habitual in at least one community, and absent without apparent ecological explanation in at least one other.

	Bossou	Kibale	Budongo	Taï Forest	Mahale	Gombe
	Guinea	Uganda		Côte d'Ivoire	Tanzania	
Pestle pound (mash palm crown with petiole)	■	⊘	⊘	○	⊘	○
Rain dance (slow display at start of rain)	○	■	●	●	■	■
Stone–stone (nut-hammer, stone hammer on stone anvil)	■	⊘	⊘	■	○	○
Stone–wood (nut-hammer, stone hammer on wood anvil)	◯	⊘	⊘	■	⊘	○
Wood–wood (nut-hammer, wood hammer on wood anvil)	○	⊘	⊘	■	⊘	○
Wood–stone (nut-hammer, wood hammer on stone anvil)	○	⊘	⊘	■	○	○
Wood–other (nut-hammer, wood hammer on earth etc.)	○	⊘	⊘	●	○	○
Pound wood (food-pound onto wood (smash food))	■	⊘	●	■	○	■
Pound other (food-pound onto other (e.g. stone))	○	⊘	○	●	○	■
Lever open (stick used to enlarge entrance)	○	○	○	●	○	■
Expel/stir (stick expels or stirs insects)	○	○	○	■	●	●
Termite fish-M (termite-fish using leaf midrib)	◯	⊘	⊘	⊘	○	○
Termite fish-S (termite-fish using non-leaf materials)	○	⊘	⊘	⊘	○	■
Fluid dip (use of probe to extract fluids)	○	●	○	■	●	■
Ant dip (one handed dip stick on ants)	■	○	○	■	○	◯
Ant fish (probe used to extract ants)	◯	○	○	○	■	◯
Ant wipe (manually wipe ants off wand)	◯	○	○	○	○	■
Aimed throw (throw object directionally)	■	◯	◯	■	■	■
Marrow pick (probe used to pick bone marrow out)	○	○	○	■	○	○
Bee probe (disable bees, flick with probe)	○	○	○	■	○	○
Index hit (squash ecto-parasite on arm)	○	○	○	■	○	◯
Fly whisk (leafy stick used to fan flies)	○	○	●	●	○	◯
Shrub bend (squash items underfoot, courtship)	●	○	■	○	■	○
Leaf groom (intense 'grooming' of leaves)	○	■	◯	○	■	■
Leaf clip mouth (rip parts off leaf, with mouth)	■	●	■	■	■	○
Leaf clip fingers (rip single leaf with fingers)	○	●	■	●	◯	○
Leaf dab (leaf dabbed on wound, examined)	○	■	○	◯	○	◯
Leaf napkin (leaves used to clean body)	○	■	■	◯	◯	■
Leaf strip (rip leaves off stem, as threat)	◯	●	○	○	◯	●
Leaf inspect (inspect ecto-parasite on leaf)	○	○	■	○	?	◯
Leaf squash (squash ecto-parasite on leaf)	○	○	○	○	?	●
Pull through (pull stems noisily to attract attention)	■	●	○	○	●	◯
Seat vegetation (large leaves as seat)	◯	◯	○	●	○	○
Self tickle (tickle self using objects)	○	○	○	○	○	●
Branch slap (slap branch, for attention)	■	○	■	■	◯	○
Hand clasp (clasp arms overhead, groom)	○	■	○	●	■	○
Knuckle knock (knock to attract attention)	◯	○	○	■	■	●
Club (strike forcefully with stick)	◯	◯	○	●	◯	●

■ **Customary:** occurs in all or most able-bodied members of at least one age-sex class (e.g. adult males)

● **Habitual:** has occurred repeatedly in several individuals, consistent with some degree of social transmission

◯ **Present:** neither customary nor habitual but clearly identified

⊘ **Not possible:** absence can be explained by local ecological features

○ **Absent:** not recorded with no apparent ecological explanation

? **Not known:** not recorded, perhaps because of inadequacy of relevant observational opportunities

a One variant was absent and another never common at the six most studied sites listed here.

 For further details, including less studied sites (Mount Assirik, Lopé), see Whiten, A., *et al.*, 1999, 2003.

based on nuclear DNA microsatellite markers and behavioral observations, found only one likely case of extragroup paternity among 14 births, an incidence of 7 percent.[259] Females may copulate with many males early in the receptive stage, but later, when the likelihood of conception is highest, they have been observed to copulate repeatedly only with high-ranking adult males.[147]

Communication

Adult male chimpanzees often give loud 'pant-hoot' calls when they arrive at fruiting trees. It is speculated that this behavior asserts the social status of the caller, rather than being of benefit to the listeners,[50, 51] or that the purpose is to rally and maintain contact with the group while recruiting allies and associates.[161] Younger males and females are generally quieter than adult males, possibly to avoid attracting feeding competition and potentially aggressive males.[49] Younger males, however, may join the alpha male in a chorus of pant-hoots and, when they do, they seem to accommodate each other by giving acoustically similar calls.[158] The pant-hoots vary somewhat geographically, which is tentatively attributed to factors such as habitat acoustics, the backdrop of sounds made by the local suite of wildlife, and body size.[49, 159] Other vocalizations include context-specific barks used in hunting and snake alarms, and combinations of barks with acoustically distinct call types or drumming.[54]

Tool use

Tool use is widespread among chimpanzee populations right across the geographical range of the species, and involves many different implements used for a variety of purposes. Like humans, chimpanzees seem to be predominantly righthanded.[108] The tools thought to be most important to chimpanzees are those used in obtaining food and inspecting their environment (including extracting, probing, and pounding). Less important are tools used for displays (including aggression against conspecifics or other species such as the leopard, and communication) and, in a minor context, for cleaning their own bodies.[28] Different communities of chimpanzees have different repertoires of tool use, some using far fewer tools than others.[154] The Taï chimpanzees have exhibited 28 out of 42 tool-use behaviors recorded throughout the range of the species, compared with 17 in Gombe and 13 in Mahale (see Box 4.3).[26, 27]

The use by chimpanzees of ant-dipping wands has been noted in Tenkere, Sierra Leone, where the average wand length was generally somewhat longer than in Guinea, Senegal, and Tanzania, and considerably longer than in Côte d'Ivoire.[1] In Guinea, however, the tool length is determined by prey attributes, with longer tools used in higher-risk contexts (e.g. at the ants' nest site or with the aggressive black *Dorylus* ants). Here, two techniques are employed: 'direct mouthing', where the ants are eaten directly from the tool, and 'pull-through', where the tool is drawn through the hand and the bundle of ants is then eaten.[114] The use of similar tools has been noted in southwest Cameroon,[117, 118] indicating a wider distribution in the general use of tools than was previously known. The use of tools to dip for driver ants was observed for the first time in the Kalinzu Forest Reserve, Uganda.[99] In Equatorial Guinea, chimpanzees were seen to use sticks to perforate termite mounds and then gather the termites with their hands.[82]

All young chimpanzees spend time playing and learning from others, but there seem to be sex-based differences in learning. A study of the acquisition of termite-dipping skills at Gombe National Park found that it took five years for young chimpanzees to develop the technique for termite-dipping.[137] However, females had learnt to fish for termites about two years before males. Females are more proficient than males after acquiring the skill, and each young female uses a technique similar to that employed by her mother, whereas young males do not.[138]

Males spent more time playing,[138] and there are cultural differences here too. Some games are known only from a particular locality, such as 'leaf-pile pulling' while descending a slope, a behavior observed only in the Mahale Mountains National Park.[197] This game involves walking backwards, raking a pile of dead leaves along with both hands and making a lot of noise.

In Sierra Leone, chimpanzees have been seen to use 'stepping sticks' (small sticks held under the feet) and 'seat sticks' (sticks for sitting on), to avoid injury from thorns encountered while feeding on the flowers of kapok (*Ceiba pentandra*, Bombacaceae).[2] In Guinea, a chimpanzee was seen sitting on a cushion made from the leaves of the carapa tree (*Carapa procera*, Meliaceae), apparently to avoid sitting on wet ground.[106]

At Mount Assirik, chimpanzees used leaf stalks to obtain termites (*Macrotermes subhyalinus*), sticks to get honey, and stones (probably as hammers), to break open the hard-shelled fruits of *Adansonia digitata* (the baobab, Bombacaceae).[15] In the CAR, a female chimpanzee used a large piece of a dead branch as a pounding tool to break into a melipone beehive and obtain honey.[68] In the Gambia, one chimpanzee was noted as sequentially using a tool set comprising four different types of tools, each with a different function, to extract honey from a bees' nest in a dead stump.[30] Similar observations have since been made in Congo.[14] In the Bwindi Impenetrable Forest, Uganda, chimpanzees used two types of tools to obtain honey: a small stick for the tree cavities and subterranean holes used by a stingless bee (*Meliponula bocandei*); and larger, thicker tools to assist in foraging for honey of the African honey bee (*Apis mellifera*).[229]

Chimpanzees also use tools to crack nuts, behavior that was documented in Sierra Leone as long ago as the 16th century.[224] At Bossou, they use tools to open oil palm nuts. Recent studies[16, 269] found that population-specific details of tool use in this area could not be explained purely on the basis of ecological differences – that is, there were cultural differences unrelated to ecology (Box 4.3). The techniques are learned when the animals are three to five years old, the age at which juveniles are also most likely to try unfamiliar foods if they are offered them.

Chimpanzees in the Odzala-Koukoua National Park used sedge (Cyperaceae) stems to scoop algae from the surface of a pond.[58] The use of a leaf sponge to drink water from tree hollows has been observed in most long-term study sites.[150] At Bossou, chimpanzees use folded leaves, most frequently from *Hybophrynium braunianum* (Marantaceae), to obtain drinking water from natural hollows in trees.[246]

In Tanzania, some individuals have been seen using a nasal probe to induce sneezing, presumably to clear out the nasal passage.[143, 195] A chimpanzee was once seen to insert a stick into a narrow hole in a tree to rouse a hiding squirrel, which was then captured and eaten.[110] Yet another Tanzanian chimpanzee was seen wearing a 'necklace' made of a piece of skin from red colobus; it had been created by a single overhand knot but it is not known whether this was tied by a chimpanzee.[156]

Nest building

Chimpanzees build nests every night, usually in trees. They use fairly substantial branches or forks to form a framework and then bend and break side branches to weave a platform, sometimes adding a lining.[70] They may also build different nests during the day for resting; these are usually in trees, but sometimes on the ground.[23] Up to 10 nests may be built in a single tree and the species of tree used varies in different regions. In West Africa, the most commonly used species are the sassy tree

Matsuzawa Tetsuro

Ilka Herbinger/Wild Chimpanzee Foundation (WCF)

Tool use for cracking nuts at Bossou, Guinea (above) and ant fishing at Taï National Park, Côte d'Ivoire (left).

Box 4.4 SEED DISPERSAL BY CHIMPANZEES

The interaction between fruiting plants and the vertebrates that disperse their seeds is increasingly attracting the attention of conservationists. This reflects the thought that to conserve tropical forests effectively, it is critical to retain the frugivores that disperse plants' seeds. Chimpanzees consistently show a year-round affinity for eating fruit, allocating up to 98 percent of their foraging time to fruit everywhere they have been studied. They ingest much larger amounts of fruit and seeds per meal than smaller frugivores, both in relative and absolute terms. Less than 2 percent of the population (around 14 percent of the biomass) of primate frugivores in Uganda is represented by chimpanzees, but they are responsible for an estimated 45.3 percent of all the seeds defecated.[277]

This high degree of frugivory is unusual in such a large-bodied mammal, but is consistent with the lineage of the species and the design of its digestive system, which, like all the apes, features a simple globular stomach with no mechanism for fermentative digestion. The digestive system is shared with the cercopithecine monkeys – macaques, guenons, mandrills, mangabeys, patas monkeys, and baboons. This limits the animal to eating foods that have relatively low concentrations of toxins or digestion inhibitors, such as tannins and fiber. In nature, such foods comprise fruits, some seeds, tender parts of plants, and animals. The larger apes, the orangutans and gorillas, can subsist on the poorer-quality end of this spectrum, such as bark and herbaceous vegetation, but chimpanzees are committed to frugivory and carnivory.

Seed dispersal in chimpanzee dung, Bafing Reserve, Mali.

Ian Redmond

The large mouth, robust dentition, and manual dexterity of a chimpanzee mean that it need not specialize in fruits of a particular size defined by its gape. Chimpanzees process fruit coarsely, swallowing the seeds of many species intact. Seeds are propelled through the gut with minimal chemical and physical damage, and may be defecated whole and in large clumps.[131] For some fruit species, the passage of seeds through the chimpanzee's gut increases their germination rate. These factors all imply that chimpanzees have excellent potential as seed dispersers, which is amplified by their behavior. Chimpanzees habitually travel widely each day, infrequently dropping seed-loaded dung on the forest floor. This foraging pattern facilitates long-distance seed transportation over a wide area. This is crucial for tree species such as *Mimusops bagshawei* (Sapotaceae), which regenerates very poorly in the immediate vicinity of the parent tree but produces viable seedlings under other tree species.[44]

Most chimpanzee-disseminated seeds are either preyed upon by rodents or insects, or succumb to pathogens (especially fungi). However, these pressures are typically even greater for seeds that are not dispersed away from the area of the parent tree. The seeds not consumed immediately may germinate *in situ*, or be dispersed secondarily – mostly by dung beetles – and germinate elsewhere. Germinating seeds face a cascade of other destructive agents, especially herbivorous animals. When a gap in the forest canopy is available, however, some fraction of the original seed cohort does finally establish to become seedlings and subsequently grow into trees.

As a by-product of their foraging behavior, chimpanzees are effective seed dispersers over long distances. This habitat-wide and year-round broadcasting of numerous seeds of multiple species is a prerequisite for the maintenance of a heterogeneous forest. This is one reason that chimpanzees have been described as keystone species in forest ecosystems. Their decline in numbers may therefore impair the composition and structure of tropical forests.

James V. Wakibara

(*Erythrophleum suaveolens*, Leguminosae) and the oil palm.[113] In the Budongo Forest Reserve, the chimpanzees particularly favored *Cynometra alexandri* (Leguminosae). Here it was found that day nests were structurally simpler than night nests, and were built in the trees used for feeding at the same height as feeding activity.[34] In the Kalinzu Forest Reserve, where there are relatively few large carnivores, many night nests are made in the low branches. In some places, nests are also quite frequently built on the ground, for example at Bili in northern DRC, and in the Nimba Mountains of Guinea.[88]

Response to habitat disturbance

Chimpanzees are robust and adaptable animals and have by far the widest geographical and ecological distribution of any ape, perhaps of any nonhuman primate in Africa apart from the commoner species of baboon. One would therefore not expect chimpanzees to be particularly sensitive to low or moderate degrees of ecosystem disturbance, such as might be caused by light selective logging or patchy settlement by low densities of farmers. Consistent with this, chimpanzee populations are known to survive well in lightly logged forest, such as at Kalinzu Forest Reserve in Uganda.[96, 97] Nevertheless, not all logging is equally benign, and many studies have shown a significant decline in chimpanzee numbers in logged forest relative to comparable unlogged areas, for example in Kalinzu,[225, 249] Ituri,[272] and Budongo Forest Reserves.[212]

Some groups are known to survive, at least for a time, in areas that have been logged and then almost totally converted to agriculture, where they travel among the few remaining small forest patches and raid crops planted by local farmers.[46] In Kibale National Park, chimpanzees were found in nine out of 20 forest fragments in 1995, but it was noted that it was their very large home range that enabled them to use these forest fragments.[206] In the Ugalla area, the survival of chimpanzees is threatened by the selective removal of *Pterocarpus tinctorius* (Leguminosae) trees, which provide very important food: flowers in March, seeds in July, and young leaves from September through November.[199, 203] In western Tanzania, shifting agricultural practices, uncontrolled bushfires, and habitat fragmentation were identified as the major threats to the survival of populations outside protected areas.[14]

Cláudia Sousa

Chimpanzee nests on palm trees at Tombali, typical of the coastal area of Guinea-Bissau.

Interactions with other animals

Chimpanzees coexist with western gorillas (*Gorilla gorilla*) in some areas (for example, the Lopé National Park in Gabon), despite having similar diets. Their keystone food resources differ and they tend mutually to avoid contact[253] (see Box 8.1).

Baboons and chimpanzees often occur together in the drier parts of the chimpanzee range. In one well studied example, at Mahale Mountains National Park, chimpanzees depend on ripe fruits, preferring those with a high calorie content,[148] which are also eaten in an unripe state by yellow baboons (*Papio cynocephalus*).[149] The baboons increased in numbers and expanded their range considerably after people moved out of the park area in 1975, reducing the ripe fruits available to the local chimpanzee group, which reacted by exploiting habitat areas and alternative food sources at a higher altitude than they had formerly.[190] Mahale chimpanzees have at least once been observed to eat baboons.

Red colobus in Taï National Park show anti-predation tactics when chimps are close, hiding high in trees where they are shielded from the forest floor, and becoming silent. In other circumstances, however, they move closer to groups of nearby Diana monkeys (*Cercopithecus diana*), presumably because the latter are efficient sentinels for predators approaching over the forest floor.[35] In Uganda, red colobus have been seen to associate with several other species of monkeys in areas of high chimpanzee density, probably for the same reason.[45] Chimpanzees also hunt guereza

Table 4.2 A summary of population data for the chimpanzee

	Central	Western	Eastern	Nigeria-Cameroon
Angola	200–500	0	0	0
Benin	0	extinct	0	0
Burkina Faso	0	extinct?	0	0
Burundi	0	0	200–500	0
Cameroon	31 000–39 000	0	0	3 000–5 000
CAR	800–1 000	0	unknown[a]	0
Congo	10 000	0	0	0
Côte d'Ivoire	0	8 000–12 000	0	0
DRC	extinct?	0	70 000–110 000	0
Equatorial Guinea	1 000–2 000	0	0	0
Gabon	27 000–64 000	0	0	0
Ghana	0	300–500	0	0
Guinea	0	8 100–29 000	0	0
Guinea-Bissau	0	600–1 000	0	0
Liberia	0	1 000–5 000	0	0
Mali	0	1 600–5 200	0	0
Nigeria	0	unconfirmed[a]	0	2 000–3 000
Rwanda	0	0	500	0
Senegal	0	200–400	0	0
Sierra Leone	0	1 500–2 500	0	0
Sudan	0	0	200–400	0
Togo	0	extinct	0	0
Uganda	0	0	4 000–5 700	0
United Rep. of Tanzania	0	0	1 500–2 500	0
Total	70 000–117 000	21 000–56 000	76 000–120 000	5 000–8 000

a 'Unknown' indicates that it is not clear how large the population is; 'unconfirmed' indicates that there may not be a population. As the table shows, the Nigeria-Cameroon chimpanzee is by far the most rare. The Ejagham and Takamanda Forest Reserves in Cameroon have also been referred to as having a 'significant population' of the Nigeria-Cameroon chimpanzee.[257]

Data compiled by Butynski, T.M. (2003);[37] see Chapter 16 for later estimates for Cameroon, Ghana, Mali, and Nigeria as well as further details of national populations.

(*Colobus guereza*) in Uganda, and the density of the latter is significantly lower in chimpanzee activity centers than outside them.[129] Guerezas appear to react to the presence of chimpanzees in a different way from red colobus: they remain quiet and make their escape along the ground.[116]

In Guinea, chimpanzees were observed capturing and toying with western tree hyraxes (*Dendrohyrax dorsalis*), but did not eat them and appeared not to regard the hyrax as a prey animal.[107] Similarly, a Mahale chimpanzee treated a squirrel as a toy, making play faces during the encounter and giving up shortly after the animal was dead.[282] A more benign association between chimpanzees and Thomas's galago (*Galago thomasi*) was noted in Uganda, with the latter found nesting inside a night nest of the former.[135]

In Taï National Park, 29 interactions between leopards and chimpanzees were observed from 1985 to 1990. In these, at least four chimpanzees were killed, with the leopards apparently being the main cause of mortality in the area.[20] A dead chimpanzee found in the Petit Loango Reserve, Gabon had also probably been killed by a leopard.[71] However, the attacks do not all run one way. In Tanzania, a group of about 33 chimpanzees surrounded a mother leopard and her cub in their den, and dragged out and killed the cub,[105] while in Uganda a small group of chimpanzees chased off a leopard.[213] In the Mahale Mountains National Park, evidence of lions eating chimpanzees was found in 1989.[247]

Unless habituated, chimpanzees usually react with fear towards human beings. On occasion, most recently in Gombe National Park (2002) and near Kibale National Park (2000), chimpanzees have attacked and killed human infants. Most primatologists believe that the chimpanzees are driven by predatory instincts rather than by the same infanticidal urges that sometimes lead to attacks on the offspring of rival males. Whatever the motivation, these rare events are devastating to the families affected and damaging to local conservation efforts.

POPULATION

Status

It is difficult to estimate the current number of chimpanzees because recent information is lacking from many areas and nothing at all is known about some. Data are summarized in Table 4.2, and indicate an estimated total population size of between 172 000 and 301 000 chimpanzees in 2003.[37]

Trends

As with other forest animals, it is difficult to assess population size and monitor trends in chimpanzee populations. Attempts have been made to estimate overall numbers by applying population density values at known sites to the remaining area of suitable habitat in the species' range. The total number of chimpanzees in 1987 was estimated at 151 000–235 000,[244] and in 1989 at 145 000–228 000.[245] The figures in Table 4.2 for 2003 suggest that these previous totals may have been underestimates. It has been argued that there were some 2 million chimpanzees around 1900, and more than 1 million as recently as 1960.[84] A decline on this scale is consistent with much else that happened in Africa during the 20th century, including widespread deforestation, the expansion of farming and infrastructure at all scales, increased access to military firearms, and human population growth.

Trends in individual countries are similarly difficult to assess because many previous estimates were probably underestimates. However, an example of decline quoted by Butynski[37] is notable: it was claimed that in Sierra Leone the population dropped from 20 000 in the late 19th century to 2 000 in 1987.[244] In Gabon, the combined population of chimpanzees and gorillas declined by more than 50 percent between 1983 and 2000 due to the increase

Ilka Herbinger/Wild Chimpanzee Foundation (WCF)

Western chimpanzees climbing, Côte d'Ivoire.

Elizabeth A. Williamson

This infant central chimpanzee is strapped in the traditional back pack of the poacher who killed its mother, Cameroon.

Box 4.5 REINTRODUCTION OF ORPHAN CHIMPANZEES

In 1989, Aliette Jamart set up the project *Habitat Ecologique et Liberté des Primates* (HELP Congo), the aim of which was to reintroduce chimpanzees rescued from the bushmeat trade to their natural environment. These activities are based in the Conkouati-Douli National Park, Congo.

From the start, the project aimed to follow the IUCN Primate Reintroduction Guidelines, and to evaluate the reasons for successes and/or failures in the short, medium and long term. The main stages of the HELP project are summarized below.

- **1989**: three islands were identified in the Conkouati lagoon as suitable for the establishment of the HELP orphan sanctuary;
- **1990**: the sanctuary received its first group of chimpanzees from Pointe Noire;
- **1992**: preliminary medical checks of captive and semicaptive groups of chimpanzees from the sanctuary were made by Marc Ancrenaz in collaboration with the Centre International de Recherches Médicales de Franceville (CIRMF), Gabon;
- **1994**: an evaluation by Caroline Tutin on behalf of IUCN was carried out to establish the release possibilities for the chimpanzees in the Conkouati area (two sites were identified);
- **1995**: a second medical examination was carried out on the group, again by Marc Ancrenaz of CIRMF, the results of which showed that they were free of disease and ready for release;
- **1996**: a second IUCN evaluation, by Caroline Tutin and the botanist Paul Sita, was carried out in the potential release sites, resulting in selection of the 'Triangle', an area surrounded by rivers adjoining Conkouati National Park;

Joanna Setchell and Benoît Goossens/HELP International

A six year old male central chimpanzee leaving the cage during his release.

- **1996**: the first group of chimpanzees was released, after a final medical evaluation and the fitting of radiotransmitters;
- **1996–2000**: a total of seven releases of 37 individual chimpanzees (10 males and 27 females) have taken place (see table).

HELP prefers to concentrate on releasing female chimpanzees, as they are well received by wild males and can go on to reproduce successfully in the wild. The capture of chimpanzees at the sanctuary, their transport to the release site, and their introduction into a new and unknown habitat can be sources of intense stress for the animals. To reduce this stress, a combination of the anesthetic agents ketamine and medetomidine are given, so that the animals are unconscious through the journey to the release site and awake in their cage at Triangle Island. The cage door is opened only when chimpanzees that have already been released are seen to be close to the cage and the new chimpanzees are fully conscious.

in logging and hunting, and the spread of Ebola hemorrhagic fever in the country.[262] Also in Gabon, a 99 percent decline in chimpanzees was recorded in Minkébé Forest, in the northeast, since 1994, when there was an Ebola epidemic.[112] A sudden population decline was also noted in Gombe National Park, Tanzania since the mid-1990s, owing to a combination of hunting and disease.[85, 86] In Sudan, it was noted in 1964 that the species "appeared to thrive particularly well in SW Sudan from where they were reported to move around in bands of 30 to 40 individuals, sometimes more,"[128] but by 1988 it was stated that the "species could be considered highly endangered if not already extinct in the country."[134]

Records at individual sites show a varied picture, ranging from local extinctions (e.g. at the Kilum-Ijim forest in northwest Cameroon in 1987 or 1988)[141] to stability, or even recovery. In the Monte

Post-release monitoring

Following their release, the chimpanzees are monitored daily from nest to nest for a period of acclimatization that varies from several weeks to months, depending on the individual. To date, the death rate is 14 percent, the survival rate is 62 percent, and the disappearance rate is 24 percent. If they have been attacked by wild males, the released males are then followed from morning through to evening. The females are located on a daily basis, but tend to leave the monitoring area when in 'estrus'. It is thought that they socialize with wild males during these periods. Bonnie, for example, was absent for six months but then returned with her baby; Rosette was absent for 18 months but then spent 15 days with the monitored chimpanzees; Matalila left for 11 months and then returned to stay with her childhood group; Massabi and Mossendjo were found two and a half months after release in a marshy area, where they were ranging over quite a wide territory. Only Massabi was fitted with a radio collar; when it was lost in May 2003, the pair could no longer be found. Massabi was reobserved in January 2004, in the company of a wild male.

A follow-up study of the released chimpanzees has highlighted that male chimpanzees cannot be released where wild chimpanzees exist, as they are likely to be killed by the existing population. Chimpanzee release was smoother when animals were anesthetized for transport, and released soon after recovering consciousness.

While reintroduction is not the only possible solution to the overpopulation of chimpanzees in sanctuaries, it has proved to be a potentially useful tool. Nevertheless, the main priority for chimpanzee conservation is to protect their habitat and reduce the pressures of hunting.

Aliette Jamart and Benoît Goossens

Release methods and problems encountered

Date	Number released (male, M; female, F)	Transport to point of release	Situation prior to release	Problems following release
1996	5 (1M, 4F)	floating cage; animals conscious	floating cage	none
1997	2 (2F)	floating cage; animals anesthetized	direct release	animals panicked and escaped
1997	8 (2M, 6F)	boat; animals anesthetized	release cage	animals stressed and panicked by captive conditions
1999	5 (2M, 3F)	boat; animals anesthetized	release cage	animals stressed and panicked by captive conditions
2000	4 (1M, 3F)	boat; animals anesthetized	floating cage until conscious at release site	none
2001	1 (M)	boat	released directly into an existing group	none
2001	12 (3M, 9F)	boat	floating cage until awake at launch site	none

Alén National Park, Equatorial Guinea, in 1994, for example, chimpanzees were found to be common all over the park and were not apparently threatened by hunting as they had been in previous years.[76] In Kibale National Park, Uganda, two sets of censuses were carried out in old-growth forests during 1975–1976 and 1997–1998; it was found that chimpanzee populations had declined only insignificantly over this period.[162]

Threats

Estimates of extinction risk for chimpanzees are largely based on the observed loss or modification of their habitats, on rates of exploitation, and also, in the case of geographically restricted populations, on the risks that are inherent to a small range size. In 2000, the Species Survival Commission of IUCN–The World Conservation Union categorized the chimpanzee species as

Box 4.6 CHIMPANZEE HABITUATION FOR TOURISM

Kibale National Park in western Uganda was gazetted in 1993 and is managed by the Uganda Wildlife Authority (UWA). It comprises about 795 km^2 of moist evergreen forest, colonizing forest, papyrus swamp, and exotic softwood plantation,[232] and is surrounded by a dense human population. It is the most important habitat for eastern chimpanzees in Uganda and, with over 1 400 individuals, homes over a quarter of the country's population.[211] Three of the communities are habituated: Kanyawara and Ngogo, for behavioral research purposes; and Kanyanchu, for tourism.

The potential for primate tourism within Kibale National Park was recognized during the late 1980s. A trail system was established within the chimpanzees' core home range to allow chimpanzees to be located and the presence of fruit in their preferred food trees to be monitored. In 1991, the Kanyanchu Visitors Centre opened to tourists, offering forest walks with the chance of viewing chimpanzees and other primates. As the success rate of finding chimpanzees varied considerably, however, a project was initiated in 1997 to raise the level of habituation and gain a more consistent viewing rate for tourists. These were the aims of the Kibale Primate Habituation Project, a joint venture of UWA and the Jane Goodall Institute-Uganda.

Habituation for tourism

The project aimed to achieve a level of habituation that balanced the need for chimpanzees to be relaxed enough in the presence of humans to behave naturally while maintaining a sufficient degree of wariness to prevent aggressive encounters. The level of habituation is of critical importance to the tourist experience, as wild chimpanzees typically disappear quickly at the sight of humans. Observations of some of the most interesting aspects of chimpanzee behavior, such as the use of tools, is correlated with the degree of habituation and the length of time that the animals can be kept in view.[83] Habituating chimpanzees, however, is not a quick or easy task, due to their ranging behavior and their fission–fusion society. The size of chimpanzee parties can vary from two to over 50 individuals of different age-sex classes.[220] During habituation and tourist visits the same individuals are not consistently followed, and contact with a large number of individuals within a chimpanzee community is inevitably intermittent.

Kibale Primate Habituation Project

From 1997 to 2001, the project worked alongside 12 Ugandan rangers to maintain dawn-to-dusk contact with the chimpanzees. A team of two rangers typically located the chimpanzees early in the morning by returning to their nest site or popular feeding trees, or by being guided by their calls. Habituators stayed with the chimpanzees, collecting baseline data on party composition and interactions, feeding ecology, and range use. Observer protocols designed for gorilla tourism were adopted and habituators took care not to show threatening behavior such as staring, sudden movement, or close proximity. The well maintained trail system was expanded to allow access deeper into the home range. Rangers were trained in chimpanzee behavior, ecology, data collection, and guiding, and were well equipped with binoculars, compasses, backpacks, uniforms, boots, and rain gear. All rangers carried radios to enable habituators to communicate the location of the chimpanzees to those guiding visitors.

Project achievements

After four years, the majority of chimpanzee individuals reached a level of habituation that would permit them to be followed all day, often at close quarters. The monthly average success rate for visitor groups viewing chimpanzees rose from 61 percent in 1997 to 88 percent in 2001. Over 60 individual chimpanzees were identified and named. Kanyanchu is one of the largest known communities in the wild, with 22 adult males. The project also compiled a database with information on demography, range use, feeding patterns, social

Endangered, i.e. facing a very high risk of extinction in the wild in the near future.

Threats multiply one another's impacts on chimpanzee populations. Light selective logging causes only temporary disturbance and may not greatly reduce the forest's carrying capacity for chimpanzees unless particularly favored tree species are lost from the biota. Increasingly intense logging and repeated re-entry logging, however, will cause mounting disruption to the forest ecosystem

hierarchy, and associations. The long-term local field staff can identify the majority of the named individual chimpanzees and provide interesting interpretations of chimpanzee behavior to their visitors. With a stunning forest, a large habituated community, and skilled rangers, chimpanzee tracking in Kanyanchu is a rewarding experience.

The project also helped diversify tourism activities, offering the unique opportunity to join the habituation team in following chimpanzees all day and witnessing the chimpanzees' full range of daily activities. Kanyanchu has become the most popular site in Uganda for viewing wild chimpanzees. With the subsequent increase in tourist numbers to the region, community-run tourism and conservation enterprises outside the park boundaries, such as the Bigodi Wetland Sanctuary, have flourished.

Conservation and challenges for the future

The benefits accruing from chimpanzee tourism include revenue generation for park management and local communities, and reducing levels of poverty, poaching, and forest encroachment. Despite such important benefits, challenges remain for the future, notably the successful coordination of different tourism activities, adherence to observer protocols, and monitoring of habituated chimpanzees. A high-quality tourism experience requires a substantial level of chimpanzee habituation; if chimpanzees lose all sense of fear of humans, however, they could pose a threat to rangers, visitors, and local people. Once the optimum level of habituation is reached, the chimpanzees must be followed daily to allow them to be located and monitored, which requires numerous well trained staff and effective on-site management.

The use of chimpanzees in tourism is a sensitive issue due to their Endangered status and close evolutionary relationship to humans. The original regulations for chimpanzee tourism were based on those for gorilla tourism, but chimpanzee tourism offers its own set of challenges and pitfalls,[136] and its long-term impact on chimpanzees is not yet known. As with gorilla-focused tourism, it is important to manage human–chimpanzee contact so as to minimize stress to the apes, to reduce disease transmission in each direction, and to avoid aggression.[278] The goal is to maximize positive effects and minimize the negative impacts on the chimpanzees, the environment, and the local people. The impact of tourism on chimpanzee behavior, ecology, or population viability is currently little understood, so the optimum number of tourists per day, duration of visits, or appropriate observer behavior are all a matter of educated guesswork. The UWA and Makerere University are currently collaborating to monitor the impacts of tourism on the Kanyanchu chimpanzee community. The results will be used to revise observer protocols, to encourage their adherence, and to design long-term monitoring systems aimed at detecting changes in the behavior, health, and environment of the chimpanzees before any irretrievable damage is done.

Julia Lloyd and Lilly Ajarova

A tree house used for habituation in Kibale National Park, Uganda.

Julia Lloyd

of which chimpanzees are a part, degrading its integrity, opening it up to drying winds and sunshine, and increasing its vulnerability to forest fires.

Increased access to the logged-over area along logging roads will encourage entry by hunters, especially where there is a commercial trade in bushmeat (as there is in much of West and Central Africa), so the effects of hunting almost inevitably compound those of logging. Mining and oil extraction can have similar effects on access, as

The boundary of Kibira National Park, Kabarore, Burundi. Here, as elsewhere, chimpanzee habitat is threatened by encroaching agriculture.

Geoffroy Citegetse

well as locally destroying natural ecosystems. Increased access also leaves the forest open to invasion by settlers, leading to further hunting and also to the fragmentation of the forest by an expanding and eventually coalescing mosaic of farms and villages. Chimpanzees are bound to find it increasingly difficult to survive in such a landscape, the more so as the reduced and fragmented populations come into more frequent contact with humans and become increasingly vulnerable to human diseases. Hence the survival of chimpanzees, like that of many other species, is threatened by the whole process of advancing human use of tropical moist forests.

With deforestation so far advanced in West Africa, only remnant tracts of primary rain forest persist. The fragmented populations of the eastern and western subspecies of chimpanzee are primarily located in remnant forest, game reserves, and national parks; unauthorized hunting, logging, mining, and farming are common in many nominally protected areas.

Hunting of adults for bushmeat has an impact on populations and is an important and increasing threat to the species. A report released in 2004 has estimated that for the Nigeria-Cameroon chimpanzee, bushmeat hunting alone is sufficient to threaten it with extinction within 17–23 years. This conclusion was based on the increasing number of orphaned chimpanzees arriving at sanctuaries in the region, and the assumption that 10 chimpanzees have been killed for each of these orphans.[207]

Bushmeat is often a major source of dietary protein in the meat-eating cultures of West and Central Africa, and sometimes also has perceived magical or medicinal benefits. Although hunting may occur at sustainable levels locally, it increases with logging and mining because food is required to maintain large labor forces, and because colonizing human communities often favor bushmeat. Civil conflict also tends to increase hunting, often by people from other regions. The impact of bushmeat hunting is now widespread, increasing rapidly in parallel with greater access to remote areas. New markets are being developed to serve rising demand from urban populations, chimpanzee products are widely sold in local and regional markets, and trade in infant chimps is often associated with hunting of adults. In some areas, however, for example the Kouilou Basin in Congo, the chimpanzee is not hunted for meat and is consequently less threatened.[63] Not all hunting is intentional, however; trapping can injure chimpanzees even where they are not the target prey species. Limb deformities were found in 11 of 52 chimpanzees living in the Budongo Forest Reserve, nearly all of them attributed to wire snares and leg-hold traps set in the forest.[260]

The dual impacts of the Ebola virus and bushmeat hunting in the heartlands of the western gorilla and chimpanzee range in the Congo Basin are unquantified and may have already greatly reduced populations of both species.[262] An outbreak of a new Ebola strain in the Taï Forest, Côte d'Ivoire killed 12 chimpanzees, about a quarter of the group under study.[69, 168] It is not known how many of the park's chimpanzees died in total. Ebola is by far the most virulent disease affecting the African great apes, but others have also had significant impacts. Wallis and Lee[261] summarized the occurrence of disease in chimpanzees in Gombe National Park, Tanzania, noting incidences of a 'polio-like' virus, pneumonia and other respiratory diseases, and scabies. They discussed the possibility that these diseases had been contracted from humans and suggested various improvements in health standards to help combat the problem. Also in Tanzania, chimpanzees are affected seasonally by intestinal nematodes, particularly *Oesophagostomum stephanostomum*, which can result in secondary bacterial infection, diarrhea, severe abdominal pain, and weakness, sometimes leading to death. Disease transmission and Ebola are discussed further in Chapter 13.

Finally, the live-animal trade involves capture of infants for the pet trade, the entertainment industry, and international biomedical business. Like hunting, this is illegal in all range states. The capture of an infant chimpanzee typically involves the deaths of all other chimpanzees present in the party targeted by the hunters. Although much concern has been expressed in the past over such uses and the possible impact on wild populations, in most areas this is thought to be a lesser threat than habitat loss and the bushmeat trade. A reasonable amount of conservation attention is focused on the rescue and rehabilitation of the orphans themselves, both for their intrinsic value and in order to create an educational opportunity out of a conservation disaster. Release schemes for rescued chimpanzees are not as far advanced as they are for orangutans, but are more advanced than for gorillas. One successful trial is ongoing in the Conkouati-Douli National Park, Congo (see Box 4.5).

CONSERVATION

Chimpanzees are the most abundant and widespread of the great apes, with a total wild population estimated to be up to 300 000 individuals. Most live outside protected areas, where they are vulnerable to disturbance of their forest habitat by logging; to habitat destruction by settlement, fire, and farming; and to hunting that supplies the increasingly entrenched and powerful bushmeat trade. Meanwhile, their fragmented populations are becoming increasingly subject to disease outbreaks as they come more often into contact with people. The detail of

Table 4.3 Conservation success scores for protected areas

	Protected area	Chimpanzee subspecies	Conservation success score[a]
Cameroon	Korup National Park	central	3.0
Cameroon	Dja Faunal Reserve	central	3.0
CAR	Dzanga-Sangha	eastern	3.5
Congo	Odzala National Park	central	5.0
Côte d'Ivoire	Marahoue National Park	western	2.0
Côte d'Ivoire	Taï National Park	western	3.8
DRC	Ituri Forest Reserve	eastern	3.5
Equatorial Guinea	Monte Alén National Park	central	5.0
Gabon	Lopé National Park	central	3.0
Ghana	Bia National Park	western	2.0
Nigeria	Cross River National Park	Nigeria-Cameroon	3.0
Uganda	Kibale National Park	eastern	3.9
United Rep. of Tanzania	Mahale Mountains National Park	eastern	4.0

a Conservation success scores derive from qualitative questionnaire assessments in which 1 indicates failure and 5 indicates very successful.

Data compiled by Struhsaker, T.T., *et al.* (2005)[233]

conservation efforts for chimpanzees varies over their broad range, as outlined in the relevant country profiles in Chapter 16 of this volume.

As these profiles show, populations of all four chimpanzee subspecies occur in protected areas across their range. These protected areas occupy tens of thousands of square kilometers of forest that range-state governments have chosen to set aside for wildlife protection. The central issues that emerge are the effectiveness with which these areas are managed, the challenges that they must overcome, and ultimately the security of the public investment in conservation that they represent. A scheme is being developed in Kibale National Park to supplement public investment with funds accruing from tourism (Box 4.6).

Kibale National Park is one of an Africa-wide sample of 13 protected areas containing chimpanzees that have been analyzed from the point of view of conservation effectiveness.[233] These are listed in Table 4.3, along with a 'conservation success score' for each, which represents the qualitative opinion of scientists and protected area managers familiar with the area concerned. Respondents were asked to score the protected area on a scale of 1 to 5, with 1 being a failure and 5 being very successful. This qualitative approach was used in the absence of protected area-wide monitoring programs that would have allowed a quantitative evaluation. The mean score for all 13 protected areas was 3.4, and the median was 3.5; this suggests that, overall, informed observers were reasonably confident that their protected areas were working.

In decreasing order of significance, the main factors contributing to an increase in conservation success scores were considered to be:

1. a positive public attitude;
2. effective law enforcement;
3. large protected area size;
4. low human population density in the vicinity;
5. the presence of nongovernmental organizations; and
6. ecological continuity.

On average, the lowest scores were obtained for West African protected areas, reflecting their greater ecological isolation, greater accessibility, and the presence of dense human populations possessing a bushmeat-eating culture. This is generally consistent with the lower abundance of the two West African subspecies, the greater deforestation in their area of distribution, and their overall more dismal conservation prospects. Elsewhere, there may be grounds for patchy optimism, although with human populations rising overall, intractable poverty in many locations, and public investment compromised by corruption and debt, the outlook for chimpanzees remains uncertain at best.

FURTHER READING

Chapman, C.A., Onderdonk, D.A. (1998) Forests without primates: primate/plant codependency. *American Journal of Primatology* **45** (1): 127–141.

Dominy, N.J., Svenning, J-C., Li, W-H. (2003) Historical contingency in the evolution of primate colour vision. *Journal of Human Evolution* **44**: 25–45.

Goodall, J. (1986) *The Chimpanzees of Gombe: Patterns of Behavior.* Harvard University Press, Cambridge, Massachusetts.

Goodall, J. (1990) *Through a Window: My Thirty Years with the Chimpanzees of Gombe.* Houghton Mifflin Company, Boston.

Grubb, P., Butynski, T.M., Oates, J.F., Bearder, S.K., Disotell, T.R., Groves, C.P., Struhsaker, T.T. (2003) Assessment of the diversity of African primates. *International Journal of Primatology* **24** (6): 1301–1357.

Kormos, R., Boesch, C. (2003) *Regional Action Plan for the Conservation of Chimpanzees in West Africa.* Conservation International, Washington, DC.

Kormos, R., Boesch, C., Bakarr, M.I., Butynski, T.M., eds (2003) *West African Chimpanzees: Status Survey and Conservation Action Plan.* IUCN/SSC Primate Specialist Group. IUCN, Gland, Switzerland.

Kortlandt, A. (1983) Marginal habitats of chimpanzees. *Journal of Human Evolution* **12** (3): 231–278.

Lonsdorf, E.V., Eberly, L.E., Pusey, A.E. (2004) Sex differences in learning in chimpanzees. *Nature* **428**: 715–716.

McGrew, W.C., Marchant, L.F., Nishida, T., eds (1996) *Great Ape Societies.* Cambridge University Press, Cambridge, UK.

Mitani, J.C., Watts, D. (1999) Demographic influences on the hunting behavior of chimpanzees. *American Journal of Physical Anthropology* **109**: 439–454.

Parr, L.A., de Waal, F.B.M. (1999) Visual kin recognition in chimpanzees. *Nature* **399**: 647–648.

Reynolds, V., Reynolds, F. (1965) Chimpanzees of the Budongo Forest. In: DeVore, I., ed., *Primate Behaviour. Field Studies of Monkeys and Apes.* Holt, Rinehart and Winston, New York. pp. 368–424.

Stanford, C.B. (1998) *Chimpanzee and Red Colobus: The Ecology of Predator and Prey.* Harvard University Press, Cambridge, Massachusetts.

Stanford, C.B., Wallis, J., Matama, H., Goodall, J. (1994) Patterns of predation by chimpanzees on red colobus monkeys in Gombe National Park, Tanzania, 1982–1991. *American Journal of Physical Anthropology* **94**: 213–228.

Struhsaker, T.T., Struhsaker, P.J., Siex, K.S. (2005) Conserving Africa's rain forests: problems in protected areas and possible solutions. *Biological Conservation* **123** (1): 45–54.

Whiten, A., Goodall, J., McGrew, W.C., Nishida, T., Reynolds, V., Sugiyama, Y., Tutin, C.E.G., Wrangham, R.W., Boesch, C. (1999) Cultures in chimpanzees. *Nature* **399**: 682–685.

Whiten, A., Horner, V., Litchfield, C., Marshall-Pescini, S.R.J. (2003) Cultural panthropology. *Evolutionary Anthropology* **12** (2): 92–105.

Wrangham, R.W., Chapman, C.A., Chapman, L.J. (1994) Seed dispersal by forest chimpanzees in Uganda. *Journal of Tropical Ecology* **10**: 355–368.

Wrangham, R.W., McGrew, W.C., de Waal, F.B.M., Heltne, P.G., eds (1994) *Chimpanzee Cultures.* Harvard University Press, Cambridge, Massachusetts.

MAP DATA SOURCES

Map 4.1 Chimpanzee data are based on the following sources, with updates as in the country profiles in Chapter 16:

Butynski, T.M. (2001) Africa's great apes. In: Beck, B.B., Stoinski, T.S., Hutchins, M., Maple, T.L., Norton, B., Rowan, A., Stevens, E.F., Arluke, A., eds, *Great Apes and Humans: The Ethics of Coexistence.* Smithsonian Institution Press, Washington, DC. pp. 3–56.

Butynski, T.M. (2003) The chimpanzee *Pan troglodytes*: taxonomy, distribution, abundance, and conservation status. In: Kormos, R., Boesch, C., Bakarr, M.I., Butynski, T.M., eds, *West African Chimpanzees: Status Survey and Conservation Action Plan.* IUCN/SSC Primate Specialist Group. IUCN, Gland, Switzerland. pp. 5–12.

For protected area and other data, see 'Using the maps'.

ACKNOWLEDGMENTS

Many thanks to Colin Groves (Australian National University), Phyllis Lee (Cambridge University), John F. Oates (Hunter College, City University of New York), and David Woodruff (University of California, San Diego) for their valuable comments on the draft of this chapter.

HELP Congo, discussed in Box 4.5, works in collaboration with the authorities of Congo at the Ministry of Forestry and the Environment and with the Wildlife Conservation Society (WCS-Congo). HELP Congo wishes to thank its sponsors: US Fish and Wildlife Service; Cleveland Zoological Society; Columbus Zoo and Aquarium; Lincoln Park Zoo; International Primate Protection League; Arcus Foundation; Pan African Sanctuaries Alliance; Fondation Brigitte Bardot; Fondation Bourdon; Société Protectrice des Animaux; One Voice; Beauval Zoo; Amneville Zoo; La Barben Zoo; Gorilla; Air Gabon; and Cardiff University; as well as the Congolese assistants who daily follow the chimpanzees in the forest of the Conkouati-Douli National Park.

The chimpanzee habituation project in the Kibale National Park, discussed in Box 4.6, was funded by the USAID Grants Management Unit, the Jane Goodall Institute-Uganda, Cleveland Zoo, North Carolina Zoo, Barclays Bank (Kampala), British Airways, Discovery Initiatives, and TOTAL (UG) Ltd.

AUTHORS

Tim Inskipp, UNEP World Conservation Monitoring Centre

Box 4.1 Craig Stanford, Jane Goodall Research Center, University of Southern California

Box 4.2 Alison Surridge, School of Biological Sciences, University of East Anglia

Box 4.3 Andrew Whiten, Scottish Primate Research Group, University of St Andrews

Box 4.4 James V. Wakibara, Tanzania National Parks/Kyoto University

Box 4.5 Aliette Jamart, Habitat Ecologique et Liberté des Primates, Congo and Benoît Goossens, Cardiff School of Biosciences, Cardiff University

Box 4.6 Julia Lloyd, Jane Goodall Institute-Uganda and Lilly Ajarova, Uganda Wildlife Authority

David W. Liggett (www.daveliggett.com)

CHAPTER 5

Bonobo (*Pan paniscus*)

CARMEN LACAMBRA, JO THOMPSON, TAKESHI FURUICHI, HILDE VERVAECKE, AND JEROEN STEVENS

The bonobo (*Pan paniscus* Schwarz, 1929), also known as the 'pygmy' or 'gracile' chimpanzee, occurs only in the inner Congo Basin of the Democratic Republic of the Congo (DRC), in Central Africa, where it is known as *chimpanzé nain* or *chimpanzé noir*.[54, 76]

Bonobos have black fur, arms as long as their legs, and a tailtuft but no tail. Perhaps the most obvious physical differences from chimpanzees are visible in the head and face. Generally speaking, adult bonobos have a short and rounded skull with a black face, red lips, sidewhiskers, and hardly any beard. The hair on their heads is long and distinctively parted in the middle.

The Congo River and the mountains of the Albertine Rift isolate the bonobo from all other apes (including chimpanzees), and all other large-bodied primates (including baboons). Although the species was not scientifically described until 1929, the existence of the bonobo throughout its modern range has been well documented since the 1880s via explorers' journals, naturalists' photographs, missionaries' reports, and colonial administrative records.[75, 77]

Field surveys of bonobos began in the early 1970s and have continued to date.[7, 50, 73, 75, 83] It has proved very difficult to obtain a clear view of either the total geographical area occupied by bonobos, or the likely number of individuals within it (estimates range from 10 000 to over 100 000).[15, 82] This uncertainty reflects the scale, challenging observational environment, and inherent ecological patchiness of the inner Congo Basin, combined with the impacts of human disturbance and (to the south) the complex gradation between dry forest and savanna.

DISTRIBUTION

The geographic range of bonobos is limited in the east, north, and west by the south bank (conventionally the 'left' bank) of the Congo River and its major tributary the Lualaba; and, in the south, by the Kasai/Sankuru river system. Their range is low lying, between 300 and 750 m in elevation, and is dominated by moist forest, swamp forest, and mosaics of grassland and dry forest.

Although bonobos occupy a mosaic habitat of forest and grassland, their area of actual occurrence is far less than their maximum geographical range and may be less than 30 percent of it. Population densities may be as low as 0.25 individuals per square kilometer.[75] The dotted range polygon shown in Map 5.1 and based on 2004 distribution data represents an area of 373 585 km^2. The total bonobo population, assuming a 30 percent occupation of this range at a density of 0.25/km^2, is 28 019 animals. This indicative figure makes significant assumptions about all the relevant parameters. It is possible to place more confidence in the estimate[33] that in the well studied sites of Wamba (in the northern Luo Reserve for Scientific Research), Ilongo (in the southern Luo Reserve), Lomako, Lilungu, and Yalosidi, a total of 4 421 bonobos occurred, although the Yalosidi population has now been lost. Recorded densities at these five sites range from 0.35/km^2 (at Ilongo) to 3.46/km^2 (at Lomako).[33]

Bonobos have been recorded in the Lomako, Kokolopori, Wamba, Ilongo, Lomami-Lualaba, Salonga, and Lukuru regions, and in small areas to the west of Lake Tumba.[57, 58, 78, 88] Bonobos seem to be absent or at low density in the central parts of the Salonga National Park, and absent from much

Map 5.1 Bonobo distribution

Data sources are provided at the end of this chapter

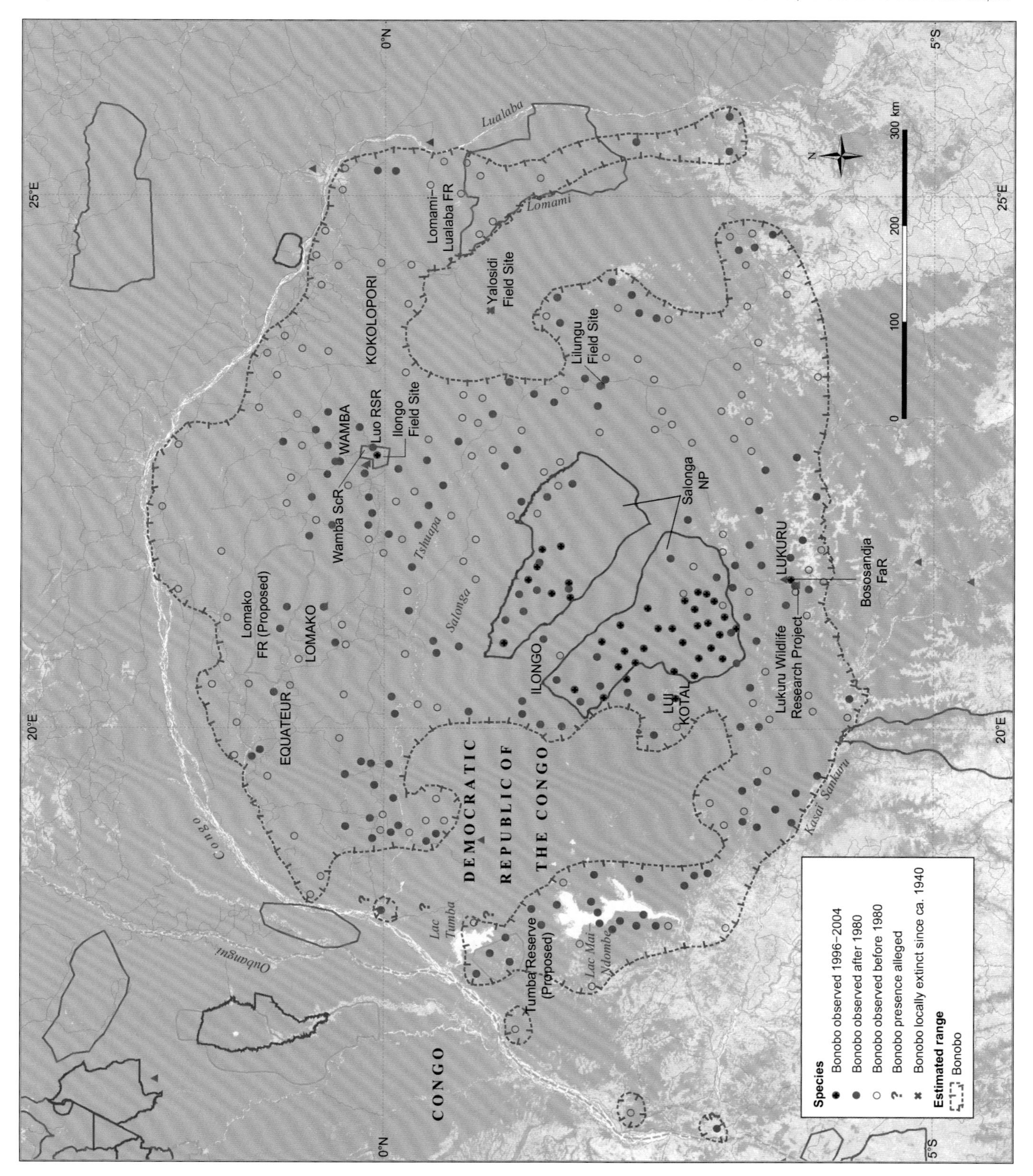

of the area between Lac Mai-Ndombe and Salonga. The other area of absence falls between the Tshuapa and Lomami Rivers (Map 5.1).

BEHAVIOR AND ECOLOGY

Habitat and diet

Most studies of bonobos in the wild have been carried out in primary forest in the northern part of their distribution, but recent work has confirmed that they also use open savanna and secondary forests. High densities are found, especially in the latter, and in patchwork and edge-habitat mixtures.[82, 83]

Bonobo groups with access to dry forest, swamp forest, and disturbed forest have been studied in the Wamba region.[34] A large part of their home range comprises dry forest, where they prefer to sleep. They also use swamp and disturbed forests, which apparently contain a higher proportion of protein-rich food that is available all year round. A tendency to use drier habitats during the rainy season is noted.

Bonobos do most of their traveling on the ground. Their trails are difficult to recognize, but usually lead either to the base of trees from which they feed, or to nesting sites. When moving in trees, they employ a method that can be described as 'quadrupedal scrambling', an inefficient means of movement that has not been seen to be sustained over more than about 40 m. The most common posture in trees is sitting, for both feeding and resting.[40, 49]

Bonobo groups number between 10 and 120 animals[27] and travel around 2 km on average each day,[24] foraging mainly for fruit. Their diet also includes leaves, pith, flowers, seeds, nuts, shoots, mushrooms, and algae. Additional food sources include high-quality terrestrial herbaceous vegetation, earthworms, larvae, termites, ants, honey, truffles, and aquatic plants.[1, 40, 55]

Subgroups or 'parties' are formed by two to 30 individuals and are composed of males, females, and their offspring. They do not usually forage together with parties from other groups.[21, 59, 95] As a presumed adaptation to the greater feeding competition in small patches, the size of feeding parties may vary with patch size.[91] Bonobos occasionally consume small mammals,[19, 42, 59, 95] including flying squirrels, infant duikers, and bats. There is little information on their hunting methods, and hunting does not seem to be a frequent practice. When meat is available, it is treated as a valuable resource; bonobos have been observed to beg the meat holder for a share.[42] It has been suggested that bonobos are more able to source dietary protein from non-reproductive plant parts than are chimpanzees, so may have less need to invest energy in hunting.[75]

Bonobos at Yalosidi, Lomako, and Wamba have been seen foraging for food in streams or marshlands, and at Lukuru in pools.[38, 78] One group at Yalosidi was observed frequently to visit a marsh grassland within the rain forest to feed on the stems of certain aquatic or amphibious herbs and grasses.[84] Among the species consumed throughout the year, the most common were *Ranalisma humile* (Alismataceae), *Pycreus vanderysti* (Cyperaceae), and *Aframomum* spp. (Zingiberaceae). Other species eaten included *Cyclosorus dentatus* (Thelypteridaceae), *Panicum brevifolium* (Poaceae), *Renealmia africana* (Zingiberaceae), *Marantochloa congensis,* and *Sarcophrynium schweinfurthianum* (both Marantaceae), as well as *Gambeya lacourtiana* (Sapotaceae).[40, 84] Bonobos have also been seen slapping the water in streams, and scooping up handfuls of dead leaves, probably to catch invertebrates and small fish.[9, 31]

Although bonobo diets are generally similar across their range, differences have been noted between populations studied for long periods of time.[2, 82] The extent to which such differences are cultural rather than dependent upon resource availability is unclear.[78]

Social behavior

Bonobo social organization is characterized by fission and fusion of small temporary subgroups (parties) within a larger, more stable, community or group. Our knowledge of bonobo social systems comes largely from two field sites, both in the Equateur province of DRC: Lomako, an unprovisioned site, and Wamba, where provisioning used to occur. Researchers provided food to some bonobo groups at Wamba prior to the cessation of studies in 1996; when research began again after the civil war, this provisioning was not resumed.[28]

The bonobos at the two sites show a number of similarities in social organization, but differences have also been noted. As in chimpanzees, the community is the largest mixed-sex social unit of individuals that maintain a closed social network. A single community's members share a discrete, relatively large, home range; extensive overlap between communities may occur and there may be seasonal and yearly variations in home ranges.[86]

Frances White

Grooming between an adult male, adult female, and her infant, from the Bakumba community of the Lomako forest.

Communities contain between 10 and 22 individuals in Lomako and between 30 and 120 individuals in Wamba.[52] There are almost equal numbers of adult males and females in Wamba,[51] whereas in at least one Lomako community, the adult sex ratio is strongly female biased.[18, 39]

Entire communities are observed together much less often at Lomako than they are at Wamba.[18, 51, 92] The smallest functional unit of bonobo daily life is the party, defined variously in terms of the individuals that remain in sustained proximity to one another, or within earshot of each other,[32] or that travel and forage together.[86] Through fission and fusion, membership of parties can change to varying degrees over days, hours, or even minutes. By contrast, membership of communities changes only with the birth or death of members, or by their permanent intergroup transfer. Larger, more stable parties are seen in Wamba, with an average of 13 members;[51] in Lomako, parties contain about five individuals on average, with a range of between one and 16.[37] Parties usually contain mature individuals of both sexes, with more females than males.[18, 37, 48, 91] A community of bonobos contains subgroups of individuals that more often form parties with each other than with others. These subgroups tend to share specific parts of the community's home range. Discernment of such subgroups requires prolonged observation and detailed analysis of abundant data on individual associations. Their occurrence is not necessarily obvious in the field.

As in chimpanzees, maturing males tend to remain within their natal (birth) community, while maturing females leave and move from community to community before settling down to breed. The result is that both chimpanzee and bonobo communities are made up of unrelated females and males that are more likely to be related to one another. This similarity masks a number of differences, however, such as the much longer time period (years rather than months) during which young female bonobos move between groups before settling down, and the very strong bonds that exist between bonobo mothers and sons.[22, 29, 93]

Bonobo social structure is dominated by female coalitions that influence mating strategies and food allocation. Females are smaller than males, but maintain their social status through cooperation with each other. Female bonobos are very skilful in establishing and maintaining strong bonds with unrelated females.[61, 62, 72, 89] The strategies employed include controlling access to desirable food, sharing food with other females more often than with males, interacting sexually with other females, and forming alliances against males when necessary.

The net result is that adult female bonobos have a social status roughly equal to that of adult males. Though males may give charging displays when they are excited, females sometimes displace males to get into preferred feeding positions. Female status is related to age, whereas the status of individual males relates to that of their mother.[51] Young adult males have been observed to rise in status through the support of their mother; males in their prime have been observed to fall in status after the death of their mother.

The high status of females in bonobo society is thought to be related to their sustained sexual attractiveness. This is independent of their menstrual cycle and is maintained into pregnancy and lactation, while further conception is impossible.[21, 23] As a result, at any given time in a bonobo community, there are on average many more females interested in mating than there are in a chimpanzee community. In these circumstances, it would be much more difficult for a high-status male to monopolize mating opportunities, so male status is less important to individuals.[25, 51] Male bonobos are rarely observed to compete or fight over access to females. Males freely access receptive and proceptive females, and treat them in a friendly manner; it is the female that determines whether copulation occurs. There is some evidence, however, that high-ranking males have more success in mating,[29, 53, 71] suggesting that competition is not entirely absent.

In this system, no male can tell whether he is likely to have sired any particular offspring. This lack of clarity over paternity is consistent with the observed generalized paternalism: adult male bonobos are extremely caring and affectionate with infants, sharing both food and nesting spaces. Amongst macaques, paternalism is known to be associated with promiscuous female sexuality, single-mount ejaculation and an even intragroup sex ratio;[4, 5] it seems that something very similar occurs in bonobos. Relatively food-poor environments, however, seem to give rise to the opposite social system among macaques, of harem-like sexual control, multiple-mount ejaculation (repeated sexual encounters between the same pair before ejaculation happens), near-certain paternity, and a lack of paternalism. The bonobo's relaxed social and sexual system has therefore been attributed to their diverse diet and resource-rich environment.

Bonobo females indulge in collaborative genital rubbing, genital–genital contact, and numerous related behaviors. The bonobo clitoris is large and shifted ventrally compared to that of the chimpanzee.[8, 51] Genital contact is a common part of bonobo social interaction, but is more frequent after an episode of aggression or when food is monopolized by an individual. Hence, it is thought that sexual activity among females may serve to promote reconciliation and the relief of social tension, thereby serving to restore and maintain coalition relationships. The frequency of genital contact is also related to female status, with low-ranking females initiating contact more often than high-ranking ones.[36] Together with the lack of restriction in heterosexual mating activity, this suite of sexual behaviors has led to an unusual amount of research and popular interest in bonobo sexuality.

The high social status of females may also be related to the difference between chimpanzees and bonobos in intergroup relationships. Among bonobos, intergroup interactions are frequent and are characterized by high-pitched excitement rather than conflict.[51] Chimpanzees are known to be antagonistic towards and sometimes kill members of other groups, while aggressive intergroup encounters are rare in bonobos. In contrast, different bonobo groups sometimes come together to feed or rest in a peaceful atmosphere. Male bonobos become excited and tend to stay behind the line of contact between the groups during such encounters, but females willingly enter a different group, and will copulate with unfamiliar males.[41]

Compared to male chimpanzees, male bono-

Box 5.1 SEED DISPERSAL BY BONOBOS AND THE SURVIVAL OF RAIN FOREST

As specialized frugivores, bonobos are essential for the long-term survival of the rain forests in which they live. In the Lomako forest, bonobos occur together with seven other primate species, but they are the only one to ingest regularly and disperse the whole seeds of a wide variety of species of rain forest tree and liana. Bonobos are excellent seed dispersers for a number of reasons.

First, they are primarily frugivorous (up to 70 percent or more of the diet is ripe fruit) and rarely damage the seeds consumed.

Second, they are large bodied and have simple guts, so that even large seeds can be swallowed whole and passed undigested via the feces. The seeds of some fruit species are very large, among the biggest being those of *Anonidium mannii* (Annonaceae). These fruit weigh 3 kg (or more) and contain over 50 seeds that each measure 3 cm in length and weigh about 10 g.

Third, bonobos travel long distances and maintain large core areas. Individuals cover over half their community range each year, and spend more than 90 percent of their time within primary rain forest, thus providing long-distance dispersal within a suitable habitat for rain-forest trees.

Fourth, they often carry fruits long distances before sharing and eating them, such as the fruits of *Treculia africana* (Moaceae) that weigh 10 kg or more. The seeds may later be dispersed even further before being defecated.

Fifth, bonobos do not sleep where they have been feeding, but move away to build nests and sleep in trees elsewhere.

Many species of tree and liana appear to have evolved with bonobos and rely on them for dispersal. *Carpodinus gentilii* (Apocynaceae) has fruits that weigh about 1 kg with a hard rind, 2.5 cm thick, that smaller-bodied monkeys are unable to open. Others, such as *Pancovia laurentii* (Sapindaceae), have seeds that germinate readily after passing through a bonobo gut, but not at all if the fruits fall uneaten beneath the parent tree or are artificially planted, even at a distance from it.

The dietary diversity of the bonobo means that it is the most important disperser of many rain forest tree species in DRC, and may be the only disperser of some of them. Of 130 fruit species collected and measured in a study, bonobos have been seen to eat 63. The list of fruit species known to be eaten by bonobos increases with each year of investigation,[94] so it seems likely that bonobos are involved in the dispersal of half or more of all fruiting trees in the inner Congo Basin. Without bonobos, therefore, major changes in this ecosystem would be likely to occur within very few generations.

Frances White

Bonobo infants are nursed until they are five years old.

David W. Liggett (www.daveliggett.com)

bos participate in less physical competition for copulation opportunities and interact less aggressively with males of other groups. They do not participate in raids on neighboring communities.[10, 29, 62] The male philopatric social organization common to the two species, in which young males stay with the natal group and young females migrate, clearly does not predict these distinctive aspects of their social behavior. More generally, bonobos of both sexes show much more frequent and varied sexual behavior than chimpanzees.[14] Grooming is more evenly dispersed between individuals among bonobo than chimpanzee communities, and grooming between individuals of the opposite sex is more frequent and occurs for longer periods of time than grooming between females or males only.[56]

It has been proposed that the differences between the two species may be less intrinsic than had been believed, and could be explained partly by environmental conditions including food supply and distribution, party size, and sexual opportunity.[2, 70] According to this view, reduced competition between females enables more stable parties to be maintained, with more female sociability than occurs in chimpanzees. Some scientists also consider that the genetic relationship between bonobos and chimpanzees may be closer than the evidence from comparison of their mitochondrial DNA suggests.[14, 37, 78]

It is not known why bonobo social interactions are so different from those of the chimpanzee. The general view is that the high level of bonobo female sexuality associated with their being receptive for extended times relieves a chief cause of male–male friction through the abundance of mating opportunities.[10, 11] The use by females of their sexuality for maintaining effective coalitions among themselves may result in a sexually egalitarian society, within which male possessiveness would be ineffective, even if attempted. Frequent grooming between the sexes reinforces social bonds and contributes to a relaxed social system.

Nonreproductive sexual behavior, like all social interaction, is potentially costly in terms of energy expenditure and reduced foraging time, so it cannot be ruled out that a food-rich environment is a necessary enabling factor for the bonobo social system. On the other hand, while there is much more sexual activity in a bonobo group than in a chimpanzee one, among bonobos "instead of an endless orgy, we see a social life peppered by brief moments of sexual activity,"[11] so the energetic costs may not be very great.

Development and reproduction

Details of reproductive development, and of hormonal and behavioral events during the menstrual cycle, are given for both chimpanzees and bonobos in Chapter 3. In brief: the first genital swelling occurs at seven years of age, sexual maturity at nine, and full adult size is reached at 16. At eight years of age, young females start to move between groups; settling in a new group occurs between nine and 13.[22]

The first offspring are born when the female is between 13 and 15. Only one infant is usually produced per pregnancy, often during a birth peak from March to May during the light rainy season.[27] The menstrual cycle lasts 36–46 days,[90] and the gestation period has been estimated at 220–230 days. Infants are nursed until they are five years old, and the mean birth interval is 4.6 years. It is not clear whether menopause ever happens, as continued menstrual cycling has been observed in females that are 45 or more years old.[47] The typical life expectancy of between 50 and 55 years leads to an average of five or six young being produced during a lifetime.[27, 59]

Observations during 1976–1996 in the Wamba region concluded that bonobos there have an infant mortality rate that is much lower than is recorded for chimpanzees. This is thought to result from some combination of the abundant fruits and herbaceous foods at Wamba, larger food patch size, better female access to prime feeding sites, male paternalism, and absence of infanticide.[27]

Vocal behavior

Bonobos are much more vocal than chimpanzees or any other great ape (see Box 5.2). They use numerous calls that are audible over long and short distances, including synchronized choruses that end up sounding like echoes. Among the most noticeable are the 'high-hoots' that are the commonest long-distance call and can be heard at all times of the day and night. Hooting occurs most frequently when the bonobos arrive at feeding sites in the early morning and while they occupy a prospective nesting site in the afternoon. Other vocal sounds have been identified during feeding and copulation, and in response to danger.[35]

Tool use

Little tool use by bonobos has been observed in the wild. In the northern sector of the Salonga National Park, bonobos have been seen digging with sticks in termite mounds,[45] and males have used branches in displays. In captivity, bonobos use various objects: rope swinging; self-wiping with leaves; and using sticks as ladders or weapons have all been reported.[30, 59, 69] Captive juvenile bonobos have also been seen using leaves in play, covering their eyes and feeling their way around while blindfolded.[10] It is likely that some of the same behaviors are found in wild bonobos.

Nest building

Both day and night nests are built by bonobos; they are used for sleeping, grooming, feeding, and playing. Built afresh every day, the nests are circular in shape and can measure up to 1.3 m in diameter. Night nests are the more elaborate and take longer to construct. They are usually built in the middle canopy (15–30 m above the ground), while day nests are usually higher up. Bonobos may gather materials from up to six trees in the construction of their night nests, whereas chimpanzees ordinarily use the branches and foliage of only one tree for this purpose.[20] There are also reports of ground nests, possibly built for purposes other than resting.[20, 40, 49]

Females build higher nests, do so more frequently during the day, and use them for longer periods of time than do males. Bonobos from different locations have been noted to have different preferences in the type and location of trees chosen for nesting.[19, 20]

Interactions with other animals

Various diurnal species of monkey share the bonobo range, including the Allen's swamp monkey (*Allenopithecus nigroviridis*), black mangabey (*Lophocebus aterrimus*), golden-bellied mangabey (*Cercocebus galeritus chrysogaster*), redtailed monkey (*Cercopithecus ascanius*), Congo Basin Wolf's monkey (*Cercopithecus pogonias wolfi*), De Brazza's monkey (*Cercopithecus neglectus*), dryad monkey or Salonga guenon (*Cercopithecus dryas*), guereza or black-and-white colobus (*Colobus guereza*), and Thollon's red colobus (*Procolobus* sp. *tholloni*). At Yalosidi, it was concluded that although there was dietary overlap between the bonobo and

William H. Calvin (www.williamcalvin.com)

Bonobos in captivity commonly make use of implements, suggesting tool use in the wild.

Box 5.2 BONOBO COMMUNICATION

Bonobo communication differs from that of other great apes in a number of interesting ways, all of which appear to be related to the fact that bonobos have adopted social strategies unlike those of other apes. Bonobos are often found in large stable communities of up to 120 individuals that move and feed together, and break up into smaller parties when they come to the ground for long-distance travel. By contrast with all other great apes, bonobos are extremely vocal, both in captivity and in the wild, where their sound is limited mainly to the canopy. Bonobos also engage in frequent exchanges of glances and gestures, drag branches, ostentatiously break multiple branches, and pound on tree buttresses as acts of communication. Preliminary observations suggest that they are able to leave messages using crushed vegetation to indicate direction of travel.[68]

The need to coordinate group travel between two to four fruiting sources per day requires high-level communication. It is not possible for a large community of bonobos to find sufficient food by traveling randomly about the forest. They must locate and arrive at ripening fruit resources throughout their environment each day. When a decision is made for the community to travel, all must agree and end up at the same location several hours later, even though they neither see each other nor vocalize as they travel on the forest floor.

Because bonobo communities are large but travel on the ground in small quiet parties, each out of sight of the other, their daily lives require high levels of symbolic communication. If, for any reason, the decision about their destination needs to change once the bonobos have come down to the forest floor (for example, if they encounter traps on the way to a feeding site), this must be conveyed quietly to one another.

Much more needs to be understood about bonobo communication systems in the wild. Like other great apes, bonobos spontaneously begin to understand spoken human speech and to pair written symbols with that speech in captive settings (see Box 3.1). These competencies require neither training nor rewards. They emerge intuitively, especially when communicative activities are embedded within the daily activity of traveling from place to place in the forest to locate food resources. Captive bonobos as young as two years of age can easily mentally map a 20 ha forest, travel to food resources by previous or novel routes, communicate their travel intentions through symbolic means, and even guide human companions new to their forest to designations the bonobos select and specify through symbolic means.[67] By four years of age, bonobos can answer questions about their travel intentions, plan two or three destinations in advance, and begin to disagree with each other about their intended destinations of travel. By eight years of age, they can decide where others should travel, so inform them, and then wait for their return. Local trackers in DRC report that wild bonobos send scouts to check out nearby food resources as they travel from point A to B. While this activity has yet to be verified among wild bonobos, it does appear among captive bonobos.

the four monkey species found there, the monkeys were all much more arboreal than the bonobo, and their ecological niches were narrower.[49]

Interactions between bonobos and other species of primate have been observed, in particular with the Angolan colobus (*Colobus angolensis*) and redtailed monkey.[64] The bonobos mostly treated these monkeys as they did their own species but, in some of the encounters, rough treatment killed the monkeys. The dead monkeys were not seen to be eaten. Interactions that did not result in harm to any of the individuals involved, including grooming between young bonobos and red colobus (*Procolobus badius*), have also been observed in Wamba.[42]

POPULATION

Status and trends

Little population trend information is available for bonobos, and the impacts of the war are so far largely unknown. The loss of the population at the former research site of Yalosidi has been confirmed,[80] and it is thought that numbers have declined at the Wamba site. The total population is thought to be much reduced because of human activities, particularly the spread of firearms (including, more recently, powerful military weapons), together with habitat clearance.

The patchy distribution of the species hinders the estimation of population numbers. Within their known range, local population densities range

Captive bonobos can acquire productive vocabularies of several hundred words, expressed via a lexigram keyboard (see Box 3.1), and are believed to be able to comprehend several thousand spoken words. Bonobos combine symbols without being taught to do so and use a simple grammar that is partially of their own construction. Their ability to understand complex grammatical structures far exceeds what they produce with lexical symbols, but this may be an artifact of the unnatural symbol boards rather than a reflection of a limit to their grammatical ability. Bonobos engage in symbolic dialogs that may run for 20 or 30 minutes and span several topics, each topic in turn giving way to another, and may return to the dialog at a later point in time. They have no difficulty in leaving one topic, moving to another, and then picking up the former topic, without any need to recreate the former conversations that led up to it.[65, 66]

Their ability to acquire language in a captive setting is paralleled by a similar capacity to acquire the rudiments of stone-tool manufacture.[69] While both symbolic language and stone-tool manufacture were initially demonstrated by human companions, the bonobos acquired these abilities through skilled imitation and observation. Even more intriguing: once one bonobo had acquired these skills, they were transferred to other adult bonobos and to their offspring, without the need for human modeling, often with far greater efficiency than that associated with the initial skill acquisition. For example, the first bonobo to acquire the techniques for stone-tool manufacturing went through phases of horizontal knapping and throwing, before finally developing the use of glancing downward 'throw-like' blows, while holding the core in the left hand and the hammer in the right. The second bonobo to develop this skill immediately adopted the developed form without the intermediate stages, after observing the first bonobo.[17]

Tools, language, and culture appear to develop in a coordinated manner among captive bonobos reared in an appropriate environment; the same is likely to be true of wild bonobos. Vocalizations in the wild are complex, frequent, exchanged as lengthy dialogs, and accompanied by pointing gestures. Distinct cultures may occur at different sites, and significant discoveries about bonobo culture and communication doubtless remain to be made.

Susan Savage-Rumbaugh

A young bonobo hooting.

David W. Liggett (www.daveliggett.com)

from 0.25 to 3.7/km^2, but the species is nowhere common.[16, 33, 50, 59, 75, 83, 87, 88] The bonobo is classified as Endangered in the 2004 *Red List*[3] of IUCN–The World Conservation Union, indicating that it has a very high risk of extinction in the wild in the near future. The bonobo is also included in Appendix I of the Convention on International Trade in Endangered Species of Wild Fauna and Flora (CITES), which DRC joined in 1976.

As the only national park within the bonobo range, Salonga is important to the conservation of bonobos. A systematic line-transect and reconnaissance survey was completed across about 61 percent of the 36 560 km^2 park in 2004, under the Monitoring of the Illegal Killing of Elephants (MIKE) program of CITES.[44] This survey was coordinated by the Wildlife Conservation Society and the Congolese Institute for Nature Conservation (ICCN) and funded by WWF International, the United States Fish and Wildlife Service, the European Community, USAID-CARPE, and the Lukuru Wildlife Research Project. The Lukuru Wildlife Research Project and the Max Planck Institute were involved in fieldwork. Bonobos were found to be distributed patchily; in some sectors, none were encountered. Relatively high densities were found in parts of both blocks of the park: the north and northwest of the northern block and the southeast, west, and northwest of the southern block, as well as in the corridor that separates the two blocks.

Threats

Bonobos were probably always patchily distributed in the vast area of the inner Congo Basin, and human pressure and forest fragmentation have reduced their distribution,[74] while killing for bushmeat has eroded their numbers.[28, 88] The latter threat has escalated with the loss of much of the agricultural economy as a result of war, and commercial hunting has intensified in areas such as the Lomako forest.[16] Although bonobos seem to do quite well in secondary forests, logging has been identified as the most important long-term threat to the species because of the widespread overlap of logging concessions – currently only partly active – with bonobo distributions.[63] The direct impacts of logging are difficult to disentangle from the simultaneous increase in hunting, as both access to the forest and demands from local markets for bushmeat accelerate. Other threats to bonobo habitats include cultivation and mining.

DRC's human population is increasing at almost 3 percent per year, the highest annual growth rate in Africa. In 1999, there were 60 million people; at current growth rates, this number is expected to double within 25 years.[63] Conditions of life in DRC are very difficult and much of the population still relies on forest products for food, shelter, and fuel. Pressure on all forest resources is increasing rapidly. Where access along rivers is possible, human immigration and land-use change is more frequent. In the absence of protective taboos, an increase in hunting pressure is likely. Where many people settle with only poor sanitation and rudimentary healthcare provision, human diseases or parasites may also be transmitted to bonobos.[74]

In some parts of the Congo Basin, bonobos are hunted and eaten by local people. Commercial hunting of bonobos is a growing threat, although it is so far thought to be absent from the Salonga National Park, Luo Reserve for Scientific Research, Kokolopori, Wamba, and Lukuru, to name a few. In the 2004 reconnaissance survey at Salonga, 339 snares and 97 hunting camps were found in 1 700 km.[44] There was little direct evidence of bonobo hunting – one skull was found, and no bonobo meat was present in around 50 loads of bushmeat examined [81] – but snares are not selective and can injure or kill bonobos that encounter them. Elephants, by contrast, are targeted for their meat by commercial hunters using semiautomatic weapons, and are viewed as under severe threat.

There are also reports of trade in live bonobos, mainly to supply private collections.[74] Where hunting does occur, females with young are particularly vulnerable, as a threatened mother will carry her offspring even when it has grown to half her size.[50] This slows her down and makes her an easy target.

Since the start of the armed conflict in 1996, some bonobos have been killed by soldiers, including some at the Luo Reserve for Scientific Research.[82] Conservation and research programs have also been disrupted, jeopardizing ongoing studies of bonobo life history and evolution, as well as conservation programs involving bonobo communities. These activities have generally been scaled down rather than halted altogether.[79] Where the war has penetrated the forest, it has led to increased local reliance on wild products, including bushmeat.[63] Adult bonobos have been killed for their meat, with juveniles being sold as pets. For the 12 infant bonobos seen in the Kinshasa market over a five month period at the height of the troubles, 60–120 bonobos are estimated to have been killed.[63] Most bonobo populations, however, are thought to have been unaffected by the war, due to their remoteness from the conflict area.[79]

A bonobo being carried from the forest after it was killed by local hunters in response to an order from a hawker (middle man) from Kinshasa.

Jo Thompson/Lukuru Wildlife Research Project

HUMAN ATTITUDES AND TRADITIONS

History and tradition

Humans are thought to have inhabited the Congo Basin for at least 100 000 years.[6] The great migration of Bantu-speaking people from the area now known as southeastern Nigeria began around 1000 BC.

They dispersed across the forests and savannas of Africa, including the current DRC.[13] As a result, DRC's population is composed of more than 250 different ethnic groups, most of which speak languages of the Bantu family. The largest are the Luba, Kongo, and Mongo.[85] Between 100 BC and 200 AD, the non-Bantu peoples sometimes known as 'pygmies' had been driven into the central Congo Basin, and vanished elsewhere through interbreeding, depopulation, and cultural dominance by the Bantu peoples. In the late 19th century, the country was colonized by Belgium and was administered for many years as the personal possession of Belgium's King Leopold II. Prior to this, human density was low and livelihoods depended on farming and hunting, probably in rough balance with land and forest resources. With colonization, however, new technologies and domesticated species were introduced, and production methods changed. Europeans conquered and to some extent settled the area, exploiting its inhabitants and resources, and using its major rivers for transport. DRC became independent of Belgian rule in 1960.

The Bantu Ndegense people now dominate the federation of four ethnic groups that occupy the Lukuru area, an important area for bonobo research and conservation.[74] Throughout the rest of the bonobo range, the Bantu Mongo people dominate.[80] Wild animals have been hunted by local people for generations.[74] Traditional hunting techniques for domestic consumption use bow-and-arrow and nets, but guns are also commonly used, especially since the beginning of the civil war.

In northern parts of the bonobos' range such as Wamba, there is a belief that humans are descendants of a younger brother of a family of bonobos that lived in the forest; in some southern parts, such as Lukuru, bonobos are believed to represent a 'fallen brother' who is trying to become human again. These beliefs support local hunting taboos. Elsewhere, and even increasingly in taboo zones, when a bonobo is killed there is demand for its by-products: the brain is considered a delicacy; the ashes of the bones are thought to confer great strength on men; and crushed bones are used to wash and strengthen babies.[46]

Recent events

Timber concessions in DRC can now be allocated on a 25–99 year lease; some have been awarded within the geographic distribution of the bonobo. According to 2003 figures from DRC's Service Permanent d'Inventaire d'Aménagement Forestier, approximately 24 percent of the bonobo's range lies within areas designated as logging concessions.[82]

Modern tools such as guns and steel wire increase hunting efficiency; even though bonobos might not be the target prey, they may be caught in traps and snares.[74] Although local human populations are familiar with bonobos, people often fear them.[75] People and bonobos frequently share the same sources of wild food, and bonobos occasionally raid gardens but are not considered to be serious agricultural pests.

Warfare in DRC during the 1990s forced the large-scale movement of human populations to areas of relative safety. These movements have become a threat to bonobos because people with no taboos against bonobo consumption have moved into bonobo range areas.

CONSERVATION AND RESEARCH

Protected areas

The minimum area needed to support a viable population of bonobos has been estimated to be 300–600 km², the exact figure depending on the bonobo population density, levels and types of threat, and other local factors.[82] Two protected areas, both over 300 km² in area, cover parts of the bonobo range: Salonga National Park and the Luo Reserve for Scientific Research. The Lomami-Lualaba Forest Reserve also contains bonobos, and there are moves to upgrade it to full protected area status. An expedition to Lomami-Lualaba, supported by the Wildlife Conservation Society and led by Mwinyihali, confirmed the presence there of bonobos in 2003.[80]

The Salonga National Park in the center of the bonobo range area was created in 1970 largely to protect the bonobo.[88] The park is still intact and covers an area of 36 560 km² in two blocks of almost equal size (*Secteur Nord* and *Secteur Sud*) separated by an unprotected corridor 40–45 km wide. The park encompasses a low plateau covered by swamp forest, river terraces with associated riverine forests, and high plateaus with dry forest cover. It has been reported that, although the park does not appear to hold good numbers of bonobo compared to Lomako and Wamba, there are significant numbers in the northeastern part.[96] Government involvement and application of laws are poor in this area, however,[63] and hunting is a current threat.[95]

The Luo Reserve for Scientific Research

Jo Thompson/Lukuru Wildlife Research Project

The relatively dry Lukuru area is important for bonobo research and conservation.

(358 km^2) is a bonobo research site, the northern section of which is in Wamba village, and the southern section in Ilongo village. This area was identified in 1973 as an ideal place to study bonobos; since then, local people have been involved as field assistants and laborers in fieldwork and conservation projects. The Primate Research Institute (Inuyama, Japan), the Wamba Committee for Bonobo Research, and the Centre de Recherche en Ecologie et Forestrie together run a project based at Wamba village. Agricultural expansion and logging are the main threats in the area. Although much research stopped during the civil war, scientists returned in 2002.[28, 82]

The Lomako forest in the north of the bonobo range occupies an area of 3 100 km^2 bounded by rivers, has good forest cover, and contains few people. It has been reported that there is a viable population of bonobos living relatively free of hunting pressure in the south-central part of the forest, but that the northern population has been affected by hunting.[16] Although local people traditionally hunt for bushmeat, bonobos are protected by taboo. This is changing, however, as the agricultural economy continues to decline, and the area is becoming more accessible. Efforts to establish a national park in the area have not yet been successful.

Recent studies have also been undertaken in the Lake Tumba and Kokolopori areas, but no protected area has been established here.[57] The Bososandja Community Forest to the south of the range area has also been proposed for official protection.[82] A protected area was proposed at Tumba by staff of the DRC's Centre de Recherche en Ecologie et Forestrie, but the status of this proposal is unclear.[80]

Conservation and research activities

Wild bonobos can be observed only in DRC, so field research on the species has been greatly affected by political events in this country. Scientists began to explore the area's biodiversity in 1973; academic research and conservation interests prospered during the 1970s and 1980s, with the encouragement of the former President Mobutu. There was later a hiatus in most field research, due to the civil war that overthrew President Mobutu in 1996/1997, affecting much of the country over the following years.

Most field research on bonobos has been done in the Wamba and Lomako areas. It is hoped that studies from Lukuru, a mosaic of dry forest and savanna, will provide insights into behavioral ecology not available from forested research sites.[78] Other research sites have included Yalosidi, Lake Tumba, and Lilungu. Bonobo research has tended to focus on social behavior, being driven by comparison with chimpanzees and also by public and academic interest in bonobo sexuality and female coalition building, concurrent with the growth of the feminist movement during the 1970s and 1980s.[70] The behavioral ecology of bonobos has also received some attention, and the range of habitats they occupy has been identified. Meanwhile, communication, language, and tool use have been investigated mainly in captive bonobos, with follow-on fieldwork being undertaken more recently.

The MIKE forest survey program, with technical and administrative coordination provided by the Wildlife Conservation Society, aims to build institutional capacity in range states for managing elephant and ape populations. Through this program, bonobo populations are being identified and surveyed in parts of the elephant range that have not been previously or recently researched.[43] Assessments of bonobo populations have been carried out in Lomako, Lukuru, and Salonga National Park, and a conservation infrastructure has been built at the latter two of these sites.[82]

The ICCN is responsible for managing the country's protected areas and related research. Supported by the United Nations Educational, Scientific and Cultural Organization (UNESCO) and the United Nations Foundation, it has worked to conserve bonobos and other great apes in protected areas during the conflict in DRC.[12] Many national and international nongovernmental organizations have also been involved in conservation efforts. The Zoological Society of Milwaukee started a project in 1997 to assess bonobos and other large mammals in the Salonga National Park. This fieldwork was suspended because of the armed conflict, but the Zoological Society of Milwaukee and the US Agency for International Development are among those supporting public awareness activities in DRC. Artists, educators, nongovernmental organizations, and government officials have been involved in research and education, and in producing booklets and magazines.[63] Other key agencies include the Wamba Committee for Bonobo Research and the National Geographic Society for Wamba; the Lukuru Wildlife Research Project for Lukuru; the Bonobo Conservation Initiative for Lac Tumba; Vie Sauvage for Kokolopori; and the Max Planck Institute for the Lui Kotal region. Further information on conservation activities can be found in the DRC profile in Chapter 16.

Conservation priorities

An updated action plan for African primates,[60] and a more specific bonobo-focused action plan,[83] were both published in the mid-1990s. The latter compiles information on research sites and activities over the previous 20 years. It identifies conservation priorities in each site, and recommends actions involving research (e.g. determination of the current range and population status), regulation (e.g. habitat protection), and education and training.

In November 1999, the IUCN Conservation Breeding Specialist Group conducted a meeting to assess bonobo conservation status, during which participants identified threats and set priorities for research and conservation. Recommendations targeted species-based conservation measures, but also gave attention to the need to improve human quality of life. Proposed activities included raising awareness of the species among the peoples of DRC; coordination of activities among various parties; assessment of bonobo populations; sharing of information; public education; and reopening of research sites. There were also calls for greater international attention to DRC, stronger efforts to restore peace, and investment to strengthen and maintain protected areas.[7] A further workshop on bonobos was organized in Japan during July 2003, which reviewed research carried out at Wamba, Lomako, and Lukuru. The workshop also established some priorities for the future, including infrastructure development and continued community participation at most of the research sites.[23]

The future of these extraordinary primates is far from secure. They are widespread over the inner Congo Basin, but scarce, with a total population likely to be much less than the maximum estimate of 100 000 individuals. They are increasingly hunted for food as local protective taboos erode with the movement of people within their range. This pressure is growing as warfare has damaged the agricultural economy and encouraged forest exploitation and commerce in bushmeat. As the war is brought to an end, moreover, it is feared that industrial-scale logging and mining will escalate dramatically. As areas are opened up, new transport routes will allow bushmeat to reach new markets among the rapidly expanding human population, leading to increased hunting of all wildlife.

Set against this, however, bonobos do occur in areas such as Salonga National Park and the Lomako forest, which remain remote and relatively unpeopled. The opportunity exists to put in place effective conservation, education, and management processes before pressures on these bonobos become overwhelming. Other conservation assets include the increasingly strong international constituency of interest in the bonobo. This arises partly from the perception that these primates have charming and peaceful social lives, offering a redemptive contrast to the stereotype of the

Jo Thompson/Lukuru Wildlife Research Project

Jo Thompson, Director of the Lukuru Wildlife Research Project, meeting with Iyo Booto Alfonse, Grand Chef de Groupement de Isolu. The Grand Chef has directed his people to collaborate with the project in protecting bonobos.

aggressive chimpanzee as an alternative model for hominid societies.

This enthusiasm could be translated into public support for a long-term commitment by donor governments to 'adopt' the species and guarantee its survival. Such arrangements would greatly amplify the impact of conservation measures otherwise based mainly on the enthusiasm of nongovernmental organizations, researchers, and many local people.

FURTHER READING

Bermejo, M., Illera, G., Sabater Pí, J. (1994) Animals and mushrooms consumed by bonobos (*Pan paniscus*) – new records from Lilungu (Ikela), Zaire. *International Journal of Primatology* **15** (6): 879–898.

Coxe, S., Rosen, N., Miller, P., Seal, U. (1999) *Bonobo Conservation Assessment. Workshop Report.* Kyoto University Primate Research Institute, Inuyama, Japan.

de Waal, F., Lanting, F. (1997) *Bonobo.* University of California Press, Berkeley.

Galdikas, B.M.F., Briggs, N.E., Sheeran, L.K., Shapiro, G.L., Goodall, J., eds (2001) *All Apes Great and Small, vol. 1: African Apes.* Kluwer Academic/Plenum Publishers, Boston, Dordrecht, London, Moscow, New York.

Hohmann, G., Fruth, B. (2003) Culture in bonobos? Between-species and within-species behavior. *Current Anthropology* **44** (4): 563–570.

Horn, A. (1980) Some observations on the ecology of the bonobo chimpanzee (*Pan paniscus*, Schwarz 1929) near Lake Tumba, Zaire. *Folia Primatologica* **34**: 145–169.

Kano, T. (1992) *The Last Ape: Pygmy Chimpanzee Behavior and Ecology.* Stanford University Press, Stanford.

Kortlandt, A. (1996) A survey of the geographical range, habitats and conservation of the pygmy chimpanzee (*Pan paniscus*): an ecological perspective. *Primate Conservation* **16**: 21–36.

McGrew, W.C., Marchant, L.F., Nishida, T., eds (1996) *Great Ape Societies.* Cambridge University Press, Cambridge, UK.

Sussman, R.L., ed. (1984) *The Pygmy Chimpanzee, Evolutionary Biology and Behavior.* Plenum Press, New York.

Thompson, J., Hohmann, G., Furuichi, T., eds (2003) *Bonobo Workshop: Behaviour, Ecology and Conservation of Wild Bonobos.* Inuyama, Japan.

Thompson, J.A.M. (1997) *The History, Taxonomy and Ecology of the Bonobo* Pan paniscus *(Schwarz, 1929), with a First Description of a Wild Population Living in a Forest/savanna Mosaic Habitat.* PhD dissertation, University of Oxford.

Thompson-Handler, N., Malenky, R.K., Reinartz, G.E. (1995) *Action Plan for* Pan paniscus: *Report on Free-ranging Populations and Proposals for their Preservation.* Zoological Society of Milwaukee County in cooperation with the IUCN/SSC Primate Specialist Group, Milwaukee.

MAP DATA SOURCES

Map 5.1 Great apes data are based on the following source, with updates as cited in the DRC country profile in Chapter 16:

Butynski, T.M. (2001) Africa's great apes. In: Beck, B.B., Stoinski, T.S., Hutchins, M., Maple, T.L., Norton, B., Rowan, A., Stevens, E.F., Arluke, A., eds, *Great Apes and Humans: The Ethics of Coexistence.* Smithsonian Institution Press, Washington, DC. pp. 3–56.

For protected area and other data, see 'Using the maps'.

ACKNOWLEDGMENTS

Thanks to Colin Groves (Australian National University) for valuable comments on the draft of this chapter. Thanks also to Stephen Blake, John Hart, and colleagues (Wildlife Conservation Society) for information about the MIKE bonobo surveys, and to Els Cornelissen (Royal Museum for Central Africa, Belgium) for archeological advice.

AUTHORS

Carmen Lacambra, UNEP World Conservation Monitoring Centre
Jo Thompson, Lukuru Wildlife Research Project
Takeshi Furuichi, Meiji-Gakuin University
Hilde Vervaecke, University of Antwerp
Jeroen Stevens, University of Antwerp
Box 5.1 Frances White, University of Oregon
Box 5.2 Susan Savage-Rumbaugh, Georgia State University

CHAPTER 6

Gorilla overview

JULIAN CALDECOTT AND SARAH FERRISS

There are two species of gorilla, separated from one another by the inner Congo Basin, that region of Central Africa to the south of the Congo River that is inhabited by bonobos. Each species has two subspecies:[12]

- the eastern gorilla (*Gorilla beringei* Matschie, 1903) is divided into the eastern lowland gorilla (*G. b. graueri* Matschie, 1914) and the mountain gorilla (*G. b. beringei* Matschie, 1903); and
- the western gorilla (*Gorilla gorilla* Savage, 1847) is divided into the western lowland gorilla (*G. g. gorilla* Savage, 1847) and the Cross River gorilla (*G. g. diehli* Matschie, 1904).

Mitochondrial DNA (mtDNA) research suggests that the lineage of the western gorilla diverged from that of the eastern gorilla around 2 million years ago (mya).[43] This does not necessarily indicate that the populations had already separated. Tropical Africa had a drier and cooler climate at this time,[19] fragmenting the forests in which the common ancestor of the gorillas lived, and possibly accelerating a divergence already underway as a result of the wide geographical range of the species.

The two gorilla species have numerous similarities[23, 38] and, until 2001, were recognized only as subspecies.[11] Both are large and sexually dimorphic, with adult males weighing up to about 200 kg and adult females around half that. Both species have broad chests and shoulders, large heads, and hairless, shiny black faces. In both, maturing males develop a silvering of the hair on their backs and sagittal crests, which when completed causes them to be known as 'silverbacks'. The eastern gorilla tends to be somewhat larger than the western. Diagnostic differences between the two species include:

- the eastern has longer, blacker hair than the western, which has sleeker and grayer or browner hair; the head hair of western gorillas

Elizabeth A. Williamson

Eastern lowland gorilla, a silverback male, Democratic Republic of the Congo.

Ian Redmond/UNESCO

An infant western lowland gorilla seeks comfort riding 'piggyback' on an older gorilla at the Lefini rehabilitation site, Congo.

tends to have red tones, with the crest and nape hair of adult males usually being a striking chestnut color;[27]

- the eastern gorilla has a more developed sagittal crest along the midline of the skull than the western, suggesting a more powerful jaw musculature; and
- the western silverback's saddle of white hair often extends onto the thigh, and grades more into the body color than the eastern silverback's, in which it tends to be more clearly delineated and to stand out more against the dark hair.

ECOLOGY

All gorillas eat much the same kinds of foods, with the precise selection varying according to their absolute and relative availability. Their larger body size enables gorillas to consume a somewhat poorer-quality diet than that of other great apes. Dietary choice is constrained by their stomachs, which are simple and nonfermenting and so preclude eating too many mature leaves.

Mountain gorillas have been studied high in the three national parks of the Virungas (Virunga, Volcanoes, and Mgahinga), and at lower elevation in the forest of the Bwindi Impenetrable National Park. In the Virungas, the diet is overwhelmingly dominated by herbaceous leaves and shoots (which are abundant there), while at Bwindi the diet contains far more fruits (which are very scarce in the Virungas, but common at Bwindi).[10, 34] In both places, the diet reflects seasonal influences, and the gorillas gorge on 'seasonal specials' such as bamboo shoots, or consume a more diverse diet including herbaceous vegetation, bark, and twigs, as availability allows.[50]

Consistent with the low quality and poor digestibility of their diet, mountain gorillas in the Virungas spend much of their daylight time feeding, and otherwise rest.[46] Their groups occupy a small area for a day or two and then move on, seldom returning for several months while the vegetation recovers from being trampled and harvested. As herbaceous vegetation is abundant, widespread, and of low dietary quality, there is little ecological need to defend home ranges against other gorillas. In some cases, mountain gorilla groups in the Virungas have completely overlapping ranges.[45] Those living at Bwindi are more mobile than in the Virungas; a group at Bwindi may use up to 40 km^2 in a year,[34] compared with 5–11 km^2 typically used in the Virungas.[42, 45, 48] This reflects the higher availability of seasonal fruit at Bwindi, and the corresponding increased travel to obtain such preferred foods.

Eastern lowland gorillas have been studied at various altitudes (some 'lowland' gorillas live at higher altitudes than some mountain gorillas) in the Kahuzi-Biega, Maiko, and Itombwe forests of the eastern Democratic Republic of the Congo (DRC). Groups occupy home ranges of 13–17 km^2 in montane forest,[54] but the size of their home range in lowland tropical forest is unknown. Like mountain gorillas, they travel less in montane than in lower-altitude forests. Their diet is made up of fruit, seeds, leaves, stems, and bark as well as ants, termites, and other insects. Fruits are preferred but, when these are scarce, eastern lowland gorillas eat more herbaceous vegetation. Large quantities of bamboo shoots are eaten seasonally.

Western gorilla diet also varies seasonally.[6, 30, 35] When fruit is abundant, it may constitute most of the diet but at other times shoots, young leaves, and bark are eaten instead. Terrestrial herbaceous vegetation, aquatic herbs, and insects are eaten year round as availability and opportunity permit. Western lowland gorillas have been studied at several sites in the Central African Republic (CAR), Congo, and Gabon but, despite their greater numbers, they are less well known than either of the eastern subspecies. They occupy a diverse range of habitat types, including lowland, swamp, and montane forests; forests with open or closed canopies; forests with dense or sparse understorys; and forests that have been disturbed and are regenerating. The rare Cross River gorilla is even less well studied, although its diet is known to include fruit, leaves, stems, piths, invertebrates, and soil.[25]

It is generally considered that fruit is more widely available in western gorilla than in eastern gorilla habitat, and that this accounts for their greater frugivory and the associated more mobile lifestyle. Typically exceeding 20 km^2, western gorilla home ranges are larger than those of mountain gorillas (except at Bwindi), and there may be extensive overlap between the ranges of neighboring groups.[9, 30, 39] These overlaps lead to gorilla groups sometimes encountering one another. Such occasions involve vocalizations and chestbeats from both groups or only one of the groups, and can lead to one group moving away. Encounters are sometimes violent and may involve lethal wounding.

SOCIETY AND PSYCHOLOGY

Gorillas are considered infants until they are weaned at about three years,[8] or possibly later in western lowland gorillas.[24, 37] Young or immature animals fall into a number of categories:[51] juveniles (3–6 years), subadults (6–8 years), and young mature males or adolescents, commonly known as blackbacks (8–11 years). The process of silvering of hair on the back and sagittal crest of mature males (age 12+, known as silverbacks) begins at 10–11 years of age and is completed by about 15–16 years. Females do not undergo this silvering as they become adult. The maximum lifespan of gorillas in the wild is unknown,[32] with the oldest known mountain gorillas at over 40 years, while the oldest gorilla to have died in a zoo reached 53 years of age.[1]

Median group sizes of between seven and 16 animals, most being typically between eight and 11, have been reported for all populations of gorillas, regardless of habitat type and prevailing diet. This median represents a dominant, silverback adult male, three or four females, and four or five of their offspring.[28, 33, 53, 54] This simple harem-like arrangement describes almost all western gorilla groups, about 90 percent of eastern lowland groups, and about 60 percent of mountain groups. The balance is made up of all-male and multimale groups. Most multimale groups result from males maturing and remaining in their natal groups. When they become silverbacks they may inherit or share mating rights within that group. Hence, multimale groups are believed to contain related adult males. It is more usual, however, especially among lowland gorillas, for maturing males to leave their natal group, either taking females with them, spending time in an all-male group, or remaining solitary until they can establish a group of their own by attracting females.

Females also transfer between groups, sometimes more than once. If a female has an infant with her at the time, there is a serious risk of the infant being killed by the dominant male of the new group.[33] It is therefore hard to see why females do this. Several factors are likely to be influential: a predisposition to leave the natal group, her preferences, and her aversions may all influence a female's choice of mate. Following the death of the adult male, either the females transfer to one or more different groups, or the harem is taken over by another male. When the dominant male of a

Martha M. Robbins

Mountain gorillas in Bwindi National Park, Uganda.

multimale group dies, the females tend to remain with the heir, with whom they are already familiar.

Gorilla social bonds, except those between mothers and their infants, do not appear to be particularly strong. As they get older, infants and juveniles spend an increasing proportion of their time close to the dominant male until they reach adolescence.[36] Interactions between adult males and females are largely limited to exchange of vocalizations between adults (see Box 8.3), aggressive displays by males towards females, appeasement of males by affected females, and interventions by males to end disputes between females (and sometimes *vice versa*). These interventions involve only moderate aggression and pose little risk to the relationships between the males and the females, but may limit the effectiveness of female coalitions and help males to maintain dominance.[49] Males in multimale groups interact little with each other, and relationships between silverbacks and blackbacks are generally minimal since the latter are subordinate and tend to spend a lot of time on the group's periphery. Affiliative (friendly or cooperative) interactions between males are therefore rare, but males in a multimale group occasionally cooperate to prevent females from leaving.[52]

Gordon Miller/IRF

Gorillas leave a trail of trampled and broken vegetation behind them.

The reproduction and sexual behavior of western gorillas is little understood; this is one of the topics that have been better studied in mountain gorillas than the other subspecies.[16, 17, 47] Female mountain gorillas reach sexual maturity at six or seven years, although between the first bout of estrus-like behavior and the first conception there is a two year period of adolescent sterility; as in chimpanzees and bonobos, this allows for a certain amount of experimentation among potential group and partner situations. The menstrual cycle among adults has a median length of 28 days, during which females are most receptive and attractive at around ovulation, that is for one to four days at mid-cycle, and mating or mating attempts occur at or near peak estrogen concentrations during menstrual cycles and pregnancy.[5, 17, 47]

In single-male mountain gorilla groups, that male sires all the offspring.[2] In multimale groups, subordinate males also mate, although often with less fertile subadult females.[31] They are often harassed by dominant males while doing so, but still manage to sire a proportion of the offspring.[3] Mating with individuals from other groups is exceptionally rare. Female choice of mate seems to be important among mountain gorillas, and is influenced by male behavior; females either stay with a mate or leave for another group.

Eastern lowland gorillas share many reproductive characteristics with mountain gorillas, including delayed conception, age at first delivery of offspring (eight or nine years), and interbirth interval (around four years).

Like all other great apes, gorillas construct nests in which to sleep at night, and can learn to use human sign languages with some facility, as well as novel motor skills taught to them in captivity. Unlike all other great apes but the Bornean orangutan, gorillas have never been observed making or using tools in the wild. This is consistent with the notion that tool use is linked to sociability in animals of sufficient intelligence and learning ability, as it broadens the pool of potentially discoverable and learnable behaviors. Solitary animals that seldom meet (such as Bornean orangutans) and group-living animals that seldom interact with one another (such as gorillas) should be the least likely to develop the use of tools. An alternative

Gordon Miller/IRF

The history of conservation is peppered with conflict, and park rangers work hard to maintain the good will of their communities, as here in Uganda.

explanation is that gorillas are not typically challenged by their foods, so have had less need to develop tool use.

CONSERVATION CONCERNS

Of the eastern gorillas, the mountain subspecies has very small but stable populations in several well managed national parks in the Virungas (about 380 gorillas in 400 km^2 of forest)[18] and Bwindi (about 320 gorillas in 200 km^2 of forest).[21] The parks are well supported by both international nongovernmental organizations and the scientific community, by profitable gorilla-based tourism programs, and by the governments of the region. These populations are too small to meet some theoretical criteria for genetic health, are vulnerable to catastrophic events such as outbreaks of disease, and would quickly be reduced by poaching if the vigilance of conservationists were to be relaxed. Nevertheless, they are being well cared for.

The eastern lowland gorilla is of much greater immediate concern; its population was estimated to be around 17 000 in the mid-1990s,[13] but it is feared that thousands had been killed by hunters by 2004.[29] Warfare engulfed the whole range of the eastern lowland gorilla during this period, while armies, rebels, refugees, and miners all lived off the land. Bushmeat, including that of the gorilla, is still consumed in great quantities. In May–June 2004, the rebel military occupation of Bukavu, and the accompanying destruction of equipment at the Tshivanga field station in Kahuzi-Biega National Park, showed that the situation was not yet stable.

Of the western gorillas, the Cross River gorilla has an estimated total surviving population of 250–280 individuals, fragmented across more than 10 highland areas.[25] It is difficult to assess population status and trends among the much more widespread and abundant western lowland gorilla, as censuses have not yet been made across large areas, and new areas of habitat have recently been identified. Together with the mountain gorilla and Sumatran orangutan, it has been listed by Conservation International and the Primate Specialist Group of IUCN–The World Conservation Union as one of the world's 25 most endangered primate taxa.[20]

Western lowland gorillas are widely distributed across a large forested region and occur in numerous protected areas,[40] but they nevertheless

Martha M. Robbins

Martha M. Robbins

Mountain gorilla infants, Virunga National Park, Democratic Republic of the Congo (top), and a silverback male, Bwindi Impenetrable National Park, Uganda.

face an uncertain future simply because of the increasing scale and cumulative nature of the threats operating upon them. These include forest clearance for farming, forest fragmentation due to clearance and the building of roads, forest degradation by logging, hunting for food, and disease. Hunting and disease are increasing as risk factors because human access to formerly remote forest areas is expanding through logging and settlement. The estimated halving of the great ape population between 1983 and 2000 in Gabon, the former stronghold of the western gorilla,[44] as a result of hunting and the Ebola virus, shows how dangerous this combination of factors can be.

Large numbers of western gorillas may remain in the Congo Basin. The presence of western gorillas living at substantial densities in swamp forests, a widespread habitat that was previously considered unsuitable, was only confirmed in the 1990s[7, 22] after first reports in 1983. In terms of national boundaries, Gabon is thought to hold the largest populations of western gorillas, probably followed by Congo. In the early 1980s, there were estimated to be 40 000 western lowland gorillas[14] of which about 35 000 were in Gabon.[41] After the discovery that western gorillas also inhabit swamp forest in significant numbers, subsequent total population estimates were revised upwards to 94 500–110 000.[4, 15, 26] However, these estimates were made prior to the significant recent impact of both bushmeat hunting and the Ebola virus.

While additional knowledge will help to guide conservation action, the long-term survival of the western gorilla in an increasingly disturbed and human-dominated landscape must depend on the attitudes of local people and the partnerships they establish with government and conservationists. Modern approaches to conservation therefore focus on community engagement, education, and empowerment, as well as global monitoring; this is all implemented in partnership with governments whose policy priorities are advanced and promoted accordingly. The basic concept is that conservation and sustainable development initiatives will be more successful where communities are stronger. Great ape conservation will suffer if communities are weak or fragmented, and their interests are either overwhelmed by outside factors or ignored in the decision-making processes of people far away. Current conservation projects therefore typically propose simultaneously starting with communities to work 'outwards' into the rest of society, and with governments to work 'inwards' towards the community level. As has been written of African conservation:

... the primary need is for the opening of friendly dialogues, partnerships and a sharing of knowledge and enthusiasm with the people, especially the poorer inhabitants ... Conservation here is primarily a social, political and human problem ... it poses problems in communication, in education and in values, because there are huge dislocations in understanding. Starting from the bottom, links have to be made between the various tiers of rural communities, old and new, national citizenries and

an external public that wants to help reconcile conservation with development.[19]

WHAT WE DO NOT KNOW

For the eastern lowland gorilla, the principal gaps in knowledge relate to the actual numbers and distribution of the species in the war-torn eastern part of DRC, and to the actions that might realistically be taken to promote their survival in the wild. From a more scientific perspective, very little is known about the ecology or demography of mountain gorillas in Bwindi, the annual home range of eastern lowland gorillas in lowland tropical forest, the lifespan in the wild, and the behavior of solitary males.

For the western gorilla, fundamental questions also remain regarding their numbers and distribution, and the actual mechanisms of population decline. There is a dearth of information on the detail of western gorilla life from a social and ecological perspective, the social bonds that maintain the species' societies, the reproductive development of individuals, and demographic information such as lifespan in the wild and age at first birth. This information is crucial in assessing population viability.

FURTHER READING

Beck, B.B., Stoinski, T.S., Hutchins, M., Maple, T.L., Norton, B., Rowan, A., Stevens, E.F., Arluke, A., eds (2001) *Great Apes and Humans: The Ethics of Coexistence.* Smithsonian Institution Press, Washington, DC.

Harcourt, A.H. (1986) Gorilla conservation: anatomy of a campaign. In: Benirschke, K., ed., *Primates: The Road to Self-sustaining Populations.* Springer-Verlag, New York. pp. 31–46.

Kingdon, J. (1990) *Island Africa: The Evolution of Africa's Rare Animals and Plants.* Collins, London.

Oates, J.F., McFarland, K.L., Groves, J.L., Bergl, R.A., Linder, J.M., Disotell, T.R. (2002) The Cross River gorilla: natural history and status of a neglected and critically endangered subspecies. In: Taylor, A.B., Goldsmith, M.L., eds, *Gorilla Biology: A Multidisciplinary Perspective.* Cambridge University Press, Cambridge, UK. pp 472–497.

Olejniczak, C. (2001) The 21st century gorilla: progress or perish? In: Brookfield Zoo, *The Apes: Challenges for the 21st Century.* Conference proceedings. Chicago Zoological Society, Brookfield, Illinois. http://www.brookfieldzoo.org/content0.asp?pageID=773. pp. 36–42.

Robbins, M.M. (2001) Variation in the social system of mountain gorillas: the male perspective. In: Robbins, M.M., Sicotte, P., Stewart, K.J., eds, *Mountain Gorillas: Three Decades of Research at Karisoke.* Cambridge University Press, Cambridge, UK. pp. 29–56.

Robbins, M.M., Sicotte, P., Stewart, K.J., eds (2001) *Mountain Gorillas: Three Decades of Research at Karisoke.* Cambridge University Press, Cambridge, UK.

Taylor, A.B., Goldsmith, M.L., eds (2002) *Gorilla Biology: A Multidisciplinary Perspective.* Cambridge University Press, Cambridge, UK.

Tutin, C.E.G. (2001) Saving the gorillas (*Gorilla g. gorilla*) and chimpanzees (*Pan t. troglodytes*) of the Congo Basin. *Reproduction, Fertility and Development* **13**: 469–476.

Vigilant, L., Bradley, B.J. (2004) Genetic variation in gorillas. *American Journal of Primatology* **64**: 161–172.

ACKNOWLEDGMENTS

Many thanks to Colin Groves (Australian National University), Alexander Harcourt (University of California, Davis), Martha M. Robbins (Max Planck Institute for Evolutionary Anthropology), Elizabeth A. Williamson (University of Stirling), and David Woodruff (University of California, San Diego) for their valuable comments on the draft of this chapter.

AUTHORS

Julian Caldecott, UNEP World Conservation Monitoring Centre
Sarah Ferriss, UNEP World Conservation Monitoring Centre

T.J. Rich/naturepl.com

CHAPTER 7

Western gorilla (*Gorilla gorilla*)

SARAH FERRISS

The western gorilla (*Gorilla gorilla* Savage, 1847) is a large animal, with a broad chest and shoulders, a large head, and a hairless, shiny black face.[55, 83, 86] Full-grown adult males weigh up to about 180 kg, about twice the weight of adult females. Two subspecies have been described:[57] the western lowland gorilla (*G. g. gorilla* Savage, 1847) and the Cross River gorilla (*G. g. diehli* Matschie, 1904). The western lowland gorilla is much more widespread and numerous than the Cross River gorilla, which is restricted to a relatively small area on the Nigeria-Cameroon boundary. This chapter will focus on the western lowland gorilla; the Cross River gorilla is discussed in Box 7.1.

DISTRIBUTION

Western lowland gorillas are widespread throughout West and Central Africa (see Map 7.1). The Congo/Oubangui River seems to delimit the eastern boundary of their distribution, and the northern boundary is essentially defined by the course of the Sanaga River and the northern limits of the closed forest. The western boundary of their distribution is formed by the Atlantic coast, and the southern edge of their distribution is defined by the forest–savanna boundary, as shown on Map 7.1. Western gorillas are found in Gabon, the Cabinda province of Angola, the western part of Congo, the extreme southwestern part of the Central African Republic (CAR), south-central and southern Cameroon, and in mainland Equatorial Guinea. They used to occur in the extreme western tip of the Democratic Republic of the Congo (DRC), but are now probably extinct in that country.

BEHAVIOR AND ECOLOGY

Observational challenges

Most research on gorilla ecology and behavior has focused on eastern gorillas (*Gorilla beringei*), particularly the mountain gorillas (*G. b. beringei*) of the Virunga Volcanoes in DRC, Rwanda, and Uganda; there has been comparatively little research on western gorillas (*G. gorilla*). Recent research has identified many differences between the two species, but many questions remain. By the beginning of this century, researchers at only three study sites had succeeded in habituating western lowland gorillas for study.[7, 28, 34] Habituation is difficult because of the limited mobility and visibility in the dense forest, the large home ranges of western gorillas, and because the gorillas often flee at the approach of humans due to having experienced hunting in the past. However, some excellent observations of less habituated gorillas have been made in marshy forest clearings (locally called *bais*), where visibility is good. Not all behavior in *bais* is typical of that seen in forested environments.[91]

Habitat

Western lowland gorillas occur in primary (old growth) and secondary (regenerating) forests (including forest swamps) as well as in both submontane[49] and lowland areas.[9] Overall, western lowland gorilla occurrence, biomass, and density seem to be positively correlated with terrestrial or aquatic herbaceous vegetation, particularly monocotyledonous plants (including gingers and palms).[8, 12, 43, 136]

During the 1980s, Tutin and Fernandez found western gorillas in seven of 15 habitat types surveyed in Gabon:[130] dense primary forest; dense

Map 7.1 Western lowland gorilla distribution (see Box 7.1 for Cross River gorilla)

Data sources are provided at the end of this chapter

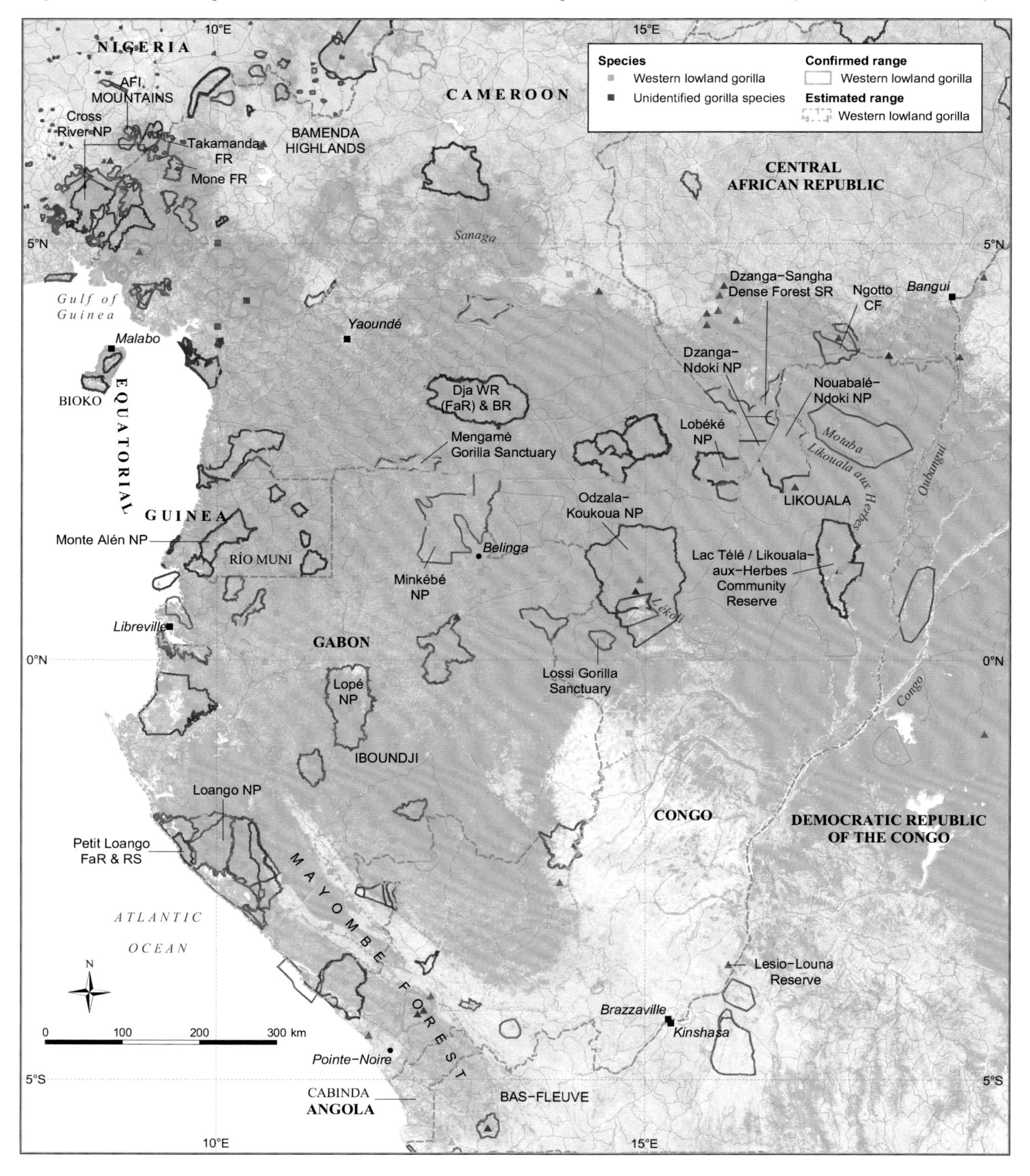

inundated (swamp) forest; thicket; undisturbed secondary vegetation; exploited forest (one to two years after logging); exploited forest (two to six years after logging); and coastal scrub. Western gorillas appeared to be absent from areas of human settlement and disturbed secondary forests; they avoided roads and plantations, but were observed in recovering secondary forests. In the Petit Loango Reserve (now part of Loango National Park), where the herbaceous vegetation favored by gorillas is generally scarce, western gorilla nests were found mainly in secondary forest with more edible herbs.[48] The fairly high density of western gorillas seen in the Dzanga Sector of the Dzanga-Ndoki National Park in CAR have been attributed to the presence of moderately disturbed or secondary forest, which is rich in nutritious folivore food such as herbs.[16]

Western lowland gorillas have been observed occasionally nesting along savanna–forest edges or in the savanna itself.[26, 135, 145] Western lowland gorillas sometimes,[137] but not always,[127] make use of forest fragments within the savanna. They do not live permanently in these habitats, however, perhaps because the forests do not provide either sufficient preferred nesting materials or a constant food supply.[137]

The Odzala-Koukoua National Park in Congo provides a good example of the variety of habitats occupied by western lowland gorillas. Here, they primarily live in open-canopy Marantaceae forest, in which sufficient light reaches the forest floor to allow plentiful understory vegetation to develop.[9] This forest type is dominant to the north of the Lékoli River, particularly in the northeastern part of the park. The ground vegetation is dominated by an almost impenetrable thicket of Marantaceae species, including *Haumania liebrechtsiana, Megaphrynium macrostachyum,* and *Sarcophrynium* spp. Western gorillas are also found in the closed-canopy primary forests of the park, which have a much more continuous canopy and a sparser understory.

In and around the northern part of the Odzala-Koukoua National Park there are more than 100 forest clearings. Those that have been investigated have a particularly sodium-rich marshy herbaceous vegetation and are known as 'salines'[76] or *bais* (see Box 7.2). Western gorillas are known to visit these clearings daily to feed on plants from families such as Cyperaceae and Asteraceae, which here are rich in sodium absorbed from the soil.[76] *Bais* therefore provide a unique opportunity for researchers to observe the animals in the open.

Although it was originally thought that western gorillas avoided water, swamp forests are now considered important habitats and feeding areas for western gorillas, supporting them in high densities.[44, 85] Western gorillas have been observed in swamp forests both in the wet and the dry season.[44] The soils of these swamps tend to be waterlogged or permanently flooded; the aquatic herbs found here, such as *Hydrocharis* spp. (Hydrocharitaceae), can provide important nutrient sources for western gorillas.[85] Species common to swamp forest include those belonging to the genera *Xylopia* (Annonaceae), *Raphia* (Arecaceae), *Klaineanthus* (Euphorbiaceae), *Trichilia* (Meliaceae), *Lophira* (Ochnaceae), *Guibourtia* (Leguminosae-Caesalpinioideae), and *Aframomum* (gingers, Zingiberaceae).[9, 44] In northern Congo, a study found that western gorillas favored those swamp forests where *Raphia* was common, a palm genus used both for food and nest construction.[12] In south-western CAR, the distribution of western gorillas seems to be influenced by the availability of *Aframomum* spp.[26]

Diet

There are two major differences in food availability between the habitat of the western gorilla and that of the best-studied gorilla subspecies, the mount-

Richard Parnell

A male western lowland gorilla feeding, Nouabalé-Ndoki National Park, Congo.

ain gorilla, affecting both diet and foraging behavior. First, 'high-quality' herbs that are easily digestible and rich in proteins and minerals are much less abundant and more patchily distributed in western gorilla habitat, outside swampy areas. Second, fruit is much more widely available in the habitats of western gorillas, so they eat significantly more fruit than do their eastern counterparts.[25, 34, 35, 37, 105, 112, 126, 146]

In the absence of direct observations of feeding, the methods used to identify the diet of a particular gorilla group include fecal analysis, and the monitoring of food types that show signs of having been processed by western gorillas and are left along gorilla trails.[34] Western gorillas consume large amounts of fiber, and they eat leaves, stems, fruit, piths, invertebrates, and soil. There are seasonal, annual, and spatial variations in the frequency of consumption of different food items[104, 106] (see Table 7.1).

The seasonal importance of fruit and herbs in the diet of the western gorilla has been the subject of much debate.[34, 73, 103] The availability of seasonal fruit appears to shape the foraging and ranging patterns of western gorillas.[103] When fruit is abundant seasonally, it may constitute most of the diet. High-quality herbs that are rich in minerals and proteins are eaten all year round, while low-quality herbs are eaten only when fruit is scarce.[36, 71, 85, 109] More leaves and woody vegetation are consumed during the dry season of January–March, when few fleshy fruits are available; more fruit is eaten at other times.[6, 34, 51, 85, 102] Favored tree fruits include those of the genera *Tetrapleura* (Leguminosae-Mimosoideae), *Chrysophyllum* (*Gambeya*) (Sapotaceae), *Dialium* (Leguminosae-Caesalpinioideae), and *Landolphia* (Apocynaceae).[34, 85] The fruits of terrestrial herbs such as species of *Aframomum*, *Nauclea* (Rubiaceae), and *Megaphrynium* (Marantaceae) are also eaten when available.[34, 91]

Some habitats of the western lowland gorilla are dominated by the leguminous tree *Gilbertiodendron dewevrei* (Leguminosae-Caesalpinioideae); at roughly five year intervals this tree produces especially nutritious seeds that contain high levels of nitrogen. The western gorillas feed heavily on the seeds during these mast fruiting events, and are willing to travel some distance to congregate in stands of *G. dewevrei*.[11]

High-quality herbs, where available, are important for western gorillas. Plants from the mono-

Table 7.1 Western gorilla diet

Study site	Country	Species (plant or animal)	Plant species: Fruit incl. seeds	Leaf	Stem/pith	Flowers	Bark	Invertebrate species	Other foods[a]
Mondika[34]	CAR and Congo	100	70	33	14	2	8	yes	–
Bai Hokóu[103]	CAR	138	77	84	14	4	–	9	dirt from termite mounds and *bais*[b]
Nouabalé-Ndoki National Park[85]	Congo	152	133	29	10	2	2	–	6 (includes roots and shoots)
Belinga[131]	Gabon	89	72	7	18	1	2	–	4 (nonplant foods)
Lopé National Park[132, 133]	Gabon	not known	91 + seeds of 21 species	up to 49	17	3	9	5	8 (includes roots, galls, and fungi)
Lopé National Park[134]	Gabon	not known	100 + seeds of 21 species	48	16	3	included in 'other'	10	22 (includes bark, roots, wood, soil, and fungi)
Southeastern Cameroon[33]	Cameroon	> 22	not known	not known	not known	not known	not known	22	not known
Afi Mountain Wildlife Sanctuary[c]	Nigeria-Cameroon	168	100	36	22	2	53	0[d]	3 (roots)

a Wood, shoots, buds, tubers, rhizomes, and fungi.
b See Box 7.2.
c Cross River gorillas.
d No evidence found in fecal samples, feeding trails, etc.

Box 7.1 THE CROSS RIVER GORILLA (*Gorilla gorilla diehli*)

The gorillas inhabiting the mountainous landscape that straddles the border between Nigeria and Cameroon at the headwaters of the Cross River were recently recognized as the subspecies *Gorilla gorilla diehli*. These Cross River gorillas have an estimated total surviving population of 250–280 individuals distributed across more than 10 fragmented highland areas.[89] Despite the relatively dense human population in this region of West Africa, these gorillas have persisted, protected by their adaptability and relative inaccessibility. As human-development activities increase within the region and gorilla habitat is further eroded, the future survival of this ape depends on urgent conservation action.

continued overleaf

Cross River gorilla distribution

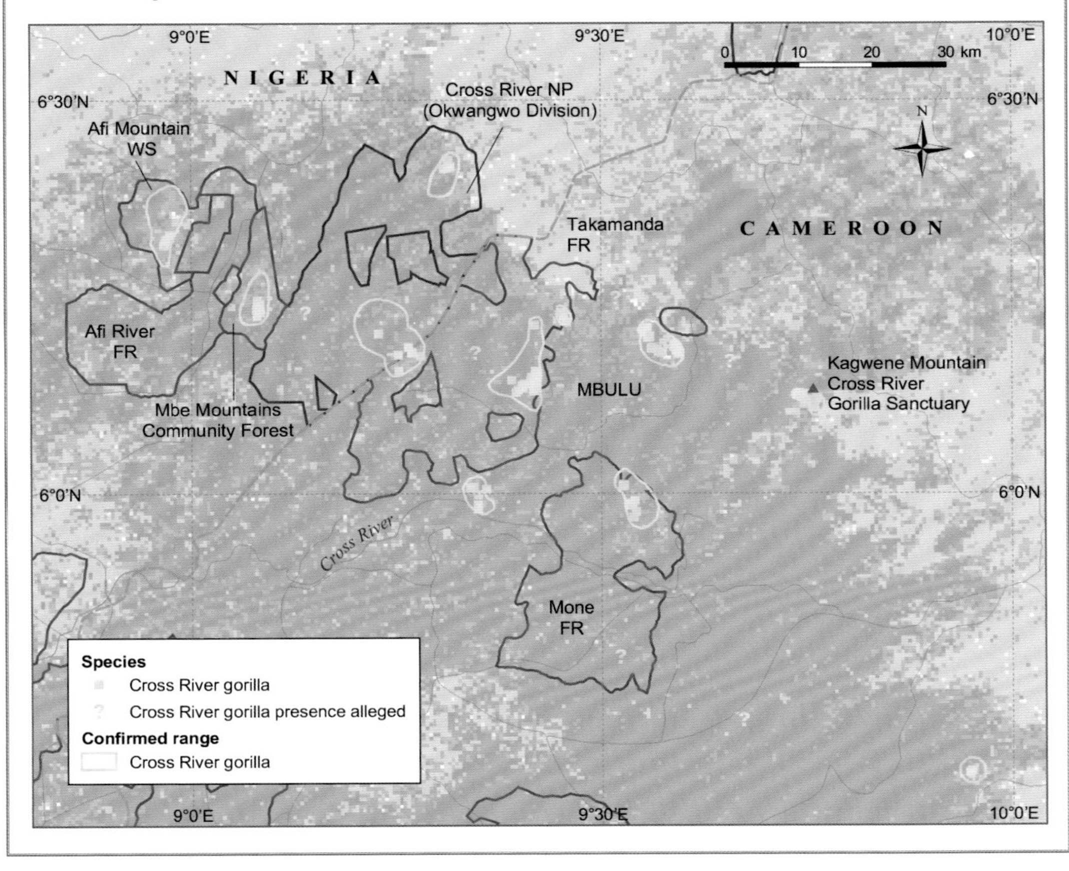

cotyledonous Marantaceae family, for example, can provide a dependable supply of food all year round. Marantaceae genera that are frequently eaten include *Megaphrynium* and *Haumania*. If available within a group's range, western lowland gorillas often feed on aquatic and semiaquatic sedges (Cyperaceae) and herbs such as *Marantochloa cordifolia*, *M. purpurea*, and *Halopegia azurea* (all Marantaceae), visiting streams, *bais*, and riverine swamps to do so.[37, 91, 145] At Mbeli Bai in Nouabalé-Ndoki National Park (Congo), preferred food plants include species of *Hydrocharis* and the sedges *Rhynchospora* and *Fimbristylis* (both Cyperaceae).[91] In the swamp forests of the Likouala region of northern Congo, the fronds of *Raphia* palm are consumed, along with species of *Pandanus* (Pandanaceae) and *Aframomum*.[12]

Outside fruiting seasons, more fibrous vegetative matter is eaten, including shoots, young leaves, and bark.[85] In the absence of preferred foods, western gorillas eat leaves, bark, low-quality herbs such as *Palisota* (Commelinaceae) and *Aframomum*,

Geographical distribution

Cross River gorillas are the most northern and western of all gorilla populations. Separated from the nearest gorilla population to the south by approximately 200 km, they occur in the Mbe Mountains Community Forest, the Afi Mountain Wildlife Sanctuary, and the Okwangwo Division of the Cross River National Park, all in Nigeria; and the Takamanda Forest Reserve (contiguous with Okwangwo), the Mone Forest Reserve, and the Mbulu Hills Community Forest, all in southwest Cameroon[89] (see map).

Taxonomic history and status

In 1904, the gorillas of this region were described as a new species: *Gorilla diehli* Matschie, 1904. Subsequent revisions of gorilla systematics amended their status to that of the subspecies *Gorilla gorilla diehli* Rothschild, 1904, 1908. In 1929, they were amalgamated with all other western gorillas in the subspecies *Gorilla gorilla gorilla* Coolidge, 1929. In the last years of the 20th century, craniometric (skull measurement) studies found that Cross River (or 'Nigerian') gorillas differed significantly from other western gorillas;[114, 123, 124] their subspecies status was then reinstated by Groves in his review of primate taxonomy.[56]

Despite new conservation efforts in recent years, the habitat of the Cross River gorilla is still being lost, and the animals are still being hunted at a low level. Given their small and fragmented population and the continuing threats to their survival, they have been listed as Critically Endangered in the *Red List* of IUCN–The World Conservation Union.[68]

Population

From the early 1930s to the late 1960s there were scattered reports on the distribution and abundance of Cross River gorillas.[2, 3, 32, 77, 122] The 1966–1970 Nigerian civil war and lack of information meant that, by the late 1970s, it was assumed that gorillas were probably extinct in Nigeria.[30] During the 1980s, however, reports appeared on the persistence of gorillas there;[40, 63] this led to renewed surveys both in Nigeria[62, 90] and in Cameroon.[41] In 1968, Critchley had estimated that 25–50 gorillas remained in Takamanda, Cameroon.[32, 62] In 1989, Harcourt and coworkers estimated that a further 100–300 gorillas remained in Nigeria. In 1996, the first long-term study on Cross River gorillas began in Nigeria, followed in late 1997 by a study in Cameroon. These studies are ongoing, and have been accompanied by extended survey efforts. The most recent estimate of the Cross River gorilla population, published in 2003, is that there are between 205 and 250 weaned individuals, with 70–90 individuals in Nigeria and about 150 in Cameroon.[89] Surveys in progress suggest that a few previously unknown subpopulations remain to be confirmed, producing a tentative total population estimate of 250–280.

Ecology

Cross River gorillas inhabit lowland semi-deciduous and evergreen submontane forests from about 200 m above sea level to at least 1 500 m. Although people have lived in and around this forest area for very many generations, there remain large tracts of primary forest throughout the habitat of the Cross River gorilla, particularly within Cameroon. Most subpopulations of Cross River gorilla exist in ridge forests above 400 m; these are typically more difficult for hunters to access due to the steep terrain. At the highest altitudes across the Cross River gorilla range, farming, burning, and cattle grazing have produced a mosaic of forest and grassland; here, especially on the edge of the Bamenda Highlands in Cameroon, the gorilla

and less-favored fruit such as *Duboscia* (Tiliaceae) and *Klainedoxa* (Simaroubaceae).[9, 34]

Western lowland gorillas have also been seen to eat at least 20 species of invertebrate.[33] In one study in Lopé National Park in Gabon, insect remains were found in about one third of feces.[132] Most of the insects eaten are termites and ants;[33] one study found the remnants of the termite *Cubitermes sulcifrons* in 30.5 percent of western gorilla feces. Three species of ant (including the weaver ant, *Oecophylla longinoda*) and, more rarely, caterpillars, grubs, and larvae from dead wood, are also known to be eaten.[131, 132]

It is possible that western gorillas have a food culture, with learned preferences passed on between individuals and generations.[85] For example, the species of insect eaten appears to vary culturally. Weaver ants (*O. longinoda*) are taken in large quantities at a study site in Lopé National Park in Gabon, but in the Belinga study site 250 km away,

Kelley McFarland

Kelley McFarland

The Cross River gorilla is extremely difficult to catch on camera.

habitat consists of relatively small isolated patches of forest, sometimes tenuously connected by gallery forest.

The only long-term field study yet completed on Cross River gorillas, by McFarland at Afi Mountain Wildlife Sanctuary,[89] suggests that, like other western gorillas, *G. g. diehli* preferentially feeds on fruit when available. But the Cross River gorilla habitat is notable for strong seasonality, with a dry season that lasts four to five months. From the late wet season through early dry season at Afi (August–January), fruit is scarce and the gorilla diet is dominated by the pith of terrestrial herbs, bark, and leaves.

Cross River gorillas are rare and shy as a result of hunting, so most information on their ecology and behavior has come from an examination of nest sites, feces, and feeding trails. Nest clusters suggest that group size is typically small (fewer than six weaned individuals), although occasional much larger groups do occur. Nesting patterns at Afi Mountain Wildlife Sanctuary suggest that a large group of about 20 individuals will sometimes divide into smaller foraging parties.[89]

Conservation prospects

The survival prospects for Cross River gorillas are promising if they and their habitat can be adequately protected. Large areas of unoccupied potential gorilla habitat remain; these provide connectivity between many of the subpopulations. To maintain these forest corridors, new protected areas need to be created and local law enforcement needs to be strengthened. Along with other conservation organizations, the Wildlife Conservation Society is working on both sides of the Nigerian–Cameroon border, in collaboration with state and national government agencies, to improve conservation of the gorillas.

Jacqueline L. Sunderland-Groves, John F. Oates, and Richard Bergl

a termite species (*C. sulcifrons*) was eaten exclusively instead, although weaver ants occur at both sites.[132] Similarly, some plant species are eaten by some populations and not by others, despite being available in the habitat of both.

Day ranges

Among primates there is a strong relationship between diet and foraging behavior. Species or populations feeding on high-energy foods that vary spatially and seasonally tend to have greater day ranges (average distance traveled by a group per day) than those feeding on lower-quality but more consistently available foods.[70] Western gorillas fit this pattern, as they travel farther when more fruit and termites are available in the forest, and have shorter day ranges when they must rely on leaves and woody vegetation.[51] They also fit this pattern in comparison with eastern gorillas, as their day range is much greater than that of either mountain

Iroko Foundation

The buttress of a large tree (*Parkia bicolor*) growing in the Oban Division of Cross River National Park.

gorillas amid abundant herbaceous foods or of eastern lowland gorillas living in places with poor fruit availability; there is no such difference between the day ranges of eastern and western gorillas when both live in areas rich in fruit.[51, 102]

At Lopé National Park, Gabon, the western lowland gorillas travel about 1 km each day.[126] At Bai Hoköu in CAR, the distance traveled varies between about 3 km/day during frugivorous months, and 2 km/day during folivorous months, with an overall mean of 2.3–2.6 km/day.[51, 102] Larger groups travel greater distances in order to obtain sufficient food.[102] Human presence, and the degree to which the western gorillas are accustomed to it, also affects daily travel distance.[28] Both human hunters and leopards (*Panthera pardus*) have been observed to influence the movement patterns of some western gorilla groups. Very long or very short distances are traveled when predators are in the vicinity, presumably as a result of efforts to avoid detection or to evade them entirely.[51]

Home ranges

The annual home ranges (the area used by a group over a year) of western gorillas are larger than those of mountain and eastern lowland gorillas;[102] the home ranges of different groups overlap quite extensively.[126] A study at Bai Hoköu, CAR, over a two year period found that western gorilla ranges varied in size between years, with the estimated minimum home range being 22.9 km^2.[102] A western gorilla group observed at Lopé National Park, Gabon, visited 21.7 km^2 over 10 years, but probably did not cover the entire area each year.[126] The home ranges of habituated western gorillas are less affected by human presence, so caution is needed when comparing data from different studies.[28] Temporary shifts in both the size and position of the home range also occur, probably in response to the availability of seasonal fruit,[126] and ranges may be much smaller during the dry season than during the wet season, when the gorillas feed on dispersed but nutritious fruits.[102]

Ecological role

Western gorillas are large, dexterous animals that affect the structure and composition of vegetation by feeding on plants and building nests with them. Trees can be badly damaged by gorillas climbing and feeding in them.[103] Like other frugivores, they both consume and disperse seeds, reducing the reproductive success of some plant species, and increasing that of others.[27, 74] Western gorillas are reliable visitors to certain trees,[139] and swallow most fruits with their seeds intact, leaving only those seeds that are too hard or too large.[85] They are effective seed dispersers, with up to 99 percent of all fecal samples containing seeds; samples typically contain many seeds, most of them intact.[98, 103] Seeds deposited at nest sites have higher germination and survival rates than those deposited elsewhere, such as on paths.[140] Western gorillas appear to be particularly important dispersers of some species, for example the endemic *Cola lizae* (Sterculiaceae) of Gabon, the fruits of which are consumed avidly and have a high germination rate after deposition.[138] Over time, by these various means, western gorillas might have a significant influence on the forest.

A number of other species use some of the same food resources as the western gorilla and share its habitat, so competition between them may occur. The relationship between western gorillas and chimpanzees is particularly interesting; although they have overlapping diets and are sympatric in CAR, Congo, Equatorial Guinea, and Gabon,[130] aggressive interactions between them have never been observed.[52] When this issue was studied in the Belinga area of Gabon, and in the Nouabalé-Ndoki National Park in Congo, 60 percent and 42 percent

Box 7.2 FOREST CLEARINGS: A WINDOW INTO THE WORLD OF GORILLAS

It is difficult to observe western lowland gorillas in their rain-forest habitat, as little light filters through the canopy, and the understory is often choked with dense vegetation. Meeting gorillas under such circumstances can be alarming for humans and gorillas alike; contacts often induce aggression or flight, so collecting unbiased behavioral data is almost impossible. Yet, until the early 1990s, our knowledge of western gorilla social behavior was based almost entirely on such observations.

Then primatologists discovered marshy clearings in the forest, which local people called *bais*; these appeared to attract western lowland gorillas into the open, and offered a fresh view of their world. Several studies of *bais* were subsequently undertaken in Congo, CAR, and Gabon. Together these have contributed greatly to our understanding of the lives of western lowland gorillas.

The classic gorilla *bai* is a treeless clearing situated around a watercourse. The size may vary from 0.04 km^2 (Iboundji, Congo) to about 0.18 km^2 (Maya Nord, Congo); most are either roughly circular or linear in form. The substrate is often extremely swampy, with water-saturated soils loosely held together by a mat of floating vegetation. Small streams and pools are often present, although the area of relatively dry ground varies considerably from *bai* to *bai*. Western lowland gorillas share these areas with many other species. Residents may include Congo swamp otters (*Aonyx congica*), spot-necked otters (*Lutra maculicollis*), sitatunga antelopes (*Tragelaphus spekei*), fish eagles (*Haliaeetus vocifer*), hammerkops (*Scopus umbretta*), and lily trotters (*Actophilornis africanus*). *Bai* visitors may include forest elephants (*Loxodonta cyclotis*), forest buffalo (*Syncerus caffer nanus*), red river hogs (*Potamochoerus porcus*), giant forest hogs (*Hylochoerus meinertzhageni*), and guereza colobus monkeys (*Colobus guereza*).

The primary activity of western gorillas while at *bais* is feeding. This generally begins as soon as an individual or group enters the *bai*, and ends shortly before they leave. It is easy to imagine, therefore, that food availability or quality is the primary reason for western gorillas to visit *bais*. Aquatic and semiaquatic vegetation is usually abundant, and dominated by plants from the Hydrocharitaceae, Cyperaceae, and Gramineae families. Many *bai* plants are highly digestible, high in protein, and contain elevated levels of salt and other minerals (attributes that make them attractive to western gorillas). Despite these qualities, the high water content and low dry weight of many *bai* plants is likely to limit the quantity that can be ingested at a single sitting. At Mbeli Bai in Nouabalé-Ndoki National Park (Congo), the average length of visits by groups was only two hours.

continued overleaf

An openly exposed family group of western lowland gorillas and a silverback male.

Richard Parnell Richard Parnell

Much beyond this, and the gorillas may simply be too 'full' to continue.

When some species of *bai* plant (such as the aquatic *Hydrocharis chevalieri*, of the Hydrocharitaceae) are picked, they are often lightly coated with sediment. Western lowland gorillas wash off this sediment fastidiously by waving each handful of plants back and forth in the water before eating it. To reach this succulent food source, western gorillas will walk through swamps and wade bipedally across streams that are more than 1 m deep, using their outstretched arms for balance. Until quite recently, it was thought that gorillas made a point of avoiding contact with water, but western lowland gorillas seem at home in a semiaquatic environment. When feeding at a *bai*, a western gorilla may sit in water up to its chest for more than two hours at a time. Even infants, although clinging to their mothers' backs during stream crossings, will happily paddle and play in shallow water while their mothers feed. Not only do western gorillas tolerate water while feeding but, in some places use water as a communicative 'tool' via dramatic splash displays'. During aggressive or playful displays, western gorillas may leap bodily into streams, generating spray and a large bow wave, or may slap the water's surface with one or both hands. At Mbeli Bai, all ages of western gorillas, except adult females, have been seen to use splash displays; this has prompted the suggestion that the displays are used primarily by males to intimidate rivals.[94]

The visitation rates of western gorillas to *bais* vary widely. There does not appear to be any 'ownership' of a specific *bai* by a particular group of gorillas; groups or solitary males with home ranges close to a *bai* are likely to be more frequent visitors, but no group has exclusive access. Visit frequency is therefore likely to be a function of the proximity of a *bai* to the center of a group's home range. Some groups may visit a particular *bai* several times a week, while others may pass through that *bai* only twice a year.

Bais have finally permitted primatologists to study aspects of western gorilla life that would otherwise probably have remained undocumented for much longer.[93] *Bais* can attract many different groups of western gorillas. For instance, at Iboundji Bai, researchers have identified 47 different groups and 25 solitary males. Sample sizes such as these have permitted a more thorough study of the dynamics of western gorilla social life, including the formation, evolution, and fission of groups, as well as how group size influences the fate of individual gorillas. We are now able to study social dynamics within a group, exploring whether there is a dominance hierarchy between females, whether young males are pushed out of their groups or leave voluntarily, and how a silverback retains his harem of females.

Bais also allow us to examine intergroup behavior. One surprise has been the degree of tolerance shown by groups toward each other. At Mbeli Bai, only 30 percent of shared uses of *bais* by different groups led to an aggressive response; this almost always came from the silverbacks. In 58 percent of cases, groups appeared to ignore each other completely. Where aggression occurred, actual physical contact was never witnessed; only display behavior was observed. By contrast, mountain gorillas have been seen to engage in contact aggression in 17 percent of group encounters; their groups were six times less likely to engage in peaceful mingling than those at Mbeli.[116] It appears that western gorilla groups are generally more relaxed when close together in *bais* than in the forest. The unparalleled visibility offered by the *bai* habitat permits a silverback to survey his group's females closely while monitoring the behavior of nearby silverbacks. The resulting reduction in tension permits other group members to interact in ways that may be unthinkable in the forest. Young males practice displaying in front of females from other groups without risk of attack, and juveniles from different groups wrestle and play with each other, potentially making contacts that will last them into adult life, and shape their interactions in the future.

Study of western gorillas in *bais* is in its early stages, and we still have a great deal to learn. *Bais* themselves, however, may provoke or permit behaviors that are rare or absent in the more usual forest habitat of the western gorilla. At Mbeli Bai, it is estimated that on average, western gorillas spend only 1 percent of their daylight hours in the clearing. *Bais* offer us perhaps the best opportunity we have to view the intricacies of western gorilla social life; however we must also be cautious in extrapolating from what is learned to behavior in other habitats.

Richard Parnell

respectively of all foods recorded for western gorillas were also eaten by chimpanzees.[73, 131] Many of the shared foods are seasonal fruits, which are relatively plentiful at the time of occurrence. It is possible that competition between western gorillas and chimpanzees is also limited by other forms of niche differentiation and by the different strategies applied by each species when there is a shortage of food[133] (see Box 8.1).

Western gorillas have few natural predators, as their juveniles are well protected and adults are large and strong. Leopards have on several occasions been suspected of having killed gorillas;[45, 108] western gorillas have been observed fleeing leopards,[45] and gorilla 'fear odor' (the scent of scared gorillas) has been detected in the air after encounters between western gorillas and leopards.[45, 142] It seems likely, therefore, that western gorillas perceive leopards as dangerous, and that this predator may represent a real threat to them.

Social behavior

Groups of reproductively active western gorillas almost always contain only one dominant, silverback adult, plus three or four females, and four or five offspring.[61, 92] Groups that contain more than one silverback have only very occasionally been reported among western lowland gorillas,[52, 76, 92, 102, 108] and all-male groups have been reported at only some study sites.[48, 50, 76, 92, 108]

Western lowland gorillas generally form stable cohesive groups.[76, 126] The takeover of a group in which its silverback is ousted by another from outside the group has never been reported, and group fission has been reported on only one occasion.[108] However, western gorillas do not appear to be as cohesive on a daily basis as their eastern counterparts.[36, 91] In some groups, members spread out with distances of over 500 m between individuals; other groups split up during the day and then reunite at the nest site.[8, 126] Where there is more than one silverback, these subgroups sometimes remain apart overnight.[36, 92, 102]

In general, group size among western lowland gorillas is similar not only to that of both eastern subspecies, but is also similar across the range of habitats occupied.[52, 92, 108, 150] The very large groups sometimes observed among eastern gorillas, however, rarely occur in western gorillas.[8, 92, 126] Group size in western gorillas appears to be influenced by the size of the foraging patch, fruit abundance,[102] and the degree of competition for food that occurs within the group.[126] Western gorillas eat considerably more fruit than eastern gorillas; this dependence on clumped food resources may constrain their group size. Pressure from predators

Ian Redmond/UNESCO

Young gorillas interact through play.

Box 7.3 POTENTIAL MEDICINAL VALUE OF GORILLA FOODS

A growing body of literature suggests that primates use certain plants for the control of illness and the regulation of reproduction.[65] In Africa, chimpanzees and humans show strong similarities in the plants they use to treat similar symptoms.[66] Few equivalent data are yet available for western or eastern gorillas. An ethnographic and pharmacological study of the properties of plants ingested by western and eastern gorillas evaluated their potential medicinal value.[31] A list was compiled of 118 plant species from 59 families, these plants being known both to be used by humans in traditional African medicine[69] and to be eaten by western and eastern gorillas in the wild.[31] Major reported pharmacological properties in humans included antiparasitic, antifungal, antibacterial, antiviral, cardiotonic (heart tonic), hallucinogenic, and stimulatory, or a utility in treating respiratory ailments.

Western gorillas ingest the fruit and seeds of a number of *Cola* species (Sterculiaceae), notably those of *C. pachycarpa* which is known locally in Gabon as 'cola of the gorillas'. The same seeds are widely chewed by people in West and Central Africa, for example by long-distance truck drivers in order to stay awake, and are used commercially in cola drinks. They contain caffeine (2–2.5 percent dry weight) and theobromine, but only small amounts of protein, suggesting that western gorillas eat them mainly for the stimulating effect of caffeine.

In Gabon, western gorillas have been reported to eat the fruit, stem, and root of *Tabernanthe iboga* (Apocynaceae). The iboga shrub is perhaps the best documented hallucinogenic plant in Africa.[107] The active principle in *T. iboga* is ibogaine, with the highest concentration being found in the roots. Ibogaine affects the central nervous system as a hallucinogenic, as well has having a stamina-boosting effect on the cardiovascular system similar to that induced by caffeine. *T. iboga* is an integral component in certain African religious cults and rites. The inhabitants of the Petit Loango region are among those to use *T. iboga* for this purpose. According to Bwiti (the religion practiced by the Mitsogos and the Fangs, two ethnic groups in Gabon) legend, the forest-dwelling 'pygmy' peoples discovered iboga. If correct, this may have been a result of their having watched wild animals ingest it. Information from more than one Gabonese source describes boars, porcupines, and western gorillas digging up and eating the roots, and afterwards going into a wild frenzy, jumping around, and fleeing as though seeing frightening images.[97, 100] Two other hallucinogenic plants eaten by western gorillas and chimpanzees in Equatorial Guinea are *Alchornea floribunda* and *A. cordifolia* (Euphorbiaceae). Both species contain the alkaloids alchornine and alchornidine,[112] and local people use the pith and leaves as an antiseptic and cough remedy.[125]

The Virunga mountain gorillas periodically visit upper mountain slopes characterized by giant *Senecio* (Compositae) plants. In 1963, George Schaller followed a group to an altitude of 4 100 m on Mount Mikeno, DRC.[115] Here the animals fed infrequently on *Senecio alticola* and *S. erici-rosenii*, preferring to eat the pith. *Senecio* species are important in ethnomedicine in the treatment of pulmonary complaints and head colds. Gorillas also occasionally feed upon the giant *Lobelia* plants that grow at these high altitudes, such as *L. giberroa* and *L. wollastonii* (Campanulaceae). All members of this genus contain bitter-tasting alkaloids that have stimulating effects lasting up to a quarter of an hour. One of these alkaloids, lobeline, is also a respiratory stimulant. The willingness of these eastern gorillas to travel to higher altitudes, expending considerable energy in the process, to reach a zone with fewer foods, less oxygen, and colder temperatures, suggests that the pharmacologically active plants that they consume once reaching this destination are of some special value to them.

More detailed field studies are required to allow a critical evaluation of the possibility that gorillas are self-medicating, and to explore the significance of the consumption of these plants in their daily lives. Observations already recorded provide an important starting point, paving the way to a greater understanding of the dietary needs and adaptations of gorillas in their various habitats. This field of research may help us to assess habitat quality and population viability, as well as providing new insights into the control or cure of diseases that afflict both apes and humans.

Michael A. Huffman and Don Cousins

(including poachers) is likely to have the opposite effect, because larger groups are likely to be able to be more alert and therefore have a better chance of self defense or escape.[1]

Total group size ranges from two to 32 individuals, with an average four to six adults, although up to 52 nests have been recorded as belonging to a single group[8, 26, 92] (see the table in Box 7.4). Larger groups typically contain a higher proportion of adult females,[92] as most groups are single-male harems. In the Maya Nord Bai (Odzala-Koukoua National Park), the adult sex ratio was sometimes as high as nine females per adult male.[76] It is assumed that a female will choose whether or not to stay with a particular male on the basis of her assessment of his fitness as a breeding partner and his ability to protect her offspring from predation or infanticide. The number of adult females in a group would therefore indicate female perception of the defensive 'quality' of the group's male.[119]

Upon reaching maturity, some western lowland gorillas remain in the group in which they were born (their natal group), but most emigrate from it.[37] Some females have been observed to transfer between groups up to three times.[108, 121] Emigrating females tend to transfer into another group or join a solitary male.[121] Male quality is likely to be an important factor in influencing female dispersal patterns.[117, 121] In smaller groups, the foraging costs associated with within-group competition may be lower.[121] No correlation has been found between group size or the number of females in a group and the reproductive success of western gorillas, although these results should be treated with caution as the sample sizes for comparison are small.[121]

Virtually all males leave the natal group in which they were born (e.g. seven of eight males observed at Mbeli Bai in Congo),[108] and generally remain solitary until they can establish their own group.[126] Sometimes immature males transfer between mixed sex groups;[108] occasionally, all-male groups form.

The mutual dependence between silverbacks and females, based on infant protection, is important in maintaining the integrity of both eastern and western gorilla groups.[118] When the silverback in a single-male group dies, the group typically disintegrates and the remaining group members transfer to other groups. Transfer of all group members as a unit sometimes occurs; in these circumstances, the new silverback sometimes kills infants.[108] Infanticide has been inferred on two

William H. Calvin (www.williamcalvin.com)

The chest-thumping display of a western lowland gorilla (here in captivity).

occasions when mother and infant transferred to a new group, but did not occur on two other occasions.[121] Silverbacks gain reproductive benefits from infanticide, which affords a more immediate opportunity to impregnate the mother, and eliminates a potential competitor to his own offspring.

Grooming and other forms of social support are rarely observed between adult females, and dominance hierarchies also seem to be weak, with female–female relationships being individualistic and ephemeral.[118, 121] Social bonds between western gorillas are not thought typically to be strong, except between mothers and their offspring. Studies in Mbeli Bai show that offspring suckle for a longer period than mountain gorillas and remain in close proximity to their mothers until weaned.[87, 88] Infants seem to use this time to learn about appropriate foods and to develop the food-processing techniques that are important for an animal that exploits such a diverse range of habitats.

Young western lowland gorillas appear to invest less time than Virunga mountain gorillas in developing relationships with other group members, perhaps because they are more likely to leave their natal group. Young gorillas spend more time developing skills that can be used later in life through play, interacting with younger gorillas, and by learning how to respond to negative behavior from other group members, albeit without forming

Preliminary studies suggest that, where an infant survives, the interbirth interval is four to six years.

Richard Parnell

long-term cooperative relationships. An alternative explanation is that western gorillas have less access to high-quality food than eastern gorillas have, and therefore may have less time to devote to developing relationships.

Intergroup encounters

Western gorilla ranges often overlap extensively, so different groups meet often, and more frequently than do mountain gorillas.[20] Several groups often make use of *bais* at the same time.[37] Serious aggression between the members of different groups is rarely seen, however; at Lopé National Park, during 43 intergroup interactions observed from 1984 to 1993, fighting occurred on only three occasions.[126] Some encounters involve exchanges of vocalizations and chestbeats but, in others, only one of the groups vocalizes. As vocalizations accompany only half of observed encounters, many encounters may go undetected. Many exchanges lead to one group remaining in the area and the other group moving away. Most intergroup encounters were related to access to or defense of either large fruiting trees or groves of trees of uncommon species. Where a resource in the core of one group's range had attracted other gorillas, the silverback of the 'resident' group would display to deter feeding competition.

Nest building

Western lowland gorillas are active only during the day. They usually rest from dusk to dawn in nests constructed each evening by pulling, bending, and breaking the stems of vegetation; these are arranged around and under their bodies.[47] Tree nests may be built by folding branches toward the crown of the tree, producing a bed of leaves at the center.[44] Nests may also be at ground level or low in the trees.[48, 130] Western lowland gorillas sometimes sleep on bare ground, without using vegetation at all; such sleeping sites represented 44 percent of 1 231 sites observed at Bai Hoköu in CAR,[102] and 45 percent of 3 725 at Mondika (Dzanga-Ndoki National Park). Gorillas are more likely to build nests in cool or wet weather.[21, 79] The type of nest constructed is influenced by the availability of suitable nesting materials, season, group size, microhabitat, and the level of disturbance by other animals.[21, 101] Some western lowland gorillas select nesting sites in particularly dense vegetation, probably in response to hunting pressure.[39]

The average nesting height varies with the availability of herbaceous undergrowth. Over four fifths of nests are found on the ground at Odzala-Koukoua National Park in Congo, where suitable plants like *Haumania liebrechtsiana* (Marantaceae) are common; a similar proportion is found in the trees in the Ngotto forest in CAR, where there is little herbaceous undergrowth.[9, 16, 21] Western gorillas do not sleep in water or on wet ground, so nests in swamp forests are built quite high up in the trees; at Likouala in the northern Congo, for example, *Raphia* (Arecaceae) fronds are a favored building material.[12]

Solitary nests can be used as an indicator of population health; a disproportionate number of solitary nests is a cause for concern. On average, about 30 percent of nests are solitary,[67] but the percentage rose to 60 percent in Minkébé National Park (Gabon) following the collapse of the local western gorilla population due to an outbreak of Ebola hemorrhagic fever, discussed later in this chapter.

Response to disturbance

Western gorillas respond in various ways to human disturbance. Contact with people prior to full habituation has been reported to produce an increase in aggression, fearful reactions, and vocalizations, as well as longer daily travel distances, but these reactions diminish as habituation proceeds.[15, 28] Western gorillas leave areas to avoid active logging; they return swiftly unless, as so often occurs, they are also hunted.[128]

Reproduction

Reproductive development has not been studied over long periods in the western gorilla, and the age at first birth is unknown. Preliminary data from two sites in Congo, Lossi Gorilla Sanctuary (Odzala-Koukoua National Park) and Mbeli (Nouabalé-Ndoki National Park), indicate that there are almost 0.2 births per adult female per year, which is similar to the birth rate of mountain gorillas.[108] It has been hypothesized that increased seasonality and spatial variation in food availability could have resulted in the western gorilla having a later age at first birth and longer interbirth interval than the mountain gorilla. The same initial reports suggested that when the infant survives, the interbirth interval is four to six years. The mean number of infants per female at the Maya Nord Bai was reported to be 0.62, more than has been seen in any other western or eastern gorilla group.[76]

Using a very small sample size (12 births involving 68 females at Lossi Gorilla Sanctuary, and 32 births involving 162 females at Mbeli Bai), mortality of infants under one year old was observed to be about 8 percent at Lossi and 43 percent at Mbeli. The death rate increased to 22 percent and 65 percent respectively, when considering infants up to three years of age.[108]

Communication

All gorillas communicate with grunts, barks, screams, hoots, and facial expressions, but the vocal repertoire of western lowland gorillas has not been fully investigated.[82] Gorillas running towards fruit trees often call out excitedly, possibly in anticipation of the imminent scramble competition for the limited number of feeding spots in fruit trees.[126] Hoots are heard more frequently and may be straightforward contact calls to communicate location in the forest.[120] No empirical studies have been conducted to test this hypothesis. Western gorillas have a larger day range and their foraging groups are more widely spread than those of eastern gorillas; this may mean that the vocal mechanisms for maintaining mutual contact are more important.

POPULATION

Status and trends

Most population surveys of western gorillas are carried out on a site or country basis rather than with reference to contiguous populations. It is difficult to assess population status and trends among western lowland gorillas because censuses have not yet been conducted across large areas, and areas of swamp habitat have only recently been identified. Distribution maps often indicate likely rather than confirmed presence. Large numbers of western lowland gorillas may remain in the Congo Basin,[128] and in many areas numbers may be higher than previously thought.[18] The presence of western gorillas in swamp forests, a widespread habitat that was previously considered unsuitable for them, was reported only as recently as 1989[42] and confirmed to be a general pattern in the 1990s.[12, 91] On the other hand, the impacts of Ebola and hunting have not been comprehensively quantified, leading many researchers to caution against overoptimism.[141] The western lowland gorilla is classified as Endangered, indicating that it faces a very high risk of extinction in the wild.[23]

Little information is available on the numbers, status, or trends for populations of western lowland gorillas at most sites in most countries. In the early 1980s, it was thought that there were only 40 000 western lowland gorillas worldwide.[59] More recent estimates have ranged from 94 500[22] to over 100 000 individuals;[60, 91, 96] several reports have, however, indicated that numbers are in decline. An informal and optimistic estimate based on the country profiles in this volume suggests that at most 82 000 remain. This figure is based on mean reported country figures, adjusted to reflect an estimated

Charlie Semeli-Botarba/UNEP/Topham

Making charcoal, Congo.

Table 7.2 Western lowland gorilla populations

Country	Date of estimate	Estimated population size
Gabon	1980–1983[130]	35 000 ± 7 000[a]
Congo	1990[22, 43, 60]	34 000–44 000
Cameroon	2000[22]	15 000
Central African Republic	1985[60]	9 000
Equatorial Guinea	1989–1990[53]	950–2 450
Democratic Republic of the Congo	2000[22]	0
Angola (Cabinda province)	2000[22]	unknown

a From 1983 to 2000 half of all great apes in Gabon were feared to have been lost.[141]

loss of half of all Gabon's great apes from 1983 to 2000.[141] Published estimates are summarized in Table 7.2; many western gorilla populations have, however, since suffered declines.

On a national level, the largest populations of western gorillas are thought to be found in Gabon and Congo,[13, 43, 60] although recent losses are unquantified. Considering data on both western lowland gorilla and chimpanzee populations together, Gabonese ape populations are thought to have declined by more than half between 1983 and 2000.[141] A census in 1989–1990 concluded that western gorillas were widespread and common in northern Congo. Areas supporting high densities included swamp forests,[9, 43] such as the forests of Odzala-Koukoua National Park and adjacent regions to the north and east.[9, 76] However, an outbreak of Ebola during 2002 and 2003 seriously affected western gorilla populations at the Lossi Gorilla Sanctuary, a community forest 50 km southwest of Odzala-Koukoua National Park;[141, 143] in 2004, an 80 percent decline in sightings of western gorillas in the park's Lokoue Bai led to fears that Ebola had begun to kill western lowland gorillas in the Odzala-Koukoua National Park itself.[54] Recent surveys have confirmed still-healthy populations of western gorillas and chimpanzees in the Lac Télé/Likouala-aux-Herbes Community Reserve in eastern Congo.[99]

There is much uncertainty about western lowland gorilla populations at most locations in Cameroon.[80] In CAR in 1996, western lowland gorillas were assessed as Vulnerable rather than Endangered at a national level.[60]

In Equatorial Guinea during 1989–1990, a census estimated between 950 and 2 450 western lowland gorillas to be present,[53] mainly in areas of plantation and secondary forest.[49] Western lowland gorillas living outside the reserves of Equatorial Guinea have been considered Critically Endangered due to threats from hunting, forest clearance for logging, and/or agriculture and the oil industry.[60] The western lowland gorilla is probably now absent from its former range in the Bas-Fleuve region in the extreme southwest of DRC, north of the Congo River.[22] Its decline and likely local extinction is probably the result of the combined effects of habitat loss, fragmentation, and hunting.[119] There is no estimate for western lowland gorilla numbers in the Cabinda province of Angola.

Threats

Western lowland gorillas are widely distributed in a large forested region and their range includes numerous protected areas. They nevertheless face an uncertain future, simply because of the increasing scale and interactive nature of the threats operating against them.[23] These include forest clearance for farming, forest fragmentation by clearance and road-building, forest degradation by logging, hunting for food, and disease. Hunting and disease are increasing as risk factors because human access to formerly remote forest area is being improved by logging, road building, and settlement. The estimated halving of the great ape population in Gabon illustrates the danger of this combination of factors[141] (see Gabon country profile, Chapter 16).

Forest clearance, fragmentation, degradation

Until recently, there has been relatively little habitat degradation over much of the Congo Basin, with little conversion to agricultural land.[128] As late as the 1980s, West and Central African timber was considered to be of low commercial value, which limited the pressure posed by selective logging;[130] this changed dramatically during the 1990s. Forest products now account for more than 10 percent of all trade recorded in Cameroon, CAR, Congo, Côte d'Ivoire, Equatorial Guinea, Gabon, and Liberia.[84]

Most of the forests in the Congo Basin are under the control of companies that are based in the European Union, operating either as concession holders or as subcontractors.[46] Many conservationists therefore consider the fate of the forest and its wildlife to be a responsibility shared by Europe. By 2000, more than half of Gabon's forests had been allocated as logging concessions,[24] and log production had increased to some 3 million m^3/year.[4] In Cameroon, over 170 000 km^2 (76 percent) of the country's forests had either been logged or allocated for logging concessions by then; satellite images have revealed that networks of new logging roads have now spread into what had previously been considered the least accessible forests in the country.[81] Other parts of the range of the western lowland gorilla to have undergone extensive logging include the mainland of Equatorial Guinea and the Congolese sections of the Mayombe forest.[53, 111, 129] Although logging occurs in some of the protected areas in Cameroon, CAR, Gabon, and Congo that are home to western lowland gorillas,[60, 80, 119] many others have escaped intact so far.

Logging roads and access routes fragment forest, as well as improving access for hunters. Forest fragmentation poses a potential threat to western lowland gorillas, in that it can block access to food sources and prevent transfer between groups. It is unclear what degree of fragmentation constitutes a barrier to western gorillas,[128] but in the continuous forest and savanna of the Lopé National Park in Gabon, western gorillas were reported to be reluctant to cross gaps in the forest that were wider than 50 m.[137]

Hunting

Western lowland gorillas are hunted for their meat, for sale to private collections (particularly as infants), for trophies, and for traditional medical or ritual purposes. Although this is illegal according to the national laws of every range state, the regulations are often poorly enforced at all levels of the legal and judicial system.[95] Hunting of great apes has been reported in the Cabinda province of Angola, Cameroon, CAR, Congo, DRC, Equatorial Guinea, Gabon, and Nigeria.[18, 19, 43, 80, 110, 130, 147]

Bushmeat hunting for subsistence is widespread in Africa, both for protein and for sale for income generation.[19] In the forested regions of Central Africa, hunting is the main threat to western gorillas, in the absence of any tradition of livestock rearing.[128] Increasing human populations and commercialization of markets have encouraged bushmeat hunting.[19]

Gorilla meat has been reported to be a popular food (where available) in northern Congo. Hunting of great apes for meat is widespread in Congo; it occurs, for example, in every part of the Motaba River area in the northeast.[72] Here, about 5 percent

Ian Redmond/UNESCO

Logging of the western lowland gorilla's habitat has increased dramatically across its range since the 1970s, including here in Congo.

of the western lowland gorilla population was estimated to be killed by hunters each year,[72] despite the sparse local human population. This level of hunting is unsustainable for slow-breeding mammals like gorillas.[96]

The intensity of hunting of western lowland gorillas varies throughout their range. Factors affecting the intensity of bushmeat hunting are local taboos, legislation (and its enforcement), the availability of ammunition and guns, and the ease of hunting under local seasonal climates.[18] Logging roads contribute both by promoting greater access to remote areas and by bringing a hungry workforce into the forests.[19] Western lowland gorilla populations have declined where timber extraction has occurred.[43] Civil wars in DRC and civil unrest in Congo and CAR have also increased hunting levels by exacerbating poverty and dependence on wild resources, particularly among displaced peoples and refugees.

There are no estimates of either the overall western gorilla population losses specifically due to hunting, or of their impact on population trends,[18, 19] although the negative impact of hunting on gorilla populations is well known.[72] As early as the 1980s, hunting was believed to be the primary threat to western lowland gorillas and chimpanzees in Gabon, as their population density was lower in areas where the animals were hunted, with density reductions of 17 percent and 72 percent observed in areas of light and heavy hunting respectively.[130] Furthermore, in many areas of West–Central Africa, gorillas are killed or injured by snares set for other species.[18]

Kelley McFarland

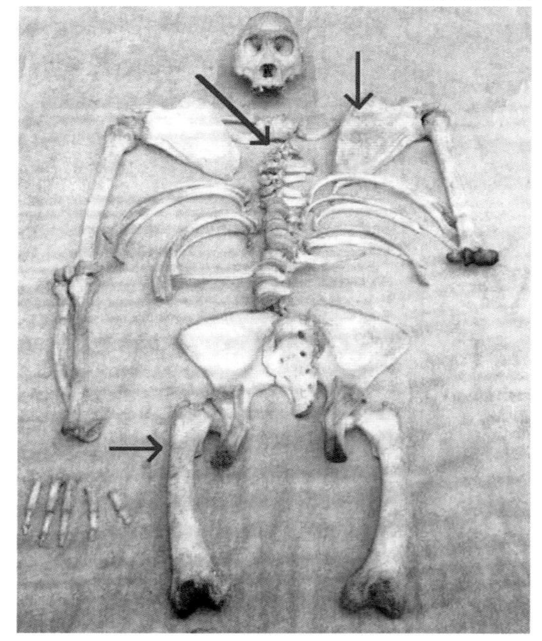

A female Cross River gorilla skeleton with gunshot hole and embedded lead shot (indicated by arrows).

Hunting and logging are considered to be threats over much of the western lowland gorilla habitat of Cameroon.[80] Hunting is a particular threat to the small population of western lowland gorillas to the north of the Sanaga River.[38] Although no logging has taken place within the Dja Faunal Reserve, hunting does occur, and timber extraction continues in the areas surrounding the reserve.[10]

In Equatorial Guinea, at least 63 live western lowland gorillas were removed from the wild between September 1966 and February 1969, many destined for zoological parks and research centers.[113] The capture of western lowland gorillas for sale and export has declined due to national and international conservation efforts, but live infants are still sometimes traded within the region.[128] In Angola, western lowland gorillas have been hunted in recent years for the bushmeat trade, and live infants sold in the capital Luanda and across the border.[110] This is largely a result of the conflict there, which has led to the immigration of people without taboos on eating apes moving into areas where these taboos have traditionally operated.

In contrast, effective controls on hunting are in place in Nouabalé-Ndoki National Park and its buffer zone (Congo), in Odzala-Koukoua National Park (Congo), and in the Dzanga Sector of the Dzanga-Ndoki National Park (CAR), where apes are rarely hunted.[16, 119] The controls are the result of successful collaborations between the national governments and outside agencies: the Wildlife Conservation Society in Nouabalé-Ndoki; the European Union program, Conservation and Rational Use of Forest Ecosystems in Central Africa (ECOFAC) in Odzala-Koukoua; and WWF–The Global Conservation Organization, along with the German overseas development agency GTZ, in Dzanga-Ndoki National Park.

Disease

Disease is a potentially devastating threat to great apes. Western gorillas are susceptible to many of the same diseases as humans, such as Ebola virus,[67, 128] the common cold, pneumonia, smallpox, chicken pox, bacterial meningitis, tuberculosis, measles, rubella, mumps, yellow fever, encephalomyocarditis, and paralytic poliomyelitis[64, 148] Of the identified western lowland gorillas in the population at Maya Nord Bai, for example, 5.7 percent were reported to be affected by yaws (frambesia tropica), while some others showed signs of the onset of this disease.[76] There are several varieties of yaws, but the pathogen in gorillas is probably *Treponema pertenue*.[78] The disease causes tissue necrosis; it is common in humans in Central Africa and can be treated with antibiotics.

It has been assumed that already small or fragmented populations are most vulnerable to disease,[64] but Ebola hemorrhagic fever has shown otherwise. Ebola is best known as an incurable human disease that kills about 80 percent of its victims. This virus has an even higher mortality rate of 95–99 percent among chimpanzees and western gorillas. Recent Ebola epidemics in West Africa have affected the western lowland gorilla in Gabon and Congo. Ebola outbreaks are thought to have contributed strongly to decline of great ape populations in Gabon, where four outbreaks are known to have occurred, two of which originated in the Minkébé National Park.[67] Farther east, declines in western gorilla populations attributed to Ebola have also been reported in the Lossi Gorilla Sanctuary of Congo, with fears that the disease may have spread into the Odzala-Koukoua National Park.[5, 54, 141] Many of the recent human outbreaks in western equatorial Africa appear to have been initiated when people handled the meat of infected great apes.

Habituation, the process whereby apes become tolerant of the presence of humans, allows regular and consistent observations by researchers and by tourists. The western lowland gorilla has proved difficult to habituate, particularly as the dense vegetation of its habitat does not allow it to be tracked easily.[144] Gorilla tourism is therefore not as well established as it is for eastern gorillas. The discovery that western lowland gorillas could easily be seen at *bais*, however, has increased the likelihood of successful gorilla tourism. Tourism can provide significant revenue that can be channeled into ape conservation, but it also increases the number of people in daily contact with gorillas. This increases the chance of disease transmission, and the stress experienced by gorillas while in contact with humans and undergoing habituation may lower their resistance to disease.[148] Although regulations help protect eastern gorillas from disease transmission from tourists and their guides,[64] no such regulations yet exist for western gorillas. Guidelines have been developed for Mbeli Bai and Bai Hokŏu.

Patricia Reed

Village outreach: education efforts in Congo teach villagers about the disease risks of contact with great apes.

However, it seems that it was Ebola (a disease not associated with tourism) that killed eight groups of habituated western gorillas at Lossi Gorilla Sanctuary between October 2002 and January 2003.[75]

CONSERVATION AND RESEARCH

Habitat protection and law enforcement

Western gorillas are legally protected from persecution in all range states. There are protected areas within the range of both western gorilla subspecies, but most gorillas live outside these.[60, 96] Due to the sparse human population in much of its range, the western lowland gorilla has so far fared relatively well in many areas that are not formally protected, although this is changing rapidly as logging spreads and hunting takes its toll. The protected areas that host western lowland gorillas include a World Heritage Site in Cameroon (the Dja Faunal Reserve, an area of 5 260 km^2), various national parks including Dzanga-Ndoki (CAR), Lopé (Gabon), Monte Alén (Equatorial Guinea), Odzala-Koukoua, and Nouabalé-Ndoki (Congo), and several other categories of reserve. The Cross River gorilla occurs in several reserves including Cross River National Park (Nigeria; see Box 7.1).

The level of nominal and actual protection afforded by the different protection categories varies between countries, reflecting their different histories and economies. Nowhere in the gorilla range states are resources for conservation abundant, however; all the countries involved are among

Box 7.4 GORILLA CENSUSES

In most western gorilla habitats, nest counts are the most practical method of estimating gorilla number and density. A common method is to count the number of nests seen per kilometer of transect walked. The size of the associated gorilla groups can be estimated by examining all nests at a nesting site (see table). Western lowland gorillas are typically found at a density of about 0.25 weaned individuals per square kilometer. At some sites, this can be as many as three or, exceptionally, five gorillas per square kilometer; poor habitat may host as few as 0.1/km^2. Western lowland gorilla densities differ between forests of similar structure and species composition, and are affected by subtle variation in the abundance and distribution of key resources such as herbaceous plants.[43, 48, 136] Interpretation of comparisons of population density are further complicated by differences in field methodology and level of effort.[99]

The number of weaned individuals in an area is estimated on the basis of nest counts, group sizes, and established local nest-decay rates (the rate at which nests disintegrate), often using a computerized transect-analysis program, such as DISTANCE.[21] Nests can also provide information on the age structure and gender balance of populations, but their results must be analyzed with care for several reasons. First, nests vary in their durability depending on how they are built and the weather to which they are exposed; vegetation and climatic differences between sites can therefore influence the results. The nest-decay rate used in population calculations is intended to correct for this.[16, 21] In the absence of further data, many studies rely on the nest-decay rate of Tutin and Fernandez,[130] introducing an additional source of uncertainty. Second, western gorilla and chimpanzee nests can be confused, as both species often nest in trees, and eat many of the same foods. Third, some western gorillas sleep on bare ground, which means that the number of western gorillas present can be underestimated.[21, 101, 102, 126] At Lopé National Park, for example, the number of gorilla nests corresponded to the number of weaned individuals in the group at only one third of fresh nest sites.[135] Large sample sizes are required to minimize the impact of these possible errors.[126]

At *bais*, it is possible to observe western gorillas directly and obtain very accurate information on the size and composition of groups. Typically, the mean size of groups seen at *bais* is larger than that recorded in nest counts elsewhere. However, these findings relate only to those western gorilla groups with access to these clearings; supplementary research is required to discover the local population density and the distance traveled to reach a *bai*.

Sarah Ferriss and Lera Miles

Group size in western lowland gorillas

Location and study method	Number of groups	Number of nests	Weaned gorillas per group[a]
Northwestern Gabon[130] (782.8 km of transects in 15 habitat types)	136	540	4 (1–19)
Dzanga-Sangha region, CAR[26] (783 km of transects)	261	1 323	5.1 (1–52)[b]
Ngotto forest, CAR[21] (94 km of transects)	–[c]	145	5.7 (2–11)[d]
Likouala swamps, Congo[12] (80 km of transects, wet season)	38	213	5.6 (2–10)[d]
Likouala swamps, Congo[43, 44] (401 km of transects, dry season)	36	–	5[d]
Bai Hoköu, CAR[102] (observation and nest counts)	1[e]	–	12–15[f]
Mbeli Bai, Congo[92] (observation from viewing platform)	14	–	8.4 ± 4.3[d]
Maya Nord Bai, Odzala-Koukoua National Park, Congo[76] (observation from viewing platform)	36	–	11.2 (2–22)[d]

a Data are expressed as mean plus range, where available.
b Including 73 solitary males.
c Where data are not available, this is indicated by a dash.
d Excluding solitary males.
e This group, unusually, had two silverbacks.
f The range observed over a 27 month period.

the world's poorest. This puts a premium on relieving constraints on conservation resources through partnerships between range state governments, official donor agencies, and nongovernmental organizations. For example, since the gazetting of the Dzanga-Sangha Special Dense Forest Reserve and the Dzanga-Ndoki National Park, both have been managed by the the Dzanga-Sangha project, a collaboration between the CAR government, GTZ (acting through the German consulting firm LUSO Consult), and WWF.[14] Partnerships with the private sector can also help; Congo is putting an innovative new law into place, which will require all logging concessions to provide patrols to discourage poaching.[119]

Some blocks of western gorilla habitat or protected areas straddle international borders, requiring cooperation between two or more countries for effective conservation. One such cross-border regional collaboration has been implemented in southern CAR, northern Congo, and southeast Cameroon, establishing Trinationale de la Sangha in 1998. This conservation initiative covers the contiguous Dzanga-Ndoki, Nouabalé-Ndoki, and Lobéké National Parks, and divides the area into regions in which human activity is managed or restricted. It allows for joint patrols by rangers from the three countries and has resulted in some successful missions in the ongoing effort to discourage poaching.[17]

Although parts of the Mayombe forest are protected in Angola, DRC, and Congo, it has been much degraded. Dialog with local communities is underway in Cabinda province, aiming to promote forest and biodiversity conservation, and to relieve poverty via a proposed system of transfrontier protected areas involving all three countries.[111]

The Mengamé Gorilla Sanctuary comprises a 1 000 km^2 biodiversity corridor in Cameroon, along its border with Gabon. It will contribute to a transborder protected area by linking with the Minkébé National Park in Gabon, as well as to an emerging tri-national initiative between Cameroon (Dja Faunal Reserve), Gabon (Minkébé National Park), and Congo (Odzala-Koukoua National Park). In 2002, the Jane Goodall Institutes signed an agreement with the Cameroon Ministry of Environment and Forests (MINEF) to establish a community-based conservation and wildlife research program in the Mengamé Gorilla Sanctuary.[58]

A key constraint on the success of western gorilla conservation is the quality of protective legislation and the degree of enforcement. Efforts made by many range states in establishing protected areas, especially those that straddle frontiers, and the protection of western gorillas in national law, show official commitment to the conservation of gorilla habitats. In all range states, however, lack of resources and financial constraints impede many efforts to enforce existing legislation effectively.

Conservation and research activities

The large range of the western lowland gorilla encourages some confidence in its survival chances, yet brings its own challenges of coordination between the multiple governments and other stakeholders. Many international, regional, and national organizations are working to safeguard the western lowland gorilla's future through conser-

John Oates

John Oates

Forested mountains at the headwaters of Asache River, a Cross River gorilla survey area, and a field survey camp in the same area.

vation and research programs (more than can be mentioned here). The country profiles in Chapter 16 provide some further details about organizations active in each of the range states.

Research often goes hand-in-hand with conservation. One of the most sustained initiatives is the Station d'Etudes des Gorilles et des Chimpanzés in Lopé National Park, Gabon, which was established by Tutin and Fernandez with initial funding from the Centre International de Recherches Médicales de Franceville (CIRMF), and with later support from other donors. Long-term studies of western lowland gorilla and chimpanzee have been conducted there since 1983. The regional ECOFAC program has assisted in the development of ecotourism at Lopé National Park, as well as supporting ecological and sociological studies in the area. ECOFAC was initiated to ensure biodiversity conservation through management of protected areas and the development of sustainable local activities throughout Central Africa.[9] Since 1992, ECOFAC has undertaken biodiversity surveys, including the collection of information on primate populations at sites such as Odzala-Koukoua National Park, Dja Faunal Reserve, Monte Alén National Park, and Ngotto forest.

Sanctuaries for captive western gorillas have also been established. Attempts to reintroduce orphans to the wild are at an early stage. In Congo, the Projet Protection des Gorilles successfully reintroduced a number of orphaned western lowland gorillas into the Lesio-Louna Reserve.[29, 91] Cameroon, Nigeria, and Gabon also have at least one sanctuary that accepts orphaned western lowland gorillas; Angola, CAR, and Equatorial Guinea do not.

There are several efforts to establish tourism operations based around sightings of western gorillas, in Congo, Cameroon, Gabon, and CAR. The salt clearings (*bais*) just north of Odzala-Koukoua National Park, which are regularly visited by western lowland gorillas, are potentially suitable. Tourism can generate significant revenue that can be channeled into ape conservation. However, experience with the eastern gorilla has shown that rules designed to regulate tourism and protect apes from human disease are often poorly enforced, mainly due to a lack of staff training and education.[148] Viewing at *bais* offers the option of using hides or observation platforms, from which tourists could observe the wildlife without the need to habituate the gorillas or for there to be any contact with them. Experience from zoos has taught us that gorillas become anxious when watched from above,[149] so where a platform might be used, a hide would be worth constructing.

In conclusion, the range states of the western gorilla are taking action to protect the species and parts of its habitat, representing a significant investment of scarce public resources and the setting aside of large areas of forest land. Threats are nevertheless increasing, and originate largely in the unplanned and unregulated infrastructure development associated particularly with the timber industry. This renders very large areas accessible to hunters and encourages development of the bushmeat market. Habitat fragmentation and disease, including outbreaks of the Ebola virus, have also played their part in further endangering already vulnerable western gorilla populations. Our collective ability to mitigate these pressures is limited by ignorance of both the status and trends among populations of the western gorilla across much of its range. Further research on the distribution, abundance, and status of western gorillas in all their range states is therefore urgently needed, with a focus on the Cross River gorilla being a matter of particular priority. Better understanding of western gorilla ecology and behavior would improve the likelihood of conservation success. Long-term studies would be the ideal way to meet both needs, while also being a proven way to promote conservation in and around the study sites.

FURTHER READING

Cipolletta, C. (2003) Ranging patterns of a western gorilla group during habituation to humans in the Dzanga-Ndoki National Park, Central African Republic. *International Journal of Primatology* **24**: 1207–1226.

Cousins, D., Huffman, M.A. (2002) Medicinal properties in the diet of gorillas: an ethnopharmacological evaluation. *African Study Monographs* **23**: 65–89.

Doran, D.M., McNeilage, A., Greer, D., Bocian, C., Mehlman, P., Shah, N. (2002) Western lowland gorilla diet and resource availability: new evidence, cross-site comparisons, and reflections on indirect sampling methods. *American Journal of Primatology* **58**: 91–116.

Harcourt, A.H. (1986) Gorilla conservation: anatomy of a campaign. In: Benirschke, K., ed., *Primates: The Road to Self-sustaining Populations*. Springer-Verlag, New York. pp. 31–46.

Harcourt, A.H. (1996) Is the gorilla a threatened species? How should we judge? *Biological Conservation* **75** (2): 165–176.

Iwu, M.M. (1993) *Handbook of African Medicinal Plants*. CRC Press, London.

Magliocca, F., Querouil, S., Gautier-Hion, A. (1999) Population structure and group composition of western lowland gorillas in north-western Republic of Congo. *American Journal of Primatology* **48** (1): 1–14.

Nowell, A.A., Fletcher, A.W. (2004) Behavioral development in wild western lowland gorillas and a comparison with mountain gorillas. *Folia Primatologica* **75 S1**: 314.

Parnell, R.J. (2002) Group size and structure in western lowland gorillas (*Gorilla gorilla gorilla*) at Mbeli Bai, Republic of Congo. *American Journal of Primatology* **56**: 193–206.

Parnell, R.J., Buchanan-Smith, H.M. (2001) An unusual social display by gorillas. *Nature* **412**: 294.

Peterson, D. (2003) *Eating Apes*. California Studies in Food and Culture **6**. University of California Press, Berkeley.

Robbins, M.M., Bermejo, M., Cipolletta, C., Magliocca, F., Parnell, R.J., Stokes, E. (2004) Social structure and life history patterns in western gorillas (*Gorilla gorilla gorilla*). *American Journal of Primatology* **64**: 145–159.

Rogers, M.E., Abernethy, K., Bermejo, M., Fernandez, M., Tutin, C.E.G. (2004) Western gorilla diet: a synthesis from six sites. *American Journal of Primatology* **64**: 173–192.

Sarmiento, E.E., Oates, J.F. (2000) The Cross River gorillas: a distinct subspecies, *Gorilla gorilla diehli* Matschie 1904. *American Museum Novitates* **3304**: 1–55.

Stokes, E.J., Parnell, R.J., Olejniczak, C. (2003) Female dispersal and reproductive success in wild western lowland gorillas (*Gorilla gorilla gorilla*). *Behavioural Ecology and Sociobiology* **54**: 329–339.

Taylor, A.B., Goldsmith, M.L., eds (2002) *Gorilla Biology: A Multidisciplinary Perspective*. Cambridge University Press, Cambridge, UK.

Tutin, C.E.G., Fernandez, M. (1984) Nationwide census of gorilla (*Gorilla g. gorilla*) and chimpanzee (*Pan t. troglodytes*) populations in Gabon. *American Journal of Primatology* **6**: 313–336.

MAP SOURCES

Map 7.1 Gorilla data are based on the following source, with updates as cited in the relevant country profiles in Chapter 16:

Butynski, T.M. (2001) Africa's great apes. In: Beck, B.B., Stoinski, T.S., Hutchins, M., Maple, T.L., Norton, B., Rowan, A., Stevens, E.F., Arluke, A., eds, *Great Apes and Humans: The Ethics of Coexistence*. Smithsonian Institution Press, Washington, DC. pp. 3–56.

Box 7.1 Cross River gorilla data are based on unpublished data from Richard Bergl and Jacqueline L. Sunderland-Groves, with additional data as cited in the Cameroon and Nigeria country profiles.

For protected area and other data see 'Using the maps'.

ACKNOWLEDGMENTS

Many thanks to Kelley McFarland (City University of New York) for information on Cross River gorilla diet, and to Alexander Harcourt (University of California, Davis), Michael Huffman (Kyoto University), John F. Oates (Hunter College, City University of New York), Richard Parnell (University of Stirling), Emma Stokes (Wildlife Conservation Society), Jacqueline L. Sunderland-Groves (Wildlife Conservation Society), and Elizabeth A. Williamson (University of Stirling) for their valuable comments on the draft of this chapter, and to Angela Nowell (University College Chester) for information on gorilla development. Thanks also to Muhammad Akhlas (UNEP-WCMC) for research into the literature.

AUTHORS

Sarah Ferriss, UNEP World Conservation Monitoring Centre

Box 7.1 Jacqueline L. Sunderland-Groves, Wildlife Conservation Society, John F. Oates, Wildlife Conservation Society and Hunter College, City University of New York, and Richard Bergl, Hunter College, City University of New York

Box 7.2 Richard Parnell, Scottish Primate Research Group, University of Stirling

Box 7.3 Michael A. Huffman and Don Cousins, Primate Research Institute, Kyoto University

Box 7.4 Sarah Ferriss and Lera Miles, UNEP World Conservation Monitoring Centre

Martha M. Robbins

CHAPTER 8

Eastern gorilla (*Gorilla beringei*)

SARAH FERRISS, MARTHA M. ROBBINS, AND ELIZABETH A. WILLIAMSON

Eastern gorillas (*Gorilla beringei* Matschie, 1903), occur in the wild more than 1 000 km from the nearest western gorillas (*G. gorilla* Savage, 1847). They are larger than the western gorilla but otherwise similar, with a broad chest and shoulders, a large head, and a hairless, shiny black face. A full-grown adult male can weigh up to about 220 kg, and a full-grown adult female about half this.[46, 97, 99] Two subspecies of eastern gorilla are currently recognized by the Primate Specialist Group of IUCN–The World Conservation Union:[48] the eastern lowland or Grauer's gorilla (*G. b. graueri* Matschie, 1914); and the mountain gorilla (*G. b. beringei* Matschie, 1903).

A very small population of unusually large eastern lowland gorillas occurs on Mount Tshiaberimu, in the Virunga National Park of the Democratic Republic of the Congo (DRC).[82] One population of mountain gorillas, that found in the Bwindi Impenetrable National Park in Uganda, has such distinctive morphology, ecology, and behavior that some have suggested that it should be considered a third subspecies.[73, 123] This is a contentious issue, as it is probable that the mountain gorilla populations have been separated for only the relatively short period during which intensive agriculture has occupied the area between them. The small size of the mountain gorilla populations and the small number of samples available for examination make it more difficult than usual to determine whether the variation between populations is greater than the variation within them. Debate continues on this issue,[14, 47, 48, 92, 135] but here we recognize only two subspecies of the eastern gorilla: the eastern lowland gorilla and the mountain gorilla.

There are few if any absolute physical differences between these two subspecies, although the mountain gorilla tends to have a larger body and longer hair, and is distinguished by its larger cranium and wider facial skeleton, as well as less rounded and more angular nostrils.[47] Genetic divergence of these subspecies was apparently confirmed by comparison of their mitochondrial DNA (mtDNA);[76] however, the high frequency of incorporation of mitochondrial into nuclear DNA recently reported in gorillas[146, 166] makes interpretation of the earlier results more difficult. The mtDNA of these two subspecies may turn out to be more similar (or indeed more different) than had previously been thought. It is estimated that the two subspecies diverged some 400 000 years ago.[166]

DISTRIBUTION

Mountain gorilla

The mountain gorilla occurs in two known populations within three countries: DRC, Rwanda, and Uganda (Map 8.1). These populations occur almost entirely within national parks. One is found among the extinct volcanoes of the Virunga Massif. These gorillas are officially protected by the Virunga National Park of DRC, the Volcanoes National Park in Rwanda (Parc National des Volcans), and the Mgahinga Gorilla National Park in Uganda; all of these are contiguous and so protect a single area of gorilla habitat (the Virungas). The other population is found mainly in Bwindi Impenetrable National Park in southwest Uganda on the border with DRC. Mountain gorillas occupy about 375 km^2 in the Virungas and 215 km^2 in Bwindi,[14] these areas being separated from each other by 25 km of settled farmland.[123]

Map 8.1 Eastern gorilla distribution

Data sources are provided at the end of this chapter

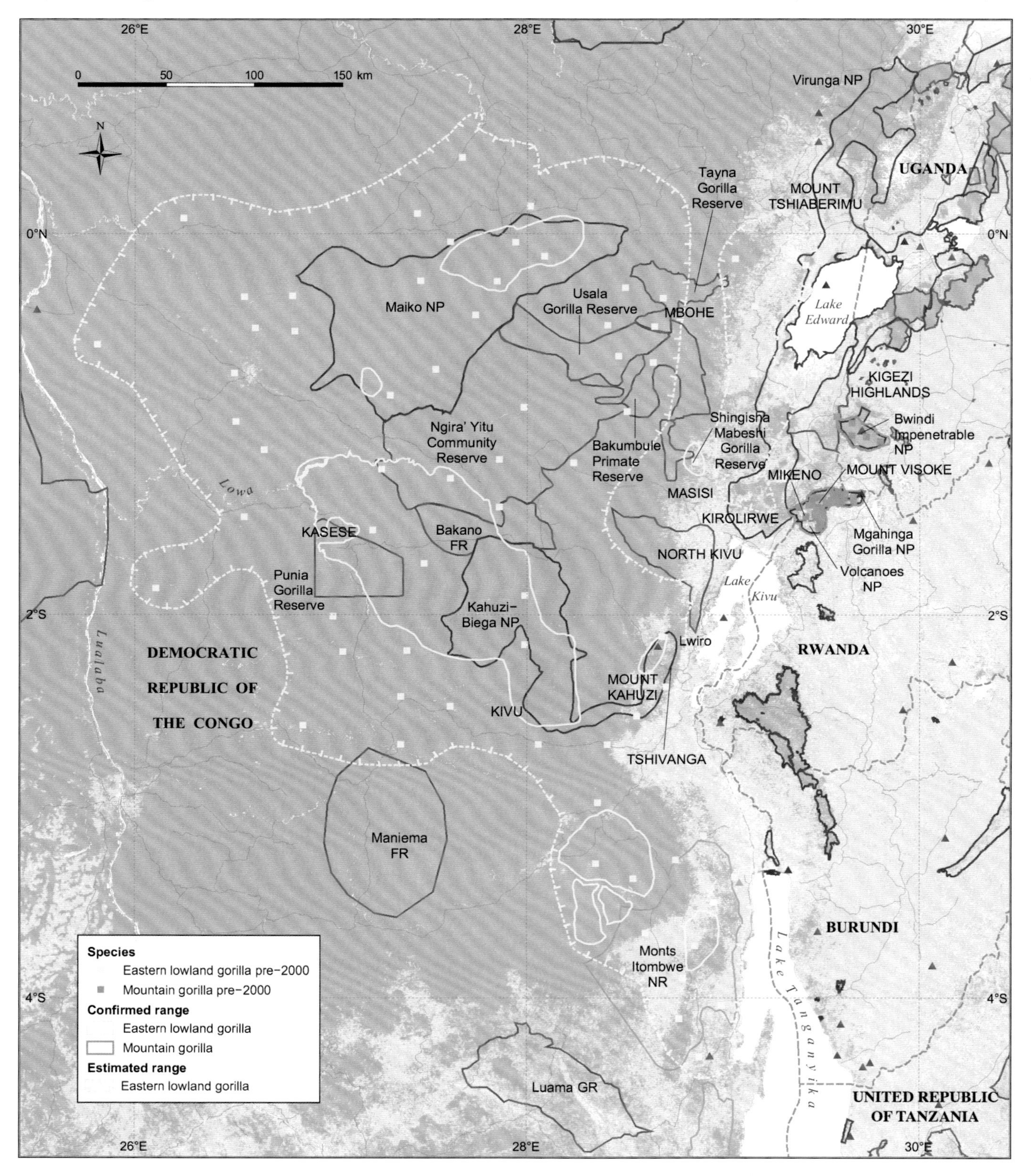

The three national parks of the Virungas contain most of the surviving forests in the region, extending to about 440 km^2. The forested area ranges in elevation from 2 000 m to 4 500 m, and therefore contains a range of ecosystems, including various montane rain forest formations and bamboo stands, as well as areas with little tree cover and abundant herbaceous vegetation.[90] The area has a high rainfall, and complex topography and drainage combine to create lakes, marshes, swamps, and peat bogs at various altitudes. The volcanic history means that there are high eroding peaks and lava plains, and that soil fertility is generally high; this factor is principally responsible for the high density of the surrounding human population.

The Virunga National Park is located in north-eastern DRC, is 7 900 km^2 in area, and has a boundary 650 km in length. It ranges in altitude from almost 800 m to over 5 100 m, so contains non-forested areas at low and high elevation. About 95 percent of the park is in North Kivu (Kivu Nord) Province and the remainder is in Orientale (formerly Haut-Zaire).[152] The park has four sectors, with gorillas found only in the southern sector.

The Volcanoes National Park of Rwanda is about 160 km^2 in area and ranges in altitude from 2 400 m to 4 507 m.[108, 153] Its boundary is located some 15 km northwest of the town of Ruhengeri in the Virunga Massif on the Ugandan and DRC borders.

The Mgahinga Gorilla National Park is 33.7 km^2 in area, ranges in altitude from around 2 400 m to 4 127 m, and is located in the extreme southwest of Uganda, on the borders with DRC and Rwanda. The park was established specifically for the protection of mountain gorillas in 1991, having been a designated but unprotected reserve prior to that.[15, 16, 17, 154]

The second population of mountain gorillas is mainly found in Bwindi National Park, which is located in the Kigezi Highlands of southwest Uganda, on the edge of the Albertine Rift Valley and bordering DRC to the west.[151] Some of these gorillas also occur across the border in DRC itself. Bwindi National Park covers approximately 331 km^2 and ranges in altitude from 1 160 m to 2 607 m.[98]

Mountain gorillas occur at a density of 0.85–1.00/km^2 in both Bwindi and the Virungas.[53, 80, 92, 186, 189]

Eastern lowland gorilla

The eastern lowland gorilla occurs only in eastern DRC, between the Lualaba River and the Burundi–Rwanda–Uganda border. Its distribution encompasses an area of about 90 000 km^2, within which it is thought to occupy an estimated 15 000 km^2 in four broad regions: the Kahuzi-Biega National Park and the adjacent Kasese region; the Maiko National Park and adjacent forest; the Itombwe Forest; and North Kivu.[49]

Kahuzi-Biega covers an area of 6 000 km^2, ranging in altitude from 600 m to 3 400 m. The park is divided into two parts, a mountain sector (600 km^2) and a lowland sector (5 400 km^2), connected by a forested corridor. Gorillas occur in both, in the region of Lake Kivu and Mount Kahuzi in the mountain sector, and the Kasese region of the lowland sector.[49, 155] Gorilla populations in this park are thought to have been devastated during the DRC civil war during the late 1990s, falling from an estimated 8 000 to an unknown number, perhaps as few as 1 000 individuals.[113]

Maiko National Park and nearby forests are located in the upland region between the central DRC river basin and the mountain ranges of the west side of the Rift Valley. The park has an area of about 10 800 km^2 and ranges in altitude from 700 m to 1 300 m.[83, 156] It is unclear how heavily the war in DRC has impacted Maiko's gorilla population. There are also several developing community reserves around the area of the Maiko, Virunga, and Kahuzi-Biega National Parks; together, these are thought to host between 700 and 1 400 gorillas as well as an unknown number of chimpanzees.[78, 93] One of these encompasses the Itombwe Forest, an area of montane, transitional, and lowland tropical forest to the west of Lake Tanganyika,[102] and includes

Pierre Kakulé Vwirasihikya

Habitat of the eastern lowland gorilla, Tayna Gorilla Reserve, Democratic Republic of the Congo.

Gordon Miller/IRF

Habitat of the mountain gorilla, Bwindi Impenetrable National Park.

protected areas of several different designations. It covers an area of 11 000 km^2, in which gorillas are found in four separate populations.[49] A community reserve is also being developed in the Masisi region in the North Kivu area, to the southwest of the Virunga National Park.

BEHAVIOR AND ECOLOGY

Much of the information on the ecology of the eastern gorilla comes from studies on mountain gorillas in the Virunga Massif.[35, 122, 124] Research in the Virungas was initiated by George Schaller in the late 1950s; since 1967, three to four habituated groups have been followed regularly by researchers from the Karisoke Research Center in Rwanda. Less is known about the ecology or demography of mountain gorillas in Bwindi,[92] or of the eastern lowland gorilla, but research is ongoing. Most studies of the eastern lowland gorilla have been carried out in Kahuzi-Biega, under the auspices of bodies such as the Congolese Institute for Nature Conservation (ICCN, Kinshasa), the Center of Natural Science Research (CRSN, Lwiro), and Kyoto University.[21, 69, 197, 207] Research findings on the ecological and behavioral differences between and within the eastern gorilla subspecies, and between western and eastern gorillas, are accumulating.[26, 38, 90, 120, 121, 201, 203]

Habitat

Mountain gorillas in the Virungas occur at altitudes from 2 000 m to 3 600 m, with occasional excursions to as high as 4 100 m, while those at Bwindi live between 1 160 m and 2 600 m. Eastern lowland gorillas occur at altitudes between 600 m and 2 900 m.[14, 90, 203, 208] Although the altitudinal ranges of the eastern lowland and mountain gorillas overlap, their populations are geographically separated.

The Virungas mountain gorilla habitat

The Virungas are a forested, mountainous volcanic region that contains a number of vegetation zones. The most widespread type is a *Hagenia abyssinica* (Rosaceae) and *Hypericum revolutum* (Clusiaceae) woodland, with a relatively open canopy and extremely dense herbaceous or, less frequently, grassy understory.[90, 180, 181] Other areas frequented by mountain gorillas in the Virungas include open herbaceous areas, often dominated by *Mimulopsis excellens* (Acanthaceae), in the flat saddle between Mounts Visoke and Sabinyo; monospecific stands of bamboo; dense ridge vegetation with abundant *Hypericum revolutum* and shrubby growth of *Senecio mariettae* (Asteraceae); and high-altitude vegetation with a stature of 4–5 m.[90]

Bwindi mountain gorilla habitat

Bwindi gorillas live at lower elevations, and are more arboreal than gorillas of the Virungas.[123] They occur in a range of vegetation types which include open forest with a discontinuous canopy, sometimes dominated by *Mimulopsis arborescens* (Acanthaceae); mixed forest dominated by understory and canopy trees and shrubs, usually interspersed with lianas and woody vines, especially *Mimulopsis* spp.; riverine forest, along permanent or temporary rivers or streams, with an open or continuous canopy; and regenerating forest that has been disturbed previously, for example by logging.[98] There is a greater density of fruit-bearing trees in the gorilla habitats of Bwindi National Park than in the Virungas.[38, 98]

Eastern lowland gorilla habitat

The eastern lowland gorilla has the widest altitudinal and geographical range of any of the eastern gorillas, living in montane, transitional, and lowland tropical forests. They have been reported at a range of densities: 0.25/km^2 in Maiko National Park, 0.55/km^2 at Mount Tshiaberimu, and 1.03–1.26/km^2 in Kahuzi-Biega.[50, 53, 111, 207] One of the best-studied populations of eastern lowland gorilla occupies the mountain region of Kahuzi-Biega. Here habitats vary from dense primary forest intermixed with bamboo, to mesophytic (moderately moist) woodland, to areas of *Cyperus* (Cyperaceae) swamp and peat bog, with alpine and subalpine grassland at higher altitudes; patches of open vegetation also occur at lower elevations.[155]

Diet

Mountain gorillas

Mountain gorillas are large-bodied herbivores; in the Virungas, they feed almost exclusively on the leaves and stems of herbs, vines, and shrubs harvested in the dense herbaceous understory, supplementing this with bark and roots.[90, 168] In contrast, the Bwindi mountain gorillas live in a more fruit-rich habitat, and take advantage of this. Around the Karisoke Research Center in the Volcanoes National Park, a study recorded mountain gorillas eating 38 plant species from 18 families.[168, 170] These included the stems and roots of *Peucedanum linderi* (celery, Umbelliferae); the stems and roots of *Laportea alatipes* (nettle, Urticaceae); and the stems and roots of *Urtica massaica* (stinging nettle, Urticaceae); as well as the leaves of *L. alatipes*, *Carduus nyassanus* (thistle, Asteraceae), and the leaves of *Galium ruwenzoriense* (galium vine, Rubiaceae). Mountain gorillas have a preference for:

- the leaves of *G. ruwenzoriense, Arundinaria alpina* (bamboo, Poaceae), and *Rubus* spp. (berry, Rosaceae);
- the stems of *P. linderi*; and
- (especially) bamboo shoots.[106, 163]

Bamboo is high in protein. Its availability fluctuates seasonally and it is consumed heavily by mountain gorillas when it is abundant.[182] A number of feeding techniques have been observed; these are interpreted as measures to avoid injury from leaves bearing stings or sharp hooks.[18, 19] As a result of the low quality and poor digestibility of much of their diet, mountain gorillas in the Virungas spend at least half of their daylight hours feeding, and much of the remainder resting.[171]

The gorillas of the Virungas and of Bwindi both require abundant quantities of easily harvestable plant material.[106] The gorilla habitat around the Karisoke Research Center contains little edible fruit,[182] as is reflected in gorilla diets there, while in Bwindi fruit is an important component of gorilla diet.[38, 121, 136] The most important fibrous foods consumed by the Bwindi gorillas also differ from those consumed in the Virungas, and include species of *Basella* (Basellaceae), *Brillantaisia* (Acanthaceae), *Clitandra* (Apocynaceae), *Ipomea* (Convolvulaceae), *Laportea* (Urticaceae), *Mimulopsis* (Acanthaceae), *Mormodica* (Curcurbitaceae), *Myrianthus* (Moraceae), *Palisota* (Commelinaceae), *Triumfetta* (Tiliaceae), and *Urera* (Urticaceae).[38]

Occasional items that mountain gorillas have been seen eating, all of which are speculated to have a nutritional function, include insects (ants and cocoons of unspecified origin);[37, 168, 172] at Karisoke Research Center, subsoil sediments five to six times per year, possibly as a source of sodium or iron;[88] dung;[40, 57, 168] and rotting wood.[36, 136]

Mountain gorillas show dietary flexibility; within both the Virungas and Bwindi their diet has been shown to vary according to the distribution and abundance of food resources that, in turn, vary according to altitudinal and climatic factors.[38, 90] For example, at Bwindi, groups of gorillas living at lower altitudes consumed more species of fibrous food (140 versus 62) and fruit (36 versus 11) than those living at higher altitudes. There is little seasonal variation in the diet of the mountain gorilla in parts of the Virungas, probably because most of their food is available throughout the year,[168] while the fruit component of the diet of the Bwindi gorilla varies over the course of a year.[38] The total number of species eaten and the degree of frugivory are more similar between eastern lowland gorillas at Kahuzi-Biega and mountain gorillas at Bwindi than between the populations of mountain gorillas at Bwindi and the Virungas.[38]

Ian Redmond

A silverback male eating *Myrianthus* fruit.

Martha M. Robbins

A gorilla infant, Bwindi Impenetrable National Park, Uganda.

Mountain gorillas appear to visit feeding areas that have received little recent use and those that produce nutritious food.[180] Foraging areas with less abundant high-quality food, or where renewal rates are lowest, are visited less frequently than other areas. When bamboo shoots are available in large quantities, mountain gorillas feed almost exclusively on bamboo. As bamboo declines in abundance, the gorillas move away from the bamboo areas and consume other herbaceous foods. When all preferred foods are scarce, the gorillas alter their diet and expand the foraging area covered each day.[163, 196]

Nutrient supply does not appear to be a limiting factor for mountain gorillas in the Volcanoes National Park.[106] While food abundance varies over the range, no areas are so productive that it would be advantageous for gorilla groups to establish and defend exclusive foraging zones.[168] Instead, home ranges may overlap by up to 100 percent,[170, 180] with the groups tending to avoid one another.[32] The availability of abundant, evenly distributed food resources means that overall feeding competition within groups is also rather low, and so the costs of social foraging are also likely to be low,[169, 182] although some competitive disputes do occur, especially in larger groups.[169] Silverbacks have priority access to food, and there are weak dominance hierarchies among females.[51, 169, 176, 178] Hence, group living may carry some costs for lower-status blackbacks, females, and juveniles.[182]

Eastern lowland gorillas

The varied diet of the eastern lowland gorilla includes a wide range of plants, their fruit, seeds, leaves, stems, and bark as well as ants, termites, and other insects.[208] Seasonality in diet and habitat use is greater for eastern lowland gorillas in low-altitude forests than for mountain gorillas.[182] Eastern lowland gorillas eat more fruit than do Bwindi mountain gorillas, but not as much as western gorillas.[38, 182, 203, 208] When fruits are scarce, eastern lowland gorillas travel less and increase their consumption of herbaceous vegetation.[206, 208] Large quantities of bamboo shoots, as well as several types of fruit, are eaten seasonally by eastern lowland gorillas of the upper altitudinal reaches of Kahuzi-Biega.[21, 39] These gorillas also occasionally feed on ants, but have not been observed eating insects as often as have eastern gorillas in lowland forests. Insects are never more than a minor part of the diet for any gorillas.[208] The ant-feeding sites have all been found in primary or ancient secondary forests on ridges or slopes. Most plant parts are eaten on the ground, although leaves, bark, and fruit are sometimes eaten in trees. Signs of feeding activity have often been observed along gorilla trails in valleys and swamps.[208]

Ranging behavior

In the Virungas, the typical annual home range of a mountain gorilla group is about 5.5–11.1 km^2.[163, 170, 175] Bwindi gorillas may use 20–40 km^2 in a year.[121] The ranging behavior of gorilla groups is mainly determined by the distribution and abundance of fruit and herbaceous vegetation in the environment,[121] but may also be influenced by social factors such as competition for mates or the mate-guarding tactics of silverbacks.[175] These complex and changing factors are reflected in diverse ranging behavior, with groups generally spending more time in food-rich areas.[90, 163] Solitary males in the Virungas have larger home ranges than would be expected for a single individual;[175, 194] there are no equivalent published data for Bwindi gorillas. Food is an important influence on the movement patterns of lone males, and other gorillas are not always avoided.[20, 175, 194, 204]

Eastern lowland gorilla groups in montane forest have home ranges of 13–17 km^2.[203] Although the size of their home range in lowland tropical forest is unknown, they are known to have shorter average day journeys in montane forest than in lowland forests.

Ecological role

Not only is gorilla behavior adapted to the ecosystems in which they live, but gorillas also help to shape these ecosystems. As large, heavy, and dexterous animals that consume a lot of foliage, they also change the structure of vegetation by trampling it. This can stimulate regrowth and productivity; stem densities of some herbaceous foods increase in the aftermath of gorilla feeding.[170] It is not certain whether there is a positive-feedback mechanism through which gorilla activity leads to a more edible plant community.[105, 180]

In many forest communities, primates act both as seed predators and as seed dispersers; they are likely to have an important impact on patterns of forest regeneration and on the diversity of tree species[86] (see also Boxes 4.4 and 5.1). Western gorillas have been reported to disperse seeds,[150] and the consumption of fruit by eastern lowland and Bwindi mountain gorillas suggests that they may also play this role.

Eastern gorillas share their habitat with other large mammalian herbivores, and so might be expected to compete with them for food. Mammalian herbivores in the Virungas include buffalo (*Syncerus caffer*) and bushbuck (*Tragelaphus scriptus*), but these are not thought to have a significant impact on the mountain gorilla population.[106, 107, 110] Other herbivores, such as black-fronted duiker (*Cephalophus nigrifrons*) and the African forest elephant (*Loxodonta cyclotis*), show little dietary overlap with the mountain gorilla.[106] Elephants have considerable potential to impact the food supply of the mountain gorilla, but their numbers are so low as to have little real effect.[107] Mountain and eastern lowland gorillas are sympatric (occur together) with chimpanzees in some areas, and their diets are known to overlap.[136, 205] Although one competitive encounter between chimpanzees and the Bwindi mountain gorillas has been observed, different foraging strategies are employed by these species and there is little evidence of feeding competition between them.[136] It has been suggested that sympatry with chimpanzees may have promoted a leaf-eating strategy in gorillas, moving their feeding niche away from that occupied by chimpanzees[203] (see Box 8.1).

The only known predators of gorillas are humans and leopards (*Panthera pardus*).[124] Evidence of attacks by leopards on western gorillas is outlined in Chapter 7, but these cats no longer occur in the Virungas, and may also have been lost from Bwindi.

Social behavior

More than 30 years of research at the Karisoke Research Center established by Dian Fossey has made mountain gorillas one of the best-studied primate species.[122] Much is known about their social behavior, feeding ecology, life history patterns, and demography.[115, 118, 134, 140, 178, 183] Given the ecological variability between gorillas in different habitats, one important question is the extent to which the information available from Karisoke applies to other gorilla populations.

Groups of eastern gorillas may contain only one mature male, several mature males (in a 'multimale' group), or may consist of males only.[116, 195] Most comprise a single dominant adult male or silverback, typically with three or four females and four or five offspring.[53, 186] Over the past three decades in the Virungas, between 10 percent and 50 percent of mountain gorilla groups have been multimale,[80] while at Bwindi about 50 percent of groups are multimale.[92] About 10 percent of eastern lowland gorilla groups are multimale.[203] If the dominant male mountain gorilla dies in a one-male group, the group may disintegrate; should this happen in a multimale group, however, one of the subordinate males can take over leadership and the group may then stay

Elizabeth A. Williamson

A male silverback eastern lowland gorilla, Democratic Republic of the Congo.

intact.[115] This pattern is in marked contrast to that seen in western gorillas, among which multimale groups are extremely rare.[120]

Group size is variable among eastern gorillas; groups ranging from two to 53 individuals have been observed.[80, 189, 196] In general, median group size is similar for both eastern and western gorillas, across various habitat types and the different diets associated with them.[103, 120, 200, 203] In the Virungas, median and mean group size are eight and 11 individuals respectively (see Table 8.2).[80] At Bwindi, a mean group size of about 10 has been reported.[92] In the area surrounding Tshivanga in Kahuzi-Biega, the mean group size of eastern lowland gorillas (excluding solitary males) is almost 10.[69] Mean group size in the highland sector of Kahuzi-Biega decreased from about 16 per group in 1978, to 11 in 1990, to 10 in 1996.[201] Other studies indicate a mean group size of seven animals in Kahuzi-Biega but only three in the adjacent Kasese region.[50]

The sex ratio at birth in both the Virungas and Kahuzi-Biega is approximately 1:1.[174, 201] Upon reaching maturity, most males and females leave the group in which they were born (their natal group). Males that emigrate usually remain solitary until they can attract females and establish their own groups; occasionally, males form all-male groups. After emigration from the natal group, some males spend a large proportion of their time alone, although in the home range of their natal group.[20, 53] It is very unusual for fully adult males to migrate into other groups.[53, 115, 118, 195] Young males may also stay within the natal group and eventually inherit its leadership.[58, 115, 118] Most multimale groups, but not all, may be the result of males maturing and remaining in their natal groups,[115] and are therefore believed to contain several related adult males. Genetic studies confirm that this is often, but not always, the case.[100]

Whether a young male remains in his natal group or emigrates could be determined by a range of factors including changes in social relationships and demographic structure, such as the availability of mating opportunities within the group, the death of a parent, or disintegration of the natal group.[28, 115, 118] Males that develop strong affiliative (friendly) relationships with the dominant silverback while they are infants are more likely to be close to the leading male during adolescence, and are therefore more likely to remain in their natal group.[58] Male eastern lowland gorillas in Kahuzi-Biega rarely stay with their putative fathers but instead form their own groups, sometimes taking females with them from the natal group.[201]

Both natal dispersal (leaving the natal group and transferring to a new group), and secondary dispersal (subsequent transfer to yet another group), occur among female eastern gorillas. Females have also been known to remain and reproduce within their natal group.[61, 134, 169, 183] Female mountain gorillas of the Virungas usually transfer from their natal group alone, while female eastern lowland gorillas sometimes transfer with another female and their offspring.[201] If a female is pregnant or has an infant when she transfers to a new group, there is a risk that the new silverback will kill the infant.[173] Infanticide has been observed occasionally in eastern gorillas,[173, 202] although not in every instance of transfer with an infant[133] (see Box 8.2).

Female transfer could offer a number of possible advantages, such as the opportunity of higher social rank,[35] especially following migration into a small or new group; avoidance of inbreeding; increased choice of mates; improved reproductive success; reduced feeding competition; or improved protection against infanticide.[134]

Females may have preferences with regard to mates, and this choice may be influenced by male behavior.[134] From the male point of view, good relationships with females are important to mating access and breeding success, as a female is free to leave the group. Although gorilla groups are essentially controlled harems, males cannot therefore afford to make them unduly oppressive ones.

Both aggressive and affiliative interactions between males and females have been observed. Males have been seen to direct aggressive displays toward females, and females to appease those males (see Box 8.3), although the reasons for these displays and their impact on female mate choice remain unclear.[134] Males may also vocalize and engage in nonaggressive behaviors toward females, possibly to maintain proximity with females.[134] Females may sometimes intervene in an attempt to end aggressive interactions between adult males.[132]

Silverback males in mixed-sex groups do not interact much with each other but, when they do, the behavior tends to be more competitive and aggressive than affiliative, presumably as a result of competition over access to mates.[118] Affiliative interactions are rarely seen,[116] but occasional cooperation by males within the same group has been observed, apparently to prevent females from leaving the group.[131] Relations between silverbacks

Box 8.1 COEXISTENCE OF GORILLAS AND CHIMPANZEES

Gorillas and chimpanzees live together in the same forests in many parts of equatorial Africa, a co-existence known as sympatry. As they are so similar to one another, how do they manage to coexist without one species displacing the other? Earlier studies[77, 124] suggested that their different diets and ranging behaviors reduced competition through 'niche differentiation'. Fruit-eating (frugivorous) chimpanzees tended to range in primary forests and stay on the dry ridges, while leaf-eating (folivorous) gorillas tended to range in secondary regenerating forests and stay in the wet valleys. These ecological differences were thought to affect their societies, and to determine their densities in different types of habitat. The dynamic 'fission–fusion' social structure of chimpanzees was therefore thought to be caused by their frugivory, while the more cohesive groups of gorillas were associated with their folivory.

More recent studies, however, have shown that there is actually extensive overlap of gorillas and chimpanzees in both diet and ranging. Western and eastern lowland gorillas include fruits and insects in their diet, and range in primary forests in close proximity to chimpanzees.[25, 85, 114, 148, 149, 195, 205, 208] Western lowland gorillas consume plant foods as diverse as those eaten by sympatric chimpanzees. Of the fruit species eaten by western lowland gorillas at Lopé National Park, Gabon, 79 percent are also consumed by chimpanzees in the same forest.[149] All fruit species eaten by eastern lowland gorillas at Kahuzi-Biega are also eaten by sympatric chimpanzees.[205] However, analysis of fecal samples at Kahuzi-Biega and Bwindi shows that there are marked differences between the two species in their reliance on particular fruit species, such as *Ficus* spp. (Moraceae), *Syzygium* sp. (Myrtaceae), *Bridelia* sp. and *Drypetes* sp. (both Euphorbiaceae).[136, 205]

The presence of gorillas is thought to influence the choice of nesting trees by chimpanzees. In secondary forests at Kahuzi-Biega, eastern chimpanzees (*Pan troglodytes schweinfurthii*) tend to avoid nesting in those trees with ripe fruits of the type preferred by gorillas.[8] Eastern lowland gorillas tend to extend their day-journey length during the fruiting season in both lowland and montane forests, while sympatric chimpanzees tend to stay in a small area, continually revisiting particular fruiting trees.[9, 205, 206] Such differences in diet, ranging patterns, and nesting-site choice may limit competition between sympatric gorillas and chimpanzees. Gorillas and chimpanzees occasionally encounter each other in the same fruiting trees at Kahuzi-Biega and Bwindi, with most encounters being tense but peaceful.[136, 205] At Ndoki, in Congo, typical encounters between western lowland gorillas and chimpanzees are even more peaceful.[144]

Many aspects of foraging behavior seen among gorillas and chimpanzees may vary with environmental conditions, and the true extent of this variability is still unknown. This is likely to be important in predicting how gorillas and chimpanzees will react to habitat change wrought by humans, which is fundamental to wise conservation planning.[187, 196] Continuing research on eastern lowland gorillas and sympatric chimpanzees at Kahuzi-Biega and Bwindi will help to clarify the scope for improving the survival of sympatric great ape populations.

Juichi Yamagiwa

A female gorilla and infant, Kahuzi-Biega National Park.

Juichi Yamagiwa

and blackbacks tend to be weak.[116, 118] Blackbacks are subordinate to silverbacks, and generally spend a lot of time on the periphery of the group.[118]

Young, unrelated males that form all-male groups are thought to do so to develop social skills, and perhaps to increase safety from predators.[116, 118] Relations between males in all-male groups tend to be more affiliative than among males in mixed-sex groups, as measured by the occurrence of playing, grooming, and time spent in close proximity.[116, 118] Homosexual behavior has also been observed.[195] Aggression is more frequent in all-male groups, but

Box 8.2 INFANTICIDE IN GORILLAS

Few behaviors observed in the animal kingdom have led to more heated debate concerning its function than infanticide (the killing of young from the same species). Why would the killing of dependent young evolve as an adaptive strategy? The prevailing view is that infanticide by males is related to competition over access to females, in line with the sexual-selection hypothesis.[66] Specifically, if a male kills unweaned offspring of other males, and thus shortens the time that he must wait to impregnate their mothers, he will increase his own reproductive success compared to that of other males who do not follow this strategy. Given the cost to females in lost reproductive effort, infanticide results in a conflict between the sexes. Infanticide occurs rarely, but over the past four decades has been suspected or observed in more than 40 species of primates, including gorillas.

Known or probable infanticide and attempted infanticide in mountain gorillas were recorded 13 times at Karisoke between 1967 and 1988; these comprised three observed cases, nine inferred cases, and one unsuccessful attack inferred from wounds.[173] If all were indeed cases of infanticide, this would have accounted for at least 37 percent of infant mortality during this period.[173] The majority of these cases occurred when the mothers of the infants were not accompanied by the group's silverback, typically because he had died. This suggests that an important motivation for females to form long-term associations with males is to obtain protection against infanticide.

Is infanticide universal among gorillas? Interestingly, in Kahuzi-Biega, female eastern lowland gorillas with dependent young have been observed unaccompanied by silverbacks for many months. Females have also transferred between social units with unweaned infants that were not killed, but three cases of infanticide have been observed at Kahuzi-Biega.[201] On the other hand, two cases of infanticide have been inferred in western gorillas following group disintegrations.[143]

The risk of infanticide is thought to have played a large role in shaping the social behavior and group structures observed in many primate species.[162] Where there is only one male per group, females can exert mate choice by transferring between social units; because of the risk of infanticide, the opportunity for a female to transfer without risk is limited to the brief time window when she does not have a dependent offspring.[134] A multimale group structure is advantageous because, in the event of the death of the leading silverback, another (often related) adult male is likely to take over the leadership of the group; this prevents group disintegration and infanticide by an outsider male.[115, 118] Since the late 1980s, while the gorilla groups studied at Karisoke have been almost exclusively multimale, no group disintegrations have occurred, and neither have any infanticides by males been observed or suspected.[80] Infanticide has rarely been observed during encounters between groups, and male eviction and group takeovers by extragroup males have not been observed in gorillas.[118, 134, 173] Recently, at Kahuzi-Biega, following the simultaneous transfer of several females, the new silverback killed one unrelated infant at the time of transfer, and (despite the efforts of the females in the group to intervene) killed two other infants shortly after their births, which occurred only a few months after the transfers.[202]

In addition to its impact on sociality, infanticide has implications for population dynamics. The death of a silverback represents initially the loss of only one individual in a population. If, however, he was the leader of a one-male group, his death is likely to lead to the deaths of all his unweaned offspring. This impacts overall infant mortality, future births, group age structure, and the rate of population growth, which can be critical for small populations such as those of the mountain gorilla.

Martha M. Robbins

disputes between males in mixed-sex groups (when they occur) are more serious and more likely to result in wounds.[116] This difference is probably a result of competition between males over mating, an issue that does not arise in all-male groups.

Apart from those between mothers and their offspring, social bonds between females tend not to be well developed. Females commonly leave their natal group, so complex social networks between females do not occur. The female coalitions that do emerge, allowing common defense against aggressors, are thought to be more common among related than unrelated individuals.[179, 183] Males frequently intervene in conflicts between females, thus limiting the effectiveness of female coalitions. Such interventions involve only moderate aggression, pose little risk to social relationships with females, and may help males to retain mates by maintaining their own status and control over the group.[179]

Immature gorillas often receive defensive support from their mothers, but rarely from unrelated adult females. Juveniles rarely receive consistent support even from their mothers, however, if they behave aggressively toward larger opponents.[179] During infancy, gorillas often develop an attraction to the leading male of the group, who may buffer the young animals against aggression from others, serve as a spatial focus for young animals, and provide an attachment figure as the maternal bonds weaken.[139] The behavior of the male toward infants and juveniles is paternalistic, though no great effort is put into this.[139] Adult males may protect immature gorillas against larger opponents but provide little support to immature individuals who behave aggressively, intervening mostly in conflicts between immature peers only to maintain control.[179]

Reproduction

Successful gorilla males typically mate with more than one female (are polygynous). Generalizing from the Virunga gorillas, it seems that female mountain gorillas reach sexual maturity around the age of six and a half years (5.8–7.1 years). Between the first bout of estrus-like behavior and the first conception there is a phase of adolescent sterility that lasts two years.[53, 174] Although less regular among young females, the menstrual cycle among adults has a median length of 28 days; females are most receptive and attractive to males at around mid-cycle, for one to four days.[24] The gestation period lasts about eight and a half months.[23, 34, 54, 147] Mating or mating attempts occur at times during both the menstrual cycle and pregnancy when estrogen concentrations are highest.[24, 54, 174]

Mountain gorillas do not have a birth season, presumably because of the lack of seasonality in food availability. Infant mortality rates at Karisoke are highest during the wettest months (April and May), when the animals are colder and more susceptible to respiratory infections.[182] The interbirth interval lasts approximately four years, as gorillas are not fertile while still suckling young (lactational amenorrhea). The recorded interbirth interval for the eastern lowland gorilla is slightly longer than that of mountain gorillas of the Virungas (4.6 versus 3.9 years).[174, 201] Should an infant die, this interval is shortened, allowing its mother to conceive again within three to six months. Infants are typically weaned at three or four years,[33, 138] but there is variation in both directions.[30]

Gordon Miller/IRF

Part of the Mapuwa group of mountain gorillas, Virunga National Park.

Social rank and group composition may change during an individual's lifetime. Assuming a different relative position within the group can be expected to change that individual's reproductive strategies. Although mountain gorillas are considered to have a one-male mating system, many multimale groups exist. In one-male groups, the only male present does all the mating. In multimale groups, subordinate males do mate, including at times when conception is likely to occur, although dominant males tend to participate in more matings with adult females, and subordinate males with subadult females.[117] Genetic studies reveal that subordinate males do sire a proportion of offspring.[12, 100, 101] Mating with individuals from other groups is exceptionally rare in mountain gorillas.[134]

In multimale groups, males often try to remain in proximity to females at mid-cycle.[134] Females sometimes mate with more than one male, sometimes even in the same mid-cycle period. This may be voluntary or the result of male coercion.[119] Harassment of copulating males can occur, and is often but not always practiced by dominant males.[117]

Eastern lowland gorillas share many reproductive characteristics with mountain gorillas, including a sterile subadult period in females, the

Table 8.1 Eastern lowland and mountain gorilla populations

Subspecies	Approximate population size	Approximate area of occupancy (km^2)[14]
Mountain gorilla (Virungas)[a]	380[44, 80]	375
Mountain gorilla (Bwindi)	320[91, 92]	215
Eastern lowland gorilla[b]	?[c] 17 000 ± 8 000[d]	15 000

a See also Table 8.2.
b See also Table 8.3.
c No data; fieldwork was being undertaken in 2005 to estimate the extent of the decline.
d Estimate based on 1998 survey data, obtained prior to outbreak of war in the area.[49]

Table 8.2 Mountain gorilla populations of the Virungas (1971–2003)

Census years	Total gorillas counted	Estimated population size	Number of social groups	Mean group size	Number of solitary males	Multimale groups (percent)	Immature individuals (percent)
1971–1973[45, 55]	261	274	31	7.9	15	42.0	39.8
1976–1978[186]	252	268	28	8.8	6	39.0	35.8
1981[5]	242	254	28	8.5	5	40.0	39.7
1986[165]	279	293	29	9.2	11	8.0	48.2
1989[129]	309	324	32	9.2	6	28.0	45.5
2000[80]	359	359–395	32	10.9	10	52.9	44.7
2003[68]	380	–	–	–	–	–	–

Adapted from Kalpers, J., *et al.* (2003).

Table 8.3 Eastern lowland gorilla populations

Geographic region	Estimated population size (2001–2004)	Estimated population size (1994–2000)
Kahuzi-Biega National Park and adjoining Kasese region	present (2005)	15 703 (7 655–22 491) (1994–1995)[49, 164]
Tayna and other proposed community reserves	1 050 (700–1 400) (2004)[93]	?[a]
Maiko National Park	assumed present (2005)	859 (462–1 135) (1996)[64]
Itombwe Forest	present (2005)	1 155 (516–1 796) (1999)[102]
Northern bank of Lowa River (north of Kasese region)	?	13 (0–26) (1998)[49]
Mount Tshiaberimu, Virunga National Park	20 (2004)[13]	?
Masisi (including Shingisha Mabeshi)	present[199]	28 (0–33) (1988–1998)[96]
Mbohe, North Kivu	?	small[49]

a '?' indicates that no data are available.

Adapted from Hall, J.S., *et al.* (1998) and later sources, as cited in the table.

age at first parturition (giving birth), interbirth interval, and infant mortality rates.[201 203]

Nest building

Adults and weaned immature gorillas build nests each night, in which they sleep. Unweaned offspring share the nests of their mothers; otherwise, gorillas sleep alone. The gorilla defecates either in or next to the nest, and the size of the dung is directly proportional to the age of the gorilla.[53, 124] Counting and measuring nests and dung can therefore provide information on the number of gorillas in a group and the age class of the individual using each nest, so it is a commonly used census method. In the Virungas, mountain gorillas almost always make nests on the ground, while about half of the nests of eastern lowland gorillas in the lowland tropical forest of Kahuzi-Biega are constructed in trees.[208] In the montane forest of Kahuzi-Biega, most nests are made on the ground but, even here, immature gorillas tend to make nests in trees more frequently than do adults; more immature and female gorillas tend to nest in trees if the group's silverback has died. This is thought to be a result of their vulnerability to large terrestrial predators.[197]

POPULATION

Status and trends

The population of mountain gorillas of the Virungas has been monitored since the 1970s. Fewer data are available on the status and trends of mountain gorillas at Bwindi, or of eastern lowland gorillas. Recent estimates of overall numbers of eastern gorillas are given in Table 8.1.

Mountain gorillas in the Virungas

The mountain gorillas of the Virungas have been studied for over 40 years. A summary of selected population estimates can be seen in Table 8.2. These data show a decline through the 1970s and into the early 1980s, with most reduction occurring in the DRC section.[70, 186] The population was estimated to contain about 450 gorillas in the late 1950s,[124] 275 in 1973,[53] and 254 in 1981.[5, 70] The 1989 census of mountain gorillas in the Virungas counted 309 animals and estimated 324 to be present.[129] A population estimate in 2000, based on repeated observation of 17 habituated groups and information on 15 unhabituated groups, suggested that the Virunga population of the mountain gorilla had further increased to between 359 and 395.[80] In the DRC parts of the Virunga Massif, seven habituated gorilla groups had declined from a total number of 103 individuals to 66 between 1995 and 1998, but showed an overall increase from 66 to 86 between 1998 and 2002.[11] The most recent census of the Virunga gorillas recorded 380 animals.[68]

The increased numbers of mountain gorillas revealed by these censuses should be viewed with some caution because nearly all of the population growth can be attributed to the Research/Susa section of the Volcanoes National Park, an area that is relatively well protected, and which is believed to be a particularly good gorilla habitat. Other sectors are known to have experienced a decline in the number of gorillas,[80] so there is still conservation work to do.

Mountain gorillas in Bwindi

The small Bwindi mountain gorilla population also appears to be stable. A survey in the early 1990s found about 300 animals,[15, 16] which was confirmed by a complete census of the entire park in the late 1990s,[92] and raised to about 320 by another census in 2002.[91]

Eastern lowland gorillas

The total area known to be occupied by eastern lowland gorillas declined from about 21 000 km^2 in 1963 to 15 000 km^2 by the early 1990s. The overall geographic range, calculated by Butynski from historical locality data, was 112 000 km^2.[14] This illustrates the degree of fragmentation of populations at that time. By the mid-1990s, there were estimated to be about 17 000 (± 8 000) eastern lowland gorillas in at least 11 subpopulations, with 86 percent living in Kahuzi-Biega and the adjacent Kasese region of DRC.[49, 111]

More recent events in Kahuzi-Biega and the surrounding region, however, indicate that the species has undergone a substantial decline in numbers[104, 113] (see Table 8.3). Access to much of the gorilla range has been difficult in recent years, and is only just becoming possible again. The available information is very limited, but there is consensus among field workers that a drastic decline in total population has occurred. This is attributed to the combined effects of the rise in demand for 'coltan' ore (discussed in more detail below) and the warfare that engulfed the whole of the eastern lowland gorilla range from the late 1990s onwards; armies, rebels, refugees, and miners all lived off the land and consumed bushmeat.[113]

Box 8.3 THE VOCAL BEHAVIOR OF MOUNTAIN GORILLAS

Mountain gorillas use a variety of vocalizations to communicate, both within and beyond their social group. Calls aimed outside the group are given primarily by adult males in response to potential danger, such as a human hunter or a rival silverback. These calls convey alarm/threat and include various types of 'barks', more intimidating roars, and screams; they are sometimes accompanied by a charge. When encountering another group or a lone male, adult males also give a form of 'long call', which is a series of loud, resonant hoots, usually combined with displays such as chestbeating or ground thumping.[31, 124]

Intragroup vocalizations are quieter and less energetic, but far more frequent and varied. Some of these signals occur in specific contexts and often evoke specific responses. To human observers, the meaning of the calls is often quite clear, as for the mildly aggressive 'cough–grunt', the whimpering of an infant that has lost its mother, the breathy 'chuckles' given only during play, or the staccato whimpers that accompany copulation.[31]

Far more mysterious are the frequent, quiet, 'close calls' that gorillas give throughout the day in various nonspecific contexts. The most common of these signals are atonal, belch-like grunts, usually of one or two syllables, that sound much like a human male clearing his throat. Other 'close calls' include syllable-free grumbles, and higher-pitched tonal calls, similar to human humming and singing.[62, 127]

If vocal communication is viewed as a form of social behavior, then 'close calls' are the most frequent social interaction between gorillas. In two study groups at Karisoke, adult gorillas vocalized about once every eight minutes. Over half of these calls occurred as part of an exchange, in which a vocalization was 'answered' by a call from another individual. A key feature of this vocal behavior is that gorillas usually give and exchange calls when other individuals are nearby, within 2–5 m.[60]

The vocal habits of gorillas correlate with other aspects of their social behavior. For example, the nature and frequency of 'close calls' are related to age and dominance status. Adult males, who dominate other group members, vocalize more frequently than do adult females, who are in turn more vocal than younger and more subordinate immature animals. The adult vocal repertoire consists mainly of syllabled grunts, whereas younger gorillas do more humming and singing.[59, 62, 127] One obvious, but not exclusive, context in which adult females grumble or hum intensely is when they are near an adult male who has just displayed. In this case, the vocalizations seem to signify subordinance and act as appeasement signals.[62, 134, 177] Most of the time, however, it is not clear what prompts an animal to vocalize, or what purpose the signal might serve. The syllabled grunts are particularly enigmatic. The animals grunt most frequently during feeding, while traveling, or resting. Calls evoke either no discernible response or, at most, a vocal answer from another animal.[31, 62]

While acoustical analyses indicate that many grunts are individually distinctive (suggesting that gorillas can recognize each other from their grunts), few features of the sounds relate to behavior.[127] As far as we can tell, grunts given during feeding are the same, acoustically, as those given during resting. It is possible that these signals convey a

The best-documented example of this decline in population is in the mountain sector of Kahuzi-Biega; here only 130 eastern lowland gorillas remained in 1999, down from 245 in the same location in 1996.[112, 184, 198] The eastern lowland gorilla population in the lowland sector of Kahuzi-Biega is believed by the park wardens to have suffered even greater casualties; a crash in all populations of large mammals is inferred from the reported lack of meat of these formerly abundant species in bushmeat supplies sold by hunters to coltan miners.[113] At the beginning of the coltan rush, the miners in the lowland sector of Kahuzi-Biega mostly ate large mammals; toward the end, they relied upon small mammals, birds, and turtles. The conflict situation has prevented field surveys, but the Wildlife Conservation Society was coordinating a gorilla survey in 2004–2005; it is hoped this will offer a more solid estimate of remaining numbers.

In summary, about 700 mountain gorillas and thousands of eastern lowland gorillas still survive. Both subspecies have declined significantly in numbers. This process is ongoing (perhaps catastrophically so) for eastern lowland gorillas, while the mountain gorillas have been slowly increasing since the early 1980s (Table 8.2). Both the Virunga

general message such as "I am about to change activity", or simply, "I am here". The function of this communication will then depend on the context. For example, during feeding periods, vocalizations might be important in interindividual spacing and the avoidance of feeding competition.[59] In other situations, 'close calls' seem to play a role in co-ordinating group movement and activity. Toward the end of a midday rest period, resting gorillas increase the frequency of their grunting, as if to indicate that they are ready to end the siesta. They seem to be signaling their 'intent' to move on, but wait to do so until they have heard from the rest of the group. Even when the animals are doing nothing but lying still, an observer can often tell when the rest period is about to end, just from the increase in 'conversation'.[141]

All our data on vocal communication in the wild comes from the mountain gorillas of the Virunga Volcanoes. Studies of western gorillas in captivity and preliminary observations in the wild suggest that the vocal repertoires of other populations of gorillas are generally similar. We still have much more to learn, however, about gorilla vocal communication.

Kelly J. Stewart

A.H. Harcourt/Anthrophoto

Above: A young silverback hooting during a chest-beating display. Below: An adult female and silverback playing; they have just sat back from some gentle wrestling. The female is beating her chest. They both have the open-mouthed 'play face' that accompanies the breathy pants known as play chuckles. These vocalizations are characteristic of, and very specific to, play. They are given by young infants upwards.

and Bwindi populations of mountain gorillas were classified (separately, because of the uncertainty over their taxonomic status) by IUCN as Critically Endangered, on the basis of their small population sizes, with fewer than 250 adults in each case; eastern lowland gorillas were classified as Endangered, albeit on the basis of the 1998 estimate of population numbers.[73]

Threats

Hunting

Gorillas are hunted for their meat, as specimens (particularly infants) for collections, and as trophies. The hunting of gorillas for sale as trophies (skins, heads, skulls, feet, and hands – sold, for example, as ash-trays) emerged in the mid-1970s, and continued until quite recently.[112, 142] Occasionally individual gorillas that raid the crops of local people are killed.[49] For example, a young mountain gorilla was stoned to death in January 2003 when his group damaged fields near the border of the Virunga National Park;[84] Rugendo, the previous silverback of this habituated group, had been killed in 2001 in crossfire.[11]

Infant gorillas have been captured for sale, or attempted sale, to public or private collections,

Gordon Miller/IRF

A plaque in Bwindi National Park commemorates seven of those who died at the hands of militia in March 1999.

and many adults have been killed while trying to protect their infants from this fate.[112] The capture of infant mountain gorillas in the Virungas was a serious problem in the 1970s, although it declined greatly through the 1980s and into the 1990s. In 1995, however, four adult gorillas were killed in Bwindi,[3] and there have been reports of infant gorillas being taken for sale to private collectors.[151] Poaching leading to the deaths of at least seven gorillas occurred in 2002 in the Virungas;[80] in 2003, nine Rwandan poachers were fined and imprisoned for two to four years each for stealing a baby gorilla in Volcanoes National Park, and for killing two adult gorillas that had been protecting it.[7] Hunting remains a threat in the Virungas.

In response to the situation in DRC from the late 1990s onwards, the United Nations Security Council established an expert panel on the illegal exploitation of natural resources in DRC. It concluded that the various armies active in DRC were systematically exploiting five natural resources either to finance themselves or to exchange for weapons; these were diamonds, copper, cobalt, gold, and coltan.[160, 161] Coltan is an alluvial ore of niobium (columbium) and tantalum, metals that are used in the manufacture of mobile telephones and computer equipment. The ore has a ready market, and its high value has attracted miners to locations where it is abundant, including rivers in Kahuzi-Biega.[6, 72, 128] Professional hunters joined the miners to provide meat for them, and the eastern lowland gorillas of Kahuzi-Biega were severely affected.[111, 113] More information on the decline of eastern lowland gorillas can be found in the DRC country profile in Chapter 16.

Traditionally, gorillas were rarely eaten in the eastern Congo Basin, which has given the eastern gorilla a certain amount of protection. These traditions are weakest in areas inhabited by the eastern lowland gorilla and, as seen in Kahuzi-Biega, are fast becoming a thing of the past. They were and remain strongest, however, around the Virungas and Bwindi, providing continued protection to the mountain gorillas there.[112]

War and political unrest

Wars kill gorillas as well as people, and death can disrupt gorilla groups as effectively as it does human communities. Gorilla groups may disintegrate in response to losses, particularly of the dominant silverback, which can result in additional mortality and declining populations.[80] Armed conflict and political unrest have taken their toll on both the eastern lowland gorilla and on the mountain gorilla populations, with a series of conflicts and wars which have affected the people, landscapes, and wildlife of DRC, Rwanda, and Uganda.

The early 1990s saw the outbreak of fighting in Rwanda, including within the Virungas; by April 1994, this had expanded into DRC and resulted in a stream of refugees pouring into the gorilla habitat and its surrounds. About half of Rwanda's civilian population was displaced during this conflict, with 860 000 refugees being concentrated in the vicinity of the Virunga National Park, and a further 332 000 having fled into DRC near Kahuzi-Biega.[27] Soon after the 1994–1995 influx of Rwandan refugees into DRC came the 1996 war in DRC; fighting broke out again in 1998.

Refugees can put massive pressure on gorillas and their habitats through uncontrolled harvesting of wood for fuel, increased hunting, and disruption of migration patterns. During the war in Rwanda, three of the four refugee camps in North Kivu were located in or near to the Virunga National Park buffer zone; much of the park has been affected by wood harvesting or poaching.[152] Subsequent conflict in DRC led to looting and destruction of the park's infrastructure, and the deaths of about 5 percent of the mountain gorilla population in the

Virungas.[80] These factors led to the Virunga National Park being placed on the 'World Heritage in Danger' list in 1994.[157] As described above, hunting for gorilla meat in Kahuzi-Biega has increased as a result of war and displacement.[111, 113]

In addition to the influx of refugees, the forests that are home to gorillas have served as hiding places and retreats for rebel forces, leading to disturbance and hunting. This is a common phenomenon at times of war in forests that straddle international borders.[112]

The long-term impacts of the recent wars in Central Africa are unclear, and the civil wars in Rwanda and DRC have made it difficult to assess how the mountain gorillas have fared,[111] although some censuses have been carried out.[68, 80] One hopeful sign relates to the mountain gorillas in the eastern Virungas. This small and somewhat isolated subpopulation numbered about 57 in 1989 and, despite intense military activity in the early 1990s, there appeared to be at least 57 gorillas remaining in 2000.[80] The lowland protected areas of DRC, where most of the eastern lowland gorillas occurred during the 1990s, remain inaccessible to researchers so it is difficult to assess their status.[111] The population in the area around Tshivanga in Kahuzi-Biega was relatively stable between 1990 and 1996[69] but, since then, two rebellions have occurred, with large numbers of eastern lowland gorillas being killed.[113, 201] Over just four years, the highland sector of Kahuzi-Biega lost more than 95 percent of its elephant population and about 50 percent of its gorilla population. Local resentment toward the park and its authorities may have contributed to this illegal exploitation of wildlife resources.[198]

Conflict can also deter international conservation organizations, aid agencies, and governments from investing in affected areas, leading to frozen budgets, withdrawal of staff, reduction in antipoaching efforts, and the closure of projects. Nevertheless, some organizations continued to support park authorities in the Virungas throughout the war,[80] even though research programs were interrupted. Protection of the gorillas in many areas has proved extremely difficult and often hazardous in recent years, and many national conservationists take tremendous risks in the course of their work, sometimes with fatal consequences. Ten staff and assistants of ICCN, for example, were murdered, apparently by militiamen who had been hiding in DRC since the genocide in Rwanda, while surveying Kahuzi-Biega boundaries to reestablish the park limits.[22, 63] These were not the first or the only park-service employees to be kidnapped or killed while they were attempting to protect the area and its wildlife.[69, 74, 75, 125, 126] In all, 92 Congolese park staff are reported to have been killed between 1996 and 2004.[67] During the conflicts in Rwanda, several workers from Karisoke lost their lives, others were imprisoned, and the center itself was destroyed;[142, 191, 192] much more international attention was drawn by the killing of eight tourists and four guides at Bwindi by Interahamwe militia in March 1999.[10] Without the determination and commitment of park rangers, it would be impossible to imagine the long-term survival of the eastern gorilla.

Habitat loss or modification

The mountain and eastern lowland gorillas live surrounded by some of the densest rural human populations in Africa, with up to 300–600 people per square kilometer, and a correspondingly high demand for land and food.[112, 145] As a result, gorillas are increasingly confined to smaller and more isolated forest fragments as human populations increase.[14]

Gordon Miller/IRF

Cattle herding in Uganda.

Habitat loss, specifically forest clearance for agriculture, was one of the main causes of population decline among mountain gorillas during the 1970s.[14, 142] In 1968, more than one third of the Rwandan Volcanoes National Park was excised for an agricultural project.[112] Little forest cover now remains in Rwanda, and virtually no forest habitable by gorillas remains outside protected areas.[83] The boundaries of protected areas are generally respected by planners and farmers, so there has been very little further habitat loss in Rwanda, although disturbance from increased human presence, social instability, genocide, and war has occurred.[111, 112] The forest has also been used as a source of wood for building and fuel, and is accessed both for water and to graze cattle.[112]

In Mgahinga Gorilla National Park, Uganda, agricultural and pastoral activities and hunting were major pressures; incursions by local people and their livestock used to be common.[70] The park has a complex history of designation changes, having originally been defined both as a game reserve and a forest reserve, established in 1930 and 1939 respectively. The boundaries for each reserve were defined by the contour line running at 2 425 m, on the lower slopes of the three volcanoes in what is now Mgahinga Gorilla National Park. In 1951, the forest reserve boundary was raised to 2 730 m, thereby significantly reducing its area and removing some important gorilla habitats.[1, 154] After the Mgahinga Gorilla National Park was designated in 1991, people living in this area were evicted. Meanwhile, the game reserve boundary was lowered to the 2 280 m contour in 1964, significantly increasing its area and including land that was already settled. The designated national park encompasses part of this additional game reserve area,[81] which means that a large community with a tradition of extractive use of park resources occurs both within and adjacent to the park. A community-based conservation program is now attempting to balance the needs of the people and the wildlife.

In DRC, demand for fuelwood by Rwandan refugees affected 105 km^2 (1.3 percent) of Virunga National Park by 1997, of which 35 km^2 had been completely cleared.[152] Since 2001, much of the Kirolirwe sector has been cleared by refugees returning to DRC from Rwanda, who were settled there by the Rassemblement Congolais pour la Democratie, an armed opposition movement.[159] Another 15 km^2 of land was cleared by Rwandan farmers in May 2004 in the Mikeno sector, also on the DRC side of the park.[67] After international protest, Rwandan soldiers removed the 6 000 loggers and farmers, killing two,[4] and the park's drystone boundary wall was rebuilt. As DRC becomes more stable, it is likely that commercial logging companies will quickly move into its forests.[111] This could well impact eastern lowland gorillas, but it is unlikely that large-scale logging would occur in the high-altitude forests of the Virungas. Gorillas often favor areas of secondary vegetation, and so might be able to coexist with logging, if it were not for the associated hunting.[111]

The Bwindi population of mountain gorillas is relatively well protected. Prior to the 1980s, manual felling and head-load extraction (i.e. the removal of no more than the quantity of wood, usually branches, that can be carried on one's head) of timber was permitted throughout the area, which was then a forest reserve.[70] These nonmechanical techniques made for very selective and environmentally benign logging. Nevertheless, only about 10 percent of the forest in Bwindi is entirely free of past human disturbance.[151] No data are available on the intensity and distribution of habitat disturbance since Bwindi was declared a national park, since when antipoaching and other enforcement efforts are thought to have led to much reduced levels of disturbance.[92]

Eastern lowland gorillas and their habitats face similar problems of habitat loss, which add to the impacts of hunting that have been noted above. The increasing human population and the corresponding need for land is a serious and ongoing pressure.[49] The boundaries of Kahuzi-Biega were altered in 1974, resulting in the loss of an important area of gorilla habitat.[49] It has been suggested that the rate of loss of habitat for the eastern lowland gorillas is probably the highest for any gorilla subspecies, but the lack of clarity about the situation in DRC means that no absolute figures are available.[111] The fuelwood reserves outside Kahuzi-Biega have been severely depleted by refugees, so fuelwood collection within the park is an ongoing threat.[111, 113]

Disease transmission from humans

Gorillas are susceptible to many human diseases, as detailed in Chapter 7; increased exposure of gorillas to humans or to human feces is occurring as more people live in or around the forests, or enter them more often because they are displaced by conflict.[190] Disease may be carried by park guards, researchers, tourists, tour guides, loggers, hunters, or by local people using nearby roads. Data on the impacts of disease among eastern gorillas, particularly outside the Virungas,[111] are limited, but the Ebola virus has not affected eastern gorilla populations.

Some eastern gorillas carry parasites including protozoans (e.g. *Cryptosporidium* spp.)[42] and nematodes (e.g. *Capillaria hepatica*),[43] but these parasite loads might be unrelated to human presence.[95] Mountain gorillas are also susceptible to the skin mites that cause scabies or mange (*Sarcoptes* spp.), an outbreak of which, in a habituated group in Bwindi in 1996, led to the death of an infant male, probably from secondary infection of scratch abrasions.[41, 79] The source of this disease is unknown, but is suspected to have originated among the people and livestock living around the park, where it is prevalent.[79] Another outbreak of mites occurred in Bwindi in 2000, but did not result in any deaths.[94] Much more seriously, an outbreak of pneumonia in Rwanda's Volcanoes National Park in 1988, possibly with an acute viral infection such as measles as the primary infection, claimed the lives of six gorillas, but 27 others were treated successfully.[130, 193] The high rate of infection (81 percent) suggested that the disease was new to these gorillas.[14] Vaccination against measles was subsequently given to 65 habituated gorillas from this population.[130, 167, 193]

While tourism can make a vital contribution to conservation by generating funds and through education, it does represent a potential source of disease[14, 16] that could threaten small populations.[95] In addition, disturbance through contact with humans may increase stress and thereby susceptibility to disease.[89] The expansion of gorilla tourism exposes more gorillas to diseases that they may never have encountered before and against which they may have no natural immunity, while encouraging protection of the gorillas from habitat loss and hunting. Healthy, fee-paying tourists who contribute strongly to financing conservation and to building political support, deter poachers by their mere presence, so in most circumstances these tourists are likely to be on balance beneficial to gorillas. A survey in 1981 compared reproductive success in 'guarded' gorilla groups exposed to tourism with that of unguarded groups. The latter were found to have a smaller proportion of immature animals.[56] Infants are not only often direct targets of hunting but, as discussed above, are likely to suffer disproportionately when groups are broken up.

Alastair McNeilage

Tourists at Mgahinga Gorilla National Park, Uganda.

In the Virungas and Bwindi, strict rules regulating tourism are in place (though not necessarily always obeyed). These limit tourist visits to one hour per day, set a maximum group size of eight tourists, and require tourists to maintain a minimum distance from gorillas of 7 m.[112] Other disease prevention measures include burying human excrement deeper than 30 cm and chasing gorillas away from private lands that surround the parks.[65, 79] Veterinary assistance is also available at these mountain gorilla centers. In Uganda, veterinary intervention is limited to diseases caused by human beings and to life-threatening conditions that could affect a substantial number of gorillas in a group.[79] The Mountain Gorilla Veterinary Project in Rwanda has a similar nonintervention policy, with restrictions on emergency treatment to illnesses that could threaten the group or population.[95]

Other threats

Gorillas can easily be caught in wire snares set for ungulates; this can result in the loss of a hand or foot.[95, 112] The three research groups in Volcanoes National Park reported 50 snare injuries to gorillas between 1971 and 1998, four of which had fatal consequences. Snares set for medium-sized mammals such as antelopes also wounded many eastern lowland gorillas in Kahuzi-Biega.[49, 207] Of the groups habituated for tourism in the montane sector of the park, at least one individual per group had lost a hand in a snare.[49] Snares are therefore considered an important threat to eastern gorillas in the Volcanoes National Park and elsewhere.[109]

The isolation and small size of mountain gorilla populations has given rise to concerns about inbreeding. However, two studies have suggested that the Virunga population, which is of much the same size and composition as the Bwindi population, is likely to be safe from genetic problems for 400 years or more.[2, 29] A comparison of a sample of Bwindi gorillas and western lowland gorillas shows only minimal reduction of genetic variability (heterozygosity) in the Bwindi gorillas, despite their small population size.[87] Nevertheless, every effort should be made to maintain or restore habitat connectivity and gene flow between gorilla populations, wherever the risks of disease transmission between the reconnected populations are exceeded by the benefits of expanding the gene pool.

Elizabeth A. Williamson

A silverback sits with a young female, whose foot (just visible) was injured by a snare, Virunga National Park, Democratic Republic of the Congo.

CONSERVATION AND RESEARCH

Conservation activities and research focused on eastern gorillas have been underway for many years. These prolonged efforts have met with much success although many problems persist. Despite the significant threats associated with warfare in the region, mountain gorilla population numbers – although small – appear to be stable and, in some cases, increasing. Eastern lowland gorilla populations are, however, declining – possibly very quickly. Increased numbers of mountain gorillas in the Virungas are probably a direct result of protection efforts, and are concentrated in one or two areas.[80, 137] These findings indicate that, with local commitment and sufficient investment, it is possible to protect gorilla populations.

The eastern gorilla is protected by national legislation in all three of its range states, and most known populations live in protected areas that are not all, or not only, 'paper parks' (areas protected in law, but not in practice). Where park rangers are present and local residents supportive, gorilla populations have a good chance of survival. A park ranger's work may include monitoring gorilla populations, patrolling for poachers, law enforcement, and community development work. That gorilla parks can make a real difference to local attitudes is illustrated by the commitment of staff members, who have been known to risk their lives in defense of their parks, even when pay has not always been forthcoming. Cooperation and coordinated efforts in park management involving the governments of Rwanda, Uganda, and DRC, supported by researchers and national and international nongovernmental organizations, have contributed to the conservation of the mountain gorilla throughout its range, and will continue to do so.

Conservation and research activities

Our growing understanding of gorilla biology (including such features as group structure and dynamics, ranging behavior, habitat requirements, and population densities), has contributed in many ways to the selection of protected areas and the design of conservation action. It has also contributed indirectly to the raising of global public and political awareness, and of much-needed funds. Population monitoring reports are particularly helpful in management, because they provide feedback on what is working and what is not, as well as early warning of new kinds of threat. This allows gorilla conservation to adapt over time, to become increasingly effective.

The Karisoke Research Center, managed by the Dian Fossey Gorilla Fund International, has sustained studies of mountain gorillas since 1967. These have included long- and short-term census work, as well as studies on social structures, group dynamics, feeding behavior, habitat use, and reproduction.[122] Because of Karisoke, the only period without regular monitoring of mountain gorillas was 15 months during 1997–1998, a time when armed conflict prevented personnel from entering the park.[142] In addition, the Mountain Gorilla Veterinary Project established a veterinary center to monitor the health of the gorillas and act in emergency situations, including the removal of snares from gorillas and dealing with disease outbreaks. Eastern gorilla studies have more recently been extended to the Bwindi mountain gorillas and the eastern lowland gorillas of Kahuzi-Biega and elsewhere.[38, 121, 200]

The Impenetrable Forest Conservation Project led to the establishment of the Bwindi Impenetrable National Park and a buffer zone in 1992.[71] The Institute of Tropical Forest Conservation, part of Uganda's Mbarara University, is the successor institution of this project. It has an active ecological monitoring program that is studying water quality, the impact of forest fires, and forest-gap dynamics. Other research includes work on barriers to crop-raiding by gorillas and a long-term project on the ecology, behavior, and population dynamics of the Bwindi mountain gorillas. This research supported the preparation of a management plan for the park, which was updated in 2001 to guide actions for tourism development, biological inventories, and other measures that are now in place.

In Kahuzi-Biega, a long-term community-based conservation project was established in 1985 with the support of the German overseas development agency GTZ, with community-focused economic development as one of its primary objectives.[15, 69] Managers at Kahuzi-Biega and GTZ developed an emergency plan for, among other things, collecting and distributing fuelwood in response to the refugee crisis of the late 1990s. GTZ has also helped to fund gorilla population censuses, including one in Kahuzi-Biega that was also supported by the Wildlife Conservation Society and others.[184] In the same region, local people, including park guards and guides, established a nongovernmental organization that helped to spread conservation knowledge and reduce conflict among local people.[198]

The revenues created by gorilla tourism have channeled significant resources into the protection of gorillas and parts of their habitat (see Box 8.4 and Chapter 14). In Uganda, the money so generated is distributed throughout the system of national parks, not just among the gorilla parks, making a broad contribution to national needs and building political support for gorilla conservation, albeit at the cost of diluting the funds available for managing gorilla populations and habitats. Conflicts deter tourists, but during those of the 1990s, the authorities of gorilla range states (the Uganda Wildlife Authority, the Office Rwandais du Tourisme et des Parcs Nationaux, and the ICCN) did what they could to maintain conservation efforts. The decrease in revenues from tourism led, however, to huge enforcement problems. This was partially offset in the Virungas by the contribution of additional funds and other resources by outside organizations. Some of the extensive educational and outreach programs developed prior to the conflict also continued.[80] The continuity of these efforts was made possible largely by international nongovernmental organizations such as the

Elizabeth A. Williamson

Firewood collection in the region of the Kahuzi-Biega National Park, Democratic Republic of the Congo.

Box 8.4 EASTERN GORILLA TOURISM

In Rwanda, Uganda, and DRC, gorilla tourism generates significant revenue, increases public awareness, and has undoubtedly been a motivator in securing government commitments to the protection of gorillas and their habitats. It should not, however, be seen as an ideal solution to the very specific problem of gorilla conservation, as gorillas are exposed to considerable risk through the consequent increased contact with humans.

The first project to develop gorilla tourism began in Kahuzi-Biega in the 1970s. Far better known, though, is the program established in Rwanda a decade later in response to plans to clear a large area of the Volcanoes National Park for cattle grazing. Habitat loss was viewed as the greatest threat to the survival of the gorillas, so a carefully planned and well controlled tourism program began as a means of making the gorillas 'pay for themselves', and further conversion of park land was averted.

The conservation benefits of this program include increased surveillance of gorilla groups habituated for tourism, and more antipoaching patrols. Daily monitoring also facilitates rapid intervention by veterinarians when necessary, for example, to remove snares from injured gorillas. With increased protection from poachers, the mountain gorilla population began to recover. International awareness and concern for the plight of gorillas has generated funds for conservation activities and research, at the same time enhancing the profiles of the gorilla range states. The gorilla was adopted as a national symbol in both Rwanda and Zaire (now DRC), and is depicted on bank notes, stamps, postcards, carvings, and murals. Today both the Rwandese passport and visas for foreigners feature mountain gorillas.

International publicity and the advent of organized tourism have attracted many visitors and made tourism an important earner of foreign currency. Tourism stimulates the economy, not only via park fees, but also through expenditure on car hire, hotel accommodation, and restaurant meals. People from communities around the parks may gain employment as guides or porters, while in Uganda, a fixed proportion of the revenue from Mgahinga Gorilla and Bwindi Impenetrable National Parks is contributed to local schools and health centers through a trust fund.

Research has begun only recently on the impact on gorilla behavior of tourist visits, and on the risk of disease transmission between humans and gorillas. Prior to these studies, conservationists relied on speculation, extrapolation, and common sense to evaluate these. Studies of captive gorillas show them to have a definite susceptibility to human diseases, leading Homsy to warn of "the catastrophic consequences of unconscious gorilla tourism."[65] Illnesses to which the gorillas have never previously been exposed are potentially the most dangerous and international tourists may carry viruses new to the region, such as novel strains of influenza. To minimize stress and risks to both gorillas and humans, there are very important regulations regarding minimum distances to be maintained between gorillas and people, the maximum number of tourists, and the duration of their visits, as well as guidelines for appropriate visitor behavior. A tourist should never attempt to get closer than the regulation 7 meters or, worse still, to touch a gorilla.

Tourism is a lucrative business, which puts pressures on the gorillas and on park authorities, leading some people to question the continued justification for gorilla viewing. The cost of gorilla-viewing permits must be set at a level that limits demand, while maintaining the revenue that needs to be accrued by the governing authorities. Despite the dangers inherent in tourism, it provides a mechanism for ensuring that national parks and the gorillas are valued for many reasons, and has certainly contributed to their survival.

Elizabeth A. Williamson

A tourist and ranger enjoy the antics of a young gorilla in Virunga National Park.

Gordon Miller/IRF

International Gorilla Conservation Programme (IGCP) of the African Wildlife Foundation, Fauna and Flora International, and WWF–The Global Conservation Organization. IGCP has run a number of projects, is involved in population censuses, and works with national institutions and agencies to support conservation efforts, strengthen resources, and build capacity.

Other international organizations are also involved in eastern gorilla conservation, often in collaboration with local organizations. For example, the Wildlife Conservation Society has projects in all of the eastern gorilla range states:

- in DRC, it is involved in gorilla monitoring, re-establishment of park infrastructure, habitat mapping, and exploration of the lowland sector in Kahuzi-Biega;
- in Uganda, it is undertaking a biological survey of Bwindi, a census of the gorilla population, and studies on the impacts of tourism on gorilla behavior; and
- in Rwanda, it provides guard support in the Virungas and is undertaking a study of crop-raiding patterns around the Volcanoes National Park.

Habitat monitoring is complementary to population monitoring, providing early warning of potential threats to gorilla ecology, and measuring the success of conservation management. The United Nations Educational, Scientific and Cultural Organization (UNESCO) and all the international space agencies established the 'Open Initiative' project, which aims to help countries to monitor World Heritage Sites via the use of satellite images.[158] In April 2003, the European Space Agency provided significant funding and technical support for a joint project with UNESCO called Build Environment for Gorillas (BEGo). A series of maps of national parks in inaccessible mountain areas (up to 5 000 m) that are home to the mountain gorilla is being produced for Uganda, Rwanda, and DRC. Comparisons with 1992 satellite images will allow the assessment of changes in gorilla habitats in World Heritage Sites.[158] In a separate exercise, a computer simulation of the Virungas was developed for the Dian Fossey Gorilla Fund International. It aims to plot the movements of gorillas through a virtual reserve, to show habitat preferences, deduce the carrying capacity of the reserve, monitor human activities (including poaching), and to assist in the management of the national parks.[188]

In conclusion, mountain gorillas survive in small but apparently stable populations in several national parks in the Virungas, and in Bwindi. These parks are managed and otherwise supported by the governments of DRC, Rwanda and Uganda, and by conservation and research groups, with gorilla-based tourism programs yielding significant funding. These gorilla populations are potentially vulnerable to disease and hunting but, by the global standards of great ape conservation, they are relatively secure at present.

Nothing similar can be said about the eastern lowland gorilla, however, the population status of which is largely unknown following the recent spread of warfare throughout its range. Many may have been killed to provide bushmeat for armed factions, displaced people, and miners, and the entire population may have collapsed as a result. As the military and political situation remains highly unstable, it is very difficult for conservationists to undertake the fieldwork required to clarify the circumstances of these gorillas, much less to support local people in their efforts to achieve sustainable development. The fates of humans – and their needs for good governance, prosperity, and peace – are intertwined with the fate of the wildlife with which they share their environments.

FURTHER READING

Butynski, T.M., Kalina, J. (1998) Gorilla tourism: a critical look. In: Milner-Gulland, E.J., Mace, R., eds, *Conservation of Biological Resources*. Blackwell Science, Oxford. pp. 280–300.

Fossey, D. (1972) Vocalisations of the mountain gorilla (*Gorilla gorilla beringei*). *Animal Behaviour* **20**: 36–53.

Hall, J.S., Saltonstall, K., Inogwabini, B.I., Omari, I. (1998) Distribution, abundance and conservation status of Grauer's gorilla. *Oryx* **32** (2): 122–130.

Harcourt, A.H. (1986) Gorilla conservation: anatomy of a campaign. In: Benirschke, K., ed., *Primates: The Road to Self-Sustaining Populations*. Springer-Verlag, New York. pp. 31–46.

Harcourt, A.H., Stewart, K.J., Hauser, M. (1993) Functions of wild gorilla 'close' calls I. Repertoire, context, and interspecific comparison. *Behaviour* **124**: 89–122.

Hart, J., Hart, T. (2003) Rules of engagement for conservation: lessons from the Democratic Republic of the Congo. *Conservation in Practice* **4** (1). http://conbio.org/inpractice/article41RUL.cfm. Accessed July 13 2004.

Homsy, J. (1999) *Ape Tourism and Human Diseases: How Close Should We Get?* Report to the International Gorilla Conservation Programme. http://www.mountaingorillas.org/files/ourwork/Homsy_rev.pdf. Accessed February 10 2005.

Hrdy, S.B. (1979) Infanticide among animals: a review, classification, and examination of the implications for the reproductive strategies of females. *Ethology & Sociobiology* **1**: 13–40.

Kalpers, J., Williamson, E.A., Robbins, M.M., McNeilage, A., Nzamurambaho, A., Lola, N., Mugiri, G. (2003) Gorillas in the crossfire: population dynamics of the Virunga mountain gorillas over the past three decades. *Oryx* **37** (3): 326–337.

Robbins, M.M., Sicotte, P., Stewart, K.J., eds (2001) *Mountain Gorillas: Three Decades of Research at Karisoke*. Cambridge University Press, Cambridge, UK.

Schaller, G.B. (1963) *The Mountain Gorilla: Ecology and Behavior*. University of Chicago Press, Chicago.

Seyfarth, R.M., Cheney, D.L., Harcourt, A.H., Stewart, K.J. (1994) The acoustic features of gorilla double-grunts and their relation to behaviour. *American Journal of Primatology* **30**: 31–50.

Stanford, C.B., Nkurunungi, J.B. (2003) Behavioral ecology of sympatric chimpanzees and gorillas in Bwindi Impenetrable National Park, Uganda: diet. *International Journal of Primatology* **24** (4): 901–918.

Taylor, A.B., Goldsmith, M.L., eds (2002) *Gorilla Biology: A Multidisciplinary Perspective*. Cambridge University Press, Cambridge, UK.

van Schaik, C.P., Janson, C.H. (2000) *Infanticide by Males*. Cambridge University Press, Cambridge, UK.

Watts, D.P. (1995) Post-conflict social events in wild mountain gorillas (Mammalia, Hominoidea) I. Social interactions between opponents. *Ethology* **100**: 139–157.

Weber, A.W. (1993) Primate conservation and ecotourism in Africa. In: Potter, C.S., Cohen, J.I., Janczewski, D., eds, *Perspectives on Biodiversity: Case Studies of Genetic Resource Conservation and Development*. AAAS Press, Washington, DC. pp. 129–150.

Yamagiwa, J., Kahekwa, J., Basabose, A.K. (2003) Intra-specific variation in social organization of gorillas: implications for their social evolution. *Primates* **44**: 359–369.

Yamagiwa, J., Mwanza, N., Yumoto, Y., Maruhashi, T. (1994) Seasonal change in the composition of the diet of eastern lowland gorillas. *Primates* **35**: 1–14.

MAP SOURCES

Map 8.1 Eastern gorilla data are based on the following source, with updates as cited in the relevant country profiles in Chapter 16:

Butynski, T.M. (2001) Africa's great apes. In: Beck, B.B., Stoinski, T.S., Hutchins, M., Maple, T.L., Norton, B., Rowan, A., Stevens, E.F., Arluke, A., eds, *Great Apes and Humans: The Ethics of Coexistence*. Smithsonian Institution Press, Washington, DC. pp. 3–56.

For protected area and other data, see 'Using the maps'.

ACKNOWLEDGMENTS

Many thanks to Dan Bucknell (Dian Fossey Gorilla Fund), Thomas Butynski (Conservation International), Colin Groves (Australian National University), Alexander Harcourt (University of California, Davis), José Kalpers (International Gorilla Conservation Programme), Michael Wilson (Gombe Stream Research Center), and Juichi Yamagiwa (Kyoto University) for their valuable comments on the draft of this chapter.

AUTHORS

Sarah Ferriss, UNEP World Conservation Monitoring Centre
Martha M. Robbins, Max Planck Institute for Evolutionary Anthropology
Elizabeth A. Williamson, University of Stirling
Box 8.1 Juichi Yamagiwa, Kyoto University
Box 8.2 Martha M. Robbins, Max Planck Institute for Evolutionary Anthropology
Box 8.3 Kelly J. Stewart, University of California, Davis
Box 8.4 Elizabeth A. Williamson, University of Stirling

CHAPTER 9

Orangutan overview

JULIAN CALDECOTT AND KIM MCCONKEY

The lineage that led to modern orangutans is thought to have diverged from that of the African apes and humans about 11 million years ago (mya), presumably somewhere in the Asian mainland. Ancestral orangutans then inhabited the areas that would become the Malay Peninsula of mainland Asia, as well as the islands of Java, Sumatra, and Borneo. This biogeographical unit, known as Sundaland,[28] comprises the lands above and below the shallow transient seas of the Sunda continental shelf (see Map 2.1). Throughout the Cenozoic era, the main land masses in Sundaland have been joined and separated repeatedly from the mainland and from one another by changing sea levels associated with high-latitude glaciations and interglacial periods.

The forests of Sundaland are characterized by a general abundance of the tree family Dipterocarpaceae. These Sundaic dipterocarp forests typically have a poor and irregular fruit supply due to their mast fruiting behavior. Within Sundaland, the abundance of dipterocarps and scarcity of other fruit trees are strong determinants of biomass among large-bodied frugivores. Many taxa show specific adaptation to these conditions and there are, for example, specifically Sundaic forms of macaque (*Macaca nemestrina*) and pig (*Sus barbatus*).[3] Much of what we know of orangutan behavior and ecology suggests a partial adaptation along the same lines.

The Bornean orangutan lineage diverged from the Sumatran 1.1–2.3 mya.[22, 27, 29] The range of estimates for speciation is such that it is possible that the two species were genetically isolated from one another before being physically separated from one another.[22] If so, speciation must have occurred through reproductive isolation: for example, a divergence of preferred mate characteristics in the two emerging species would lead to a reduction in gene transfer between populations.

Populations of the Sumatran orangutan (*Pongo abelii*) may contain remnants of three or

Ian Singleton/SOCP

Subadult male Sumatran orangutan just a few days after release in the forests adjacent to Bukit Tiga Puluh National Park, Indonesia.

Ian Singleton/SOCP

Secondary lowland forest in an old logging area near Bukit Tiga Puluh National Park in Jambi province, Sumatra, Indonesia, where orangutans are being reintroduced.

more separate lineages. After the initial divergence of the Sumatran type, a new influx of orangutans from Borneo and the Southeast Asian mainland is thought to have entered the area during one of the periods of land bridge connection. Interbreeding may have produced three types: one linked to the Sumatran form; one linked to the Bornean; and one with closer affinities to the now-extinct mainland orangutan.[18] No subspecies of the Sumatran orangutan are recognized.[8]

The dispersal of Bornean orangutans (*Pongo pygmaeus*) through the Bornean part of Sundaland, where Sundaic ecological conditions are more pronounced than elsewhere, started in the southwest corner; dispersal was constrained by large rivers and high mountain ranges, both of which can act as barriers to these animals. The population later became divided when climate change made many earlier dispersal corridors inaccessible. This shows up in minor differences in microsatellite and mitochondrial DNA among four regional populations,[27] and three subspecies are now recognized in the different parts of Borneo:[8, 19, 21]

- the northwest Bornean orangutan, *P. p. pygmaeus*, which is medium sized and occurs in northern West Kalimantan and Sarawak;
- the central Bornean orangutan, *P. p. wurmbii*, which is the largest subspecies and is found in southern West Kalimantan and Central Kalimantan; and
- the northeast Bornean orangutan, *P. p. morio*, which is the smallest and occurs in Sabah and East Kalimantan.

Both the Bornean and Sumatran orangutans are large and have obvious sex differences in appearance and behavior, with adult males weighing about 75 kg and adult females about 40 kg.[1] Fully developed adult males have prominent cheek pads or 'flanges', the development of which is linked to the individual achieving high social status, which can take as many as 10 years of adulthood and may not happen in all males.[4, 8, 20] Both species are long lived, and may reach 45 years in the wild.[12] There are a number of physical differences between the two species:

- the Bornean is stouter and stockier, and usually has a dark red–brown coat, rather than the lighter cinnamon fur of the more gracile Sumatran;
- the Bornean has little fur around the face, while Sumatran females have a distinctive beard and males a prominent beard and moustache;
- fully developed adult Bornean males have a large, pendulous throat sac and a distinctive figure-eight-shaped face founded on the presence of a suborbital fossa that is lacking in the Sumatran species, combined with forward-facing cheek pads or flanges, while Sumatran males have flat cheek flanges that are covered with downy hair.

ECOLOGY AND DISTRIBUTION

Like all apes, orangutans have simple, globular stomachs that cannot ferment food, so they are limited to eating materials that are not excessively fibrous, toxic, or protected by digestion-inhibitors such as tannins. In a rain-forest context, this translates into a diet of sugary, ripe fruit (which orangutans strongly prefer) and undefended seeds, plus a variety of minor items and 'famine foods' such as leaf shoots, insects, flowers, and bark. Orangutans are large bodied, so they can tolerate a certain amount of mildly toxic material, and will often eat soils that may help adsorb and neutralize secondary plant metabolites. Being strong and dextrous, orangutans can gain access to edible items such as seeds and palm hearts that are physically well defended by woody material or thorns. Their intelligence enables orangutans to memorize the locations of cryptic or temporary food sources, and to use clues – such as the behavior of other animals – to find fruiting trees and lianas.

Long arms, highly mobile hip joints, and opposable toes mean that orangutans are strongly adapted to arboreality. This allows them to move through a complex three-dimensional environment with great facility; orangutans spend most of their time in the trees and are deeply familiar with conditions in the canopy in their large home range.[1] They forage in a typically zigzag way through the forest unless they happen to know where large sources of fruit are to be found, when their travel becomes much more directed. Orangutans track the seasonal changes in production in the patchy rain forest, where the timing of fruiting peaks can vary with elevation and aspect. These are all attributes of an animal that has pushed a fruit-eating niche to the limits in a fundamentally rather fruit-poor environment.

Trees that provide fruit suitable for orangutans are typically found at higher densities in Sumatran forest than in Bornean forest, and bear it more continuously, although there is much patchiness and dynamism in the forests of both islands. An important feature is the lesser dominance of dipterocarp trees in Sumatra, which are replaced by other trees that collectively fruit more steadily. A number of differences between Sumatran and Bornean orangutan behavior have been attributed to the different patterns of food supply.[26] Sumatran orangutans have the opportunity to eat more fruit and to share it among adults, making for greater sociability; among other things, this allows orangutans to acquire tool-using behavior from one another. Figs in particular occur at such high densities in some parts of Sumatra that these fruit alone are thought to have enabled greater sociality among orangutans in Sumatra compared to those in Borneo.[4] Average home-range size, day-range length, and population density all respond to differences between locations in the abundance and continuity of fruit availability, seasons, and ecosystem types (both within and between Sumatra and Borneo). Detailed data on these are still emerging from the few long-term field studies being undertaken, and are supplemented by additional survey information on orangutan populations and forest composition. Deforestation in both islands, however, may prevent the discovery of many patterns in forest ecology.

The preindustrial distribution of orangutans was discontinuous in both Borneo and Sumatra.

Orangutan Foundation

A female Bornean orangutan and her offspring. The mother–infant bond is very close in orangutans.

Orangutans were, even then, absent from very large areas of apparently suitable habitat; examples include the forests in most of the southern two thirds of Sumatra, and those between the Rajang River in central Sarawak and the Padas River in western Sabah. Prehistoric hunting may have extirpated orangutans from some forests; ancient cave sites in areas now lacking orangutans often contain their bones, along with those of other species eaten by humans. Furthermore, some locations with abundant orangutans are associated with cultural influences on people's willingness to hunt them; for example, in the strongly Muslim Aceh province in Sumatra, and in the Batang Ai catchment in Sarawak, where a hunting taboo has long been in place among local Iban people.

An alternative interpretation is based on the theory that orangutans live close to the edge of an ecological niche that can become unviable with a slight change in forest composition in favor, for example, of dipterocarps. Patchiness in the distribution of breeding populations of orangutans could therefore simply reflect patchiness of the forests on which their survival depends. Both theories may be correct for different places, with local extirpations affecting some areas and ecological absence others.

SOCIETY AND PSYCHOLOGY

Orangutans are wide-ranging animals with behavior that has been interpreted as characteristic of 'residents', 'commuters', or 'wanderers'.[20] According to this interpretation, residents occupy a defined area over many years, but may go beyond it to exploit seasonally abundant foods.[7, 11, 13] Commuters are seen regularly in a particular area for several weeks, but vanish and return each year, repeating this behavior over several years; these individuals are assumed to commute between two or more regular feeding grounds and may follow waves of fruiting across areas or altitudes. Finally, wanderers are seen very infrequently, sometimes only once, and may never return to an area.

Other authors dispute this classification of orangutan ranging behavior, and interpret all observations in terms of orangutans having very large but stable and overlapping home ranges, that extend far beyond the study areas of human observers, and are therefore simply too large to monitor completely.[23] This would account for 'wandering' and 'commuting' behavior by orangutans that arrive and depart irregularly or at long intervals; it is becoming clearer that, the longer a field study continues, the more likely it is that occasional visitors will be seen again.

Whatever the interpretation of the observations, it is clear that orangutans are not territorial, and that neither sex excludes others from areas that they use habitually. Fully adult males, however, are intolerant of each other, so they may use the home ranges of other adult males only cautiously, if they wish to avoid combat. At one study site, as many as six males ranged independently over a given area at the same time, despite ferocious battles ensuing when they came into contact with one another.[11]

When young animals of either sex first gain independence from their mothers, they often range widely for a time before settling down, in the case of females often close to the home range of their mothers. Subadult males may continue to range widely for a lengthy period and, if they do settle, do so farther away from their mothers. New individuals in an area are almost always subadult males, and all 'wandering' individuals seem to be adult or older subadult males; up to 20 percent of these may never become established in a known location.

The transition between subadult and adult is a complex one in male orangutans of both species, as they exhibit a maturation process known as 'bimaturism'.[25] This is unique among the apes and is not yet fully understood. The timing of maturation is extremely variable, with puberty beginning any-

A mature flanged male orangutan.

Cindy Fromme/BOS-USA

where between the ages of five and 16 years, with the mean being somewhat earlier in the Bornean than the Sumatran species. There seem to be two alternative developmental pathways involved thereafter.[9, 10, 14] Some males develop certain features associated with high testosterone levels, including the prominent cheek pads,[4, 8] and reach full sociosexual maturity sooner than others. These males are therefore described as 'flanged', and those of a similar age that have not yet developed such features are called 'unflanged'. Some males remain unflanged and less than fully mature for up to 20 years.[14, 26]

Unflanged males have testosterone levels intermediate between those of flanged males and juveniles. A flanged male is thought to have certain advantages over an unflanged male, notably higher status that gives him more secure access to an established home range, food sources within it, and any receptive females that may be available. There are costs, however. A flanged male is large bodied and combative, so there are metabolic challenges and a higher risk of injury in fighting, and high blood levels of testosterone may also in themselves reduce lifespan. Mobility is also more costly for fully developed males due to their larger body size. These costs may be worth bearing only if the individual has a strong chance of gaining high status and becoming a mate of choice for females. Hence it is thought that the trigger for becoming flanged must be something that relates to the balance between benefits and costs of high and low testosterone levels.

In captivity, removal of the flanged male from an enclosure will prompt unflanged males to become flanged.[15] In the wild, flanged males produce regular long calls that certainly inform females and unflanged males of their whereabouts and status. It seems likely, therefore, that the onset of flangedness is delayed in males until there is a gap in the flanged male population that would make it worthwhile for them to become flanged and accept the costs of doing so. The mechanism seems to be that the hypothalamus in young males, which regulates testosterone production, is affected by the young males hearing the long calls of flanged males.[14]

A male that remains unflanged for a time may be subordinate but he avoids combat and metabolic costs and is not without mating opportunities, even though females have a strong preference for fully adult males as mates and approach them for sex at around the time of ovulation. Females are also attractive to unflanged males at this time, however; unflanged males are occasionally able to catch them and force copulation upon them.[5, 17, 20] Hence, in both species, unflanged males sire a proportion of offspring, although each successful flanged male sires more infants than does each unflanged male. In Ketambe, northern Sumatra, unflanged males fathered half the offspring over a 15 year period,[25] with the other half being fathered by a smaller number of flanged males. In Tanjung Puting, southern Borneo, subadult males often find receptive females before adult males do, but 86 percent of matings initiated by them are resisted by the

Cyril Ruoso/BOS-USA

A young Bornean orangutan.

female.[6] This all suggests that there may be two stable male strategies at work, the one linking delayed maturation with forced copulation, and the other linking full maturation with consensual intercourse. In this context, it is relevant that early-developing males do not appear to father late-developing sons.[25]

The mother–infant bond is very close in orangutans, but it gradually weakens with age; by the time the apes are fully adult, interaction between them is often limited to glances.[26] After independence, females tend to stay near the range where they were born and maintain friendly relationships with local females, which are likely to be relatives. Hence orangutans live in loose communities that may consist of one or more clusters of related females and the adult male with whom they all prefer to mate.[23] Researchers have noted that the movements of community members are subtly coordinated, and that they may come together as a real group on some occasions.[20] Individuals in clusters of closely related females at Suaq Balimbing, Sumatra, not only share home range boundaries, but appear to coordinate their reproduction; the timing of births is similar within a cluster, but different between clusters.[23] If true, this implies a subtle but powerful organizing influence within the community.

CONSERVATION CONCERNS

The most recent estimates of total surviving numbers for the Sumatran and Bornean orangutans are around 7 300 and 57 000 respectively, in most cases distributed among small and isolated subpopulations in fragmented islands of habitat.[16, 24] Numbers continue to decline, aggravated by the secondary trade in juvenile orangutans; there are now very few locations where a viable population of at least 250–500 individuals inhabits a forest area that is both protected in law and potentially protectable in practice.[24]

The chief causes of this decline in population are logging, followed by forest fire, and the conversion of forests to farms and plantations (often of oil palm, *Elaeis guineensis*). These factors are especially potent in the lowlands, below about 500 m, where forests are more accessible and more valuable in terms of timber (especially dipterocarp timber), and the land is more amenable to farming and settlement. Even low, flat areas of deep peat-swamp forest that are completely unsuitable for farming have been cleared for settlement in

Suherry/SOCP

A continuing trade in orangutans exacerbates the decline in populations caused by logging, forest fire, and land conversion. Here, staff of the Sumatran Orangutan Conservation Programme and an officer from the Indonesian government's Conservation Department have confiscated a female orangutan infant from the village of Namo Tala in northern Sumatra.

Indonesia, reflecting the central-planning failures of the former political regime. The loss of central control over the forest management and protection systems after 1998, weak as they were before then, allowed interest groups at the provincial and regency levels to promote very rapid logging and forest clearance. This led to the worst wildfires in recorded history, and the expectation that virtually all the lowland forest would soon be lost from both islands. This situation may be stabilizing in some locations; certain local government leaders see advantages to forest conservation and have started to propose new protected areas to central government, such as the Batang Gadis National Park in Sumatra, which was designated in 2004.

WHAT WE DO NOT KNOW

After three decades of field research, large gaps remain in our understanding of both the Bornean and Sumatran orangutan. We are yet to understand fully the unusual development of male secondary sexual characteristics. Ranging patterns and social systems are better understood in Sumatra, and it is important to determine how Bornean orangutans organize their movements and sociality, apparently in a relatively fruit-poor habitat. Most importantly, it is crucial to extend our understanding of how orangutans cope with habitat disturbance, so we can better evaluate their response to the increasing habitat degradation that is occurring. Our understanding of the apes' social system and ranging patterns have been vastly improved in recent years, but the extent to which these patterns can be extrapolated to regions outside the limited study sites is not known. Almost all studies in Sumatra, for example, have been conducted in the relatively fruit-rich forests of Ketambe and Suaq Balimbing. Until studies have been extended to other areas, firm conclusions cannot be drawn about whether certain characteristics are typical of the wider population, the species, or even the genus; this makes it difficult to draw firm conclusions about differences between the two species. Lastly, studies must be extended to disturbed areas, to deepen understanding of the ecology of disturbance, adaptation, and recovery.

FURTHER READING

Bennett, E.L. (1998) *The Natural History of Orang-utan.* Natural History Publications, Kota Kinabalu, Sabah.

de Boer, L.E.M., ed. (1982) *The Orang utan: Its Biology and Conservation.* Dr W. Junk Publishers, The Hague.

Delgado, R.A., van Schaik, C.P. (2000) The behavioral ecology and conservation of the orangutan (*Pongo pygmaeus*): a tale of two islands. *Evolutionary Anthropology* **9** (5): 201–218.

MacKinnon, J. (1974) The behaviour and ecology of wild orangutans (*Pongo pygmaeus*). *Animal Behaviour* **22**: 3–74.

Maple, T. (1980) *Orang-utan Behavior.* van Nostrand Reinhold Company, New York.

Rijksen, H.D., Meijaard, E. (1999) *Our Vanishing Relative: The Status of Wild Orang-utans at the Close of the Twentieth Century.* Kluwer Academic Publishers, Dordrecht.

Singleton, I., Wich, S., Husson, S., Stephens, S., Utami Atmoko, S., Leighton, M., Rosen, N., Traylor-Holzer, K., Lacy, R., Byers, O., eds (2004) *Orangutan Population and Habitat Viability Assessment: Final Report.* IUCN/SSC Conservation Breeding Specialist Group, Apple Valley, Minnesota.

ACKNOWLEDGMENTS

Many thanks to Ian Singleton (Sumatran Orangutan Conservation Programme), Colin Groves (Australian National University), Raffaella Commitante (Cambridge University), and David Woodruff (University of California, San Diego) for their valuable comments on the draft of this chapter.

AUTHORS

Julian Caldecott, UNEP World Conservation Monitoring Centre
Kim McConkey, UNEP World Conservation Monitoring Centre

Cyril Ruoso/Still Pictures

CHAPTER 10

Bornean orangutan (*Pongo pygmaeus*)

KIM MCCONKEY

Bornean orangutans (*Pongo pygmaeus* Linnaeus, 1760) survive in 306 fragmented and increasingly isolated populations on the island of Borneo, which is about 740 000 km^2 in area (Map 10.1).[96] The populations inhabit forest blocks that are separated by impassable barriers such as rivers or areas of cultivation. They are concentrated in Central Kalimantan (Indonesia) and Sabah (Malaysia), with smaller populations in West and East Kalimantan and Sarawak (Malaysia). No permanent populations are thought to exist in the independent sultanate of Brunei Darussalam,[123] though sightings of single adult males have been reported there,[78, 112] indicating that nomadic individuals may wander through the forests on occasion. Next to Brunei is a large area of recorded absence, from the Rajang River in Sarawak to the Padas River in Sabah. This may reflect local extinctions caused by hunting pressure in the past,[123] an unsuitable environment, or a combination of both.

Bornean orangutan populations are divided into at least three subspecies. *Pongo pygmaeus pygmaeus* (the northwest Bornean orangutan) is medium sized and occurs in West Kalimantan and Sarawak. The largest subspecies, *P. p. wurmbii* (the central Bornean orangutan), is found in Central Kalimantan; the smallest, *P. p. morio* (the northeast Bornean orangutan) occurs in Sabah and East Kalimantan. There is evidence that the Sabah and East Kalimantan orangutans may also be distinct subspecies, but this is yet to be fully accepted.[51] Orangutans no longer occur in South Kalimantan, the Indonesian province that forms the southeastern part of Borneo. Rivers form the main barriers between the subspecies. In the middle of Borneo, where the rivers are sufficiently small for orangutans to cross, the three subspecies probably interbreed.[51]

BEHAVIOR AND ECOLOGY

Feeding, foraging, and ecological strategy

Bornean orangutans have a strongly expressed preference for large, succulent fruit with a high-energy content in the form of sugars, particularly when they occur in large crops.[75] The availability of such fruit influences virtually all aspects of a Bornean orangutan's life: its ranging patterns, social behavior, timing of reproduction, and health.[37, 69, 123] Fruit makes up over 50 percent of the diet (Table 10.1), but orangutans are opportunistic foragers rather than absolute frugivores. Dietary composition may therefore change markedly with time of year,[47] as orangutans make use of 'famine foods' in times of shortage. This opportunism and dietary flexibility is essential to a large-bodied forest animal that cannot digest large amounts of mature leaves because its simple, globular stomach lacks fermentation chambers.

Bornean forests are renowned for their irregular food supply for fruit-eating and seed-eating animals. They support a low frugivore biomass compared to that supported by other tropical moist forests, such as those in Africa.[10, 18, 63, 86, 146] This is largely a result of the abundance of one mast fruiting tree family, the Dipterocarpaceae.[63] Forests dominated by dipterocarps are also seen in Sumatra and Malaya, but are more widespread in Borneo. These trees produce huge crops of winged, chemically undefended, oil-rich seeds during 'masts' that occur one to six years apart, with many different dipterocarp species participating in each mast. Masting is a complex form of tree behavior

Map 10.1 Bornean orangutan distribution

Data sources are provided at the end of this chapter

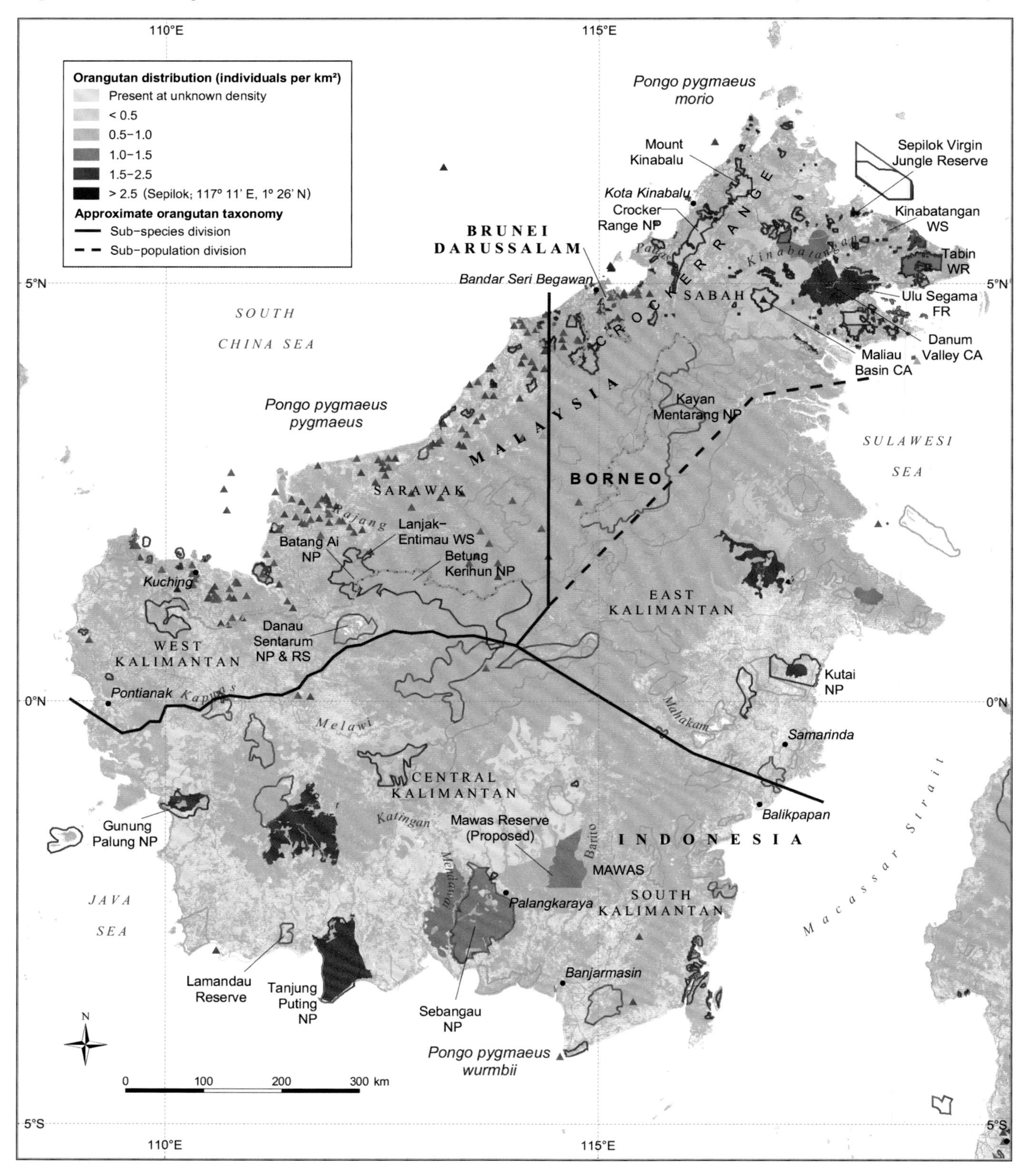

that is linked to drought, which in Borneo is strongly influenced by the El Niño Southern Oscillation cycle.[32, 149] As trees of many other families also flower and fruit in response to drought, masting events often involve a large proportion of all trees in the forest.

Frugivores in Borneo are adapted to cope with this irregular food supply. They risk being satiated during masts and starved at other times. This situation has favored mobile species with flexible diets. For example, the Bornean bearded pig (*Sus barbatus barbatus*) is very mobile; its long legs and ability to swim and even climb trees enable it to take advantage of a 'phenological mosaic' in which fruiting peaks move around over large areas.[76]

Orangutan strategy has some similarities to that of bearded pigs (Box 10.1). They gorge themselves during mast fruiting, increasing their calorie intake by 50–70 percent, eating only fruit (including dipterocarp seeds) while this is abundant.[32, 69] When explosions in insect populations occur, these are exploited in a similar way.[123] Orangutans may remain in the vicinity of a fruiting tree for several days until the crop is exhausted. The orangutans store fat during these times, which helps them to survive for a few weeks afterwards.

Their usual foraging strategy is to follow a steady zigzag pattern through the forest, feeding on many different sources.[82] Individual orangutans may be familiar with an area of over 30 km^2 in extent.[123] Several authors have concluded that orangutans track food supply across a landscape that may contain a range of elevations, topographical features, and forest types, each with differently scheduled food sources.[47, 82] Orangutans appear to have little difficulty locating available fruit sources, and can take direct routes between them; this suggests that they remember food plant seasonality and locations.[35, 82]

Observations also suggest that orangutans can interpret clues regarding the location of fruit sources, such as the strong smell of durian fruits (*Durio* spp., Bombacaceae), the noisy interactions of other feeding animals, and the mass movement of hornbills (*Buceros* spp.) and flying foxes (fruit bats, *Pteropus* spp.) toward new feeding grounds.[82, 123] The congregation of animals around a tree with ripe figs, for example, can be seen or heard for some distance and often attracts orangutans.[123] These behaviors suggest a significant capacity for making intelligent decisions based on a detailed spatial memory,[81, 130] and are suggestive of how a large-bodied, semisolitary, arboreal mammal can succeed in making a living in an ecosystem where fruit is in short and unpredictable supply.

Figs (*Ficus* spp., Moraceae), especially strangling figs, are an exception both to the pattern of mast fruiting and to the annual-to-biannual fruiting pattern of many other tree species.[75] Fig trees fruit abundantly, and may do so two or three times a

Table 10.1 Diet and range in three Bornean orangutan populations

	Tanjung Puting Kalimantan[43, 47]	Kutai Kalimantan[100]	Kutai Kalimantan[124, 125]	Kutai Kalimantan[143]	Ulu Segama Sabah[82]
Number of study years	4	1	1	5	2
Diet					
No. of types of food eaten	317	–	–	–	–
No. of fruit species eaten	169	–	–	–	–
Proportion of diet (percent)[a]					
Fruit	61 (16–92)	–	54 (13–89)	61	53 (10–97)
Young leaves	15 (0–40)	–	29 (5–57)	25	36 (7–73)
Flower	4 (0–41)	–	2 (0–11)	0	–
Bark	11 (0–47)	–	14 (0–67)	13	16 (0–37)
Insects	4 (0–27)	–	1 (0–3)	1	2 (0–8)
Ranging					
Female home range (km^2)	5.0–6.0	0.4–0.6	0.5	0.6–3.0	–
Male home range (km^2)	6.0	–	1.0	4.0–8.0	–
Female day range (m)	710	–	305	–	500
Male day range (m)	850	–	305	–	500

a Mean figures, followed in parentheses by the monthly range observed.

KOCP

An orangutan eating tree bark, a particularly important food source when fruits are scarce.

year, with each tree on a different cycle. There are peaks in abundance, but these are weak enough for an orangutan's home range to be likely to contain some figs at all times.[123] Orangutans typically eat other fruit in preference[75] but, where figs are present and there are few alternatives, they comprise an important part of the orangutan diet. Orangutan populations also occur in areas such as the swamp forests of Tanjung Puting National Park that are virtually devoid of fig trees.[47] In these forests, staples include *Tetramerista glabra* (Tetrameristaceae), which fruits throughout the year; *Xanthophyllum rufum* (Xanthophyllaceae); *Gironniera nervosa* (Ulmaceae); *Lithocarpus* spp. (Fagaceae); and *Nephelium* spp. (Sapindaceae). In the drier alluvial valleys, *Ficus* spp., *Aglaia* spp. (Meliaceae); *Baccaurea* spp.; and *Mallotus* spp. (both Euphorbiaceae) are all important dietary sources for orangutans.[47, 82] Several tree genera favored by orangutans (e.g. *Durio*, *Aglaia*, *Nephelium*) also produce fruits of dietary and commercial importance to humans.

Other orangutan adaptations to fluctuations in food availability are their flexibility and resilience to unusual food sources. Outside masting periods, orangutans regularly consume unripe fruit, seeds, and other items.[35] During periods of food stress, less than 16 percent of the orangutan diet may be fruit.[35, 43, 123] In such circumstances, they consume large quantities of bark (37 percent of diet in one study),[69] leaves, gingers (Zingiberaceae), and stems.[70] Orangutans also consume the growth layer under the bark (mainly of *Ficus* trees); their teeth show signs of being adapted to this task.[35] Several orangutans studied in Kutai, East Kalimantan, were able to persist for half a decade by consuming bark and vegetable matter, after their forest was devastated by fire in 1982–1983.[123, 144] In some areas of Borneo, the young leaves, top shoot, and inner meristematic tissue of the mature leaf stem of the palm *Borassodendron borneensis* form extremely important food sources during periods of fruit scarcity. Local people insist that orangutans can survive only in areas where this species is fairly abundant.[41, 76, 125, 127]

No confirmed observations of carnivory have been made for the Bornean orangutan, although an infant was observed biting off a rat's head before playing with its body,[70] and a formerly captive two year old was seen to bite off and eat the head of a bird before discarding the body.[31] Bird eggs may also be eaten.[82] Leaves, young shoots, flowers (notably *Madhuca* spp., Sapotaceae), mushrooms, wood pith, honey, insects (leafhoppers, crickets, bees, ants, and termites), and mineral-rich soils are certainly eaten.[82, 123, 126, 150] These soils sometimes have high concentrations of kaolin, which may help to adsorb and neutralize the large amounts of tannins and other secondary plant metabolites consumed in the Bornean vegetation, despite selective feeding. Orangutans have been seen eating sections of the tubes of soil deposited along tree trunks by termites, and may also descend to the ground to eat clumps of earth. They visit 'mineral licks' such as caves with high concentrations of important minerals (e.g. sodium, potassium, and calcium); these are visited by a wide variety of animals, including deer and gibbons.[82]

Male and female nutritional needs are somewhat different. Males eat more young leaves, bark, and termites than females do; they also spend more time on the ground where termites occur.[47] Female orangutans eat a more varied diet in general, but consume fewer calories, than males.[47, 69] Females therefore store less fat in times of plenty, and are more affected by periods of shortage.[69]

Ranging behavior

Bornean orangutans are wide-ranging animals that occupy potentially very large home ranges; being

Box 10.1 ADAPTATIONS OF BEARDED PIGS TO LIFE IN DIPTEROCARP FORESTS[18, 19, 20]

Bearded pigs (*Sus barbatus*) consume roots, fungi, small animals, turtle eggs, carrion, and items from at least 50 genera of plants. Fruit supply controls growth, fattening, and breeding, and the oil-rich seeds of dipterocarps (*Shorea*, *Dipterocarpus* spp., and others), oaks (*Lithocarpus*, *Quercus* spp.), and chestnuts (*Castanopsis* spp.) are especially important. Highland oak forests produce fruit regularly, and are important as predictable food sources, while dipterocarp forests provide large supplies of food only irregularly by masting. Dipterocarp fruiting is well documented in Sarawak because the seeds of many of these trees are exported commercially, and records have been kept for many decades. Since 1899, virtually no dipterocarp seeds were exported in one third of years, small amounts in most other years, and more than 1 000 tons in some years. Episodes of heavy fruiting usually occur every three to five years. Exceptions during 1945–1988 were four pairs of years with heavy dipterocarp fruiting (1953–1954, 1958–1959, 1982–1983, and 1986–1987). During the 1980s, these events were correlated with repeated droughts linked to El Niño Southern Oscillation phenomena. The bearded pig population rose dramatically in 1954, 1959, 1983, and 1987, but in no other year during 1945–1988, suggesting that explosive population growth occurred in response to sustained availability of dipterocarp seeds over two consecutive years.

Features that may allow this response include large average litter size (which can range from three to 12, depending on the mother pig's size); short gestation length (90–120 days), permitting up to two litters per year; variable but potentially early age at first rut and pregnancy (10–20 months); efficient conversion of dietary fat to body fat; variable but potentially high growth rates; very flexible group sizes (10s to 100s); synchronization of birth peak with fruit fall; and travel-adapted features such as long legs and swimming ability.

The synchronized birth peak is achieved by a rut, prompted by the falling of massed flowers from the forest canopy (the 'confetti effect'). Male bearded pigs bombard the females with olfactory and/or pheromonal signals from their urine.[17, 18, 20] Sadly for the male, in order to achieve a successful rut, he must go into acute urinary retention, which can destroy his kidneys. Males may delay participating in a rut (analogous to unflanged male orangutans delaying full maturation; see Chapter 9) until some combination of factors suggests that the likelihood of success is worth the risk of death. Male bearded pigs may stake the reproductive success of their entire lifetime on one rut, during which they may sire hundreds of piglets.

In the Malay Peninsula, there is historical evidence that bearded pigs migrated regularly to take advantage of predictable fruiting in camphorwood (*Dryobalanops aromatica*) forests, which have since been felled for their valuable timber. During the eruptions in bearded pig numbers, in the 1950s, Pfeffer described annual population movements in Kalimantan over distances of 250–650 km;[114] in the 1980s, Caldecott tracked herds of bearded pigs moving through the upper Baram area of Sarawak at a rate of 8–22 km/month, migrations that were sustained over four or more months.[17] Large-scale movements of populations of bearded pig have also been reported from the Malay Peninsula.[2, 54, 55, 56, 67] During these, the pigs were described as moving consistently in one direction, in scattered or condensed herds, over either a broad or a narrow front; this was observed to persist over a period of days, weeks, or months. The animals were variously described as being in good, poor, or very poor physical condition. They were sometimes accompanied by piglets and sometimes not, and regularly swam across rivers, sometimes across coastal bays, and even out to sea. In some cases, the population was thought to retrace its route later, or to follow a circular course to its point of origin.

Julian Caldecott

A bearded pig.

Cede Prudente

KOCP

Flanged males are larger and more aggressive than unflanged males, as well as being more attractive to receptive females.

also slow moving, their daily travel distances are small relative to the size of their home range. Home ranges vary considerably in area between study sites (see Table 10.1). They are not defended and overlap extensively, although fully adult males are intolerant of each other. The ranging pattern adopted by a particular individual appears to be determined by a combination of food availability, social position, and reproductive condition.[123] Females use a smaller area of forest than do males, usually in the range of 0.5–5.0 km^2, especially if they are restricted in their movements by dependent infants.[123, 126] The home ranges of males are consistently at least two to three times larger than those of females. High-status or dominant adult males, however, appear to maintain a relatively small home range during their period of dominance, within which they attempt to monopolize access to receptive females.[140] Individuals new to an area are almost always subadult males,[70] and it seems certain that males are the dispersing sex, with solitary males being encountered at times far from locations that support breeding populations (for example at high altitudes in Sabah and Sarawak).[21, 105]

Party formation

The scarcity of large, reliable food sources across most of Borneo is thought to be the key factor that reduces the opportunity for social interaction among orangutans.[150] Typically, Bornean orangutans live in very loose, dispersed communities with little interaction compared to most great apes. Mating opportunities are restricted by population density, as well as by female receptivity; this, in turn, is thought to be related to food supply.[69] Population densities vary across Borneo, with individuals from the relatively dense and social Kutai area mating much more frequently than do those from Gunung Palung National Park (West Kalimantan).[98] The same pattern is seen for all other aspects of sociality, including aggressive interactions, which are regularly observed amongst the Kutai apes but seldom at Gunung Palung.[98] The fact that Bornean orangutans have not been seen to use tools in the wild, unlike their more social Sumatran relatives, has been interpreted to be a consequence of the lack of opportunity for individuals to learn from one another.[147]

Bornean orangutans spend most of their time alone or in mother–offspring binary units. They associate loosely with others at times and are no doubt familiar with the individuals that use their home ranges fairly regularly, having been observed to coordinate their travel.[82] Familiarity and relatedness presumably go together, at least for females, as they tend to settle near their mothers.[47] Groups of two or more individuals, excluding mother–infant pairs, formed 19 percent of observations at Tanjung Puting[46] and 20 percent of observations at Ulu Segama.[82] Associations or 'parties' typically occur at large fruit sources, but some individuals may also travel between food sources together.[47]

Females are much more likely to form parties than males;[46, 100] fully adult Bornean males are the least social orangutans, spending 91 percent of their time alone.[46] Juveniles may play together, even when their mothers show no interest in each other.[82] Subadult orangutans of both genders can also be quite gregarious. In Tanjung Puting, adolescent females spend about 45 percent of their time in groups, compared to 10 percent for adult females.[46] Subadult males spend about 41 percent of their time in groups, of which 83 percent is spent with females, and only 3 percent spent exclusively with males.[46] In the presence of adult females, adult males do chase off subadult males, but usually the latter are able to move quietly away.[46]

Relations between males

The large body size of fully adult males, and their consequent need to consume large quantities of ripe fruit, makes resource competition inevitable between males in the same area. Males do not actively defend territories and aggregations of males can occur in areas with an abundance of resources, but adult males are aggressive towards each other at close range; battles can ensue, sometimes with fatal results.[46, 70, 82] Combat is rare, as adult males tend to avoid each other.

Development and reproduction

Female orangutans reach maturity between the ages of 11 and 15 years[44] and reproduce about every eight years,[42, 82] depending on ecological conditions. The processes of male maturation is a complex one that involves some males becoming fully mature, or 'flanged', earlier than others that remain 'unflanged', sometimes for many years (see Chapter 9). Flanged males have larger bodies, are more aggressive and dominant over the others, and are the preferred mates of adult females. The latter actively seek and frequently obtain attention from flanged males around the time of ovulation,[44] but do not display an obvious physical signal like the perineal swellings of the chimpanzee (although vulval swellings occur during pregnancy). Conception may require more than one cycle of mating,[146] and is most likely to occur during periods of mast fruiting, when female estrogen levels are highest.[69]

Mating can take two basic forms. In the first, the female initiates contact with a flanged male, which then actively solicits sexual intercourse by means of posturing and a penile display.[135] Intercourse is consensual and often occurs with the female 'on top';[36, 135] it may be repeated with the same male during a consortship lasting several consecutive days, or the female may mate with more than one male around the time of ovulation.

The second form of orangutan mating involves forced copulation by unflanged or, less often, by flanged males that are strangers to the female.[99] In either case, juveniles are often intolerant of males mating with their mothers and may attack the male.[82] Mating behavior by unflanged males in Tanjung Puting National Park has been documented by Biruté Galdikas.[46] Here, 86 percent of matings initiated by unflanged males were resisted. Unflanged males often locate receptive females first, but are soon displaced by flanged males. An unflanged male does not attempt copulation with a pregnant female or when a flanged male is in the vicinity.

Some flanged males maintain a circumscribed range and mate with the females whose ranges overlap with their own; others may be more mobile and mate opportunistically in the home ranges of other males, sometimes by forcing copulation. Unflanged males may be tolerated by a flanged male and spend most of their time within his range, or they may wander in search of food and females. Such males may consort with females but most copulations that follow are forced. The implication of these observations is that females prefer to mate with high-status, fully developed males; these tolerate the presence of unflanged males as they represent little competitive threat.

Orangutan pregnancy lasts for eight and a half months.[68, 85] About 90 percent of young survive their long, dependent infancy. While traveling, a mother orangutan carries her offspring in a side-ventral position until it is several years old. By 11 months, the young orangutan can find its own food,[82] but may continue to suckle until five or six years of age.[145] Mothers may play with their young, and sometimes feed them fruit directly. Offspring become fully independent at seven to 10 years of age, when they may leave the maternal home range.[46, 113]

Vocal behavior

The long call of the male orangutan is among the strangest and most penetrating sounds heard in nature. These calls are given by the highest-ranking males three or four times a day and are often accompanied by branch-shaking displays and bristling of hair.[82] Long calls are given when a male arrives at a new location; as orangutans often call in the direction in which they are traveling, their calls

Cyril Ruoso/BOS-USA

Orangutan infants have a long period of dependency relative to that of other great apes.

KOCP

This orangutan nest is a few days old: most of the leaves are still green in color.

may allow males to space themselves out and reduce the chance of violent interactions. Long calls are often given after associations with females and are more frequent in areas where there are few consortships underway.[99] They may have a role in attracting females, or at least in alerting them to the male's presence, in addition to the effect they seem to have in inhibiting the full development of other males (see Chapter 9).

Tool use

Bornean orangutans have been seen to take shelter from the sun or rain by holding twigs or leaves over their heads,[82] but have not been observed to use tools in the strict sense of using one object to influence another. Captive Bornean orangutans, however, learn tool-using behavior quickly from others, including humans; in zoos, both Sumatran and Bornean orangutans can surpass chimpanzees in their speed of invention and learning, and in their creativity.[141] Rescued orangutans at Camp Leakey, Tanjung Puting National Park, have been observed to imitate humans in many activities, including riding in canoes, 'brushing' teeth, sawing logs, hammering nails, shooting blowguns, and washing laundry.[131] In Sabah, rehabilitated apes learned to use leaves as a feeding vessel. They carefully selected leaves and arranged them into a vessel able to hold semifluid material. A mouthful of milk and bananas was spat into the vessel, then eaten slowly.[66] These observations are taken as evidence that Bornean orangutans are capable of using tools and transmitting this behavior culturally, as do Sumatran orangutans, but that they meet each other too infrequently for this capacity to be highly developed in the wild.

Nest building

Every night, just before sunset, orangutans construct night nests. These are springy platforms, formed by bending and weaving small branches. They are often covered with loose twigs and branches. An existing nest may be used, if situated close to a prime feeding tree. Orangutans usually remain in their nests until midmorning, although occasionally some individuals move during the night. Bornean orangutans often nest close to the last food tree visited that day, and feed there again the next morning; they frequently return to the night nest for a late-morning rest. Orangutan nests are considerably easier to see in the forest canopy than are the apes themselves. They can be distinguished readily from those of other animals, such as sunbears (*Helarctos malayanus*), although less easily (at least from the ground) from those of giant squirrels (*Ratufa* spp.). Nest observations can therefore be used to assess the density of orangutans; helicopters can be used to survey large areas very quickly for nests.[4]

Natural enemies

Aside from human activity, there is no evidence of predation on orangutans in Borneo, but juveniles are probably at threat from clouded leopards (*Neofelis nebulosa*), pythons (*Python reticulatus*), and black eagles (*Ictinaetus malayensis*).[82] Iban folklore has accounts of orangutans fighting crocodiles but the two species are unlikely to meet except when an orangutan falls into a river by accident.[82] Some juvenile orangutans undergoing rehabilitation have been killed by wild bearded pigs.[47]

Response to disturbance

Orangutan densities are reduced by habitat disturbance,[62, 123] with the extent of this depending on the severity of damage to the forest. One study showed that densities differed by only 7 percent between disturbed and undisturbed lowland dipterocarp forests, but by 21 percent between disturbed and undisturbed peat-swamp forests in Gunung Palung National Park. The authors suggested that

Box 10.2 ORANGUTANS IN DEGRADED HABITATS

Sabah (Malaysia) harbors more than 10 000 orangutans with most of these living outside primary forest reserves, in forests that have been selectively logged.[4] Rather than regarding orangutans living in secondary forests as ultimately doomed, the French organization Hutan, together with the Sabah Wildlife Department, conducted the first detailed long-term study on how orangutans cope with changes in their natural habitat, seeking realistic solutions to enhance their prospects of long-term survival. The Kinabatangan Orangutan Conservation Project was set up in 1998 in the Kinabatangan floodplain in eastern Sabah, a corridor of highly degraded and fragmented forests that still harbors remarkable wildlife diversity and abundance, including one of the largest orangutan populations in Malaysia.[5, 71]

Since the mid-1950s, all forests (mainly mixed lowland dipterocarp type) of the Lower Kinabatangan floodplain have been subject to increasing pressure from commercial logging and agriculture. As a result, climax (mature) forests have been greatly transformed from the towering, highly structured ecosystems of the past. These forests now have a low stature, lack a clearly recognizable division of canopy layers, and have an overall low stem density of 332 trees (i.e. trees that have a diameter over 10 cm at breast height) per hectare. There are very few large trees, small basal area, large gaps in the canopy, and a high degree of soil disruption; these attributes are characteristic of disturbed habitat.[3]

The emergent trees belong to the families Dipterocarpaceae, Leguminosae, and Moraceae. Although this is comparable to primary forest, the trees are never taller than 35 m, in contrast to the heights of 70 m or more reached in primary forests.[148] Fast-growing and light-demanding pioneer species that are tolerant of dry conditions and show opportunistic life-history characteristics are

continued overleaf

Plant families most often consumed by orangutans

	Fruits	Leaves	Bark
Feeding time allocated to item type as percent of total[a]	66	22	9
Fruit family as percent of food type			
Moraceae (fig)	27	13	5
Fagaceae (beech)	12	<1	<1
Anacardiaceae (mango)	10	<1	<1
Ebenaceae (ebony)	3	18	<1
Rubiaceae (coffee)	3	<1	11
Polygalaceae (milkwort)	<1	4	<1
Sterculiaceae (cocoa)	<1	6	32
Woody climbers (except Moraceae)	13	45	28
Other families	32	15	25

a During 1 700 hours of feeding observations at the Sukau study site.

this reflected much higher disturbance rates in the peat-swamp forest.[65] Between 1999 and 2004, the illegal logging in and around Gunung Palung National Park accelerated, with removal of large numbers of fruit trees, halting research in the park.[15, 77]

Only limited assessments are available of the behavioral response of orangutans to and their ability to cope with such degradation.[71] In the peat-swamp forests of the western Sebangau catchment, Central Kalimantan, illegal logging operations caused large shifts in orangutan distribution, often into areas of suboptimal habitat. The resultant overcrowding led to stress, increased juvenile mortality, and a decrease in fecundity.[62]

In the Kinabatangan Wildlife Sanctuary in Sabah, a research site has been established by the Kinabatangan Orangutan Conservation Project to investigate orangutan populations in logged forest. Initial results indicate the presence of a very high population density,[71] although this may be a temporary consequence of the concentration of

common in these forests: Euphorbiaceae (e.g. *Macaranga* spp.), Rubiaceae, Sterculiaceae, Moraceae, creepers (*Spatholobus* spp., *Meremia* spp. and *Uncaria* spp.), and grasses (*Paspalum* spp., *Imperata* spp.). The diversity of these disturbed Kinabatangan forests remains high by global standards, with a minimum of 1 056 tree species belonging to 129 families and 512 genera.[6]

Today, the composition and structure of these forests is a mosaic of different types of habitat, all at different stages of degradation and early regeneration. The heterogeneity of these habitats differs greatly from that of mature lowland dipterocarp forests and this is likely to affect general orangutan behavior and forest use.

KOCP

Large woody climbers are frequent in the disturbed forests of the Lower Kinabatangan floodplain.

Orangutans were studied in this environment at Sukau, where they were found to have an average active period each day of slightly more than 10 hours (612 minutes). Based on more than 5 000 hours of direct observation of nine habituated wild orangutans (four adult males, five adult females), the following proportion of time was spent on these mutually exclusive daily activities: 47 percent resting; 35 percent feeding; 11 percent moving; 1 percent nesting, and 6 percent unknown. The average daily distance traveled was about 300 m. Unflanged males travel more than other orangutans. When in consortship, unflanged males adapt their movements to the movement of adult females, so the distance they travel daily is significantly less than at other times.[1]

Orangutans at the Sukau study site were seen to eat a total of 310 species of plants, from 66 families and 156 genera; these included 210 species of trees, 87 species of vines, and 13 species of monocotyledons (see table). Among these 310 species, 135 were consumed for their fruits, 185 for their leaves, 97 for their bark, and 19 for their flowers. Fruits are the most common food, accounting for more than 66 percent of the diet of orangutans, followed by leaves (22 percent), barks (9 percent), flowers (2 percent), and insects (1 percent).

Most foods consumed by orangutans at Sukau originate from either pioneer tree species or the woody climbers that are very common in secondary forests. Woody climbers are of special importance due to the year-round availability of their leaves and bark; they are used extensively at times of fruit scarcity.

Our preliminary results suggest that orangutans possess an unexpectedly high degree of behavioral and dietary flexibility that would allow them to survive in highly logged and secondary forests. However, further longer-term observation is needed to understand fully orangutan socioecology in degraded habitat. This will be of crucial importance in the design of a sound management strategy for the orangutan population in the Lower Kinabatangan and other similar areas.

Marc Ancrenaz, Isabelle Lackman-Ancrenaz, and Ahbam Abulani

individuals away from areas of active disturbance.[128] On the other hand, the area provides a rich and constant fruit supply, due to the abundance of pioneer and climber species that have invaded the site after logging. Thus, the orangutans consume a lot of fruit (which makes up 60 percent of their diet),[1] and appear healthy. Compared to populations in primary forest, they feed and travel less and rest more. Due to the loss of large trees and canopy continuity, however, males are forced to move more along the ground. As females and young orangutans are lighter, they can still move arboreally. These orangutans showed remarkable flexibility and adaptation to the levels of disturbance encountered (see Box 10.2).

Ecological role

The importance of fruit in orangutan diets indicates the vital ecological role these animals could play as seed dispersers, possibly affecting patterns of forest regeneration and plant-species diversity.[72] The species-rich nature of tropical forests is partly maintained by frugivores moving seeds from the parent tree to sites some distance away. This not only disperses germination sites but can also increase germination and establishment rates, as predation by seed-eating animals and seedling competition are both often higher under the parent tree.[24, 61, 138] The main factor limiting the density of many plant species is the number of their seeds reaching suitable growth sites.[39]

Little more is known about the role of orangutans in seed dispersal,[30] but much can be inferred from their observed diet and habits. Given their large body size and frugivory, orangutans certainly have the potential to disperse many seeds. The mechanism of seed dispersal is also relevant, as the process that the seeds undergo can have important consequences for their fate.

Seeds that are swallowed whole by orangutans and later defecated are more likely to be dispersed away from the crown of the parent tree, and dispersed further, than seeds that are spat out or dropped after the fruit pulp has been removed.[30] Orangutans often carry fruit away from the parent tree, by as much as 200 m, before they spit out or drop the seeds.[45] This may be far enough for seeds to escape unfavorable conditions beneath the parent tree crown.[8, 27] Those seeds that are swallowed may take up to several days to be defecated.[22, 45] Such seeds would typically be moved 800 m or, conceivably, as far as 8–10 km from the parent tree.[45]

Many potential seed dispersers inhabit Borneo's forests.[30] Several are even more frugivorous than orangutans (Table 10.2), and may be more efficient at dispersing the seeds. Of 413 fecal samples obtained from gibbons (*Hylobates muelleri* X *H. agilis*), for example, 100 percent contained intact seeds.[89] Similarly, Galdikas[45] found intact seeds in 94 percent of 64 Bornean orangutan fecal samples, although this high rate may have been a seasonal effect.

Seed size is an important factor influencing seed dispersal by frugivores. The small seeds of figs, for example, may be dispersed by abundant small birds and insects, but the dispersal of large seeds relies on animals big enough to carry or swallow them. These larger birds and mammals are the chief targets of human hunters; their selective loss from an ecosystem seriously impairs the dispersal of plants with large-seeded fruits.[102]

Orangutans are the largest arboreal frugivores in Borneo's forests, and among the largest of all frugivores.[30] Their sheer size may have enabled orangutans to fulfill an important role in dispersing large seeds. Larger animals – banteng (*Bos javanicus*) and rhinoceros (*Dicerorhinus sumatrensis*) – are virtually extinct on Borneo,[113] and the identity of the seed species they disperse is not known. Even if they do disperse the same species as the orangutan, they are unable to access fruit in the tree canopy and would rely on fallen fruits (such as those dropped by an orangutan).

Despite their much smaller size, fruit bats, such as *Pteropus vampyrus*, can disperse large seeds by carrying them away from the parent tree.[30] For this to happen often there must be many bats,[117] and their populations are also dwindling in Borneo.[12, 52] Gibbons (*Hylobates* spp.), civets (Viverridae), sunbears (*Helarctos malayanus*), barbets (Megalaimidae), mouse deer (*Tragulus* spp.), and pigeons (Columbidae) also disperse fairly large seeds by defecation, although they are probably unable to disperse the largest seeds taken by orangutans (see Table 10.2).[30, 89, 92, 94] Hornbills (Bucerotidae) regurgitate large seeds,[30] and travel long distances quickly[58] so are probably among the most important seed dispersers in Borneo. Primates and hornbills eat distinctly different subsets of fruit in Cameroon.[115] There is evidence of a similarly small dietary overlap in Borneo; generally, hornbills consume fruit with high-fat arils (seed coverings), while orangutans prefer moist sugary fruit.[76]

Orangutans also function as seed predators, destroying around 30 percent of their dietary species.[45] They may destroy and disperse seeds of

KOCP Lynda Dunke

***Durio* sp., one of the favorite orangutan fruits in the forest (left), and *Ficus* sp., a less favored but highly important food source.**

Table 10.2 Bornean fauna capable of dispersing large-seeded fruit[a]

Category[b] and taxon	Common name of frugivore (English/Bahasa[c])	Degree of frugivory[d]	Most likely fate of large seeds
4 Loss of this frugivore would have a major impact on seed dispersal[b]			
Pteropus vampyrus	large flying fox	D	dropped
Hylobates spp.	gibbon (*wak-wak*)	D	defecated
Paguma, Paradoxurus spp.	civet (*musang*)	D	defecated
Arctictis binturong	bearcat (*binturung*)	D	defecated
Bucerotidae	hornbill	D	regurgitated
Ducula spp.	imperial pigeon	D	defecated/regurgitated
Megalaimidae	barbet	D	defecated/regurgitated
3 Loss of this frugivore may have a significant impact on seed dispersal[b]			
Pongo pygmaeus	orangutan (*mawas*)	C	defecated/dropped/spat
Macaca spp.	macaque (*beruk, kera*)	C–D	dropped/spat
2 Loss of this frugivore may have a minor impact on seed dispersal[b]			
Helarctos malayanus	sunbear (*beruang*)	C–D	defecated
Tragulus spp.	mouse deer (*pelanduk*)	D	defecated
Dicerorhinus sumatrensis	Sumatran rhinoceros (*badak*)	B	defecated
Bos javanicus	banteng	A–C	defecated
Muntiacus, Cervus spp.	deer (*kijang, rusa*)	B	destroyed/regurgitated
Argusianus argus	Argus pheasant	B–C	defecated?
1 Loss of this frugivore would have little impact on seed dispersal[b]			
Sus barbatus	bearded pig (*babi hutan*)	B	destroyed[e]
0 Loss of this frugivore will have no impact on seed dispersal[b]			
Presbytis spp.	langur	C–D	destroyed[e]
Nasalis larvatus	proboscis monkey	A–B	destroyed[e]
? Effect of loss of this frugivore on seed dispersal is not known[b]			
Ratufa, Callosciurus spp.	squirrel	D	dropped/destroyed[e 6]
Hystricidae	porcupine	B	destroyed[e]

a Animals that rarely eat fruit are not tabulated.
b Taxa are grouped according to Corlett's importance index.[30] This index runs from '0' to '4', with 4 being the most important; unknown effect (?) is also tabulated.
c Bahasa is a Malay dialect used as the national language in Indonesia; a sample is given in this column of a few common names from Indonesia and Malaysia.
d Degree of frugivory, also adapted from Corlett,[30] is coded: A, occasional; B, consistent; C, fruit is seasonally dominant; D, fruit forms the majority of the annual diet.
e Some seeds may be dropped, or cached and forgotten.

the same species, as a single individual may process seeds from the same species in different ways; orangutan feces can contain intact and broken seeds from the same species.[30] They are also very wasteful feeders and drop a lot of fruit, often unripe, to the ground. This provides food for terrestrial fruit feeders, such as mouse deer (*Tragulus* spp.), kijang (*Muntiacis muntjac*), sambar deer (*Cervus unicolor*), bearded pigs (*Sus barbatus*), and porcupines (Hystricidae) that are unable to access the fruit in the tree canopy.[45] These terrestrial animals may in turn disperse the seeds.[30, 45]

In much of Borneo, orangutans are declining

because their forest habitat is being destroyed, eliminating many plant and animal species. Those fragmented forests that remain, but are too small to support orangutans, may change in composition because of the loss of orangutans as seed dispersers. Without detailed research on seed dispersal, it is impossible to determine exactly how these forests are changing. It is unlikely but possible that there are plant species that rely solely on orangutans for dispersal which would face extinction without them.[30] More likely is a loss of genetic variability within plant species that are dispersed by orangutans and the development of a more clumped – and therefore more ecologically vulnerable – spatial distribution. A similar effect has been noted in other areas following the loss or significant decline of an important disperser.[106, 138] A shift to smaller-seeded plants in areas where orangutans are absent may also occur.[25]

KOCP

A young Bornean orangutan whose progress has been followed at the Kinabatangan Orangutan Conservation Project intensive study site.

Interactions with other animals

Food, and especially fruit, abundance limits most orangutan populations. They would therefore be expected to compete with other fruit-eating animals for supplies. Borneo's most frugivorous animals are the gibbons, hornbills, squirrels, leaf monkeys (Colobinae), macaques (*Macaca nemestrina, M. fascicularis*), sunbears, and arboreal civets. Orangutans show little response to these animals in fruit trees, suggesting they are rarely perceived as competitors.[82] Competition among fruit-eating animals is alleviated to some extent by the differences in their food tolerances and preferences. The orangutans' closest dietary overlap is probably with that of other primates.[122]

Several features of orangutan behavior give them advantages over other animals. Their bulk permits orangutans to consume the largest proportion of the fruit crop, and their flexible feeding and foraging strategies mean that they can take advantage of local differences in fruit availability. This flexibility may allow them to be more wasteful of a limited resource, which must significantly reduce food availability for other animals. In one study in Sabah, for example, gibbons fed daily in a liana (*Combretum nigrens*); the fruit of this plant ripens slowly, so only a few ripe fruit could be found each day, and on each visit gibbons plucked only the ripest fruits.[82] Orangutans visited the liana on two consecutive days, removing whole bunches of fruit and dropping the unripe fruit. Orangutans also consume the unripe fruit of some species, reducing future supplies for other animals.[82] While orangutans certainly compete with other frugivores for limited resources, it is the other animals that may suffer the effects of this competition.

KOCP

A proboscis monkey living along the River Kinabatangan.

Niche separation between primates is achieved by differences in fruit and habitat preference. Orangutans can tolerate very sour or bitter fruit that sympatric primates may avoid.[82] In Borneo, orangutans share this tolerance with the five species of

Sheena Hynd

Tabin Wildlife Reserve in eastern Sabah.

leaf monkeys (*Presbytis* spp., *Semnopithecus* spp., *Trachypithecus cristatus*); however, the diet of the leaf monkeys is dominated by these types of fruit and leaves (supported by their fermentative digestive system), while orangutans prefer other fruit types.[34, 75] The soft juicy fruit that orangutans prefer[75] are shared more with the macaques and gibbons.[74, 90]

Niche separation from the macaque species may be achieved primarily through differences in habitat. Long-tailed macaque populations are concentrated in riverine areas, while pig-tailed macaques are found in highest numbers in dry, upland areas.[16, 122] Of all the primates, the greatest dietary overlap is between gibbons and orangutans. Gibbons are very selective feeders; they manage to feed in the presence of orangutans by rising earlier and reaching the main food supplies before the larger apes. Gibbons also travel much longer distances each day. Their energy-efficient brachiation (rapid movement between tree branches, achieved by swinging from handgrip to handgrip) makes more food sources available to them.[122]

HABITAT REQUIREMENTS

Bornean orangutans have a patchy distribution, and so the broad areas shown in the distribution map (Map 10.1) may not accurately represent their true occurrence within an area of forest. In an uninterrupted expanse of forest, their numbers vary in space and time according to the availability of food resources and, possibly, the availability of dietary minerals.[113] Often there is an intermediate zone of low density between areas of presence and absence, occupied by mobile single males rather than a resident breeding population.[111] Bornean orangutans are most abundant in peat-swamp forest and flood plains (see Table 10.3). Peat swamps appear to provide a particularly rich habitat and some regions may support quite an even distribution of animals.[103] In general, orangutans are more commonly found close to streams, rivers, and swamps, probably because of the higher incidence of preferred fruit trees there. Rivers also act to limit distribution, as rivers that are more than about 10 m wide and 60 cm deep cannot be crossed by orangutans.[49, 123]

The main reason that orangutans are absent from large areas of Borneo may be past hunting pressure. Alternatively, this absence and the large-scale distributional patchiness may simply be a reflection of habitat suitability. Bornean forests are relatively inhospitable to large-bodied frugivores and, even in the better parts, orangutans find it difficult to thrive during times of poor fruit availability. A slight change in forest composition, for example in favor of dipterocarps (which are particularly dominant in the areas of Sarawak and Sabah not historically occupied by orangutans), could make it impossible for a breeding population to survive. The hunter–gatherer Penan peoples demonstrate a comparable example of ecological factors influencing distribution; they are absent from areas such as eastern Sabah, where the forests lack abundant hill sago palm (*Eugeissona utilis,* Arecaceae), their staple starch.

Rijksen and Meijaard estimated[123] that 35 percent of a typical peat-swamp forest would be suitable as orangutan habitat, but this may be conservative. More than 90 percent of the Sebangau catchment area is used by orangutans.[103] There are several types of peat forest (see Table 10.3), with the highest orangutan densities being found in shallow coastal peat swamps (e.g. at Tanjung Puting and Gunung Palung National Parks), with lower densities in the deep peat swamps of the Sebangau catchment area, even though orangutan distribution is less patchy and more widespread in these latter areas.[103]

Orangutans are found at intermediate densities in alluvial and lowland dipterocarp forest and are least abundant in hill dipterocarp forest, in forest on ultrabasic soils, and in both submontane and montane forests. Orangutans persist in secondary forest and selectively logged lowland forest, but their population density may be as much as halved, depending on the extent of degradation; long-term impacts of logging are not known.[111, 123] In Gunung Palung, some disturbed peat forest contained higher ape densities than primary lowland forest,[65] but areas that have been burnt can support very few, if any, orangutans.[123]

The drier forests away from rivers have few food trees and only 25 percent (flat lowland) to 30 percent (hill country up to 500 m) of these areas have habitats suitable for orangutans.[123] The least suitable natural habitat is heath forest, with its low-stature vegetation growing on extremely infertile soils, often over white sand. The low density of primates, and virtual absence of orangutans, in the forests of Barito Ulu (Central Kalimantan) have been attributed to: the high incidence of dipterocarps;[88] the low incidence of large fruiting trees;[14] historical hunting; or the separation of Barito Ulu from other suitable habitat by rivers and heath forest.[91]

Bornean orangutans are rarely found at elevations above 500 m,[50] although exceptions are increasingly being reported. Gunung Palung National Park hosts relatively large numbers of orangutans in mountain habitats, although the presence of a breeding population has not been confirmed.[65] Two sets of sightings suggested that populations may occur between 700 and 1 300 m in the Mount Kinabalu and Crocker Range National Park of Sabah,[111] but there have been no sightings in either location since about 1988.[77] The Kinabalu sightings were thought to be of transient individuals. Tracks of a solitary individual were found in 2002 in the Maliau Basin Conservation Area, in central Sabah, at an elevation of about 1 000 m.[21] Similarly, wandering individuals have been reported at high elevations in the Tamabu Range (Pulong Tau National Park) in northern interior Sarawak.[105]

The characteristic altitude limit of 500 m primarily reflects the distribution of food types favored by orangutans, as there is a sharp decline above this altitude in the abundance of soft-pulped fruit.[37] The presence of orangutans at Mount Kinabalu and in the Crocker Range at higher

Table 10.3 Bornean orangutan densities[a] supported by different habitats

	Borneo (1999 review)[123]		Danau-Sentarum National Park[133]	Sebangau peat-swamp forest[103]	Gunung Palung National Park[65]
Habitat type	Density	No. of sites	Density	Density	Density
Flood plain and peat forest	3.0 (2.2–3.5)	4	3.29–4.09	0.91–4.20	4.09–5.87
Shallow peat forest	–	–	–	–	4.09–5.87
Mixed swamp (deep) forest	–	–	–	1.92–2.42	–
Low pole[b] (deep peat) forest	–	–	–	0.91–1.15	–
Transitional (low pole/tall interior) forest	–	–	–	2.04–2.57	–
Tall interior (deep peat) forest	–	–	–	2.04–2.57	–
Alluvial and lowland dipterocarp forest	2.2 (1.2–3.5)	6	–	–	3.22
Secondary and selectively logged forest	0.75 (0.5–1.0)	2	1.28	0.85–4.20	3.00–4.59
Mixed swamp (old disturbance) forest	–	–	–	3.33–4.20	–
Mixed swamp (recent disturbance) forest	–	–	–	0.85–1.08	–
Disturbed peat (shallow) forest	–	–	–	–	3.20–4.59
Disturbed lowland forest	–	–	–	–	3.00
Upland (hill and dipterocarp) forest	0.5 (0.1–1.1)	6	1.80	–	2.60–2.77
Submontane and montane forest	–	–	–	–	2.13
Disturbed peat forest	–	–	–	–	3.20

a The mean number of individuals per km^2, and/or range.
b 'Low pole forest' is a description used in the region for a forest with many small trees.

Adapted from Rijksen, H.D., Meijaard, E. (1999), and later sources (see numbered citations).

Box 10.3 PEATLANDS IN SOUTHEAST ASIA AS HABITAT FOR ORANGUTANS

About 70 percent of all tropical peatlands, some 260 000 km^2, occur in Southeast Asia (especially in Borneo and Sumatra) and in Indonesian New Guinea (West Papua). In these areas, lowland rain forests grow over soils that are flooded or sufficiently waterlogged to inhibit the decomposition of fallen vegetation, so peat accumulates to various depths. This creates a particularly demanding environment: not all species of lowland forest trees can tolerate these conditions. As a result, biodiversity is relatively low in peat forests, although some plant and animal species are restricted to these environments.[29, 119, 121, 139] There is significant local and regional variation within this vegetation type,[119] with differences in both hydrology and nutrient availability exerting strong influences on forest composition and structure.[108, 121, 139]

Peat-swamp forests represent a significant part of the terrestrial carbon store, in both the trees and the peat, because of the large biomass of the former and the thickness of the latter.[110] Much of this carbon is released into the atmosphere following deforestation, drainage, and development and also when the surface vegetation and peat are burned.[109]

As more information on the biodiversity of tropical peat-swamp forest accumulates, it is becoming clear that this ecosystem has been undervalued as a habitat for rare and threatened species, not all of which are wetland specialists.[73] The rivers of these peat-swamp forests were long considered to have low fish-species diversity and productivity,[63] but this view has changed as many new taxa have been found to be associated with peat-swamp habitats, such as those in North Selangor, Malaysia[104] and in the River Sebangau catchment, Central Kalimantan.[107] Meanwhile, the diversity of avian species in peat-swamp forest is considerable and the importance of this habitat as a refuge for a number of rare and threatened species has been demonstrated.[107] Several Southeast Asian bird species are considered to be peat-swamp specialists, and are not associated with any other wetland or forest habitat. Recent studies have also highlighted the role that tropical peat-swamp forests play in providing habitats for endangered, threatened, and vulnerable species of mammals.[38, 107, 116]

Of particular conservation importance are the relatively large populations of orangutans in Borneo associated with peat-swamp forests.[13, 62, 95, 103, 107] These forests provide one of the most important remaining habitats for this endangered primate. Five of the six largest extant populations of Bornean orangutans are found in habitats with deep peat-swamp forests (the Sebangau, Kapuas-Barito, and Katingan-Mendawai catchments) or in areas with large expanses of shallow, coastal peat-swamp forests (Tanjung Puting and Gunung Palung National Parks).[142] As much as 40 percent of Borneo's remaining orangutan population may occupy peat-swamp forests.

Field surveys carried out since 1995 in the River Sebangau catchment in Central Kalimantan,

Peatlands in Southeast Asia

Most peatlands occur on the islands of Borneo and Sumatra and in Peninsular Malaysia. The true extent and thickness of these peat deposits are poorly documented.

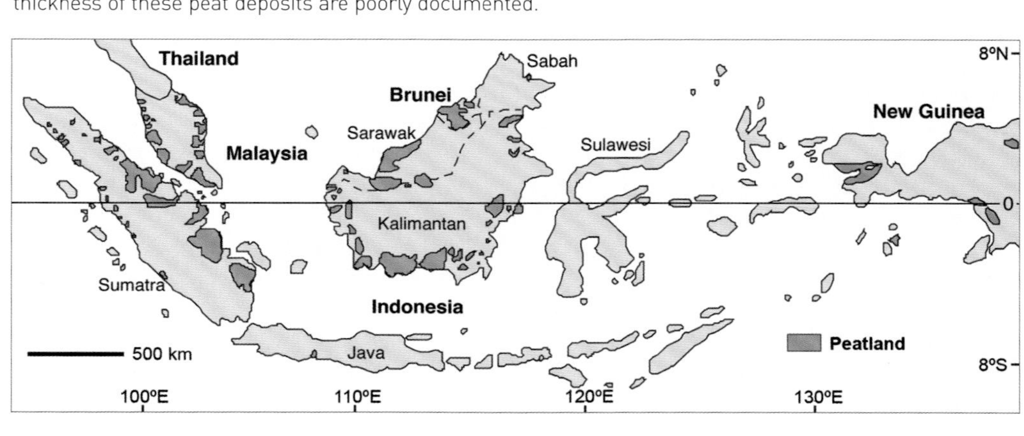

John O. Rieley

John O. Rieley

Top: Canopy of the peat-swamp forest in the Sebangau catchment.
Above: Impact on peat-swamp forest of the Proyek Lahan Gambut drainage channel.

Indonesia have shown it to be a very important habitat for orangutans.[103] In common with most forested areas in Kalimantan, however, it is experiencing large-scale, indiscriminate illegal logging.[33] In October 2004, most of the Sebangau catchment (5.68 km^2) was designated as a new national park by the government of Indonesia. This change in status alone will, however, be insufficient to save the peat-swamp forest habitat and its large orangutan population. Urgent action is required to prevent future fires and to restore the hydrological integrity of the area by negating the impact of the extraction canals associated with illegal logging that hasten the runoff of water from the peatland ecosystem.

Other threats to the Sebangau National Park include plans to increase the area of land under plantation and arable crops (such as oil palm, rice, and vegetables) to support increases in the number of people who will be settled around the park perimeter. It is strongly recommended that a buffer zone be established around the new park, to control human activities and reduce their negative impacts.

Peat-swamp forests are in decline throughout Southeast Asia and steps should be taken to increase international awareness of their global importance in the carbon cycle and in climate-change processes, as well as for the conservation of the orangutan and other wildlife.

John O. Rieley, Susan E. Page, and Suwido H. Limin

altitudes may have been due to unusually high densities of fruit trees such as mangosteens (*Garcinia* spp.) and durians (*Durio* spp.).[111]

It is difficult to quantify the continuous area required to support a healthy population of orangutans. They have a complicated and poorly understood social and ranging system of loose communities that travel long distances in response to mast seasons, presence of females, and food shortages. With our current knowledge, it is difficult to be confident of either the number of individuals required to preserve the orangutan social structure, or the area these individuals require.[123, 128]

POPULATION STATUS, TRENDS, THREATS

Human attitudes and traditions

Human settlements have existed in Borneo for at least 40 000 years.[7, 9] Archaeological evidence from caves at Niah in Sarawak and Madai in Sabah shows that orangutans were a regular food item of prehistoric people, who may have caused the extinction of orangutans from several regions of Borneo. Some tribes and clans still have a religious association with the apes, often involving hunting them. Orangutans are absent from the largest part of Sarawak's lowland forest, from north of the Rajang River, through all of Brunei, and from much of the flood plain area of the Kapuas River and the valleys of the Kayan Mentarang area in East Kalimantan.[123]

The relationship is unclear between the original prehistoric inhabitants of Borneo and the more recently arrived inhabitants known collectively as 'Dayaks'. These Austronesian peoples came to Borneo from the Malay Peninsula and Sumatra in waves starting some 4 000 years ago. This was part of the great island-hopping expansion of the Austronesians from their original base in Taiwan.[28] On Borneo, the Dayaks diversified to form societies adapted to meet a variety of conditions. In the interior, the dominant form of land use was shifting cultivation, originally of root crops, with hill rice being adopted around the 5th century. Hunter-gatherer societies, such as the nomadic Penan, arose in some areas. The Iban, now among the most populous people in parts of Sarawak and West Kalimantan, are relatively recent arrivals from Sumatra. They invaded Borneo along the Kapuas River in West Kalimantan during the 14th and 15th centuries. Resident Dayak groups were absorbed or displaced as the Iban spread.

Under traditional systems of land tenure among the Dayaks, every longhouse community in Borneo claims an extensive tract of community land. Within these territories, individual families have rights to the lands they clear for farming (areas known as *kebun* or 'gardens'); these rights are inherited. The 'gardens' are maintained in a system of cultivation followed by a fallow period when secondary vegetation reclaims the site. Community land extends beyond the cultivated areas and includes swamp and virgin forests that are conserved for the collecting of wild produce and hunting of animals.[59]

Hunting of orangutan was widespread amongst the Dayak peoples, generally using a blowpipe and poison. The apes were considered to be a prized and delicious food item and also a medicine to ensure sexual potency.[26, 123] Infant orangutans were kept as pets in longhouses or sold. Under some traditional systems, certain families or clans were restricted from eating orangutan meat; this may have spared the species from extinction in a number of areas.[66] The Iban believed that after death an ancestor becomes a member of a particular species (*tua*) and, in this form, can help his or her living descendants. This belief continues in some tribes today; one study found the orangutan to be a *tua* species among seven households in the Batang Ai region of Sarawak,[59] though it was permissible to hunt the apes to obtain a head on special occasions.[123]

Orangutans were also hunted for religious reasons. To some Dayak peoples, they possessed a soul[122] and this 'soul substance' was considered higher than that of humans. Some Iban groups regarded the orangutan as being a representative of their war god.[60] Related to this belief was the custom of the Kelabit and Kayan peoples that involved rubbing dirt and ape hair into the breast of a newborn child, to prevent it being stolen by an ape.[122] The skin and hair of orangutans have been used until the present for war cloaks, jackets, and caps, and to decorate the handles of swords.[123]

Compared to other animals, orangutans feature only irregularly in traditional mythology. Linnaeus related travelers' reports that suggested they were once human rulers, who had fallen from grace.[66] According to some Malay people, orangutans are humans who refuse to speak for fear of being enslaved.[122] The Iban seemed to view orangutans as their forest cousins; some stories suggest they made no clear distinction between humans and apes.[66] Indeed, traditionally prepared trophies of human skulls have been found mixed with orangutan skulls.[123] Many stories tell of attacks, and rapes, of women by male orangutans, and (occasionally) of men[80] by female orangutans. In these stories, women were often kept as the orangutan's lover or bride; in some accounts inter-breeding occurs. Other stories describe the great strength of orangutans in fights with pythons, crocodiles, or bears. The Kenyah and Kayan peoples are unable to look straight into the eyes of an orangutan in case they offend it, but they will still hunt and eat it.[66]

Recent history

Products such as hornbill 'ivory' and rhinoceros horn from the interior of Borneo have been traded as far afield as China for a thousand years or more.[53] Trade goods from China, such as ceramic jars and bronze gongs, have found their way deep into the interior of Borneo in return. Growth of international trade relations in the early 19th century precipitated massive cultural changes. Major movements of indigenous peoples of Borneo were induced by the head-hunting habits of the Iban around the same time. The constant threat of warfare and transition being undergone by many societies diluted the traditional mythical status of the orangutan.[123] This process was accelerated by more recent mass migrations into Borneo by Javanese, Maduranese, and other people who came as transmigrants sponsored by the government of Indonesia, or who came spontaneously to take advantage of new trading and farming opportunities. The traditional regulatory land-use system and religious constraints on hunting the ape vanished. Loss of traditional beliefs and community responsibility removed constraints on selling previously untouched forest to timber companies or illegal loggers. Old logging techniques were superseded by more technologically advanced and more destructive methods.[59]

Coupled with these changes was the arrival of European people seeking 'scientific collections' or opportunities for sport hunting. One biologist shot 217 orangutans in one small area;[137] 'bags' of around 40 individuals were common in the 1890s. The spread of the Muslim culture into much of Borneo has meant that many people no longer see orangutans as potential food items; this is *haram* (forbidden) under Islamic law. Nonetheless, some army personnel (not all of whom are Muslim in either Indonesia or Malaysia) have been reported to go on poaching expeditions.

Some foreign tourists still purchase ape skulls as traditional Dayak relics, although they may be of recent provenance. Orangutan losses due to hunting and the associated pet trade are sizeable; they increased in 1997 when an economic crisis emerged in Indonesia.[150] It has been estimated that every week two infant orangutans from Kalimantan are smuggled out of Jakarta via Batam to Singapore.[23] Orangutans are poached throughout Borneo and several other trade routes also exist.[123] The number of young orangutans confiscated by 2000 represents 1 percent of the estimated total population at that time.[87] This figure does not, of course, take into account those infant orangutans that die before being confiscated, the mothers that were killed to obtain these infants, or the infants that are not rescued from the pet trade.

The current position

The Bornean orangutan is classified as Endangered by IUCN–The World Conservation Union, indicating it has a very high risk of extinction in the wild in the near future.[40] By 2004, the total remaining area of Bornean orangutan habitat was around 86 000 km^2, supporting 45 000–69 000 individuals. Their distribution is summarized below.[96, 142]

- Central Borneo orangutans (*P. p. wurmbii*): Central Kalimantan has several forest areas large enough to support viable orangutan populations, and is thought to hold over 32 000 individuals. The peat swamps of the western Sebangau catchment may hold 6 000 and the Gunung Palung National Park nearly 2 500.

Raffaella Commitante

Young orangutans at an orphan sanctuary in Borneo.

- Northwest Borneo orangutans (*P. p. pygmaeus*): Danau-Sentarum National Park (also a Ramsar Site) and the surrounding peat forests in West Kalimantan are estimated to hold about 1 500 individuals following the 1997–1998 fires.[133] There are relatively few orangutans remaining in Sarawak, with the last viable population of 1 140 to 1 760 being in the Batang Ai National Park and adjacent Lanjak-Entimau Wildlife Sanctuary.[97]
- Northeast Borneo orangutans (*P. p. morio*): Sabah is thought to hold around 11 000 (8 000–18 000) individuals,[4] with about another 3 000 in East Kalimantan.

The total number of Bornean orangutans today is estimated to be less than 14 percent of what it was at the start of the Holocene, 10 000 years ago, when the last ice age ended, sea levels rose, and Borneo became isolated from Sumatra (Table 10.4). The decline of the species accelerated towards the end of the 20th century, with deforestation and degradation of Bornean forests since the 1970s. In the unlikely event that all human disturbance and anthropogenic mortality suddenly ceased, orangutan numbers would recover at only 0.006 percent per year due to their very low reproductive rate.[87]

Feeding time at the Sepilok rehabilitation center in Sabah.

Elaine Marshall

A Population and Habitat Viability Assessment workshop held in January 2004 identified a set of 10 habitat units that were considered to offer the greatest potential for sustaining the Bornean orangutan (Table 10.5);[142] the absence of a population from this list does not imply that it fails to merit conservation attention. It was noted that the "loss of any of these populations would seriously jeopardize the taxon's integrity as an evolutionary unit."[142] The workshop's selection procedure considered taxonomy, habitat diversity (uniqueness and peripherality), and distribution among major political units (states and provinces).

Sabah (73 371 km^2) occupies about 10 percent of the island of Borneo; about half remains under natural forests,[93] although oil palm and pulpwood plantations continue to expand. Eastern Sabah was almost completely uninhabited until about 1960, but now only 25 percent of land area remains under lowland forest, much of it heavily logged. From the late 1960s into the 1990s, Sabah's forests were managed in ways that resulted in severe depletion of the state's timber reserves,[129] and they are now virtually exhausted.[101] Close to 8 percent of Sabah's land area is included in the state's system of protected areas, but about 60 percent of the orangutans in Sabah live outside protected areas, in production forests that have been through several rounds of timber extraction and that are still being exploited for timber.[3]

Sarawak (124 500 km^2) is Malaysia's largest state and, like Sabah, around 8 percent of its land area is included or proposed for inclusion within state-protected areas. It has been a major exporter of timber since the 1960s; log production increased steadily during the 1970s and 1980s to more than 14 million m^3 per year. More than 2 000 km^2 of forest were logged each year during the 1990s, a process that continues to date. Much of Sarawak is uninhabited by orangutans due to past hunting, conversion of forest by shifting cultivation, or for ecological reasons. Significant populations of orangutans now occur only in the south-central interior, in and around the Batang Ai National Park and Lanjak-Entimau Wildlife Sanctuary, which is contiguous with Betung Kerihun National Park in West Kalimantan. These protected areas in Sarawak are thought to be subject to relatively little threat, although they are understaffed and vulnerable to illegal logging and hunting, including probable cross-border logging from Kalimantan.[142]

The four provinces of Indonesian Kalimantan

(536 000 km² in total) occupy more than 72 percent of Borneo; orangutans still occur in three of these. Threats to their populations and habitats are abundant and widespread, however, and have been building up for many years. An example is the Proyek Lahan Gambut million hectare mega-reclamation program in Central Kalimantan. This project, supported by former President Suharto, was designed to boost the faltering transmigration program and rice production in the mid-1990s. It was responsible for draining 15 000 km² of peat swamp, including up to 7 000 km² of prime orangutan habitat.[123] The project was abandoned in 1998,[103] and the land was burned in 2002; it now appears on satellite images as a circular patch of deforested land. Subsistence agriculture is an additional cause of forest loss; by 1999, slash-and-burn agriculture was reported to have affected 27 percent of land area in Kalimantan, 87 percent of which was presumed to have been prime orangutan habitat at some stage.

Indonesian forests are also being increasingly converted to plantations, especially for oil palm and *Acacia*.[23] Coal mining has long been an issue in Kutai National Park, and is becoming one in other areas, such as Barito Ulu, with mining rights also being sought inside additional national parks.[23] Furthermore, fires and droughts have ravaged Kalimantan repeatedly since the early 1980s. Significant fire damage occurred during 1997–1998, burning around 52 000 km² of forest in the worst-affected province, East Kalimantan.[57] Localized damage was extreme; in Kutai National Park, 95 percent of lowland forest was lost.[150] The disproportionate loss of peat-swamp and lowland forest was the most serious consequence as this is the richest habitat for orangutans.[65] Large numbers of orangutans were killed by people while fleeing the flames and smoke during and after the fires; the displacement of apes may have precipitated a shockwave of 'refugee crowding' in adjacent forests,[123] causing stress that reduces breeding success.[150] As a result of the fires, Borneo's orangutan population may have been reduced by 33 percent in just one year.[123] Serious fires also occurred in 2004, with the resultant haze again threatening human health in Borneo's cities.[79, 118]

The remaining Indonesian forests that are officially intended to be permanent are dedicated to timber production, watershed protection, or biodiversity conservation. Official timber concessions almost completely overlap the fragmented distribution range of the orangutan in Kalimantan, and it is generally thought that orangutans are unable to survive long term in heavily logged forests,[150] although the northeast Borneo orangutan appears to be exceptionally resilient to logging (see Box 10.2).

It is the illegal, rather than the legal, activity that is causing the current deforestation crisis in Kalimantan, however. Vast areas are being converted to unplanned, private plantations or burnt after repeated logging. Illegal logging in the national parks is rampant and has caused a huge decline in orangutan numbers. Failure to stop illegal logging in these areas will inevitably lead to a further dramatic reduction in forest cover and in orangutan population size.[142] Danau-Sentarum National Park and Betung Kerihun National Park, for example, are being destroyed by illegal logging and have almost no capacity or support to prevent this; Tanjung Puting and Gunung Palung National Parks have also been badly affected. The construction of canals for illegal logging drains and destroys deep peat-swamp forests, and represents

Table 10.4 Bornean orangutan population decline since 8000 BC

Year	Estimated number of Bornean orangutans	Decline from previous estimate
8000 BC	420 000	–
1900	230 000	45 percent
2004	57 000 (45 000–69 000)	75 (±5) percent

Based on Singleton, I., *et al.* (2004) and Rijksen, H.D., Meijaard, E. (1999).

Table 10.5 Bornean orangutan populations critical to species integrity

Habitat unit	Estimated population	Subspecies
Sebangau	6 300	*P. p. wurmbii*
Tanjung Puting	6 000	*P. p. wurmbii*
Belantikan	5 000+	*P. p. wurmbii*
Mawas	3 500	*P. p. wurmbii*
Gunung Palung	2 500	*P. p. wurmbii*
Sabah Foundation forestry concession (east)	5 320	*P. p. morio*
Kinabatangan	4 000	*P. p. morio*
Gunung Gajah/Berau/Kutai	3 000 (?)	*P. p. morio*
Batang Ai/Lanjak-Entimau/ Betung-Kerihun	> 2 500 (?)	*P. p. pygmaeus*
Danau-Sentarum (and surrounds)	500 (1 500)	*P. p. pygmaeus*

Based on Singleton, I., *et al.* (2004).

a major threat to orangutans (see Box 10.3).[103] Current trends suggest that all Indonesian lowland forest will be lost or badly degraded by 2010, with the main driver being timber extraction.[64] If this forecast is correct, it is hard to see a future for wild orangutans in Kalimantan.

CONSERVATION AND RESEARCH

The field study of Bornean orangutans began in the late 1960s, at Kutai National Park in East Kalimantan,[124, 125] and in the Ulu Kinabatangan region of Sabah, Malaysia.[82, 83] A long-term investigation began in 1971 in Central Kalimantan, in the area now designated Tanjung Puting National Park,[43, 46, 47, 48] and continues to the present day. Later studies have included work at Danum Valley, the Kinabatangan floodplain, Tanjung Puting National Park, Kutai National Park and elsewhere, on feeding behavior, adaptation to disturbance, and behavioral adaptations of rehabilitated individuals.

The chief experience of researchers in Borneo since the 1970s has been of escalating habitat damage and declining wildlife populations, with orangutans coming under increasing pressure from logging, land clearance, and forest fires. As a result, most researchers have either moved on or become more involved in conservation, often through nongovernmental organizations. For example, in Sabah, Hutan runs a community conservation and research project along the Kinabatangan River, supported by WWF-Malaysia and working closely with the Sabah Wildlife Department. In Sarawak, WWF-Malaysia was involved during the 1980s in preparing a statewide conservation strategy and in conducting field studies to support management of Batang Ai National Park and the Lanjak-Entimau Wildlife Sanctuary.

In Indonesian Borneo, the Orangutan Foundation International funds patrols in Tanjung Puting National Park, rehabilitates and releases orphan orangutans in Lamandau Strict Nature Reserve, and supports research into both conservation and forest restoration. The Borneo Orangutan Survival Foundation rehabilitates and releases orphans in the Balikpapan area as well as in other parts of Kalimantan, and is involved in proposals to protect the Mawas area, which comprises 5 000 km^2 of peat-swamp forest inhabited by orangutans. In an exciting application of modern technology, habitat in the Mawas reserve is monitored by satellite, supplemented when necessary by aerial photography. This remote sensing captures evidence of illegal land uses and wildfires so that teams may be dispatched to counter them.[136]

Slow-moving, slow-breeding orangutans, which depend largely on fruit in fruit-poor lowland Bornean forests, are very vulnerable to rapid lowland forest clearance, whether for plantations or through forest fires incidental to logging and drought. The destruction of forest ecosystems on the island of Borneo has an historical momentum that will be difficult to restrain even in most protected areas; it is largely irreversible. We cannot predict whether it will also become complete, but isolated moist forests surrounded by flammable and fire-maintained scrub, grass, and farmland may well prove difficult to protect.

The current population of the Bornean orangutan, at about 57 000, is a snapshot of a species in decline. It is symbolic of increasing habitat damage across the world's most biologically rich tropical island. Potentially viable orangutan populations and forests still exist, but to secure them will require not only focused investment in and around the areas themselves. In addition, coordinated and sustained effort will be required by all the Bornean states and territories to improve land use and local governance, and to resist the factors that dispose forests to be cleared and burned. There are signs that local governments in some areas are beginning to work more closely with local peoples to safeguard their forests, and that cross-border cooperation is beginning to become easier both bilaterally and as facilitated by the Association of Southeast Asian Nations. It can only be hoped that these efforts will continue and multiply.

FURTHER READING

Ancrenaz, M., Calaque, R., Lackman-Ancrenaz, I. (2004) Orang-utan nesting behavior in disturbed forest of Sabah, Malaysia: implications for nest census. *International Journal of Primatology* **25** (5): 983–1000.

Ancrenaz, M., Gimenez, O., Ambu, L., Ancrenaz, K., Andau, P., Goossens, B., Payne, J., Tuuga, A., Lackman-Ancrenaz, I. (2005) Aerial surveys give new estimates for orang-utans in Sabah, Malaysia. *PLoS Biology* **3** (1): e3. http://dx.doi.org/10.1371/journal.pbio.0030003. Accessed December 8 2004.

Bennett, E.L. (1998) *The Natural History of Orang-utan.* Natural History Publications, Kota Kinabalu, Sabah.

Currey, D., Doherty, F., Lawson, S., Newman, J., Ruwindrijarto, A. (2001) *Timber Trafficking. Illegal Logging in Indonesia, SE Asia and International Consumption of Illegally Sourced Timber.* EIA and Tekepak Indonesia.

Goossens, B., Chikhi, L., Jalil, M., Ancrenaz, M., Lackman-Ancrenaz, I., Mohamed, M., Andau, P., Bruford, M. (2005) Patterns of genetic diversity and migration in increasingly fragmented and declining orang-utan (*Pongo pygmaeus*) populations from Sabah, Malaysia. *Molecular Ecology* **14** (2): 441–456.

Jepson, P., Jarvie, J.K., MacKinnon, K., Monk, K.A. (2001) The end for Indonesia's lowland forests? *Science* **292**: 859–861.

Maltby, E., Immirzi, C.P., Safford, R.J., eds (1996) *Tropical Lowland Peatlands of Southeast Asia.* IUCN, Gland, Switzerland.

Morrogh-Bernard, H., Husson, S., Page, S.E., Rieley, J.O. (2003) Population status of the Bornean orang-utan (*Pongo pygmaeus*) in the Sebangau peat swamp forest, Central Kalimantan, Indonesia. *Biological Conservation* **110** (1): 141–152.

Rieley, J.O., Page, S.E., eds (1997) *Biodiversity and Sustainability of Tropical Peatlands. Proceedings of the International Symposium on Biodiversity, Environmental Importance and Sustainability of Tropical Peat and Peatlands, Palangka Raya, Central Kalimantan, 4–8 September 1995.* Samara Publishing, Cardigan, UK.

Rijksen, H.D., Meijaard, E. (1999) *Our Vanishing Relative: The Status of Wild Orang-utans at the Close of the Twentieth Century.* Kluwer Academic Publishers, Dordrecht.

Ross, M.L. (2001) *Timber Booms and Institutional Breakdown in Southeast Asia.* Cambridge University Press, Cambridge, UK.

Russon, A.E. (1998) The nature and evolution of intelligence in orangutans (*Pongo pygmaeus*). *Primates* **39** (4): 485–503.

Schwartz, J., ed. (1988) *Orangutan Biology.* Oxford University Press, Oxford.

Wright, S.J., Carrasco, C., Calderón, O., *et al.* (1999) The El Niño southern oscillation, variable fruit production, and famine in a tropical forest. *Ecology* **80**: 1632–1647.

MAP SOURCES

Map 10.1 Orangutan data are based on the following sources:

Ancrenaz, M., Lackman-Ancrenaz, I. (2004) *Orang-utan Status in Sabah: Distribution and Population Size.* Kinabatangan Orang-utan Conservation Project, Sandakan, Malaysia.

Meijaard, E., Husson, S., Lacy, R., *et al.* (forthcoming) Orangutans: out for the count or hanging on? [in preparation for *Science*].

Meijaard, E., Dennis, R., Singleton, I. (2004) Borneo Orangutan PHVA Habitat Units: Composite dataset developed by Meijaard & Dennis (2003) and amended by delegates at the Orangutan PHVA Workshop, Jakarta, January 15–18 2004.

Singleton, I., Wich, S., Husson, S., Stephens, S., Utami Atmoko, S., Leighton, M., Rosen, N., Traylor-Holzer, K., Lacy, R., Byers, O., eds (2004) *Orangutan Population and Habitat Viability Assessment: Final Report.* IUCN/SSC Conservation Breeding Specialist Group, Apple Valley, Minnesota.

For protected area and other data, see 'Using the maps'.

ACKNOWLEDGMENTS

Many thanks to Mark Attwater (Orangutan Foundation), Raffaella Commitante (University of Cambridge), Simon Husson (University of Palangkaraya), Ashley Leiman (Orangutan Foundation), Helen Morrogh-Bernard (University of Palangkaraya), and Rondang Siregar (University of Cambridge) for their valuable comments on the draft of this chapter.

AUTHORS

Kim McConkey, UNEP World Conservation Monitoring Centre

Box 10.1 Julian Caldecott, UNEP World Conservation Monitoring Centre

Box 10.2 Marc Ancrenaz, Isabelle Lackman-Ancrenaz, and Ahbam Abulani, Hutan Foundation

Box 10.3 John O. Rieley, University of Nottingham, Susan E. Page, University of Leicester, and Suwido H. Limin, University of Palangkaraya

Anup Shah/naturepl.com

CHAPTER 11

Sumatran orangutan (*Pongo abelii*)

KIM MCCONKEY

Sumatran orangutans (*Pongo abelii* Lesson, 1827) are now mainly restricted to 11–13 isolated forest units, found mostly to the north of Lake Toba on the Indonesian island of Sumatra (Map 11.1).[44] The absence of orangutans farther south is considered to be the result of hunting over many hundreds (perhaps even thousands) of years, although there are records of orangutans farther south as recently as the 1830s.[30] The majority now live in the province of Nanggroe Aceh Darussalam (Aceh). Most are found in lowland areas, below 1 000 m, although transient individuals – usually males – are sometimes seen at higher altitudes.

The most viable populations are found in northern Sumatra, including in the Leuser Ecosystem (the large landscape within which the Gunung Leuser National Park is set) of Aceh, and particularly in the coastal swamps and certain lowland parts of the Alas Valley.[64] Two highland areas in the Alas Valley area also sustain orangutan populations, these being the upper Mamas Valley and the Kapi Plateau. The soil type and ecosystems in these areas are similar to those seen in the lowlands, allowing orangutan densities of three to six individuals per square kilometer to occur.[31] Two isolated orangutan populations occur farther south, one in the Padang Sidempuan, Tarutung, and Sibolga area (known as the West Batang Toru forest block), and the other in forest to the east of there, known as East Sarulla.[70] There are also a very few orangutans in the Lumut coastal swamps, a population that is not considered viable.[40]

BEHAVIOR AND ECOLOGY

Diet and feeding behavior

Sumatran orangutans have little difficulty locating fruiting trees and appear to be familiar with the topography of a large area. They are able to recognize important food species, are thought to possess a mental map of food sources, and to be aware of fruiting seasons. There is also evidence that they read various signs indicating the presence of fruit. Orangutans observe the flights of hornbills and have been seen to follow their flight paths.[30]

Fruit availability has a pivotal influence on the behavior and distribution of orangutans. Their typical restriction to altitudes of less than 1 000 m and their patchy distribution within broad expanses of forest at lower altitudes are both thought to be governed by the availability of soft-pulped fruits.[6] Orangutans consume a variety of fruit types, which together form the largest portion of their diet in most months. Over three years at the Ketambe site in Gunung Leuser National Park, 58 percent of feeding observations were of fruit, taken from 92 different tree and liana species.[30] Several of these usually semisolitary apes may be attracted at the same time to a tree bearing fruit with soft juicy pulp, such as *Antiaris toxicaria* (Moraceae), *Cyathocalyx sumatranus* (Annonaceae), *Mallotus schaeorocarpus* (Euphorbiaceae), *Tinomiscium phytocrenoides* (Menispermaceae), *Garcinia* spp. (Clusiaceae), the rambutan *Nephelium lappaceum,* and *Xerospermum* spp. (both Sapindaceae).[30, 49] Large fruits with a volume greater than 71 cm^3, such as durians (*Durio* spp., Bombacaceae) and jackfruit (*Artocarpus elastica,* Moraceae), are also favored, although smaller fruits are usually more abundant and feature more regularly in the diet.[30, 49] In the Suaq Balimbing swamp forests of Gunung Leuser National Park, the following produce almost continuous supplies of fruit: *Tetramerista glabra* (Tetrameristaceae), *Sandoricum beccarianum*

Map 11.1 Sumatran orangutan distribution

Data sources are provided at the end of this chapter

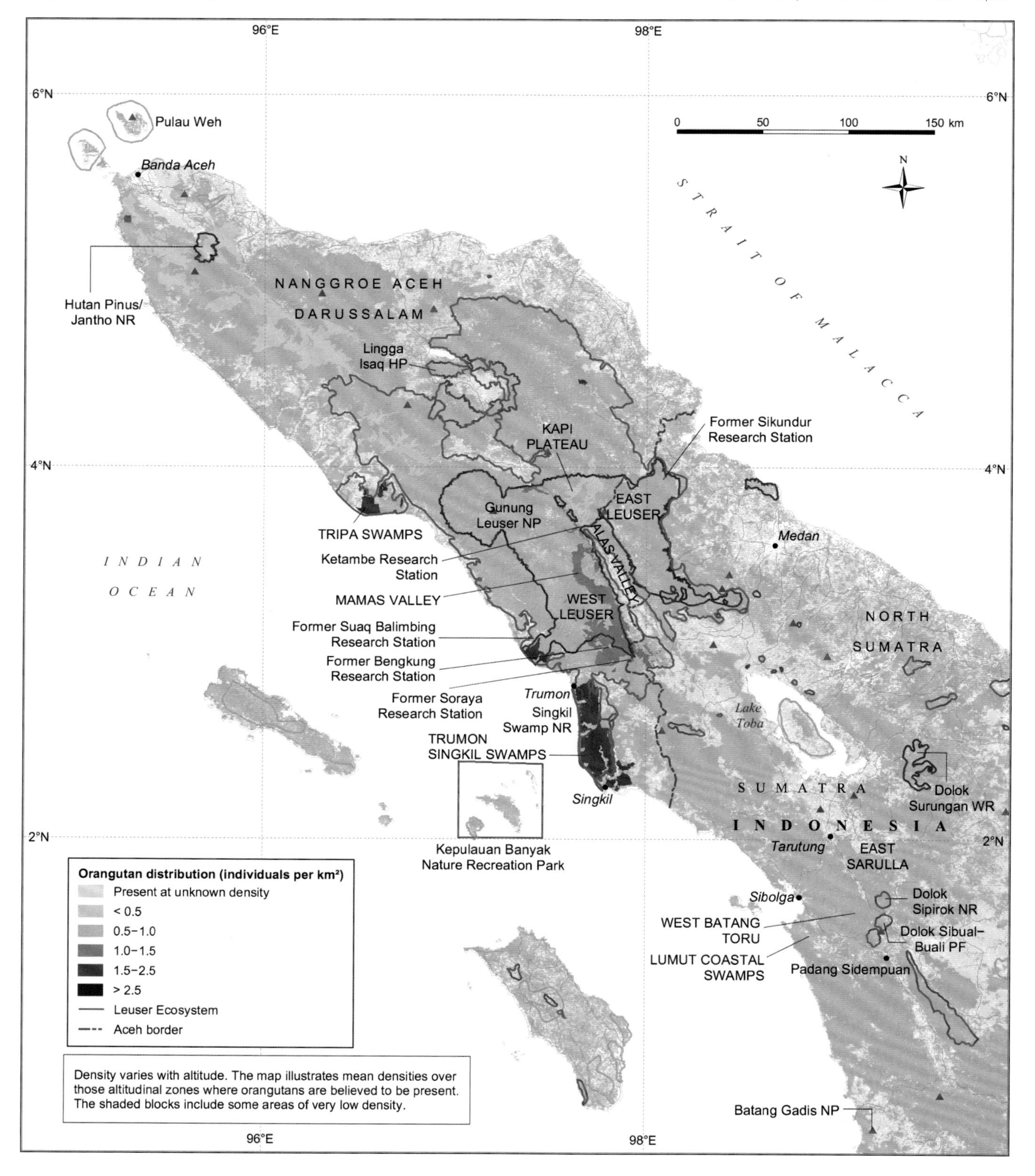

(Meliaceae), and *Neesia* cf. *glabra* (Bombacaceae). These are considered staple orangutan foods in these habitats.[56, 57]

Although there is evidence for seasonal dietary changes in Sumatran orangutans, some sites offer almost continuous availability of high-quality fruit so monthly variation in diet is accordingly slight.[56] For example, figs (*Ficus* spp., Moraceae) of at least eight strangler species were available for eight months of the year in Ketambe.[30, 45] If fruit does become scarce the apes may move to find new supplies, but there is also a tendency to switch to less-preferred foods,[42, 57] such as bark and leaves.[23, 30]

At Ketambe, figs have been recorded as providing around half of all the fruit eaten by orangutans.[30] Strangler fig trees occur at very high densities in some areas and produce large crops of easily digested fruit at relatively short intervals.[57, 64] Certain fig species are favored over others, and at Ketambe they include *Ficus annulata*, *F. benjamina*, *F. drupacea*, *F. stupenda*, and *F. subulata*.[30]

Seeds are often rich in fat with a high calorie content, to provide the seedling with early sustenance after germination. As these same attributes make seeds attractive to animals, in tropical forests seeds are often protected by hard cases or irritant hairs. Their strong teeth and jaws,[57] manual dexterity, and strong hands, allow orangutans to overcome these defenses and obtain the seeds of various species, including the favored fruit of *Heritiera elata* (Sterculiaceae).[30] Orangutans at Suaq Balimbing also use tools to extract the highly nutritious seeds of *Neesia* cf. *malayana*[58, 62] (see below). The ability to exploit these seeds, combined with a tolerance of unripe and acidic fruits, gives orangutans an advantage over other primates in the same forest.[30, 49]

Leaves are also regularly eaten by Sumatran orangutans, representing 5–25 percent of feeding observations.[23, 30] The apes usually eat new shoots or buds, but also consume the mature leaves of selected species.[30] A high tolerance of generally unpalatable species has been demonstrated. For example, orangutans eat the leaves of stinging nettles, *Dendrocnide* spp. (Urticaceae), although they take great care to prevent their lips coming into contact with the leaf surface.[30] Other items consumed include aerial roots, epiphytic fungi, orchids, the stems of climbers, grass, and leaf galls; orangutans also use their strong teeth and jaws to strip bark from trees. Even the phloem and xylem layers of the wood of certain trees are eaten.[30, 57] At least 17 species of insects are represented in the orangutan diet. Soil is also consumed, presumably to provide mineral nutrients or (in the case of kaolin-rich clay) to settle the stomach.

Carnivory is rare in Sumatran orangutans, and those incidents that have been reported have been interpreted as opportunistic rather than as the outcomes of planned hunting expeditions.[46, 53] No vertebrate remains were found in feces during a three year study in Ketambe; individual orangutans did occasionally eat bird eggs and sometimes also inspected squirrel nests, so may eat nestlings if the opportunity arises.[30] An adolescent female was observed eating an infant gibbon (*Hylobates lar*) or slow loris (*Nycticebus coucang*),[46] and seven slow lorises were also seen to be eaten by three adult females over 20 years at Ketambe and Suaq Balimbing.[53] Lorises are easy, slow-moving prey (although with sharp teeth), and orangutans are ill equipped to catch faster-moving animals.

Ranging behavior

A single area is likely to be used by several orangutans with quite diverse ranging patterns. As noted in Chapter 9, these have been interpreted by some observers in terms of three social classes linked to ranging behavior: 'resident,' 'commuter,' and 'wanderer.'[31] This conclusion was based on data from Ketambe, where 'commuters' were said to make up 60 percent of the population, with 30 percent 'residents,' and 10 percent 'wanderers'. Data from Suaq Balimbing indicate that most individuals live in large but stable and widely overlapping home ranges.[42] At Suaq Balimbing, up to 16 adult females, nine adult males, and at least 15 subadult males have been seen to use a forest patch of only 4 ha, the area bounded by a single square of a 200 m grid of survey paths. Clusters of related females sharing home range boundaries have also been identified. Some subadult males may range very widely, but there was no evidence for the presence of commuters. Ranging behavior may be better described in terms of very large lifetime home ranges, parts of which are used more persistently than others, depending on various social and ecological factors. Differences between orangutan populations in ranging behavior may be resource driven, as the Suaq Balimbing swamp forests offer more plentiful and continuous supplies of nutritious fruit.

The home ranges of male orangutans are consistently at least two to three times larger than

Ian Singleton/SOCP

Fruit makes up the larger part of an orangutan's diet through most of the year. Here, a rehabilitated orangutan in Jambi province, Sumatra, is enjoying a papaya supplied by one of the post-release monitoring staff.

those of females. High-status or dominant adult males, however, appear to maintain a relatively small home range during their period of dominance, within which they attempt to monopolize access to receptive females.[37, 42] The other adult and subadult (unflanged) males must search for females over a larger area, usually avoiding the areas occupied by the dominant males.

The overall extent to which Sumatran orangutan home range size and movements are influenced by food availability is not clear.[37, 47] In Ketambe, where the terrain is rugged, those orangutans that do not restrict themselves to a circumscribed area follow the availability of soft-pulped fruit as it varies over steep altitudinal gradients.[22] In Suaq Balimbing, which is largely flat, many of the major food trees fruit synchronously; here the orangutans are less mobile on a daily basis, but occupy very large home ranges.[37, 42] Females occupy home ranges of around 8.5 km^2 in these swamps, compared to only 3 km^2 in Ketambe, while subadult and adult male ranges are around 25 km^2 in the swamp forest and 8 km^2 in Ketambe.[5, 42] The Suaq Balimbing orangutans may be using areas of swamp and hill forest in order to make best use of the spatial and temporal patchiness of food resources.[42] The greater range overlap seen in the swamp forest probably results from orangutans sharing access to areas with a high density of favored food trees.[42]

Food-related movements do occur regularly during mast fruiting, when trees from the family Dipterocarpaceae flower in synchrony, usually along with several other species, to provide a superabundance of food. In normal fruiting seasons the lowlands tend to be more productive, but during a mast period in 1997 some orangutans in the Gunung Leuser National Park were observed to move from the swamps into the hills to exploit local increases in food abundance.[36]

Social behavior

Orangutans are remarkable among the great apes in that they appear to lack distinct social units or groups. Instead, Sumatran orangutans are described as living in 'loose' communities that consist of one or more clusters of genetically related females and the adult male with which they all prefer to mate.[42, 43] Researchers have noted that the movements of community members are subtly coordinated, and that they may come together as a real group on some occasions.[31]

The mother–infant bond is very close in orangutans, but it gradually weakens with age; by the time the apes are fully adult, they are mainly solitary. Interaction among adults is often limited to glances,[57] although juveniles may play together. After independence, females tend to stay near the range where they were born and maintain amiable relations with local females, which are likely to be close relatives.[57] Individuals in clusters of closely related females at Suaq Balimbing not only share home range boundaries, but appear to coordinate their breeding; the timing of births was similar within a cluster, but differed between clusters.[42]

After becoming independent of their mother, males move away and either settle in a large range or wander over large areas, sometimes even well beyond the forest region occupied by the breeding population. As subadults they may still travel together, but flanged adult males generally avoid encounters with each other. When flanged males meet, violent aggressive displays can ensue, and potentially fatal fights may occur.[42] Subadult and flanged males commonly occupy overlapping home ranges, however, and flanged males will tolerate subadult males provided that they keep their distance.[57]

Long calls are a prominent means by which orangutans may maintain links within their loosely knit communities. The calls carry over long distances and may enable females and juveniles to remain aware of the location of the males. Only fully developed adult males produce the long call, which "starts as a series of quiet bubbling grunts, then builds up into a full-blown gravelly roar ... often accompanied by vigorous branch shaking."[2] The throat pouch is fully inflated during calls and may act as a resonating chamber,[57] and the calling male's hair stands on end during the display. Calls are made three or four times a day. Calling frequency is higher where local orangutan density is higher.[5, 55] It has been suggested that the cheek pads of adult males may help to focus and direct the long calls, and that cheek-pad differences between Bornean and Sumatran orangutans relate to divergent long calls.[5]

An orangutan community is brought together under the influence of abundant food and sexually attractive females.[42, 51, 54] Sufficient fruit is not always available to sustain groups of orangutans, however, even should they wish to be gregarious.[5] When individuals do aggregate, it is in one of three different modes:

- consortships, in which a receptive female and male travel together for an extended period;[51]
- temporary aggregations, which occur when individuals feed together but leave the food source separately; and
- travel bands, in which all individuals feed together within a patch, leave together, and may visit the next patch together.

As single food sources must be divided among the group, aggregations are likely to be costly to participating individuals. By forming these parties, however, orangutans are thought to obtain mating opportunities, protection of females from harassment by subadult males, and opportunities for the socialization of infants.[51] At Suaq Balimbing, females with mid-sized infants are least likely to participate in groups, and adult males rarely participate in aggregations other than consortships.[54] Nevertheless, the high density of orangutans in Sumatran swamp forests, and their sociability, is thought to have facilitated the social transmission of skills including tool use[62] (see below).

Ian Singleton/SOCP

While Sumatran orangutans spend more time in groups than the Bornean species, they generally lead solitary lives, with males ranging over a much bigger area than females.

The mean daily party size at Suaq Balimbing has been seen to be about 1.7, while at Ketambe it was about 1.5; individuals have been observed with others for 68 percent of the time in the former area and 54 percent in the latter.[30, 57] Consortships were of longer duration at Suaq Balimbing, where several females have also been seen to converge on one male, creating a large mating aggregation that also attracted subadult males.[57] These differences are all attributed to the greater and more reliable availability of fruit in the swamp forests at Suaq Balimbing, which effectively frees orangutans to behave more sociably by relieving the chief constraint on proximity, that of food scarcity and competition.[5]

Tool use

Wild Sumatran orangutans have been observed to use 54 different tools for extracting insects or honey, and 20 for opening or preparing fruits.[2] Tool use has been conservatively estimated to occur in

Ian Singleton/SOCP

Although the mother–infant bond weakens as orangutans mature, females tend to stay close to the range in which they were born.

65 percent of feeding sessions in Sumatra.[62] Sticks are used to extract termites and other social insects[30] and to access the nutritious seeds of *Neesia* cf. *malayana* fruits.[62] *Neesia* seeds are rich in fat but protected within a large, very tough, five-angled woody capsule. As the fruit ripens, it dehisces (bursts open), exposing a mass of irritant hairs, among which sit the rows of seeds. To access the seeds, an orangutan holds in its mouth a short stick from which the bark has been stripped, and rubs the stick inside the fruit to dislodge the seeds. This technique can be used as soon as the first crack appears in the capsule. Later in the season, a fat-rich aril (fleshy seed coat) develops, which the orangutans remove by hand. Tool-use has only been seen in the swamp forests of Sumatra, and not to date in any dryland forests; it has been witnessed many times at Suaq Balimbing and evidence in the form of fallen fruits with the tools still embedded has been seen in the Singkil and Tripa Swamps as well[40] (see Box 11.1).

Nest building

Sumatran orangutans sleep in nests and usually build a new one every evening. They weave together branches, twigs, and leaves; this normally takes only a few minutes but sometimes takes up to 20 minutes.[2] About 5 percent of nights, however, are spent in old nests that the orangutan finds just before retiring, in which case the inner lining is renewed. The main use of a night nest is for that night's sleep but, sometimes, orangutans also play or rest in it during the following day. Infants share a nest with their mother until they are weaned.

Sumatran orangutans also often make nests in which to rest during the middle of the day. They sometimes cover their nests with leaves and twigs when it rains, and have been observed holding a leaf or collection of twigs over their head to provide shade from the sun, or as an umbrella against the rain.[30]

Development and reproduction

Sumatran orangutan males usually reach puberty at the age of 14 years, although some mature as early as five and some as late as 16. As described in Chapter 9, the males of both species of orangutan undergo a complex maturation process, with some males becoming fully mature, or 'flanged' (i.e. with fully developed cheek pads), earlier than others, which may remain 'unflanged' for many years. Flanged males have larger bodies, are more aggressive and dominant over the others, and are the preferred mates of adult females.

Female Sumatran orangutans typically reach maturity five years earlier than do males, and are thought to give birth around every eight years.[16, 48] Their menstrual cycle lasts about one month. Unlike other great apes, orangutan females do not have perineal swellings around ovulation at mid-cycle, but their marked proceptive display behavior signals their fertile status; that is, they actively seek sexual encounters with favored males.[15]

Flanged males usually mate with females during a consort relationship. The consortship tends to be initiated by the female.[35] Males actively solicit sex with fertile females via 'male presenting', by posturing and displaying the penis. Consortships may last up to three months in Sumatra and several females may converge on a single male at one time, in contrast to the shorter and more exclusive consortships in Borneo.[54, 57] Flanged males are also known to force copulations with females, but this is rare in Sumatra.[5] Consort copulations have a much higher chance of producing young, as the male is with the female during her most fertile period. Females may mate with more than one male during the menstrual cycle, and may take more than one cycle to conceive.[65]

Except among bonobos, homosexual behavior has been reported infrequently among great apes in the wild. In orangutans, such behavior usually occurs only between captive or rehabilitated individuals. Over a 9 000 hour study, two instances of same-sex genital contact or manipulation were observed at Suaq Balimbing, involving a pair of subadult males that had formed a party together.[11] The interaction was associated with 'kiss-squeak' vocalizations (usually an alarm call) and branch shaking, which were interpreted to indicate competitive social tension. Another interaction between two adolescent males was observed at Ketambe; this was associated with more friendly affiliative behavior.[11] There is too little evidence to identify any particular function for homosexual behavior in orangutans.

Unflanged subadult males try to associate with potentially receptive females but, when females are ready to conceive, they seek out a dominant male and form a consort relationship with him. The flanged male is able to prevent all but the most determined subadult males from mating with her. Unflanged males do force copulations upon available females when they can,[57] and will follow females closely for extended periods, interfering with their foraging efficiency.[10] A female that is carrying an infant may find it more difficult to escape; in any case, even a subadult male is larger than an adult female. During enforced consortships, a female will try actively to evade the adherent male and seek the protection of a flanged male if one is available, although she will not necessarily mate with him.

The orangutan gestation period is about 245 days or just over eight months.[2] Infants are carried by their mothers for several years and may continue to suckle until they are five or six years old.[48] This gives the orangutan the slowest breeding rate of any primate, and one of the slowest of any mammal.[2] The young are carried in a side-ventral position and may be played with by the mother, and sometimes share her food. By 11 months of age, they are beginning to find their own food.[23] Immature individuals become independent of their mothers and may wander away from her home range at seven to 10 years of age.

Both species of orangutan may reach 45 years of age in the wild,[20] and recent analyses of data collected over 30 years at Ketambe suggest that they may in fact live beyond 50.[40] Male-biased sex ratios at birth give way to a relative abundance of females as adults in at least some areas, suggesting a net loss of males during their dispersal.[42, 47]

Coping with disturbance

Orangutan densities are negatively affected by selective logging,[31] but there has been little assessment of behavioral change and ecological adaptation of Sumatran orangutans following disturbance. Those individuals that try to remain in their home range are susceptible to injury or death while it is being logged. If dispersal away from the logging operations is possible, then most of the apes move. However, if orangutan populations are close to the carrying capacity of the forests where they live, an influx of new individuals into an area will cause crowding, stress, and probably starvation among both refugee and resident individuals.[31] Where the original population is not at carrying capacity, this is probably because the area had previously been logged or hunted.

Comparisons of nest counts in logged and unlogged forest at Ketambe indicated that orangutan populations declined by 40 percent, matching the 40 percent decline in the availability of soft-pulped fruit.[29] The apes remaining in the disturbed area had become more folivorous, traveled more,

Ian Singleton/SOCP

Selectively logged and regenerating forest adjacent to Bukit Tiga Puluh National Park. After a couple of decades, this area had recovered enough to be suitable as a release site for rehabilitated orangutans. Staff monitor them as part of the post-release program.

Box 11.1 CULTURE AND SOCIALITY IN SUMATRAN ORANGUTANS

Orangutans, like chimpanzees (see Box 4.3),[67] appear to have simple cultures: behavioral variations between wild populations that are not easily attributable to ecological distinctions, and are likely to be the result of social learning. When researchers from all the established orangutan study sites compared the animals' behavioral repertoires, 17 behavior variants were found to be widespread in at least one site, but absent at another, despite there being no clear, relevant ecological differences.[27, 58] Each well studied orangutan population exhibited a unique suite of cultural behaviors.[27]

This comparative study considered both Sumatran and Bornean orangutans. Some of the apparently cultural behaviors were exclusive to one or the other island. For instance, 'bunk nests' (building a complete nest above, to shelter an occupied nest below from rain) and 'kiss-squeak with leaves' (holding leaves near the mouth while giving the kiss-squeak alarm call) were seen only in Borneo but 'tree-hole tool use' (using a twig to probe tree holes for insects or honey) and 'branch scoop' (using a leafy branch to get water from a tree hole) were found only in one Sumatran population. None of these cultural behaviors was diagnostic of either species – that is, no behavior was universal on one island but absent from the other. What may be most interesting are the behaviors found in one or more populations on both islands, but absent from at least one other population on each island. These behaviors include 'nest raspberries' (a 'phhhhhp' noise made by blowing through pursed lips, seen and heard as the orangutan finishes building a nest) and 'sun covers' (orangutans piling leaves and branches over both the nest and themselves for shade on warm, clear days).

Suaq Balimbing is a lowland swamp forest, and Ketambe is a primary hill and riverine forest; these research sites are both located in the Gunung Leuser National Park in the north of Sumatra. They host the best-studied Sumatran orangutan populations, representing two of the three populations that display most apparently cultural behavioral variants (seven and four, respectively).

As summarized in the table below, there are distinct cultural differences between the two populations. At Suaq Balimbing, orangutans regularly use both feeding tools (at tree-holes, and to get the seeds out of *Neesia* sp. fruits) and 'branch scoops'. They also perform 'nest raspberries', 'twig biting' (by passing the ends of twigs in front of the mouth, or actually biting them prior to inserting them into the lining of a nest), and the 'symmetric scratch' (slow, exaggerated, scratching movements

Behaviors likely to have cultural origins, observed in Sumatran orangutans at two locations

Behavior	Suaq Balimbing	Ketambe
Tree-hole tool use	customary[a]	absent (without ecological reasons)
Seed-extraction tool use	customary	absent (with ecological reasons)
Branch scoop	habitual	absent (without ecological reasons)
Nest raspberry	customary	absent (without ecological reasons)
Twig biting	customary	absent (without ecological reasons)
Symmetric scratch	customary	rare
Kiss-squeak with hand	habitual	customary
Leaf padding	absent (with ecological reasons)	habitual
Autoerotic tool	absent (without ecological reasons)	customary
Sun cover	absent (without ecological reasons)	habitual

a 'Customary' indicates a higher frequency of occurrence than 'habitual'.

and rested less and in shorter bouts. They were forced to move on the ground, because of the discontinuous canopy, which is more energetically expensive than arboreal travel for an animal that is so well adapted to life in the trees. However, if the area is left unlogged long enough, and if orangutans still persist in surrounding areas, densities can eventually recover. In the Sekundur area of the

William H. Calvin (www.williamcalvin.com) William H. Calvin (www.williamcalvin.com)

Although only Sumatran orangutans have been seen to develop tool use in the wild, both species adopt tool-use behaviors when in captivity.

of both arms, in something that resembles calisthenics or T'ai Chi). At Ketambe, orangutans use 'leaf padding' (holding leaves to protect the hands while handling spiny fruits), 'autoerotic tools' (a stick to stimulate their own genitals), and 'sun covers'. Both orangutan populations display the cultural behavior 'kiss-squeak with hand' (holding a flat or cupped hand near their mouth while giving a kiss-squeak).[27]

Over 30 years after Jane Goodall first reported regular feeding-tool use in chimpanzees at Gombe, the Suaq Balimbing orangutans were the first (and so far, the only) wild population of another ape species found to regularly use twigs as probes while feeding.[61] Primate populations with regular and widespread feeding-tool use have been proposed to show greater social tolerance.[59] A study of individual differences in tool-use frequency among Suaq Balimbing females supports this model, showing that time spent with other adult orangutans at less than 50 m distance correlates positively with tool use.[60]

A detailed comparison of social interactions at Suaq Balimbing and Ketambe has been conducted to test this model further. Orangutans at both Suaq Balimbing and Ketambe are known to be more social than any population of Bornean orangutans yet studied; initial measures of time spent with other orangutans at distances of less than 50 m showed little difference between these two sites.[26, 59] More detailed analysis revealed that the tool-using orangutans at Suaq Balimbing spent more time in close proximity to other individuals (both at a distance of less than 10 m and, especially, at less than 2 m); they also permitted a wider variety of social partners to approach so close.[26] This suggests that tolerance of social partners at close proximity is an important factor in spreading and maintaining regular feeding-tool use through social learning in orangutan populations. The overall greater gregariousness of Sumatran orangutans compared to those of Borneo can help explain why Sumatran populations exhibit more cultural behavior variations. Sumatran orangutans have many more frequent opportunities to watch and learn from one another, so behavioral innovations can spread relatively rapidly through their populations.

Michelle Merrill

Gunung Leuser National Park, logging was carried out in the 1970s at a removal rate of 11 large trees per hectare. After five years, over half the trees still showed signs of damage. By 2001, tree density, fruit availability, and orangutan densities resembled those of pristine forest elsewhere in the park.[18] It is not known whether the total number of orangutans had recovered to prelogging levels, or whether the remaining orangutans had spaced themselves out to exploit available resources, but it is clear that the habitat had recovered sufficiently to support the same number of orangutans as an unlogged forest.

Even less is known about how Sumatran orangutans cope with shifting agriculture and conversion of land to plantations. They do enter plantations or gardens that can provide food, but such areas are not used exclusively. There are some isolated populations in Sumatra surviving in rubber gardens with a very few forest trees along some of the stream valley bottoms. These animals are remnants that have been cut off by rapidly expanding oil palm plantations. Nevertheless, left alone it seems they might survive at low densities (some are still breeding), although they are gradually being exterminated as pests. Most orangutan populations are probably unable to survive long term in severely fragmented forest, as their fruit-dominated diet requires them to occupy large ranges to ensure sufficient supplies.[31]

The Ketambe River near its confluence with the Alas River, in Gunung Leuser National Park, Sumatra.

Serge Wich

Ecological role

Given the importance of fruit in the orangutan diet, their main ecological role is likely to be as seed dispersers. Virtually nothing is known, however, about this aspect of Sumatran orangutan ecology, beyond the 96 fecal samples investigated by Herman Rijksen,[30] 44 percent of which contained intact seeds. This topic is discussed in Chapters 2 and 10.

Our limited knowledge about their frugivory and seed dispersal makes it difficult to define the ecological role of Sumatran orangutans in relation to other species. Orangutans share their environment with several frugivorous species of similar or larger size, such as elephants (*Elephas maximus*), rhinoceroses (*Dicerorhinus sumatrensis*), sunbears (*Helarctos malayanus*), and, in forest patches south of Lake Toba only, tapirs (*Tapirus indicus*). These species may be less important than the orangutan in seed dispersal,[4] but practically nothing is known about their role or the degree of dietary overlap with orangutans. Elephants and rhinoceroses are able to transport much larger seeds internally than orangutans can and tapirs are considered good seed dispersers in neotropical forests,[12, 13] as are sunbears in Borneo.[25] Of the other primates, the siamang (*Symphalangus syndactylus*), which is 15–30 percent the size of an orangutan, is also probably capable of dispersing many of the same seeds.

Interactions with other animals

The food resources of a forest are limited, forcing sympatric animal species to develop their own ecological niche to reduce competition. Except when feeding from the large strangler figs that attract a great diversity of animals, Sumatran orangutans usually share most of their preferred fruit species with around five medium-sized primate species. In any given lowland forest, these are a subset of the following: pig-tailed macaques (*Macaca nemestrina*); long-tailed macaques (*M. fascicularis*); five leaf monkey species (*Presbytis* spp., *Trachypithecus* spp.); and three gibbons (two small species of *Hylobates* and the larger siamang).[30] These primates all consume figs, other fruits, and leaves; they are active in the forest canopy, but differ in food selection and range size. All but the siamang will feed peacefully with each

Table 11.1 Feeding strategies of sympatric apes in the Ketambe area

Species characteristics	*Hylobates*	*Symphalangus*	*Pongo*
Weight (kg)	6–8	10–13	30–70
Territoriality	territorial	territorial	overlapping home ranges
Home range (km^2)	0.4	0.2	2–10+
Time of rising	05.00–06.00	06.00–07.00	07.00–08.00
Speed of travel	very fast	fast	very slow
Distance/day (m)	1 250–1 500	400–800	500–900
Quantity of food consumed	small	medium	very large
Ecological strategy	selective feeder; prefers small sweet figs[49]	aggressive competitor	least selective feeder; stores excess energy as fat tissue
Fruit (percent of diet)	56	52	60
Leaves (percent of diet)	34	40	34

Adapted from Rijksen, H.D. (1978) *A Field Study on Sumatran Orangutans (*Pongo pygmaeus abelii *Lesson 1827).* H. Veenman and Sons, Wageningen.

other in the same tree. In Ketambe, female and young orangutans have been seen being attacked by siamang, but apparently only by certain individuals. Ketambe orangutans usually show no hesitation in entering trees where siamang are feeding.[30]

Leaf monkeys possess complex, fermentative stomachs that allow them to consume more mature leaves than other primates do. Although they are able to process the same hard-husked fruit as orangutans, they avoid the sweeter pulpy fruit favored by orangutans, gibbons, siamang, and macaques.[30, 49] Orangutans are also distinguished from the other primates by their greater tolerance of unripe or acidic fruits.[49]

The two macaque species found in Sumatra differ from all other primates – and from each other – in their ranging behavior. Long-tailed macaques live in large groups in small home ranges close to rivers, where they remain unless fruit is very abundant elsewhere; pig-tailed macaques have huge home ranges and travel long distances on the ground in forests away from large rivers.[3, 30, 33, 50]

The orangutans and gibbons occupy the most similar ecological niches, but still have features that ensure competition is minimized (Table 11.1). The smaller gibbons (*Hylobates* spp.) are more selective feeders, rise earlier in the day, and remain within small territories (usually 0.5 km^2 or less), which they defend against conspecifics. Orangutans consume large quantities of fruit at one time and store the excess energy as fat. Siamang show intermediate characteristics.[30]

Orangutans show little response to non-primate species and usually feed peacefully in trees with them. Fruit bats (*Pteropus* spp.) may shift their roosts periodically, causing a mass influx of bats to an area. They establish feeding territories in trees and may remain in these trees during the day, which can deter orangutans from entering them, even in the case of a fruiting fig. Bearcats (binturong, *Arctictis binturong*) show a tendency to avoid fruiting trees where orangutans are feeding.[30]

Humans appear to be the only frequent predator of adult orangutans, but the latter's near-exclusive arboreality may reflect the risk of predation by tigers (*Panthera tigris sumatrae*).[30] The body of an old male orangutan was found in 1975 that appeared to have been freshly killed by a tiger.[66] The Sumatran tiger itself is designated Critically Endangered on the *Red List of Threatened Species,* with only a few hundred individuals remaining in 2004. Clouded leopard (*Neofelis nebulosa*) and hunting dogs (*Cuon alpinus*) may occasionally take immature orangutans.[30] There are no leopards of the species *Panthera pardus* in Sumatra or Borneo.

HABITAT REQUIREMENTS

Sumatran orangutans are most abundant in the forests of flood plains, alluvial bottomlands, and freshwater and peat swamps (Table 11.2), where they may occasionally reach densities as high as 10 individuals per square kilometer.[38] They live at intermediate densities in lowland dipterocarp forest and hill dipterocarp forest, and can be found at very low density in some submontane and montane forests.[31] Orangutans also occur in some secondary

Table 11.2 Sumatran orangutan densities in different habitats

Habitat type	Density[a] in multiple locations[31]		Density[a] in the Leuser Ecosystem[64]
	No of locations	Density	
Flood plain and peat swamp	3	6.9 (4.5–7.0)	3–5 (max. 7+)
Alluvial and bottom land forest	3	3.2 (3.0–5.5)	–
Secondary and selectively logged forest	2	1.2 (1.1–1.3)	–
Upland (hill and dipterocarp forest)	6	1.1 (1.0–2.2)	1.0
Submontane and montane	5	0.7 (0.4–1.2)	3 (max. 6)[b]

a Mean orangutans per km^2, followed in parentheses by information on range, where available.
b In the Upper Alas valley.

From Rijksen, H.D., Meijaard, E. (1999)[31] and van Schaik, C.P., *et al.* (2001).[64]

forests. Selectively logged areas support few orangutans relative to undisturbed areas of comparable forest composition; the reduction depends on the intensity of the logging that has taken place.[29]

The current distribution of orangutans seems to be determined by two main factors – the availability of preferred fruits and human presence. Hunting by humans may be responsible for the complete absence of orangutans in some regions of Sumatra, and their continued decline parallels that of increasing human occupation and clearance of their forest habitat. These apes survive best in areas of low human population density.[31] Rivers more than 10 m wide and 60 cm deep also restrict orangutan movement.

Ian Singleton/SOCP

South of the provinces of North Sumatra and Aceh, the only free-living orangutans in Sumatra are in the environs of Bukit Tiga Puluh National Park, Jambi, where over 50 ex-captive orangutans have so far been reintroduced by the Sumatran Orangutan Conservation Programme, and are now breeding.

Orangutans prefer soft-pulped fruit and the availability of such food is an excellent indicator of orangutan density;[64] it explains low densities at high altitudes[6] and in recently logged forest,[29] and also their variable abundance in specific habitat types.[31] Certain fruit species are more important than others. Lianas make up 17 percent of all orangutan food plants (fruit and leaves) in Ketambe and they also provide an important means of arboreal transport.[30] Strangling figs produce large fruit crops at short intervals, providing a regular food supply for the apes, where they occur. In some Sumatran swamp forests, species of *Tetramerista, Sandoricum,* and *Neesia* fulfill this function.[56]

A tropical forest can be conceived as an "immense patchwork of plant communities of different compositions."[31] Each patch varies in its usefulness to an orangutan seeking food. The richest patches are in swamps, floodplains, and peat forests, between rivers, and in the 10–15 km area surrounding these habitats. These areas have a high diversity of productive food trees;[63] 30–50 percent of trees typically provide fruit suitable for orangutans and 10 percent of the trees fruit each month. However, low-quality patches for orangutans still occur within these otherwise good habitats. It has been estimated that perhaps 50 percent of an expanse of swamp forest constitutes suitable habitat for orangutans.[31]

In comparison, the drier areas away from rivers are dominated by wind-dispersed dipterocarps, with few regularly fruiting trees. Only around 35 percent of these areas are thought to form suitable orangutan habitat.[31] During dipterocarp mast fruiting events that occur every few years, orangutans move into these dry regions to take advantage of the masses of fat-rich seeds that are produced. Further discussion on dipterocarps and mast fruiting is given in Chapter 10.

Above 1 000 m, Sumatran orangutan densities decline markedly, although occasional pockets of higher density occur in upland valley habitats. Individual 'wandering' males are seen as high, or higher than 1 500 m, but breeding populations are not supported.[64] These altitudinal limits mainly reflect the distribution of favored food types: that is, the upper limit of many trees whose fruits are of a type preferred by orangutans.[6] The healthy orangutan populations found in the two high plateaus of the Alas Valley are thought to depend on the less acidic soils that support an unusually high abundance of figs.[64]

The area required to support a viable Sumatran orangutan population varies with habitat quality. The questions that still remain to be answered about their social and ranging systems also make it difficult to estimate it precisely.[34] Home ranges are typically of 5–25 km² or larger for males and 1–10 km² for females.[31, 42] The largest occupied home ranges incorporate areas of swamp and hill forest.[42] Range overlap varies considerably among habitats and is higher in food-rich areas.

It is assumed that 250–500 individuals are needed for a viable population,[44] which would require between 50 and 600 km² of occupied forest, depending on its suitability.[31] Larger areas are needed in practice, to allow for local patchiness in orangutan distribution. It is possible to estimate how many orangutans would be supported in 100 km² of forest of quality similar to that of Suaq Balimbing, assuming that ranging patterns are also similar.[39] If the range sizes and degree of overlap for each age-sex class yield a density of 7.25/km², then 100 km² would support a total of 725 orangutans. Of these, 229 would be adult females and 100 would be subadult males, but only 33 would be fully adult males. Of these 33, fewer than seven would be expected to be dominant at any one time, and thus contribute significantly to the gene pool.

By 2002, Sumatran orangutans occupied only 13 blocks of primary forest totaling around 20 500 km² in area. Only 9 000 km² was at a low enough altitude to sustain permanent orangutan populations.[44] Only four blocks supported over 500 individuals (see below).

Orangutan Foundation

Orangutan Foundation

Logging and mining, such as here, in and adjacent to Tanjung Puting National Park, Indonesia, have had a devastating effect on orangutan habitat throughout Southeast Asia.

POPULATION STATUS, TRENDS, AND THREATS

Human attitudes and traditions

Paleolithic human migrants may have settled along Sumatra's east coast and larger rivers as long as 80 000 years ago. Based on archeological evidence from caves in the Padang Highlands of western Sumatra, these people consumed orangutans in relatively large quantities.[31]

Seven hunter-gatherer peoples are known to have existed in Sumatra: Abung, Kubu, Mamaq, Sakai, Akit, Lubu, and Ulu.[21] These peoples lived mainly on the eastern side of the Bukit Barisan mountain range, occupying the banks and dry ridges of the extensive floodplains and peat swamps of eastern Sumatra south of Lake Toba.[31] The fact that few, if any, orangutans survive in this region today suggests that they exerted heavy hunting pressure on the species.

Ian Singleton/SOCP

A Sumatran orangutan before confiscation in December 2003, found in a cage at the back of a restaurant in Desa Petai, Riau province. He was later found to have air rifle pellets embedded in his abdomen.

Hunter-gatherers were absent from the forests of the Leuser Ecosystem in historic times, and this is likely to be the reason that great ape populations were able to persist there. Devout Muslims who do not eat orangutans (or most other wildlife) have inhabited much of the area for centuries. Local people extracted timber for subsistence needs and caused some habitat disturbance through shifting cultivation, but were primarily agriculturalists and had a very limited impact on orangutan populations.[32]

The hunter-gatherers of central Sumatra appear to have been the most effective hunters. Apes were among their favorite prey due to their "somewhat sweet, but nice taste";[28] they were hunted with dogs and spears in the lowland parts of their range, or with blowpipes and poison-tipped darts. The darts were often poisoned with the sap of the *Antiaris toxicaria* tree, ironically a favorite fruit source of the orangutan. The traditional belief system of the hunters deterred them from hunting in the higher hills, where spirits were thought to reside.[31] Certain clans or families also traditionally avoided eating orangutan meat, but did not criticize their neighbors for hunting orangutans.[31]

The Batak people believe that eating orangutan meat will make them strong, and this belief has prevailed into modern times. The staffs and wands of shamans were often decorated with orangutan hair.[31] Other reports claim that to draw the spirit into the staffs, a kidnapped child had to be fed upon an orangutan liver and then sacrificed.[21] Various other items were also adorned with orangutan skin or hair; even as late as 1971, a Batak youth in the Alas Valley was seen wearing an orangutan-skin cap.[30]

Recent history

During the 20th century, orangutans began to be viewed as having an economic value, and a considerable international trade occurred in the 1930s and 1960s. The Dutch professional animal collector, van Goens, captured at least 218 adult Sumatran orangutans to export to circuses and zoos abroad. The Gayo people were renowned for their ability to trap live orangutans.[30] Numerous infants were also sold into the national and international pet trade, with many being exported to Taiwan.

Orangutans continue to be killed by local and foreign people, despite legal protection. The meat is still eaten by some in Sumatra, but the spread of Islam to many traditional people (Gayo, Alas, and Achenese) has halted their hunting of orangutans for food. However, the apes are still killed by people of all faiths if they raid fruit crops[30] and to obtain infants as pets. There are also reports that the army elite organized hunting safaris in northern Sumatra as late as the 1990s.[31] In the recent past Rijksen and Meijaard aptly described the feeling towards orangutans – that they have "an economic, medicinal, nutritious or nuisance value – any of which warrants persecution."[31]

Indonesia is in a unique historical position, and the fates of all its wild species will be greatly influenced by the outcome of complex social and economic processes over the next few years. The country is reinventing itself, but as what is by no means clear. There are powerful forces exerting various kinds of influence, including pressures for local and participatory democracy opposed by others who seek a return to centralized 'guided' democracy. Political, religious, military, capitalist, and bureaucratic elites compete for influence and opportunity at the local, regional, and national level. The meteoric pace of Indonesian development during the 1970s to 1990s can never be repeated, as it was fuelled by the opening of virgin lands and the clearing of virgin forests that have now been all but used up, by the sale of petroleum resources that are now depleted, and by the borrowing of immense wealth that has largely fled the country leaving a legacy of debt. The chief issue for orangutans and most other species of the Sumatran forests is whether or not the habits and momentum established by this model of development will destroy the remaining frontier, before the fundamental issues of sustainability are at last addressed by the Indonesian people. In practical terms, the fate of *Pongo abelii* depends on the peoples of the Leuser Ecosystem finding a way (with the support of the European Union, Indonesian nongovernmental organizations, and their friends in government) to secure the ecological architecture of their own environment and, thereby, a future for themselves as well as their neighbors and cousins in the forest (see Box 11.2).

The current position

It is estimated that the total number of Sumatran orangutans is less than 7 percent of what it was in 1900, and 2 percent of what it was 10 000 years ago when the last ice age ended, sea levels rose, and Sumatra became isolated from the Asian mainland.[31, 44, 70] The decline of the species accelerated towards the end of the 20th century, with massive exploitation of Sumatran forest habitats occurring in the 1970s, 1980s, and 1990s, and continuing to date. In the absence of human intervention, the recovery of this species would be extremely slow (around 0.006 percent per year) due to their very slow reproductive rate.[24]

Sumatran orangutans were classified as Critically Endangered by IUCN–The World Conservation Union in 2000, indicating that the species faces an extremely high risk of extinction in the wild in the immediate future.[8] It is clear that the population is fragmented into a number of small units that, in their turn, are being split up or extinguished. The exact population sizes are unknown. Rijksen and Meijaard estimated that 12 500 Sumatran orangutans survived in 1993, and predicted that by 2020 only 7 500 individuals would remain. This was based on the fact that 45 percent of their habitat was formally protected.[31] This projection is now considered to have been optimistic. The 1993 figure of 12 500 orangutans was an overestimate, because it included unconfirmed populations in southern Sumatra which are now thought to have already died out at the time. Populations have continued to decline since 1997. At least 1 000 individuals have been estimated to have been lost each year between 1997 and 2000 from the Leuser Ecosystem alone.[44] By 2002, only around 7 300 individuals are believed to have remained[44] (see Table 11.3).

The Leuser Ecosystem is one of the best-protected sites for Sumatran orangutans, and the heart of their present range. The Leuser Ecosystem, which includes Gunung Leuser National Park, has been the focus of European Union support via the Leuser Development Programme (see Box 11.2). However, this ecosystem is now fragmented into at least four areas of forest (West Leuser, Trumon-Singkil, East Leuser, and Tripa), with the prospect of further fragmentation due to a proposed new road scheme.[44]

Recent models propose that populations of over 500 Sumatran orangutans are large enough to be demographically stable in the long term.[44] In 1993, it was thought that six such populations existed; by 2002, four remained, only one of which was outside the Leuser Ecosystem. In addition, eight smaller populations that were thought to exist to the south of Lake Toba had disappeared by 2002; two of these have been lost over the 10 year period and the others may in reality have been lost before 1993[44] (see Table 11.4).

Sumatran orangutans are unable to survive long term in severely fragmented forests. Habitat loss, fragmentation, and degradation are the major threats to orangutan survival.[34] The fragmentation of the Leuser Ecosystem probably has the greatest single impact on Sumatran orangutans.[44] In the

Table 11.3 Sumatran orangutan population decline since 8000 BC

Year	Estimated number of Sumatran orangutans	Decline from previous estimate
8000 BC	380 000	–
1900	85 000	78 percent
1997	12 500	85 percent
2001–2002	7 334	41 percent

From Rijksen, H.D., Meijaard, E. (1999),[31] and Singleton, I., *et al.*, eds (2004).[44]

Table 11.4 Estimated population structure of Sumatran orangutans

Surviving population size	1993[31]	2001–2002[44]
< 100 individuals	10	1
100–500 individuals	7	8
500–1 000 individuals	2	1
> 1 000 individuals	4	3
Total	23	13

From Rijksen, H.D., Meijaard, E. (1999),[31] and Singleton, I., *et al.*, eds (2004).[44]

Box 11.2 HISTORY OF THE LEUSER ECOSYSTEM

During the 1980s, the work of Herman Rijksen, Mike Griffiths, Carel van Schaik, and others revealed that the proposed Gunung Leuser National Park at 7 927 km^2 was inadequate to conserve a representative sample of the flora and fauna of northern Sumatra. This should not have been a surprise, as the boundaries, set in 1936 and 1976, bear little relation to the terrain.

One boundary section was simply a line drawn on a map between two mountain tops 50 km apart. Either it was thought not to be significant that a unique volcanic plateau and two important water catchments were bisected by the boundary, or there was an inadequate knowledge of the geography at the time of designation. Another boundary was a semicircle of 30 km radius, centered on the summit of Gunung Leuser itself.

Proper demarcation of boundaries on the ground is a condition for full gazettement of a national park in Indonesia, prior to which it has a weaker legal status. The initial boundaries failed to take into consideration the natural ecology of the region. Furthermore, the mountainous character of the planned park made the placing of physical boundary markers practically impossible – an issue that would need to be resolved before the national park could be legally formalized.

The Alas River, one of two major rivers flowing through the Leuser Ecosystem.

SOCP

Studies carried out in the early 1990s revealed the movements of wide-ranging Sumatran species such as elephants, fruit bats, and hornbills. Similar studies revealed that species such as the orangutan and tiger needed extensive tracts of lowland forests to maintain viable populations. Combining the habitat needs of these charismatic species with the geomorphology of the region revealed a naturally bounded area of 27 000 km^2 – roughly the same size as Haiti or Rwanda. This zone was named 'the Leuser Ecosystem', and includes samples of most Sumatran ecosystem types.

The Leuser Ecosystem stretches from the sandy beaches bordering the Indian Ocean, right across the breadth of Sumatra almost to the mangrove swamps bordering the Malacca Straits. It includes two great mountain ranges reaching over 3 000 m in altitude. These are separated by a great rift valley through which two large rivers flow – the Tripa to the northwest and the Alas to the south. The rivers flow into the Indian Ocean, after passing through extensive freshwater peat swamps that are home to the densest orangutan populations on Earth. It is the last place where viable (or potentially viable) populations of the Sumatran varieties of elephant, orangutan, tiger, and rhinoceros exist, and the only place where all these species are found together.

It was only after the scientific work in identifying the Leuser Ecosystem had been completed in 1990, that it was realized that most of this area had already been promoted for conservation by the traditional leaders of the peoples in that part of Sumatra as early as 1928. For six years, the local leaders had lobbied the Dutch colonial government to have the forests of Leuser conserved in

early 1960s, Sumatra was almost completely covered by tropical forest, but this cover was greatly reduced and fragmented by logging, infrastructure development, resettlement (transmigration), and plantation development on a massive scale during the 1970s and 1980s.[5, 68] There was a 61 percent decline in forest area in Sumatra between 1985 and 1997.[14] From 1998 onwards there was a sudden upsurge in deforestation due to the collapse of control by central government in Indonesia, following decades of poor governance and environmental neglect.[14, 17] Kalimantan (Indonesian Borneo) was also affected, and the impacts on Indonesian wildlife are immeasurable.

Given that 80 percent of current orangutan habitat is covered by timber concessions, or is vul-

perpetuity. Eventually an area approximating the Gunung Leuser National Park was granted, excluding most of the valuable lowland forests and coastal plains.

Once the Leuser Ecosystem was identified as an area of tremendous significance that required holistic conservation management, the great challenge was to translate this vision into reality. The effort required to do so has proved to be enormous, but there was one important factor that made it easier than it might otherwise have been. In the early 1990s, the European Union was keen to help conserve tropical rainforests and the Indonesian government was keen to do the same, as long as there was solid financial support from the international community. This meeting of policies culminated in a series of commitments by the European Commission and the government of Indonesia to provide about US$38 million to put the right conditions in place to conserve the Leuser Ecosystem. This commitment, which began in 1992, was due to be completed toward the end of 2004. It took the form of an Integrated Conservation and Development Project for Lowland Rainforests in Aceh, followed by the Leuser Development Programme.

These activities yielded several important achievements, the first of which was to develop a new management system for the area as a whole. This was needed because the area includes plantations and locally claimed lands, as well as natural forests under various kinds of planned use. This meant that decision making was fragmented and conservation management capacity was also weak. The government of Indonesia therefore assigned management of the Leuser Ecosystem to a private foundation, the Leuser International Foundation, which had established itself with the express aim of conserving the Leuser Ecosystem. The Leuser Development Programme then continued to provide support through the Leuser International Foundation.

Serge Wich

Ketambe Research Station in the Leuser Ecosystem.

The second important achievement was to build support for the conservation of the Leuser Ecosystem among an array of stakeholders from central and local governments, universities, the business community, local communities, and others.

The third was to gain legal recognition for the Leuser Ecosystem. Various decrees at the presidential, ministerial, and provincial levels have been issued supporting the legal status of the Leuser Ecosystem, its management, and its inclusion in all spatial plans from the local to the national level. The consolidation of this legal status required 3 000 km of boundaries to be demarcated with concrete markers every 2 km – a task

continued overleaf

nerable to illegal logging and habitat conversion, habitat loss is likely to continue at high rates. The fires and droughts that ravaged Kalimantan over the past two decades have been less of a threat to the orangutans on Sumatra, so far. There, approximately 5 percent of the 1997–1998 fire hotspots occurred within orangutan habitat.[72] Large-scale forest fires to the south of the orangutan range area, in central Sumatra, were reported in June 2004. Urgent consultations on smog control were underway between the Malaysian, Indonesian, and Singapore governments.[1]

The tsunami of December 2004 that led to the tragic destruction of many towns and villages around Aceh's coastline had little direct impact on orangutan habitat. The indirect impacts remain to

requiring massive investment of labor and funds. Those charged with the work had to row ashore through the surf of the Indian Ocean, carrying cement bags and steel rods as well as supplies to sustain them as they installed the markers on isolated coastlines. They had to climb high peaks where night temperatures hover around 0°C. In addition to the physical challenges, the demarcation was done during a time of armed conflict in Aceh, and there was a real risk of being caught in the crossfire. A further challenge was that few local people understood conservation. A lengthy dialog was therefore needed at all the settlements near to the planned boundaries before the boundary markers could be erected with local consent.

The ratification process was even more difficult than physical demarcation. This was a bottom-up affair with district (*kabupaten*) leaders signing off first, before provincial leaders and finally the central government in Jakarta. The whole task was eventually completed, however, and the Leuser Ecosystem was formally constituted in legal documents at all levels as an official conservation area. The foundations had thereby been laid for the zonation and protection of this priceless area.

Not all the Leuser Ecosystem comprised virgin forest. Some areas had been given out for logging or even plantations, and some were being drained to convert swamps to agricultural land. Plans also existed to convert lowland forests to cattle ranches and to settle people from Java to work on estates. Hence an ongoing job for the Leuser International Foundation and Leuser Development Programme has been to resolve these conflicting land uses. By June 2004, more than half of the 12 original logging concessions had been closed down, and the licenses of several problematic plantations revoked. Swamp drainage has stopped and there is no cattle ranching anywhere in Leuser.

Another, and hopefully the last, great challenge is facing the conservation of Leuser: a network of roads called Ladia Galaska is being promoted by the province of Aceh. This would pass through some of the biologically richest forests in Leuser and would open up access for logging, plantations, and settlements (legal or otherwise). The Leuser International Foundation has developed national and international alliances to oppose this plan and a decision on the future of Ladia Galaska had been expected shortly after the presidential elections in September 2004. The Leuser International Foundation will continue to argue that the road project would do little if anything to improve the overall welfare of local communities, and that alternative transport arrangements would do less harm and contribute more to local development.

The tsunami of December 2004 had few direct effects on the Leuser Ecosystem, with only the Tripa swamp area known to have experienced a temporary influx of seawater.[41] Following the devastation caused to Aceh's coastal communities, some people have suggested that the 4–8 million m^3 of logs required for rebuilding should be sourced from the Leuser Ecosystem area.[7] Others have called for the timber to be imported as a form of international aid, being sourced from sustainably managed forests in temperate countries.[71] It is not clear whether the Ladia Galaska project will be accelerated or delayed as a result of the rebuilding effort.

At stake ultimately is the long-term future of the inner core of the Leuser Ecosystem – about 21 000 km^2, mostly of forest. This area, if preserved, will safeguard the supply of water and other catchment services to 4 million people downstream. It would also mean the likely survival of the orangutan and the many other denizens of Leuser's forests.

Mike Griffiths

be seen, but it is feared that the need for firewood and construction timber is likely to lead to further loss of forest.

CONSERVATION AND RESEARCH

The first detailed studies of Sumatran orangutan behavior and ecology were carried out in the 1970s, by MacKinnon at Renun[23] and Rijksen at Ketambe.[30] Only Ketambe has been in continuous use by researchers since then; it is located in the Alas Valley within the Gunung Leuser National Park. Another long-term study site was opened in 1993 in swamp forest at Suaq Balimbing in the south of the Gunung Leuser National Park. These two sites, in lowland and relatively fruit-rich environments with exceptional densities of orangutans, have yielded almost all that is known about the Sumatran orangutan in the wild, although some additional work has also been done by Priatna at Soraya and Sikundur, and by Fox at Agusan.[9, 10]

The Sumatran Orangutan Conservation Programme conducts most of the survey and monitoring work concerning the status and distribution of wild Sumatran orangutans at present. It has established a release program in Bukit Tiga Puluh National Park in Jambi province. The Research, Monitoring, and Information Division of the Leuser Management Unit manages research activities within the Leuser Ecosystem (see Box 11.2). Initially it ran four research stations:

- Ketambe, which is still operational (lowland forest), with studies focused on habituated orangutans and Thomas leaf monkeys, *Presbytis thomasi*;
- Soraya (lowland forest), protected within a logging concession but closed down as a result of the conflict;
- Suaq Balimbing (lowland peat swamp forest), closed down as a result of the conflict; and
- Bengkung (based in a transmigration site in the middle of lowland rain forest), closed down as a result of the conflict.

Two monitoring posts have also been established, considerably extending the potential for survey work within the vast and diverse Leuser Ecosystem.[19]

Whether any of this work will be enough remains open to question, for these are our final maneuvers to preserve the Sumatran orangutan in the wild. After centuries of population contraction and decades of deforestation and logging, these animals have nowhere else to go but the already fragmented Leuser Ecosystem and one other forest patch nearby. The best estimate of the current population is about 7 300, and it is still believed to be declining despite considerable conservation investment in and around the Leuser Ecosystem. Nevertheless, the concentration of opportunities for Sumatran orangutan protection may itself make the task easier, as conservation, research, education, and enforcement efforts in close proximity can be strongly synergistic. If a breathing space can be maintained through these efforts, during which local people, governments, and businesses can learn how and why to preserve enough of these forests, then it is just possible that the Sumatran orangutan will survive within them.

Ian Singleton/SOCP

An orangutan being moved to its release site in Jambi province.

SOCP

Awareness of the orangutan's plight is being taken into schools by the Sumatran Orangutan Conservation Programme.

FURTHER READING

Corlett, R.T. (1998) Frugivory and seed dispersal by vertebrates in the Oriental (Indomalayan) region. *Biological Review* **73**: 413–448.

Djojosudharmo, S., van Schaik, C.P. (1992) Why are orangutans so rare in the highlands: altitudinal changes in a Sumatran forest. *Tropical Biodiversity* **1** (1): 11–22.

Knop, E., Ward, P.I., Wich, S.A. (2004) A comparison of orang-utan density in a logged and unlogged forest on Sumatra. *Biological Conservation* **120** (2): 187–192.

MacKinnon, J. (1974) The behaviour and ecology of wild orangutans (*Pongo pygmaeus*). *Animal Behaviour* **22**: 3–74.

Utami, S.S., Goossens, B., Bruford, M.W., de Ruiter, J.R., van Hooff, J.A.R.A.M. (2002) Male bimaturism and reproductive success in Sumatran orang-utans. *Behavioral Ecology* **13** (5): 643–652.

van Schaik, C.P., Ancrenaz, M., Borgen, G., Galdikas, B., Knott, C.D., Singleton, I., Suzuki, A., Utami, S.S., Merrill, M. (2003) Orangutan cultures and the evolution of material culture. *Science* **299**: 102–105.

van Schaik, C.P., Fox, E.A., Sitompul, A.F. (1996) Manufacture and use of tools in wild Sumatran orangutans: implications for human evolution. *Naturwissenschaften* **83**: 186–188.

van Schaik, C.P., Monk, K.A., Robertson, J.M.Y. (2001) Dramatic decline in orang-utan numbers in the Leuser Ecosystem, northern Sumatra. *Oryx* **35** (1): 14–25.

Wich, S.A., Singleton, I., Utami-Atmoko, S.S., Geurts, M.L., Rijksen, H.D., van Schaik, C.P. (2003) The status of the Sumatran orangutan *Pongo abelii*: an update. *Oryx* **37** (1): 49–54.

Wich, S.A., Utami-Atmoko, S.S., Setia, T.M., Rijksen, H.D., Schürmann, C., van Hooff, J.A., van Schaik, C.P. (2004) Life history of wild Sumatran orangutans (*Pongo abelii*). *Journal of Human Evolution* **47** (6): 385–398.

MAP DATA SOURCES

Map 11.1 Orangutan data are based on the following sources:

Dadi, R.A., Riswan (2004) Orangutan distribution polygons: developed at the Leuser Management Unit as part of the Leuser Development Programme, funded by the European Commission and the government of Indonesia. Leuser Management Unit, Sumatra, Indonesia. Based on technical criteria set by Singleton, I. Main sources of field data: van Schaik, C., Idrusman, Singleton, I., Wich, S. Additional information from Dadi, R., Griffiths, M., Priatna, D., Rijksen, H., Riswan, Robertson, Y., Universities of Bristol and Bogor Expedition to Sumatra (Burton, J., Bloxam, C., Kuswandono, Long, B., McPherson, J.), and members of the Leuser Management Unit's Antipoaching Unit.

Singleton, I., Wich, S., Husson, S., Stephens, S., Utami Atmoko, S., Leighton, M., Rosen, N., Traylor-Holzer, K., Lacy, R., Byers, O., eds (2004) *Orangutan Population and Habitat Viability Assessment: Final Report.* IUCN/SSC Conservation Breeding Specialist Group, Apple Valley, Minnesota.

For protected area and other data, see 'Using the maps'.

ACKNOWLEDGMENTS

Many thanks to Raffaella Commitante (University of Cambridge), Simon Husson (University of Palangkaraya), Helen Morrogh-Bernard (University of Palangkaraya), Ian Singleton (Sumatran Orangutan Conservation Programme), and Rondang Siregar (University of Cambridge) for their valuable comments on the draft of this chapter.

The orangutan distribution polygons were produced by Rahmadi A. Dadi and Riswan of the Leuser Management Unit as part of the Leuser Development Programme, funded by the European Commission and the government of Indonesia. They were based on technical criteria designed by Ian Singleton, in turn based on many years of field research on the Sumatran orangutan (especially on densities by habitat type and altitudinal limitations), together with the most recent information from the Leuser Management Unit on destroyed or degraded habitat (from extensive ground surveys, aerial photomosaics, and satellite images). This team was greatly assisted by Nick Jewell, who digitized most of the forest-cover data. The main sources of field data were: Carel van Schaik, Idrusman, Ian Singleton, and Serge Wich.

Additional information was provided by Rahmadi A. Dadi, Mike Griffiths, Dolly Priatna, Herman Rijksen, Riswan, Yarrow Robertson, members of the Universities of Bristol and Bogor Expedition to Sumatra (James Burton, Catherine Bloxam, Kuswandono, Barney Long, James McPherson), and members of the Antipoaching Unit of the Leuser Management Unit.

AUTHORS

Kim McConkey, UNEP World Conservation Monitoring Centre

Box 11.1 Michelle Merrill, Emergent Systems

Box 11.2 Mike Griffiths, Leuser Management Unit

CHAPTER 12

Gibbons: the small apes

DAVID J. CHIVERS

The great apes of Southeast Asia share their habitat with smaller but no less interesting apes: the gibbons (family Hylobatidae). In some cases, the gibbons are even more threatened than the great apes, but receive much less public attention. This brief overview is intended to raise the profile of the gibbon family, and offers the opportunity to contrast their unique behavior and ecology with that of the other apes.

Following Carpenter's pioneering study of lar gibbons in Thailand in the 1930s,[3] most species of the family Hylobatidae were first studied in the field in the 1960s, 1970s, and 1980s. They have been shown to be monogamous, territorial, frugivorous, and suspensory, with elaborate duets by the adult pair. The complexities of each species have been investigated in recent years, and the roles of gibbons in both seed dispersal and forest regeneration have been demonstrated. Gibbons live on the mainland of Southeast Asia, and on the islands of Java, Sumatra, Borneo (Kalimantan), and associated islands. These all sit on the Sunda shelf, which emerged from the sea as a consequence of volcanic activity about 12 million years ago (mya). It owes its uniquely rich fauna and flora to an admixture of immigrants. These came first from the Indian subcontinent (the Siva-Malayan fauna) and then later from China (the Sino-Malayan fauna).[7] The gibbon lineage diverged from that of the other apes about 15 mya somewhere in forested, tropical, or subtropical Asia.

GIBBON TAXONOMY

There are still burning issues concerning the validity of species to be resolved with regard to gibbon taxonomy, especially in the northeast of the family's range. Apart from clarifying distribution and abundance from lesser-known areas, DNA analysis is the key to resolving disputes. The gibbons are now divided into four genera. These are mainly

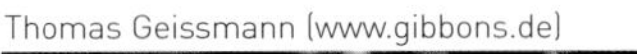

Thomas Geissmann (www.gibbons.de)

A lar or white-handed gibbon (*Hylobates lar*) with infant.

A large *Ficus* tree in the riparian forest along the river Kinabatangan, Borneo. *Ficus* are critical resources for wildlife since they produce fruits several times a year.

KOCP

allopatric (with disjunct ranges) except where the siamang range overlaps that of the lar and agile gibbons.

(1) *Symphalangus*, the siamang, *S. syndactylus*, of Sumatra and the Malay Peninsula (two subspecies);
(2) *Nomascus*, four species of crested gibbons, each with several subspecies,[11, 12, 14, 21] from southern China, Viet Nam, Cambodia, and Laos:
    1. *N. concolor* (four subspecies) and *N. nasutus* (two subspecies) in the north;
    2. *N. leucogenys* in the center (two subspecies);
    3. *N. gabriellae* in the south;
(3) *Bunopithecus*,[34] the hoolock gibbon, *B. hoolock* (two subspecies), of Assam, Bangladesh, and Myanmar (extending across northern Thailand into the southwest corner of China); and
(4) *Hylobates*, comprising five to six species, ranging from Thailand through the islands of the Sunda shelf:
    1. *H. klossii* (the Kloss gibbon), confined to the Mentawai Islands off the west coast of Sumatra;
    2. *H. pileatus* (the pileated gibbon), of southeast Thailand and west Cambodia;
    3. *H. moloch* (the Javan gibbon), of Java, now confined to the west;
    4. *H. lar* (the lar or white-handed gibbon), with two or three subspecies in Thailand (and Yunnan, China), one in the Malay Peninsula, and one in north Sumatra;
    5. *H. agilis* (the agile or black-handed gibbon), with one subspecies whose range stretches from the Malay Peninsula (between two lar subspecies) to the east of Sumatra, one in the rest of Sumatra south of Lake Toba, and one in the southwest of Borneo (West and Central Kalimantan, bounded by the Kapuas and Barito Rivers); and
    6. *H. muelleri* (Müller's gibbon), with three subspecies radiating around the rest of Borneo.[24]

In view of the extensive hybridization between the last two groups in the center of Borneo,[25] it may be necessary to sink *H. muelleri* into *H. agilis* as a fourth subspecies of the latter, but Geissmann argues that the agile is more similar to the lar gibbon.[13] He also defined the four genera, based on molecular data suggesting a split as far back as 8 mya.[9]

The calls and pelage color and markings are distinctive among the gibbons.[9] Species are either monochromatic (to the west black, and to the southeast grey), polychromatic (in the center), or asexually or sexually dichromatic (in the north, in the more open semideciduous habitat); this reveals an intriguing geographical pattern. Both calls and pelage coloration have a genetic basis; these features, with profound behavioral significance for reproduction, should therefore be taken seriously in classifying gibbons. The other key parameter for species and sexes is song, with the female 'great call' being the diagnostic feature. Family groups of gibbons tend to sing daily to advertise their territory and the strength of their pair bond. Male and female gibbons have distinctive parts;[12] it is a true duet in the majority of species, which is most unusual among primates, although more common in birds.

GIBBON EVOLUTION

Chivers has proposed a model of gibbon evolution, relating to frequent changes of sea level during the latter part of the Pleistocene.[7] As ice formed at high latitudes and altitudes, the Sunda shelf was exposed as one land mass; as the ice melted and the shelf was flooded, a number of islands were left exposed. The wholly or partially isolated gibbon populations evolved in separation, and then migrated once land bridges were restored. The key point in this model is that, after the initial spread of three ancestral lineages (or gibbon genera) into different parts of the Sunda shelf, gibbon speciation occurred within the shelf with subsequent sequential spread back to the mainland. This competes with the idea that gibbon species spread out from the Asian mainland. According to the Chivers model, the hoolock gibbon was the first to enter the Asian mainland, followed by the pileated and lar gibbons; the Kloss (Mentawai), Bornean, and Javan gibbons originated on the edges of the shelf; and the agile and lar originated in the center of the shelf. During the periods of lowest sea level, the center of the shelf dried out; at that time the key rain-forest relicts, into which gibbons and other forest animals retreated and out of which they spread when sea level rose, were in eastern Indochina and southern China, northeast Borneo, west Java, northern Sumatra, and southern Myanmar, as well as the Mentawai Islands.

ECOLOGY AND BEHAVIOR

Habitat

Gibbon habitats span the two main forest formations of this part of the Oriental biogeographical region:[8] the semideciduous monsoon rain forests of 'mainland Asia' north of the isthmus of Kra; and the evergreen rain forests of the Malay Peninsula and the islands of the Sunda shelf.[42] The evergreen rain forests of the Sunda shelf comprise the main gibbon habitat. Significant numbers of species and individuals occur in the more seasonal forests of mainland Asia, concentrated in pockets of evergreen forest, and survive in the moister areas under maritime influence in, for example, Indochina, Thailand, Myanmar, and Bangladesh.

Tine Geurts

Gibbons and orangutans eat many of the same fruits, such as *Blumeodendron* sp. (left) and *Ficus* sp. (below).

Serge Wich

Thomas Geissmann (www.gibbons.de)

Adult male lar or white-handed gibbon (*Hylobates lar*) singing.

Trees of the family Dipterocarpaceae are typical of most gibbon habitats, ranging from 1 to 43 percent of forest composition,[25] and averaging 16 percent,[1] whereas the tree family Leguminosae varies inversely in abundance from 13 to 14 percent. Between 24 and 50 tree families have been documented from gibbon habitats (averaging 37), with about 400 trees per hectare.[1] Moraceae (figs) and Euphorbiaceae are the commonest tree families used as food sources in gibbon habitats, followed by Leguminosae, Myrtaceae, Annonaceae, Rubiaceae, Clusiaceae, and Anacardiaceae. Figs are a particularly good predictor of gibbon biomass. These families present 161 species, comprising 45 percent of all known gibbon foods.[25] Gibbons prefer lowland forests, where diversity and density of fruiting trees is greatest; however, the largest species, the siamang (which, at about 10 kg, weighs twice as much as the others), has a greater tolerance for leaves in the diet and occurs more frequently in higher-altitude forests.

Feeding and activity patterns

Gibbons spend between 57 and 72 percent of their feeding time eating the reproductive parts of plants (fruit and flowers). The siamang, which is larger, is an exception at 44 percent; the moloch, Müller's, lar, and agile gibbons spend around 60 percent; and the Kloss, pileated, and hoolock spend around 70 percent. About 25 percent of the fruit intake is figs (nearly 40 percent in siamang). Young leaves are important for most gibbons, especially the siamang, but not for the Kloss gibbon in the Mentawai Islands, where the soils are poor and the leaves better defended chemically. Animal matter, mostly invertebrate, provides an important source of animal protein (about 10 percent of feeding time). The underlying preference is for the smaller sources of ripe (sugary) and pulpy (fleshy) fruit.

The gibbons, then, are fruit-pulp specialists,[8] like the chimpanzee, but they compete more than most primates with large birds such as pigeons and hornbills for small, colorful, and sugary fruit. This food focus seems to have required gibbons to learn the location of suitable food resources and to protect them. The small area that can be effectively protected does not provide enough food for many individuals, hence the gibbon's territoriality and monogamy. The focus by monogamous family groups on small fruiting trees avoids competition with the large multimale, multifemale groups of macaques (*Macaca* spp.) and the large-bodied orangutan.

For some plant species, gibbons are key seed dispersers; for others, especially those also dispersed by several bird species, they are less important. In dispersing seeds, animals encourage trees of the same species to fruit asynchronously, which prolongs food availability.

Gibbons differ from other primates in not having a markedly bimodal pattern of daily activity with feeding peaks early and late in the day, and a long midday siesta.[10] After active bouts of feeding, gibbons sustain activity through the heat of the day by foraging in the cooler lower levels of the canopy; they retire early for the night, usually several hours before sunset. This frees food sources shared with other primates, langurs (*Presbytis* and *Trachypithecus* spp.) or macaques (*Macaca* spp.), that are monopolized during the morning by gibbons.

Population density

The density of monogamous family groups of gibbons, typically comprising around four individuals, varies from 1.5 (for two species in Malaya) to 6.5 (in Thailand) groups per square kilometer. Biomass density, as indicated by gibbon biomass per square kilometer, is a more useful measure of population density than the number of individuals or groups per square kilometer, as it relates more closely to food availability, presumably at times of scarcity.[8]

The combined biomass of siamang and lar gibbons in Malaya was 126 kg/km^2, with 34 kg/km^2 for Müller's gibbon in Kalimantan and 104 kg/km^2 for lar gibbons in Thailand. Therefore, gibbons are at least as numerous in the more seasonal forests farther north.[8]

Day range

Pileated and Müller's gibbons, and siamang travel from 0.8 to 0.9 km daily on average, while other gibbons travel between 1.2 and 1.5 km. Siamang have been seen to travel as little as 0.15 km/day and as much as 2.86 km/day. The day range of the hoolock gibbon varies from 0.28 to 3.40 km; other gibbons show comparable variation, from about 0.40 to 2.50 km. These figures reflect variation in food distribution. In the Malay Peninsula, siamang ranged less and ate more leaves when fruits were scarce, increasing day ranges when fruits were abundant and energy levels in excess.[6] In the monsoon forests, however, where leaves are not such a viable alternative for the smaller gibbons, increased day ranges may reflect a wider search for sufficient fruit.[1]

Home range and territory

Home range varies between 0.16 km^2 for lar gibbons in Thailand and 0.17 km^2 for moloch gibbons in Java, to 0.45+ km^2 for hoolock gibbons in Bangladesh and 0.56 km^2 for lar gibbons in Malaya, where siamang home ranges are also large (0.3–0.4 km^2). Where there are two species of gibbon in the same area (one always being the largest species, the siamang), it is likely that the home ranges of both are larger than when alone because of competition for particular fruiting trees. Agile gibbons have smaller home ranges that are evenly used over any five day period; siamangs and lar gibbons each have larger home ranges with both more limited and more variable patterns of ranging over five day periods, although centered on the same fruiting tree or trees.[10] The proportion of the home range that is defended as territory for the exclusive use of the resident group varies: siamang, 62 percent; Kloss, 64 percent; hoolock, 86 percent; Müller's, 88 percent; moloch, 94 percent. Other species defend 75–77 percent of their home ranges.

Vertical use of forest canopy

With their suspensory behavior, gibbons exploit the high forest canopy more than most other rain forest primates, but they are equally at home among the flexible supports of the small trees of the understory. Indeed, they escape from the heat of the midday sun by foraging in small trees with fruit or new leaves in the relative cool of the understory.[6, 35] Comparison with other primates in the Malay Peninsula, for example, shows the preference of gibbons for the main canopy and emergent trees, in the high forest away from edge habitats.[10]

Social organization

Group size averages 3.8 – equivalent to an adult pair and two young – but ranges from two to seven, i.e. there may sometimes be three or four young.[1] Only the concolor gibbon has been recorded as living in polygynous groups, with two or three adult females and young, and an average group size of 7.2 in Yunnan[15] and 5.3 in Bawanglin Nature Reserve, Hainan,[20] although this requires confirmation. Infants are aged up to two or three years until wholly capable of independent travel; juveniles up to five or six

KOCP

Gibbons compete with large birds such as hornbills for their preferred sugary fruit.

Thomas Geissmann (www.gibbons.de)

Thomas Geissmann (www.gibbons.de)

Above: Juvenile male yellow-cheeked crested gibbon (*Nomascus gabriellae*).
Right: Female and male siamang (*Symphalangus syndactylus*); an adult pair during a duet, with partially inflated throat sacs.

years; and subadults, physically adult-like, to eight years or so, when they leave the natal group.

Social interactions within groups

Social interactions within groups are relatively infrequent, presumably because the family group is so cohesive and so familiar with its daily routines. Overt signals are rare, as the young watch and follow their parents. The only sounds heard, apart from the resonating group calls and the movement of branches and foliage, are squeals from an immature animal, usually a subadult, who has come too close to a parent (usually the male), or the bleats of an infant in distress as it is encouraged to move independently. Overt facial expressions are limited to open-mouth threats in aggressive/submissive interactions.

Only in the siamang does the male carry the infant during its second year of life, after it has been weaned from the female (although it may still suckle at night as it sleeps with her). In this way the infant learns first about those animals on whom it is most dependent for its survival: the female, the male, and then the subadult with whom it plays while the adults groom. It interacts least with the juvenile, with whom one might expect it to play most. The adult female usually leads the group around the home range, hence the need to stop carrying the growing infant at the earliest opportunity. The juvenile follows the female, the subadult lags behind at the rear. It is clear, however, that the adult male, from its central position, is influencing the direction of travel.[6] The smaller gibbons separate more often to forage on a broad front as they move between the main food trees.

Grooming involves either adults and subadults during rest periods, or adults and young as they settle for the night (the juvenile tends to sleep with the male, the infant with the female). Play is the other main social activity, recorded in up to 4 percent of the active day in some studies (siamang, lar, pileated, and hoolock). While the infant and juvenile spend much time playing alone – swinging, jumping, manipulating tree parts – they also swing from, grapple with, and bite at, adults or subadults, and sometimes other juveniles.

Singing and social interactions between groups

The duet in all gibbon species is believed to maintain mate and territory, specifically to advertise availability and attract a mate, to develop the pair bond (and cement other bonds within the group), and to defend the mate and the territory. Females seem to exclude other females to defend their mate, and males to exclude strange males to defend their forest space. These songs are reinforced by boundary patrols while food is being sought and during chases to and fro across the boundary.

The complex interaction of multiple factors makes it difficult to explain the observed behaviors in simple terms. These have been best clarified by playback experiments on Müller's and agile gibbons in Borneo[26, 27, 28, 29] and on lar gibbons in Thailand.[36, 37] The resident pair has been shown to respond differently to the songs of neighbors than to those of strangers; the former they expect, and the latter cause much greater agitation. The female reacts strongly to a strange female. Groups will duet in response to a lone female calling, but will approach silently a lone male that is calling. Bornean agile gibbons respond to calls of agile gibbons, even those from Sumatra, but not usually to those of Müller's gibbons.

There has been extensive discussion of the key features of gibbon sociology – monogamy and territoriality – that are thought to confer benefits as well as imposing costs.[2, 19, 22, 26] In being monogamous, the male could be said to be reducing his potential reproductive success; it is thought to be the available niche and distribution of food that leads to this sacrifice, as a result of the energetic costs of patrolling and defending a territory with a rich and predictable food supply. Most field workers believe that gibbons are monogamous because they are adapted to surviving on small fruiting trees. It has been argued by van Schaik and Dunbar that gibbons are monogamous to prevent infanticide, and that they could live in polygynous groups; this analysis was, however, based on incomplete and disparate data.[41]

Some exceptions to a strictly monogamous pattern have been observed, but typically in extreme circumstances. Palombit studied siamang and lar gibbon at Ketambe in the Gunung Leuser National Park in Sumatra.[32, 33] He observed the pair bond to be much stronger in siamang; the pair was more cohesive and equal amounts of grooming were performed by the male and female. Several observed cases of mate desertion, mate switching, and extrapair copulations were caused by a high incidence of disease and death. Ahsan has reported similar observations among the groups of hoolock gibbons in Bangladesh, because they were restricted or isolated in forest fragments of varying size.[1] Any tendency toward promiscuity or even polygyny may relate to a shortage of males in such forest fragments.

Reichard and Sommer echo the argument that females are defending their mate and males defending the territory's resources, suggesting that extrapair copulations (12 percent of those seen) help to confuse paternity and forestall infanticide; hence, kin relations extended into neighboring groups.[38, 39] They studied isolated lar populations in Thailand, near the hybrid zone inhabited by both *H. lar* and *H. pileatus*. Here, the home ranges of pairs of lar gibbons overlapped by around 64 percent. Encounters between groups were therefore common and occupied 9 percent of the active day. Such circumstances seem to be able to weaken the basic monogamous and territorial pattern.

Group formation

Given the stability of gibbon family groups over long periods, it is rare to observe dispersal of maturing young and the formation of new groups. The pattern that has emerged from observations in Malaya[9] and Mentawai[40] is of young adults, recently excluded from the natal group, acquiring a territory with or without parental help, and then obtaining a mate. Daughters tend to wander less far from the parental territory than sons and are more likely to receive parental help. A rare alternative is for a young adult to take over the natal territory when one or both parents disappear; if one parent survives, mating may occur, but this incest is usually transient and/or reproductively ineffective.

Thomas Geissmann (www.gibbons.de)

Adult male agile gibbon (*Hylobates agilis*); a dark variant with light cheek patches.

Adult female northern white-cheeked crested gibbon (*Nomascus leucogenys leucogenys*).

Thomas Geissmann (www.gibbons.de)

Juvenile male northern white-cheeked crested gibbon (*Nomascus leucogenys leucogenys*).

Thomas Geissmann (www.gibbons.de)

GIBBON CONSERVATION

Forest clearance is the greatest threat to the survival of primates and many other animals, and also undermines human wellbeing. Once the forest cover of a tropical country dips below about 50 percent, climatic changes and water and soil problems seem to escalate catastrophically.[7] As few countries seem able to afford to keep more than 10 percent of their forests totally protected, at least another 40 percent of forest area has to be managed for sustained yields of a wide variety of products.[30, 31] Managed forests provide a buffer zone for protected forests, which provide replenishment of plants and animals. The shapes, sizes, and spatial relationships of managed forest areas need to be planned carefully on the basis of systematic research, much of which still needs to be conducted. The third part of the strategy is to use to maximum efficiency the land already cleared of forest or so degraded that its role as forest cannot be redeemed.

Logging, primates, and people

Selective logging represents the compromise between human and animal needs in the long term, but it will only work if timber extraction is very light and carefully controlled. Johns studied this approach in the Sungai Tekam area of the Malay Peninsula.[16, 17, 18] Even if only 10 trees per hectare are extracted (i.e. 4 percent of the trees), 45 percent of the total stand (i.e. 68 percent of the plant biomass) is damaged during access, felling, and extraction. It is the larger and more frugivorous species that are the most vulnerable, but their populations should recover fully within 20–30 years (if there is no further disturbance). For example, gibbons and langurs adapt their foraging strategies by eating more leaves as fruit availability declines in newly logged forest. Gibbons maintain their territories, but the stress affects their breeding. Langurs may emigrate temporarily from the disturbed area, and there is increased mortality among immature monkeys (because of travel difficulties across gaps), which adds to the breeding loss.

Selective logging enhances the diversity of microhabitats characteristic of the mosaic of successional stages of the forest; it is the colonizing plants of immature forest that provide more nutritious and less chemically defended foods. Bird communities maintain much the same trophic structure, but species composition may be changed markedly: dietary generalists survive better than insect and fruit specialists, whose food supply may be highly disrupted temporarily. Mosaics of primary and logged forest can maintain viable populations of the large wide-ranging hornbills. Thus, the persistence of primary forest in an area may be crucial to the survival of certain animal species, and it is the relationships between these two types of forest that need to be investigated urgently. Additional information on the effects of selective logging is available for the Malay Peninsula from surveys of primary and variously disturbed forest[23] and from East Kalimantan.[43] In contrast to the tolerance of gibbons and langurs, orangutans and proboscis monkeys (*Nasalis larvatus*) are seriously affected by selective logging.

Shifting cultivation has been practiced in much of Southeast Asia for centuries, especially along rivers. Where population densities are low, there are long fallow periods between incidents of fire and cultivation at any given place, allowing forest recovery. Under these conditions, a landscape may be maintained indefinitely under a mosaic of forest patches of different stature and age, with little impact on biodiversity or ecological functions relative to natural conditions. When human population density rises, fallow periods become so short that the system becomes unsustainable. Forest can quickly disappear from the landscape under these conditions.

Human needs mean that much forest must be managed for a great variety of forest products, not just timber.[31, 32] What is required is the improved protection of watersheds and national parks representing all ecosystems (especially the richest, lowland ones), with the efficient, sustainable management of large buffer zones, and the more productive use of land already cleared of forest.[9] Such a strategy should ensure that viable populations of all gibbon taxa survive in perpetuity, but it will not be easy to put into effect.

Translocation, captive breeding, reintroduction

Mather[25] developed the valuable approach of analyzing gibbon food trees from all previous studies for comparison with the density of gibbons in each area.[9] He shows that there is a direct correspondence between gibbon biomass and the abundance of these preferred gibbon foods. Group size increases in localities with more fig trees. This analysis enables one to assess whether a gibbon population is at carrying capacity, or below (because of human disturbance), or above (because of immigration from nearby disturbed areas). The suitability of proposed sites for reintroduction or translocation can be assessed, and stocking density determined; where there is selective logging, the reduction in carrying capacity can be determined.

Our improved taxonomic and socioecological understanding of this diverse group of apes and of their habitats in tropical moist forests improves the chances for their effective conservation. Clearer recognition of species and subspecies, and improved quantification of the use of resources (social structure, feeding, and ranging) in relation to what is available are essential to effective protection and/or management.

The predictions of a drastic reduction in gibbon populations[7] are being realized, with the Kloss, moloch, and concolor gibbons being the most endangered. As the clear-felling of forest areas declines, however, their prospects are boosted if adequate selectively logged forest (with low extraction rate) persists, as gibbons have shown themselves to be very adaptable to such disturbance.[17, 18, 23] Little progress has been made in developing techniques for translocation – the movement of social groups from a doomed to a protected habitat – presumably because of the physical difficulties involved, and the lack of suitable destination habitat (but for real progress see Cheyne's work).[4] It remains a possible solution where populations become critically endangered, but adequate preparation, care (with veterinary supervision), and monitoring are essential.

Captive breeding worldwide provides valuable publicity (about the plight of rain forest animals) and education, with fund-raising opportunities for conservation activities. It also helps to conserve the gene pool, with meticulous studbooks. The prospects of reintroduction to the wild are gloomy, however, given the costs involved and the lack of available habitat. If habitat is available, it is much more cost effective and successful to translocate

Serge Wich

Degraded habitat in Borneo.

social groups from doomed forest fragments to any understocked protected forest. The prime effort must be to protect natural habitat and to conserve wildlife within it.

Kalaweit in the Bukit Baka National Park in Central Kalimantan offers a beam of hope. Facilities are being developed to accommodate confiscated gibbons, to support pair formation, and (when ready) to reintroduce them to protected forest. Another possible area in which to do this is being developed nearer to Palangka Raya, the provincial capital of Central Kalimantan.[4, 5]

Education is essential at various levels, as successful programs in many countries demonstrate. In the long term, education of local people (whose lives are most immediately affected by destruction of forests) and the young (the next generation) the world over is essential. Most critical, however, is the need to influence the decision makers of today. These are the governments of tropical countries (who now mostly see what has to be done) and, more importantly, the governments of 'user countries' as well as the heads of international and national commercial concerns. Policy and activities need to be changed rapidly to avert impending catastrophes. Values have to be changed and resource flows significantly altered if this planet is not to be damaged irreparably. An international network concerned with disseminating this interdisciplinary bio-environmental approach could have a critical role to play in this process.

Threatened gibbons

Indochina is key to gibbon conservation, indeed to all primate conservation (it contains about eight of the 20 most endangered primates in the world). The four crested gibbon species (*Nomascus* spp.) in northern Indochina and southern China are seriously threatened, but the most endangered are the Hainan (China) and Ca Vit (northeast Viet Nam) gibbons (*Nomascus nasutus*), with fewer than 20 individuals each. Efforts are being made to ensure that they all flourish. The rarer they are, the more effort local people are often willing to make. The northern and southern white-cheeked gibbons in Viet Nam and Laos (*Nomascus leucogenys*) are also struggling, while the yellow-cheeked gibbon (*Nomascus gabriellae*) in southern Viet Nam and Cambodia seems to be the most numerous of the genus.

The other gibbons most endangered by habitat loss are the Javan or silvery gibbon (*Hylobates moloch*), which survives only in the west of Java, and the Kloss gibbon (*H. klossii*) on the Mentawai Islands off the west coast of Sumatra. The status of the hoolock gibbon (*Bunopithecus hoolock*) is unknown in Myanmar, and perhaps a cause for serious concern; numbers in Bangladesh and eastern India are not large. The pileated gibbon (*H. pileatus*) is restricted in Thailand and its status is unknown in Cambodia. Otherwise, the more widely distributed siamang (*Symphalangus syndactylus*), lar, agile, and Bornean gibbons (*H. lar, H.agilis,* and *H. muelleri*) are present in good numbers where forests remain, even in selectively logged ones.

FURTHER READING

Brockelman, W.Y., Srikosamatara, S. (1984) Maintenance and evolution of social structure in gibbons. In: Preuschoft, H., Chivers, D.J., Brockelman, W.Y., Creel, N., eds, *The Lesser Apes: Evolutionary and Behavioural Biology.* Edinburgh University Press, Edinburgh. pp. 298–323.

Carpenter, C.R. (1940) A field study in Siam of the behaviour and social relations of the gibbon (*Hylobates lar*). *Comparative Psychology Monographs* **16** (5): 1–212.

Chivers, D.J., ed. (1980) *Malayan Forest Primates: Ten Years' Study in Tropical Rain Forest.* Plenum Press, New York.

Chivers, D.J. (2001) The swinging singing apes: fighting for food and family in far-east forests. In: Brookfield Zoo, *The Apes: Challenges for the 21st Century.* Conference proceedings. Chicago Zoological Society, Brookfield, Illinois. http://www.brookfieldzoo.org/content0.asp?pageID=773. pp.1–28.

Geissmann, T. (1995) Gibbon systematics and species identification. *International Zoo News* **42**: 65–77.

Leighton, D.R. (1987) Gibbons: territoriality and monogamy. In: Smuts, B.B., Cheney, D.L., Seyfarth, R.M., Wrangham, R.W., Struhsaker, T.T., eds, *Primate Societies.* University of Chicago Press, Chicago. pp. 135–145.

ACKNOWLEDGMENTS

Many thanks to David Woodruff (University of California, San Diego) for his valuable comments on the draft of this chapter.

AUTHOR

David J. Chivers, Wildlife Research Group, Department of Anatomy, University of Cambridge

Conserving the great apes

TOSHISADA NISHIDA

The conservation of nature can be divided into short-term, middle-range, and long-term perspectives. A short-term project aims to secure immediate protection by actually preventing harmful acts through effective law enforcement. On a simple level, a short-term project is more important than a long-term one, since if animals are extinct today, there will be no need for a long-term project tomorrow. However, even if you succeed in saving animals this decade, but then lose them in the next, the current venture is completely pointless. Our mission in the short term should be to save the great apes from imminent extinction and to formulate at least a middle-range conservation scheme by implementing measures to save them. As GRASP may be considered a middle-range scheme, its mission should be to help each range country to formulate a national great ape survival plan and to integrate these plans into a global network. Limiting road construction and logging, along with the introduction of Forest Stewardship Council principles of forest management, are perhaps the most important elements to consider.

I feel that the three chapters that follow deal excellently with short-term and middle-range conservation measures, and so I will confine myself to taking a long-term perspective for great ape conservation.

I am often asked how many chimpanzees still live in Africa, and I answer that perhaps there are only 100 000. Upon hearing this, the usual response is "Oh! There are still so many chimps!" People just do not stop to consider that even the smallest satellite cities of Tokyo or Osaka contain more than 100 000 humans. We are so anthropocentric that we do not think twice about the fact that we are overpopulating the Earth at the cost of other living things. Anthropocentrism is the driving force pushing the great apes to extinction. Therefore, if we do not succeed in educating people to abandon our current anthropocentrism there can be no hope of saving the great apes.

Of course, we humans cannot subsist without killing animals and plants. However, we can respect them and should refrain from killing them solely to seek pleasure or to satisfy our excessive appetites. Only 50 years ago, for example, the prohibition on wasting food was a universal tenet of human culture, except for rare special occasions such as a feast or potlatch; this attitude should again be enthusiastically embraced.

'Progress' appears to be currently regarded as the sole ethical purpose of human existence. However, this only became widely accepted across Europe after the 18th century Enlightenment. During most of human history, and even now among more traditional communities around the world, conservatism or respect of customs remains the dominant ethic. I often asked my Tanzanian assistants why they were doing this rather than that. Most of the time, they responded, "Oh, because my grandfather used to do this." We rarely recognize that most modern developments provide only short-term benefits by wasting materials.

Although the ultimate result of today's anthropocentrism is the endless expansion of the human race, I believe that the primary cause for the decrease in the number of great apes, and many other species of wild fauna and flora besides, is the ever increasing demand, particularly in the industrialized world, for raw materials: cheap timber, cheap agricultural products, cheap natural resources. So, perhaps those of us in the industrialized world should first be asking questions of ourselves about our current consumerist attitudes and habits. Only then, for example, can we work with any integrity with those range states where it is practiced to discourage the eating of great apes.

There are other ways forward, too. For four years now, I have been engaged in the Great Ape World Heritage Species Project (GAWHSP) as I believe that a World Heritage Species should be nominated based on its universal value, encapsulating scientific, cultural, and conservation ideals. This is one of the guiding reasons to propose granting World Heritage Status to the great apes.

Although ecotourism and even research have occasionally had negative effects on the health of great apes through the introduction of anthropozoonotic diseases, large numbers of apes have been protected by park rangers funded at least partially by ecotourism, for example in Mahale, United Republic of Tanzania. Thus, ecotourism has a part to play.

Both academic and conservation-oriented research fees can also be encouraged, with range states regarding these as one source of long-term finance for protected areas, as has been the case in some East African countries. Although research fees might be lower than tourist prices, researchers should accept some costs, since their work provides individual benefits to the researchers. Short-term success in the conservation of great apes in the last 20 years has been disrupted by war and armed conflict. However, during even the worst period, some scientists continued to visit their study sites and provide salaries to their assistants. Although the study populations suffered devastating losses, at least some survived the crises, with the support of determined scientists proving to be invaluable.

Accordingly, I would suggest that any protected area should have a long-term scientific research team that is locally based, while organized internationally. Gombe, Mahale, Karisoke, Wamba, Kahuzi-Biega, Bossou, Taï, Kibale, Budongo, Kutai, and others each have such a team of scientists. Kalinzu, Bwindi, and Moukalaba may soon join this long-term club. The list could be extended to all the protected areas containing great apes, possibly with international nongovernmental conservation organizations involved in helping to provide funds for long-term research. And as researchers who monitor great apes every day are very protective of their 'own' animals, perhaps we – as proposed by John Oates in his excellent book *Myth and Reality in the Rain Forest* – should be encouraged to establish trust funds to ensure the viability of continuing long-term research.

Toshisada Nishida
Executive Director, Japan Monkey Centre

CHAPTER 13

Challenges to great ape survival

LERA MILES, JULIAN CALDECOTT, AND CHRISTIAN NELLEMANN

Great apes are endangered because people are bringing into their world both deliberate change, such as land clearance, and accidental change, such as forest fires. This destruction of tropical moist forests puts countless numbers of species at risk of extinction. Hence, the challenge of great ape conservation cannot be disentangled from the management and future of tropical forests as a whole. That said, great apes are particularly vulnerable because it is easy and often profitable to shoot them, they reproduce slowly and are susceptible to many human diseases. The main threats to their survival are habitat loss, degradation, and fragmentation due to logging and clearance for agriculture (particularly in West Africa and Southeast Asia); forest fires (especially in Southeast Asia); and hunting (particularly in West and Central Africa). Potential sources of further risk include diseases, human conflict and mineral extraction. Demand from overseas consumers for luxury resources such as tropical timber and cheap staples such as palm oil contributes to many of these pressures. The range and intensity of threats have led many observers to conclude that great ape numbers will further decrease, and rapidly, within 10–20 years.[6, 41, 113, 131, 153]

PREDISPOSITIONS TO ENDANGERMENT

Life history and vulnerability

Great apes have relatively low reproductive rates, long lifetimes, and long 'childhoods' (Table 13.1). This combination of factors makes their populations very vulnerable to high rates of adult mortality, from which they cannot easily recover. This, combined with the requirement for a large area of natural habitat, is the ecological factor at the root of the vulnerability of apes to the impacts of humans.

Threat status classification

The *Red List of Threatened Species* of IUCN–The World Conservation Union is a guide to determining which species are in most urgent need of conservation action. Recent or expected population losses are an important criterion for the *Red List*. If a species' population has declined by 80 percent or more over ten years (or three generations, whichever is the longer), or is expected to, then it is classified as Critically Endangered; if by 50–80 percent it is classified as Endangered. All the great ape species are in the Endangered or Critically Endangered categories of the *Red List*.[77] Of particular concern are the Sumatran orangutan, the mountain gorilla, and the Cross River gorilla, all of which are classed as Critically Endangered (see Table 13.1). The *Red List* coding system also reflects

Vu Danh Viet/UNEP/Topham

Forest fires are one of many threats to ecosystems, particularly in Southeast Asia.

the type of evidence on which the classification has been based.[76] The Sumatran orangutan, for example, falls into the category 'Critically Endangered (CR) A2bcd'. These letters and numbers mean that:

- the orangutan fits criterion A (population decline);
- of type 2 (reduction of more than 80 percent over the last 10 years or three generations);
- based on evidence of types b (index of abundance), c (continuing decline in 'quality of habitat'), and d (level of exploitation).

The major threats are also categorized. For Sumatran orangutans, these are:

- 1.1.1.1 (habitat loss through shifting agriculture);
- 1.3.3 (habitat loss through wood extraction);
- 3 (harvesting/hunting).

The IUCN system therefore condenses expert opinion and scientific evidence about the status of a species, subspecies, or population. There are 352 Endangered and 162 Critically Endangered mammal species on the 2004 *Red List*, a reminder that this is a common tale of threatened extinction.[77] The endangerment of apes can be seen as an early warning of the loss of many species that are less well known. The close relationship between human beings and the great apes makes their case particularly resonant for us: for these are threats to our own kin.

Table 13.1 Great ape reproductive characteristics and *Red List* status

Taxon	Minimum age at first pregnancy	Reproductive interval in adult females	Maximum lifespan	Estimated no. of individuals	Year of estimate	IUCN Status[a] [129]
Chimpanzee (*Pan troglodytes*)	10–13[70]	4.4–6[75]	40–50[80]	172 700–299 700[24]	2003[b]	EN A3cd
Western chimpanzee (*P. t. verus*)	..	..	..	21 000–56 000[24]	2003[b]	EN A1cd+2cd
Eastern chimpanzee (*P. t. schweinfurthii*)	..	..	..	76 400–119 600[24]	2003[b]	EN A3cd
Central chimpanzee (*P. t. troglodytes*)	..	..	..	70 000–116 500[24]	2003[b]	EN A3cd
Nigeria-Cameroon chimpanzee (*P. t. vellerosus*)	..	..	..	5 000–8 000[24]	2003[b]	EN A1cd+2cd
Bonobo (*Pan paniscus*)	13–15[67]	4.8[53]	50–55[115]	10 000–50 000 (to > 100 000)[23, 38]	2001	EN A2cd
Western gorilla (*Gorilla gorilla*)	8.5[c] [64, 160]	4–6[d] [133]	35–45[c] [12, 65, 165]	94 500–110 000[e] [23, 119, 157]	2000[b]	EN A2cd
Cross River gorilla (*G. g. diehli*)	..	..	..	250–280[118]	2004	CR A2c; C2a(i)
Western lowland gorilla (*G. g. gorilla*)	..	..	..	94 500–110 000[e] [23, 119, 157]	2000[b]	EN A2cd
Eastern gorilla (*Gorilla beringei*)	..	3.9–4.6[160, 168]	..	?[f]	2005	EN A2cd
Mountain gorilla (*G. b. beringei*)	..	3.9[160]	..	700[61, 84, 101, 102]	2003	CR C2a(ii)
Eastern lowland gorilla (*G. b. graueri*)	..	4.6[168]	..	?[f, g] [129]	2005	EN A2cd+ 3cd+4cd
Bornean orangutan (*Pongo pygmaeus*)	11–15[56]	8[154]	45[94]	45 000–69 000[139]	2004	EN A2cd
Northeast Bornean orangutan (*P. p. morio*)	..	..	..	11 000–21 000[1]	2004	Not assessed
Northwest Bornean orangutan (*P. p. pygmaeus*)	..	..	..	2 640–3 260[105, 137]	2004	EN A2cd
Central Bornean orangutan (*P. p. wurmbii*)	..	..	..	> 40 500[139]	2004	EN A2cd
Sumatran orangutan (*Pongo abelii*)	..	..	..	7 334[139]	2004	CR A2bcd

a EN, Endangered; CR, Critically Endangered; for a full explanation of threat criteria see IUCN (2000) *IUCN Red List Categories and Criteria Version 3.1*. http://www.iucn.org/themes/ssc/redlists/RLcats2001booklet.html. Accessed March 28 2004.
b Chimpanzee and western gorilla estimates were collated by Thomas Butynski (2001[23], 2003[24]) and include estimates ranging from 1984 to 2003. Please consult the country profiles for further details.
c Based on eastern gorilla.
d Based on western lowland gorilla.
e Recent decline due to Ebola hemorrhagic fever not quantified.
f No data; fieldwork was being undertaken in 2005 to estimate the extent of the decline.
g Reduced from 17 000 ± 8 000 in 1998.[62]

Risk and uncertainty

As our ecological knowledge has increased since the mid-20th century, our estimates of the population sizes of great apes have also increased, even though their populations actually declined during this period. The Bornean orangutan is a typical example (see Table 13.2). This is because early estimates did not always recognize the breadth of habitats and regions occupied by the great apes, and further errors have arisen from the variety of sampling and extrapolation methods used. These often rely on nest density, and ape nests are not easily visible in the forest: radically different results may be obtained from surveys on foot or by helicopter, for example.

Nevertheless, we can be confident that the species with the smallest populations are the eastern gorilla and Sumatran orangutan (see Table 13.1), that the scarcest subspecies is the Cross River gorilla, and that the chimpanzee is the most numerous great ape species. There are, very roughly, at least twice as many chimpanzees as western gorillas, four times as many chimpanzees as Bornean orangutans or bonobos, and about 30 times as many chimpanzees as Sumatran orangutans. There may be over 90 000 each of the central and eastern chimpanzee. The only other great ape subspecies that may exist in similar numbers is the western lowland gorilla. Contrast this with the global human population in 2005, which was estimated at 6.465 billion: the equivalent of about 27 000 people for every chimpanzee.[125]

There are few figures on the actual rate of decline in great ape numbers. The 2004 *Red List* provides some estimates for Sumatran orangutans (see Table 13.3). It also states that Bornean orangutans, chimpanzees, and eastern gorillas are declining in numbers, but that there is insufficient information to judge the trend for the western gorilla and bonobo.[77] The dual impacts of Ebola and bushmeat hunting in the heartlands of the western gorilla and chimpanzee range in the Congo Basin are unquantified and may already have much reduced populations of both these 'common' species.

Table 13.2 Estimates of Bornean orangutan populations through time[a]

Year	Estimate	Trend
1960s	450–900[b] [78]	declining
1970s	250[b], 2 000–3 000[c] [78]	declining
1980s	> 3 500[c] [78]	declining
1990	30 000–50 000[43]	not stated
1995–1996	15 953–24 497[109, 132]	declining
2004	45 000–69 000[139]	declining

a See also Table 10.4.
b Sarawak only.
c Sabah only.

Table 13.3 Sumatran orangutan decline, based on 2004 IUCN *Red List* [a]

Year	Taxon	Percent decline over period	Trend
1992–1999	*Pongo abelii*	46	declining[42]
1992–2000	*Pongo abelii*	> 50	declining[42]
2000	*Pongo abelii*	17	declining[43]

a See also Table 11.3.

HABITAT LOSS AND FRAGMENTATION

The great apes are creatures of tropical moist forests, although not all are equally arboreal. Orangutans are seldom seen on the ground, whereas the more terrestrial bonobos spend relatively little time in the trees. Nevertheless, most of the time, all the apes build nests in trees and depend on arboreal food sources (although mountain gorillas forage mostly at ground level within a herbaceous 'canopy'). Even chimpanzees, the species most likely to be found in open woodland or farmland, need access to forest in which to sleep and feed. Hence, the long-term survival of the great apes will be determined by the fate of the forests in their range countries.

It is hard to find global maps of deforestation. The Food and Agriculture Organization of the United Nations (FAO) collates information about forest area by country, including estimates of the amount of change. Forests are defined by FAO as areas with canopy cover of more than 10 percent, including both natural and plantation systems. Estimates of change in natural forest are not provided separately from estimates of change in forest cover, but this information can be obtained approximately by subtracting the change in plantation forests from the total (see Table 13.4).

Deforestation is sometimes the outcome of a development decision (e.g. replacing lowland rain forest with oil palms, or peat forests with a resettlement scheme), but it can also be the consequence of many small actions that coalesce to create a deforested environment (e.g. settlers and their smallholdings multiplying along a forest highway). It can also occur when one form of disturbance (such as logging) allows the forest to dry out enough for it to burn if fires are set nearby (as happened in Borneo in 1997–1998 and in Sumatra in 2004). Deforestation can be irreversible if, for example, exposed soils are badly leached and eroded or if fire-adapted grasses, such as alang-alang (*Imperata cylindrica*), become well established and are then maintained by regular burning.

Large-scale fire is a particularly important factor in the Southeast Asian forests where orangutans live, especially in Borneo. Here fertile soils are scarce, human populations sparse, and land-use practices are highly consumptive of space and forests. Local farmers are used to clearing whole hillsides for one or two harvests before moving on, and at the same time hunt in large areas of forest. Meanwhile, central government planners often treat the interior of Borneo as more-or-less 'empty' land, for the allocation of logging or plantation permits, and the location of major infrastructure projects or resettlement schemes.

Table 13.4 Decline in natural forest cover in range states, 1990–2000[a]

Year	Natural forest 2000 (km²)	Change in natural forest 1990–2000 (km²)	Annual change (percent of 1990 figures)
Angola	696 150	-12 630	-0.18
Burundi	211	-1 279	-8.58
Cameroon	237 780	-22 820	-0.88
Central African Republic	229 027	-2 983	-0.13
Congo	219 767	-2 213	-0.10
Côte d'Ivoire	69 327	-27 703	-2.86
Democratic Republic of the Congo	1 351 103	-53 837	-0.38
Equatorial Guinea	17 520	-1 030	-0.56
Gabon	217 896	-1 164	-0.05
Ghana	62 590	-12 230	-1.64
Guinea	69 042	-3 678	-0.51
Guinea-Bissau	21 855	-2 165	-0.90
Indonesia	951 155	-168 695	-1.51
Liberia	33 625	-8 725	-2.06
Mali	131 715	-9 935	-0.70
Malaysia	175 425	-40 375	-1.87
Nigeria	128 239	-45 261	-2.61
Rwanda	462	-3 228	-8.75
Senegal	59 418	-6 012	-0.92
Sierra Leone	10 490	-3 610	-2.56
Sudan	609 865	-100 265	-1.41
United Republic of Tanzania	386 761	-8 939	-0.23
Uganda	41 472	-9 358	-1.84

a These estimates of deforestation between 1990 and 2000 were determined from FAO Forest Resource Assessment (FRA) data by subtracting the 'change in plantation area', FRA 1990 to FRA 2000,[44, 45] from the 'change in all forest area', FRA 2000.[45]

Every few years, the El Niño Southern Oscillation (ENSO) causes a delay to the onset of the rainy season in Southeast Asia. Together with the local tradition of using fire to clear land, and the effects of logging in opening and allowing the drying of forest, these droughts exacerbate the spread of fires. This is especially the case where canals are cut in peat-swamp forests to float out logs, since they are seldom filled in after use, and continue to drain the swamps and lower the local water table long after logging has ended. Major ENSO-linked fires occurred in 1982–1983, 1991, 1994, and 1997–1998.[9] The 1997–1998 fires were so widespread that millions of people were affected by atmospheric pollution,[3] and other such haze events also occurred in 2002 and 2004.

In many areas, fire and farming combine to create a stable patchwork landscape, often with moist forest retained in gullies and valleys, and fire-maintained grassland on the slopes in between. Much of the habitat of the Cross River gorilla is like this, for example on the Obudu Plateau in the Okwangwo Division of the Cross River National Park in Nigeria. Small populations can find themselves trapped within such patches of forest, and unable to disperse to other or larger areas. Corridors of farm development or housing along roads can also fragment a forest landscape, and with it a great ape population. Some of the measures that conservationists use are designed to offset this effect, for example, by linking habitat blocks with forest corridors. Fragmentation is a hazard because it reduces the size of the gene pool within a breeding population, thus reducing its genetic heterogeneity and increasing its vulnerability to the effects of inbreeding. Isolated populations are also potentially vulnerable to catastrophic or random events such as disease and forest fires.[55]

HABITAT DAMAGE

The last half of the 20th century saw both the establishment of a global market for tropical timber and the availability of capital and equipment within tropical countries to enable widespread

Box 13.1 LUCKY GORILLAS?

Large animals that reproduce slowly are usually the first to be threatened with extinction. Gorillas are the largest primates, and only the other great apes reproduce as slowly as they do. Nevertheless, neither the western nor the eastern gorilla is listed as one of the most threatened primate species (see Table 13.1). Over 6 percent of the 296 primate species recognized in the IUCN *Red List* are listed as Critically Endangered,[77] but five out of six great ape species are only Endangered. (On the other hand, one subspecies from each of the gorilla species is Critically Endangered.)

Why are the gorillas, the largest of the primates, not among the most threatened of primate species? Whether a taxon is in danger of extinction is determined first by the nature and intensity of the threats, and second by its biology. Gorillas appear to be blessed with some luck in both areas.

Lucky biology (1)

Gorillas, unlike other great apes, can survive largely on herbaceous food as opposed to relying on food with a higher energy content, such as fruit or meat.[100] In any forest, this relatively poor-quality food exists in greater abundance than does high-quality food. When fruit is in short supply, the other great apes must expend considerable energy seeking it, while gorillas start to eat more low-quality food. Despite their large size for a given amount of food, gorillas can therefore survive in smaller areas of natural habitat than can the other apes.

Lucky biology (2)

Western lowland gorillas can occur at high densities in swamp forest, as biologists have only recently discovered.[14, 46, 119] Within the gorilla's range, vast areas of swamp forest remain intact, particularly in Congo.[47] Orangutans also do well in swamp forest, but in Southeast Asia these have been greatly affected by logging and associated changes arising from drainage and subsequent fire.

Luck with threat (1)

Mammals and birds are more often threatened where human density is higher,[87] but very few humans live in the vast areas of West African swamp forest. If low human density and large geographic range both make taxa safer, there are grounds to think that western gorillas may be less threatened than the IUCN *Red List* suggests. When ranked using these criteria, the western gorilla emerges as one of the less threatened primates; over 50 percent of other primate species are more threatened.[66]

Ian Redmond/UNESCO

A lucky gorilla?

Luck with threat (2)

Not only do the interior of the swamp forests currently host very few people, but access to them is still extremely difficult. The gorillas in West African swamp forest, perhaps now the majority of gorillas in Africa, suffer less from logging or the bushmeat trade under comparable human population densities than they or other animals do elsewhere.

Luck with threat (3)

Not all parts of the range of gorillas have a low human population density. Some of the highest densities in Africa occur in the range of the eastern gorilla in eastern Central Africa.[63] The chaos of war and rebellion in this area has occasionally led to mass slaughter of gorillas, such as that of the eastern lowland gorillas of Kahuzi-Biega National Park.[167] At the same time some of the very best protected areas in Africa occur within the eastern gorilla's range; through all the vicissitudes of war and genocide, the Virunga mountain gorilla population has flourished.[63, 84]

Of course, just because two gorilla subspecies (western lowland and mountain gorillas) seem less threatened than many other primate species we cannot assume that the genus as a whole is safe. Ebola hemorrhagic fever epidemics appear to have eliminated western lowland gorillas from large areas of eastern Gabon.[157] Despite decades of conservation effort, the threats to gorillas from land-use changes and hunting are still very real.

Alexander H. Harcourt

This orangutan in Central Kalimantan is showing signs of starvation as a result of habitat disturbance.

Tine Geurts

industrial-scale logging. Even the best-managed mechanized logging has a strong impact on the local environment. Roads and log-pounds are built using heavy equipment, compacting and exposing the soil; trees must be felled, crushing their neighbors, and then dragged around, damaging other trees. Bearing in mind that this is being done in a tropical, high-rainfall environment, often on steep terrain, and frequently far from the supervision of professional foresters and forestry officials, the impact on the forest ecosystem is likely to be severe. How severe will depend on various factors, including the intensity of the operation. Southeast Asian dipterocarp forests are often harvested at a much higher rate than has been characteristic of similar operations in African forests, because the dipterocarps grow so densely and are such valuable timber trees.

Conversion of great ape habitat to agricultural land. This area lies between Bwindi and Mgahinga in Uganda.

Gordon Miller/IRF

Studies of timber-producing tropical countries have revealed a consistent pattern. Government forest services have tended to function reasonably well until a combination of new technologies and markets made it possible to make huge profits from logging.[136] At that point, multinational companies were often invited by the government in each country to harvest the forests on a profit-sharing basis. This generally required both that the government's own forest service be partially disabled so that it could no longer insist on sustainability, and that legislation protecting the forests be rewritten to allow long-term public interests to be overridden by short-term, often personal, financial interests. This occurred at various times in different places, for example during the 1970s and 1980s in Malaysian Borneo, and during the 1980s and 1990s in Indonesia.

Indonesia still held most of its forests as late as 1950, but over the following 50 years forest cover declined from 1 620 000 km^2 to 980 000 km^2.[54] The rate of forest loss is still accelerating, with lowland forests being most at risk. At current rates, lowland forests will disappear entirely from Sumatra and Kalimantan (Indonesian Borneo) within 10–20 years and perhaps even sooner.[54, 79, 113] Most Indonesian forests are allocated to export-oriented logging concessions. Barito Pacific Timber Group, the largest holder of logging concessions in Indonesia, exported over 94 percent of its production in 2001.[54] Extensive illegal logging for export is also a major problem within the industry.[120] A program funded by the UK government found that 73 percent of all logs in Indonesia came from undocumented sources,[138] while illegally exported logs have for years provided raw material for the Malaysian sawmill industry, which has far too great a capacity for its own (legal) in-country supplies. The Indonesian government, supported by the European Union, United States, and others, has been trying for several years to gain some control over this, and in July 2004 proposed introducing the death penalty for those found guilty of illegal logging.[86]

Box 13.2 SATELLITE ANALYSIS OF THREATS TO GOMBE CHIMPANZEES

On January 31 1961, Ham, a four year old male chimpanzee, flew in one of the first NASA space missions. Ham's flight was a success that helped pave the way for the United States' manned space flight program.[111] Space technologies now have the potential to contribute critical information to help save chimpanzees and other endangered great apes from extinction.

If the great apes are to survive, we need to measure objectively the success of conservation action. This means we need extensive information on habitats and land-use patterns that include sufficient detail in both time and space. Unfortunately, the great ape ranges occur in developing parts of the world where little field information is available.

A multitude of satellite-based sensors with different characteristics now make it possible to map the location, extent, and magnitude of certain types of human activity within great ape ranges. Satellite imagery has been used to improve our understanding of the threats to chimpanzees and to support conservation efforts in Gombe National Park, United Republic of Tanzania.

In Gombe, Jane Goodall's pioneering work and groundbreaking discoveries about chimpanzee behavior helped to narrow our view of the gap between human and nonhuman beings, teaching us a great deal about our own place in nature. Located on the shores of Lake Tanganyika in western Tanzania, Gombe National Park was established in 1968 and was the first park created specifically to protect chimpanzees. Gombe hosts one of the world's most longstanding sites for the study of animal behavior. Research began in 1960; for decades, staff members of the Gombe Stream Research Center have been tracking chimpanzees to collect data daily. As so much is known about each individual, the Gombe chimpanzees are a unique resource in seeking better understanding of the behavioral ecology and conservation needs of their species.

Gombe's chimpanzees face many threats and are in crisis. The park is only 35 km^2 in area; chimpanzee numbers are declining as a result of disease, poaching, and habitat loss. Once a good understanding of these threats and their causes is established, it may be possible to improve the practical strategies employed to reduce or eliminate these threats, and to monitor conservation success.

At Gombe, satellite imagery has proved to be an excellent tool in mapping threats to chimpanzee habitats at different spatial scales. Gombe chimpanzees depend on a mosaic of evergreen and deciduous forests and woodlands, so satellite imaging has been used to evaluate the extent to which these habitats have been lost from the Gombe region.

Landsat satellites have been orbiting the Earth since 1972, continuously collecting images of the Earth's surface. The accumulated data archive includes multiple images of Gombe, permitting comparison of satellite images from different dates. 'Vegetation difference indexing', which quantifies changes in green vegetation cover, was performed using satellite images captured during dry seasons in 1972 and 1999. This revealed forest destruction and conversion to oil palm plantations, along with massive clearing of miombo (dry deciduous) woodlands for farmland and charcoal. The vast majority of loss of local tree cover occurred outside the national park, on village lands as well as within forest

continued overleaf

Until the 1980s, timber from West and Central Africa was considered to be of low commercial value, which limited the pressure posed by the selective logging that was taking place. All this changed dramatically during the 1990s, as the Southeast Asian forests became depleted. By 2000, more than half of Gabon's forests were allocated as logging concessions;[32] and log production had increased to some 2.5–2.7 million m^3/year. Meanwhile, in Cameroon, more than 170 000 km^2 (76 percent) of the country's forests had either been logged or allocated for logging concessions, and satellite images have revealed that networks of new logging roads had spread into what had been considered the least accessible forests in the country.[107] Extensive logging had also occurred in the Río Muni area of Equatorial Guinea, in the DRC and Congo sections of the Mayombe Forest, and in other parts of the western gorilla's range.[59, 135, 146]

Many of the timber concessions in DRC have been awarded within the range of the bonobo. According to the national forest service, some 24 percent of the range of the bonobo is now under

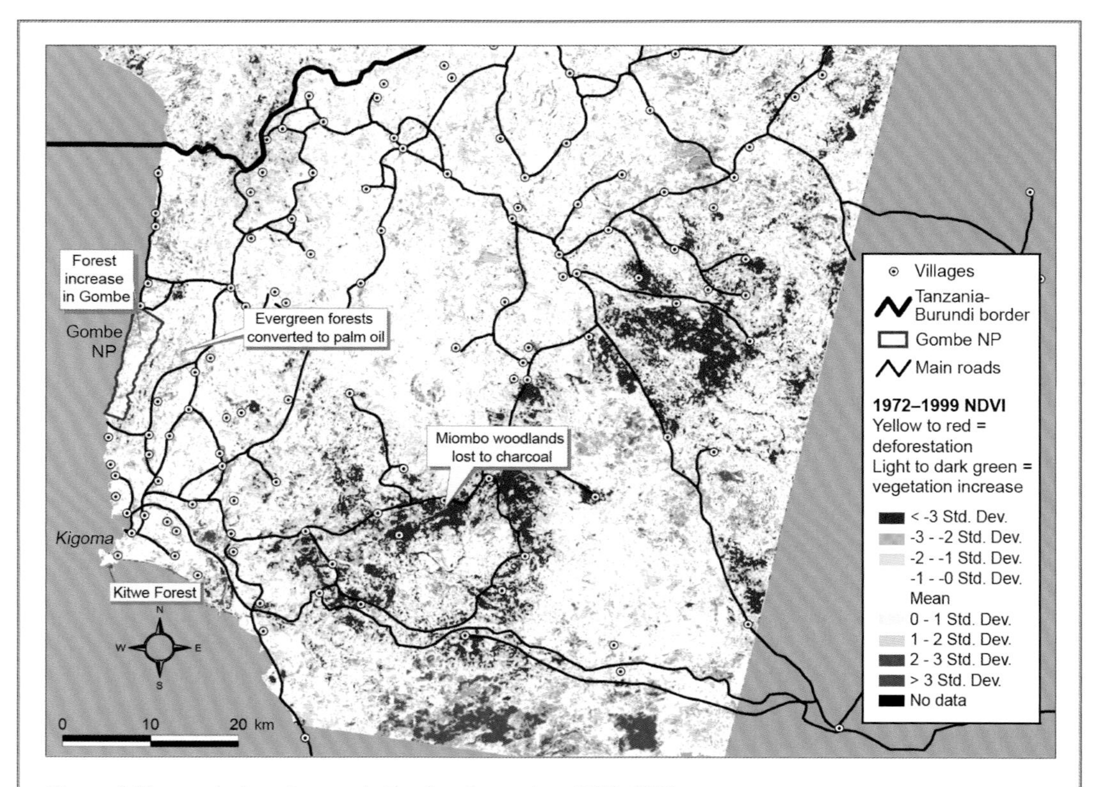

Figure A Change in forest cover in the Gombe region, 1972-1999

Forest cover is expressed in terms of the Normalized Difference Vegetation Index (NDVI).

reserves, in areas that are less well protected but known to be inhabited by chimpanzees (Figure A).

The analysis also revealed an increase in forest cover both inside Gombe National Park and within protected patches of forests such as at Kitwe (Figure A), which have been restored by the Tanganyika Catchment, Reforestation, and Education Project of the Jane Goodall Institute. This demonstrates the enormous potential for restoration of forests and miombo woodlands in the Kigoma region. If large enough patches of these habitats can be restored in strategically important places for chimpanzees and connected to existing forest remnants, then there is hope for the Gombe chimpanzee population in the long term.

Historical Landsat satellites can only map land features larger than 57 meters in length. Most human settlements, roads, and farms in western Tanzanian landscapes are smaller than that and therefore difficult or impossible to map. More precise satellite imagery has recently become available that can detect objects in the 0.6–4 m scale.

logging concession;[143] other observers calculate the figure to be as high as 55 percent.[130] Forest products now account for more than 10 percent of all trade recorded in Cameroon, CAR, Congo, Côte d'Ivoire, Equatorial Guinea, Gabon, and Liberia.[112] Many of the logging companies operating in the Congo Basin are based in European Union countries, such as Denmark (e.g. DLH Group), France (e.g. Rougier, Thanry, and Interwood), Italy (e.g. Alpi), and Germany (e.g. Danzer, Feldmeyer, and Wonnemann).[49]

Logging in tropical moist forests is a complex process that interacts with varied and diverse ecosystems, having wide-ranging effects on the various species present.[158] It is not always clear that logging at moderate intensities, considered alone, has a wholly negative impact on great apes, which are robust and mobile and have unspecialized dietary needs. Where the process of logging reduces fruit availability, however, then carrying capacity at least for the orangutan will inevitably decline. The more folivorous and terrestrial gorillas

Figure B Fifi's progress (yellow dots) through Gombe National Park
This 1-meter IKONOS image was collected in July 2000 and is draped over a digital elevation model.

The effectiveness of new technologies in mapping heterogeneous tree canopies along with human farms, settlements, and paths is demonstrated in Figure B (a 1-meter IKONOS image). The yellow dots represent the location of Fifi, a female chimpanzee, as she was being followed at 15 minute intervals in 1999. This is the fine scale at which many human activities impact directly on chimpanzees, so such images are extremely useful in planning, implementing, and evaluating conservation measures in the Gombe region.

Greater accuracy and higher-resolution imaging costs more. The dozens of sensors carried by various satellites planned in the near future will increase the range of options open to researchers. The cost-effectiveness of these tools depends on matching technological advances with high-quality questions, and the integration of the results obtained into improved efforts in great ape conservation.

Lilian Pintea

and the more adaptable chimpanzees are likely to be less affected, even by moderate logging. If they are displaced from their normal home ranges, however, this can cause stress and disruption to social interactions with neighboring groups and communities.[161]

What is very much clearer, however, is that the workers who drive the bulldozers and wield the chainsaws of a logging operation, their camp followers, and people who arrive later along newly opened logging roads, are likely to want to supplement their food supply through hunting. They may also wish to supplement their finances by selling forest products, if they can. Many will regard great apes as food, and smoked ape meat as a commodity; these additional hunting pressures are a serious threat associated with logging. Once industrial logging has run its course, moreover, plantations, farms, ranches, fires, and alien invasive species tend to enter the logged areas, creating a new ecosystem with few patches of native, closed-canopy forest, and therefore few or no great apes.

Pierre Kakulé Vwirasihikya

Bushmeat for sale at a Kasese market stall, Maniena province, Democratic Republic of the Congo.

HUNTING AND THE BUSHMEAT TRADE

Historical markets

Apes have long been valued as sources for a variety of traditional uses, including the production of 'charms'. Gorilla bones are an important ingredient of protective amulets, which are much sought after by local healers.[34] Chimpanzee and gorilla body parts are important elements in traditional medicines in West Africa[116] and in Congo.[128] In Sumatra, the staffs and wands of shamans were often decorated with orangutan hair.[131] In Borneo, where head-hunting traditions are both widespread and recent, orangutan skulls can fetch up to US$70 in towns in Kalimantan, apparently as a less illegal alternative to using human skulls (see also Box 13.3).

The trade in apes as pets and research animals has historically provided hunters with additional income. There was a booming international trade in orangutans between the 1930s and 1960s, although hunting them became illegal in 1924 in Sumatra and in 1931 in Sarawak.[131] In the 1980s, popular movies and television soap operas created a new demand for pet orangutans, especially in the Far East. An estimated 1 000 orangutans may have been imported into Taiwan for the pet trade between 1995 and 1999; the price of an individual ape in the private market was reportedly anything between US$11 000 and US$20 000. Much of this trade has now been stopped, but it is highly profitable where it does occur. There has also been a lucrative trade in infant bonobos, chimpanzees, and even gorillas as pets, for zoos, and for private collections; hundreds of chimpanzees were exported for biomedical research, a trade that continued into the 1980s on existing permits, despite restrictions under the Convention on International Trade in Endangered Species of Wild Fauna and Flora (CITES).[127] The profitability of this illegal trade continues.

Apes as food

Consumption of ape meat is limited by tradition in many places, sometimes because people regard apes as objects of reverence, or as too similar to themselves, or as unhealthy to eat.[155] Eating primates is forbidden under Islamic law, which has provided some protection in Sumatra and in parts of both Borneo and Africa. Human population movements and cultural exchanges have altered some of these traditions, however, as has the increasing market demand for bushmeat, which has turned hunting into a profitable activity (see Boxes 13.3 and 13.4).

For some communities, great apes have traditionally been a harvestable resource to be hunted for subsistence, or to meet market demands for meat and medicinal products. Prehistoric and modern indigenous peoples of Indonesia are known to have hunted orangutans extensively for meat and to have preferred ape meat to many other kinds of game.[131] The Batak people of Sumatra believe that eating the meat will make them strong, and this belief has prevailed into modern times. Throughout Central and West Africa some communities have traditionally eaten ape meat and there is a high market demand for it,[17] despite customary or legal prohibitions. Gorilla meat in particular is seen as a dish fit for powerful men in some cultures, so that a chief who did not serve it to visiting dignitaries would risk embarrassment. Gorillas are eaten by some who believe they will thereby gain strength; and chimpanzees are eaten in the hope of acquiring their luck and cunning.[103]

Scale of bushmeat trade

By 2000, the illegal bushmeat trade was estimated to be worth nearly US$1 billion annually,[11] a proportion of which involves great ape meat.[17, 85, 122, 164] It has been facilitated by the opening up of formerly remote areas by logging companies, especially in Central Africa,[5, 145, 147] where a single chimpanzee or gorilla carcass can fetch the equivalent of US$20–25.[4]

Box 13.3 HUMAN BELIEFS AND TRADITIONS

The cosmopolitan, rationalist, 'western' culture is not the only one with strong opinions about apes. The ethnographic record shows that traditions and beliefs about apes are widespread, with both positive and adverse results. In Borneo, many Iban believe that after death an ancestor becomes a member of a particular species (*tua*) and in this form can help his or her living descendants; and in the Batang Ai area of Sarawak, the orangutan is such a *tua* species.[72] Other peoples of Sumatra and Borneo saw orangutans as representatives of their war god, as former human rulers who had fallen from grace, as humans who refused to speak for fear of being enslaved, or as their forest cousins.[73] In the northern inner Congo Basin, bonobos represent a 'fallen brother' who is trying to become human again. In Liberia, the Wehdjeh clan of the Sapo people consider themselves to be relatives of the chimpanzee, from whom their knowledge of forest skills is derived,[114] while the Vili people of Congo hold that chimpanzees are reincarnations of people. Similarly, people in the Boé region of Guinea-Bissau believe that chimpanzees shelter the spirits of their elders, so killing them is taboo. In Côte d'Ivoire, many of the 60 ethnic groups have chimpanzees as their totem and refuse either to kill or to eat them.[69] The attitude of the Fang subtribes of Equatorial Guinea amounts almost to a taboo; they believe that eating apes, especially gorillas, causes infertility in women.[10] In 1891, it was reported that the Fang believed that a pregnant woman seeing a gorilla risked giving birth to one.[103] Some Fang tribes hold the gorilla to be their totem.

A skull for sale at a tourist gift shop in Bali.

Orangutan Foundation

These traditional attitudes reflect the special nature of apes, rather than their protection status. There are just as many cases of apes being killed as a result of tradition and belief as not. In Congo, the Kwélé, Kota, Mboko, and Djem peoples eat gorilla meat as part of a circumcision ritual for young men.[58, 103] In Guinea, chimpanzee blood is thought to cure epilepsy, and the meat is believed to strengthen young children.[89] In DRC, even in areas where the killing of apes is generally taboo, when a great ape is killed, the powdered bones and hair are added to the bathwater of babies to improve their strength and health. Although it is taboo to eat primate meat among many Islamic populations (in e.g. Sierra Leone, Senegal, Liberia, Mali, and northern Nigeria), ape body parts may still be used for medicinal purposes and witchcraft.[103, 124]

The former peoples of Borneo who engaged in head hunting (the Iban and other Dayaks) sometimes valued orangutan skulls as much as human ones as a source of spiritual energy for their long-house communities. The Batak people of Sumatra still believe that eating orangutan meat will make them strong.[131]

Many peoples have valued apes as simply another source of meat, which is "somewhat sweet, but with a nice taste."[108] This attitude has been spreading in the cities of Africa, where traditional belief systems have lost their power, but where the logging companies recruit their laborers. The loss of taboos and the movement into the great ape areas of people who no longer respect these traditions represents a significant threat to the great apes.

The local people near Wamba in DRC, for example, traditionally refrained from hunting bonobos for religious reasons [141] but, in the mid-1980s, poachers were recorded hunting bonobos for meat.[144] The 'bushmeat crisis' in West and Central Africa is a reflection of the spread along the new logging roads of a cultural and market system that enthusiastically consumes smoked ape meat; this is penetrating deep into areas where people once held very different beliefs. Nevertheless, many people in these regions still believe that these hominids (that can learn our languages, break sticks in the forest to indicate their direction of travel, build nests, use tools, and develop their own cultures) deserve better from the hominids with guns than to be killed, gutted, and smoked over a fire.

Julian Caldecott

Box 13.4 BUSHMEAT HUNTING AND TRADE IN SENDJE, EQUATORIAL GUINEA[91]

The village of Sendje is situated in Río Muni, or continental Equatorial Guinea, 42 km by road south of its main town, Bata, and about 10 km west of the Monte Alén National Park. During the country's time as a Spanish colony, many people in the area were employed in logging or on oil palm plantations, living in camps and villages scattered throughout the forest. After independence from Spain in 1968, these forest inhabitants were relocated forcibly to settlements along major roads, and the village of Sendje officially came into being.

Sendje's population has increased to around 400 people, but following the cessation of large-scale commercial logging and agriculture, job opportunities are scarce. However, the recent discovery of offshore oil in Equatorial Guinea's waters has meant that the country as a whole has taken an economic upturn. This oil boom has fuelled urban demand for fresh meat and fish, including bushmeat, creating conditions for a lucrative commercial trade in the meat of wild animals.[39] In addition, improved infrastructure and reduced transportation costs have meant that it is profitable to trade meat from increasingly remote locations. In Sendje, as in many other rural villages in the country, while women work in the fields cultivating the main crops of cassava and peanut for subsistence, men hunt and trap in the forest for income. The majority of the meat caught is sold in the markets of Bata.

Sendje is particularly well suited to commercial hunting, as it is the gateway to a large area of forest that is still relatively rich in wildlife. Old logging tracks lead into the forest, enabling easy access; abandoned villages and logging camps have been transformed into hunters' camps. Trapping on a subsistence level has been going on around the village for decades, but the recent increase in the scale of hunting and trapping activity (particularly in terms of number of trappers and total number of traps) has led to people traveling ever deeper into the forest in search of prey. The majority of commercial trapping and hunting now takes place well inside Monte Alén National Park, which means a trek of up to 30 km, or 10 hours, from Sendje.

Trapping is much more common than hunting with guns, as the entry costs are lower (wire versus a shotgun and expensive, unreliable cartridges) and fewer skills are needed. Around 75–80 trappers and hunters were recorded during a 15 month study. With a single exception all were male, and their ages ranged from 13 to 74 years. Only around 15 of these people used guns, and this was usually in conjunction with traps. More than 5 000 active traps were recorded in the area at any one time, with a mean of about 100 per trapper; some younger, more vigorous, and more commercial trappers operated up to 250 traps. Most said that they did not choose to trap or hunt through tradition or because they enjoyed it, but because it was their only means of earning money.

Mammal surveys show that prey has been severely depleted near Sendje. Densities of small- and medium-sized terrestrial species such as duikers (forest antelopes) and large rodents that are targeted by trapping, are low throughout the area. Primate species are also targeted by hunters with guns; primate density is much more dependent on the distance from the village and therefore the intensity of hunting with guns. Primate numbers are still quite healthy at the farthest hunting camp (inside the Monte Alén

The quantity of bushmeat harvested is related both to its availability and the cost of other protein sources (see also Box 13.4). Factory fishing by foreign vessels along the coast of West Africa, for example, has contributed to the decline of fish stocks.[18] This in turn has contributed to increased hunting pressures in nature reserves in Ghana, and to a decline of mammal populations there.[18] Conversely, the more fish that is available in local markets, the less bushmeat is sold.

The trade in bushmeat may involve about 3 million tons of wild meat annually in Central and West Africa, with consumption levels in rural areas ranging up to about 16 kg per person per year.[31] A similar consumption rate of around 12 kg per person per year was recorded in Sarawak in the mid-1980s,[26] and is generally consistent with data from other tropical locations where people eat bushmeat. However, consumption rates of up to 106 kg per person per year have been reported for hunter–gatherers in the central Congo Basin and in north Congo.[7, 88] In the Peruvian Amazon, bushmeat

National Park), but abundance is much lower near the village. Most conspicuously, the black colobus (*Colobus satanas*), a large-bodied, easily hunted species rated as Vulnerable by IUCN's Primate Specialist Group, is now virtually absent around the village.

A total of 8 396 animals were harvested in Sendje in 2003, of which the majority were ungulates (35 percent) and rodents (33 percent), belonging in particular to two species: the blue duiker (*Cephalophus monticola*, 30 percent) and the brush-tailed porcupine (*Atherurus africanus*, 28 percent). Primates made up only 10 percent of the animals taken, reflecting the relatively low and still recent use of guns. Only one chimpanzee and one gorilla were known to have been killed in 2003; over half of the primates taken were black colobus. In areas where hunting is intensive, black colobus rarely feature among animals taken or in market records, suggesting that the Monte Alén National Park area is still relatively rich in primates. The large number of hunted animals recorded overall, combined with the fact that hunters and trappers need to travel ever farther from the village in pursuit of prey, indicate that hunting and trapping at such levels cannot be sustained.

At present, gorillas and chimpanzees are hunted only opportunistically in this area. Both species are taboo for the Fang ethnic groups that live in Sendje and, although Fang people from other parts of Río Muni will eat it, gorilla meat in particular has a low value in local markets. In view of the fact that encounters with a gorilla group can be dangerous for humans, injury from a trap is currently a greater risk than the threat of being shot by a hunter for gorillas in the area.

In Bata, bushmeat is not necessarily preferred over other types of meat or fish, but fresh produce is preferred over the cheaper and more widely available frozen produce. As incomes rise with continuing economic development, more people will demand fresh meat and fish. Without greater provision of alternatives in the form of improved sustainable fisheries and livestock husbandry (or the development of other income-generating opportunities for rural people), and without the education of consumer markets, demand for bushmeat is liable only to increase. As yet, there is no active enforcement of hunting regulations (black colobus and great apes, for example, are supposed to be protected under Equatorial Guinean law) and hunting within protected areas continues unchecked. Unless urgent action is taken, neither the socio-economic nor the biological sustainability of the bushmeat trade in Sendje appears to be achievable.

Noëlle Kümpel

Smoked monkey, Equatorial Guinea.

Noëlle Kümpel

has been reported to be consumed at a rate of 19–168 kg per person per year.[35]

The national and international trade in live great apes also appears to have increased in recent years across many range states. The main causes are the availability of live orphans as a by-product of the bushmeat trade in West and Central Africa (principally affecting chimpanzees and bonobos because gorilla infants usually die before reaching a likely buyer), hunting related to logging in Southeast Asia (affecting orangutans), and poor economic conditions in many countries[131, 169] (see Chapters 16 and 17).

Great apes are also frequently injured or killed where snares are set for other species, usually medium-sized ungulates.[40, 52] Primate conservation projects sometimes include specific programs to patrol for snares, particularly in East Africa (e.g. in Kibale National Park, Uganda[124] and in the Virungas).[52] The presence of rangers dedicated to the removal of snares also deters other forms of illegal exploitation of the forest.

Nicole Simmons

A family living near Kanyanchu on the forest boundary, Kibale National Park, Uganda.

Table 13.5 Human Development Index rankings, 2002[149]

World rank[a]	Ape range state[b]
59	Malaysia
109	Equatorial Guinea
111	Indonesia
122	Gabon
131	Ghana
139	Sudan
141	Cameroon
144	Congo
146	Uganda
151	Nigeria
157	Senegal
159	Rwanda
160	Guinea
162	United Republic of Tanzania
163	Côte d'Ivoire
166	Angola
168	Democratic Republic of the Congo
169	Central African Republic
172	Guinea-Bissau
173	Burundi
174	Mali
177	Sierra Leone

a Out of 177 countries.

b Insufficient data to allow calculation for Liberia.

ECONOMIC CONSTRAINTS

The governments of the 23 great ape range states have generally been supportive of conservation efforts for the great apes, even though their resources are very limited, especially in Africa and Indonesia. This is shown dramatically by the ranking of these nations according to the Human Development Index (HDI) of the United Nations Development Programme (UNDP),[149] a widely accepted indicator of wealth and social wellbeing in national societies (Table 13.5). A low HDI rank implies a challenging development context that is affected by conflict, poverty, and demand for the extraction of natural resources, often at the expense of the environment.

The HDI can be computed for 177 countries, and of the 23 great ape range states, 21 are ranked between 109th (Equatorial Guinea) and 177th (Sierra Leone). The only exceptions are Liberia, which is not ranked, and Malaysia, which is ranked 59th. The last figure may mislead, as Malaysian Borneo (Sabah and Sarawak), where orangutans live, is significantly poorer and more rural than the urbanized Peninsular Malaysia. Despite this, Sabah and Sarawak are constitutionally responsible for managing their own forests and wildlife without necessarily receiving federal support to do so. The HDI analysis indicates that virtually all the range states have significant challenges that can make it difficult to undertake the organized, long-term social investments demanded by successful conservation, at least according to the prevailing model.

Great apes are protected by law in all their range states, and most protected areas include some great ape habitat and populations. Many of these protected areas are of world-class quality, having been selected as World Heritage Sites and/or Biosphere Reserves. They all represent the permanent setting aside in law of large areas that might otherwise have been used for logging, farming, mining, or other purposes, at significant opportunity cost to the countries concerned. This reveals the willingness of these societies to preserve their national patrimony, even where this implies additional economic hardship. To judge from events, the commitment of range states to their protected areas may far exceed that of the much wealthier donor community.

A single example is enough to make this particular point. A plan for the establishment and development of Cross River National Park in south-

Box 13.5 EBOLA AND GREAT APES IN CENTRAL AFRICA

Diseases that can be transmitted from animals to humans under natural conditions (zoonoses) have been in existence for as long as we have shared the Earth. Through the centuries, diseases such as the bubonic plague, rabies, tetanus, and measles, have crossed from animals to humans. Recently, zoonoses have captured international attention because Severe Acute Respiratory Syndrome (SARS) and Ebola outbreaks have taken their toll on human populations around the world. The Ebola hemorrhagic fever virus is only one of at least 100 infectious agents that humans and great apes have in common.

Ebola was identified in 1976.[81, 121] Since then, it has affected human populations at least a dozen times in six different countries of Equatorial Africa. In a recent epidemic in Central Africa, gorilla, chimpanzee, and human populations were hit hard.[156] In a small area in northwestern Congo, this devastating disease killed over 130 humans and was estimated to have killed half of a population of about 1 200 great apes. The absence of entire family groups of gorillas and chimpanzees during and following the outbreak was confirmed by laboratory testing of samples collected from gorilla carcasses. Some gorillas survived even after other members of the group had died of Ebola hemorrhagic fever, as had been observed in chimpanzees exposed to Ebola in Côte d'Ivoire.[51] It is now feared that Ebola may have spread into the Odzala-Koukoua National Park.[60]

The worst-case scenario for Ebola in great apes may have arisen in the Minkébé forest region of northeastern Gabon, where western lowland gorilla and chimpanzee populations came close to disappearing during outbreaks of human Ebola infection during 1994 and 1996.[74] Tens of thousands of gorillas and chimpanzees may have died from Ebola. No work was undertaken in the region during the human outbreaks on the collection of samples or on wildlife observations in the forest to determine conclusively whether or how Ebola affected the ape populations.

Prior to the 2002–2003 Ebola outbreaks in Congo, and in anticipation of an epidemic or disease event, the Wildlife Conservation Society worked together with the European Union program, Conservation and Rational Use of Forest Ecosystems in Central Africa (ECOFAC), to train management staff of national parks and protected areas in Congo and Gabon in techniques for conducting wildlife censuses, monitoring the health of wildlife, conducting postmortem examinations, and performing standardized data collections. The results demonstrated mortality of gorillas due to Ebola hemorrhagic fever; direct genetic linkages; the similarity of the virus afflicting both great apes and humans; and evidence for multiple, genetically distinct Ebola viruses circulating in the forest at the same time.[96] This suggested that there

continued overleaf

east Nigeria was prepared with the joint support of WWF–The Global Conservation Organization, the government of the United Kingdom, and the European Commission.[27, 28, 29] The park contains most of the remaining moist forests in Cross River state, and comprises two divisions: Oban (containing Nigeria-Cameroon chimpanzees, *Pan troglodytes vellerosus*) and Okwangwo (containing Nigeria-Cameroon chimpanzees and Cross River gorillas, *Gorilla gorilla diehli*). The plan was accepted by the Nigerian federal government in December 1989, which then asked the European Commission for assistance to implement it. In April 1990, a draft financing proposal was prepared by the European Commission delegation in Lagos, and was subsequently approved by the Commission in Brussels. A team from the European Commission and the German Development Credit Agency (Kreditanstalt für Wiederaufbau) visited Cross River state in June 1991 to assess the feasibility of the project and to amend its structure in preparation for putting it out to international tender. Consulting firms were shortlisted for the management contract in March 1993, and the contract was awarded in October 1993.

More than five years elapsed between the beginning of project planning and the beginning of project implementation. While this story gradually unfolded, the Nigerian government proceeded to fulfill its stated intention of creating the Cross River National Park. It achieved this despite its own political and economic difficulties, which included attempted coups d'état, factional riots, general strikes, and financial crises on an enormous scale.

might be multiple reservoir species. These trained field teams drew on essential, established relationships with local villagers and hunters, to enable them to detect and report Ebola fatalities in great apes months before the first human cases. Since the mid-1990s, the earliest cases of Ebola infection in humans, prior to wider outbreaks in Central Africa, have been analyzed. These have shown a link between the handling of Ebola-infected gorillas or chimpanzees and susceptibility to the disease. Reducing the frequency of this contact route could reduce the incidence of human

Ebola outbreaks in Central Africa

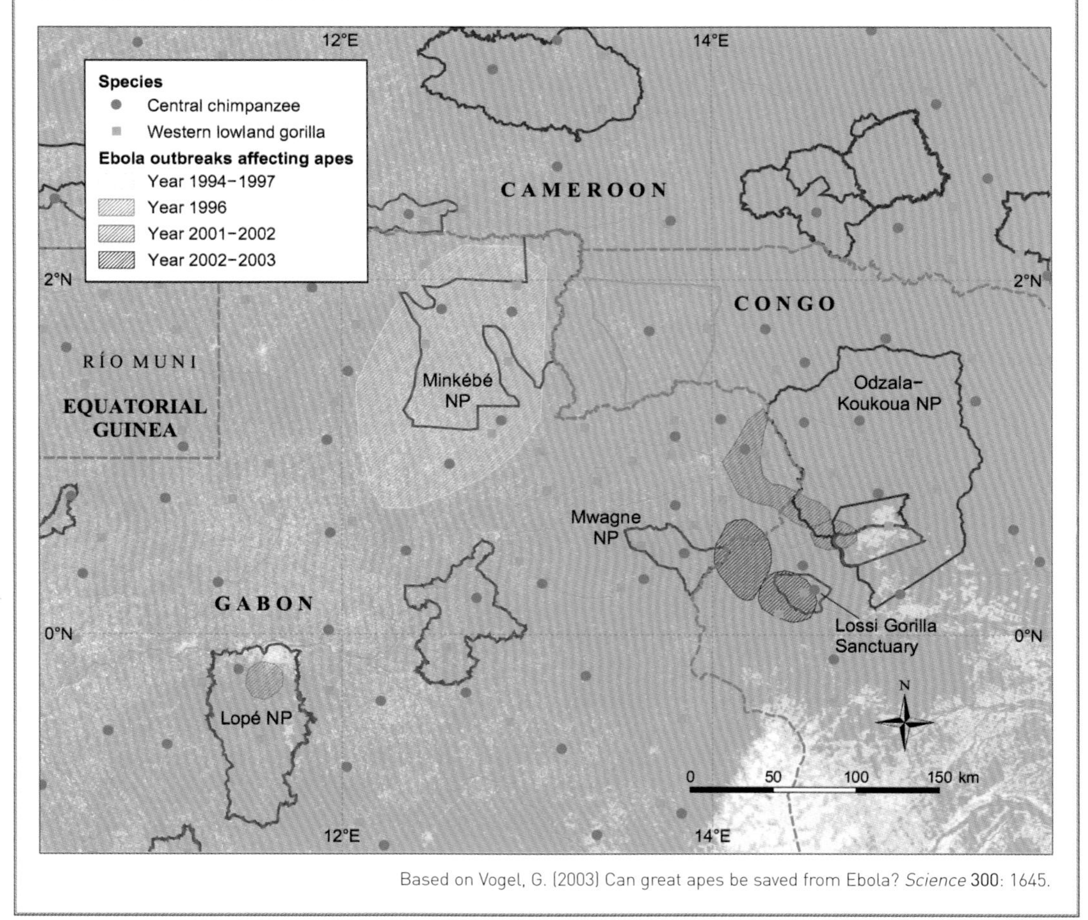

Based on Vogel, G. (2003) Can great apes be saved from Ebola? *Science* 300: 1645.

Despite these interruptions, and precisely on schedule, the Federal Nigerian Council of Ministers approved Cross River National Park in October 1989, leading to Presidential Decree No. 36 of 1991, which legally created the national park on October 2 1991.

Fortunately, the threats faced by the park were less acute than had been thought when WWF assessed them initially, and the most significant effect of the delay was to undermine local trust in the conservation process as financed by outsiders. Had WWF been better prepared for the slow pace of international activity, it would have been possible to avoid raising expectations in the project area. Rather than investing solely in planning for early implementation, a program might instead have been undertaken involving basic conservation work around the obvious priorities, which were and remain: "to protect the forest; to maintain lines of communication between the people affected by that protection and the people doing the protecting; and to help both sides understand the ecological limits of their environment and how to live the best possible lives without exceeding those limits."[27]

outbreaks, while also discouraging the hunting of gorillas and chimpanzees.

Work is still underway to identify the natural reservoir (or reservoirs) of the Ebola virus. Some species of fruit bats and insectivorous bats can survive the infection and then shed the virus in their excrement.[140] Fieldwork in CAR has demonstrated the presence of at least fragments of Ebola viral particles in a number of rodent species.[110] Similar work in Congo has found the same in some bats.[95] These findings should be interpreted with caution as the techniques used determine only the presence of the genetic material; this does not always indicate the presence of live or viable virus. Other studies have shown the presence of antibodies to Ebola in apparently healthy humans and primates in Central Africa, indicative of previous exposure having been successfully overcome by an immune response.[95, 110]

The scientific evidence to date suggests that Ebola is widespread in Equatorial Africa and persists in nature between observed outbreaks in humans. It is believed to have one or more natural hosts, in which it probably causes minimal disease problems at the population level. The conditions for the transfer of Ebola virus to other, more vulnerable, species are unknown. Rather than being a virus of deep forest refugia, Ebola may be more common at forest peripheries, in fragments, and in mosaics.[110] This could reflect the preferred habitats of the reservoir species or the type of habitat in which transmission between species is more likely to occur. Changes in climate or vegetation patterns may alter ecological relationships between animal populations and promote the transfer of the virus, as has been observed with other viral diseases.

In order to understand better the disease caused by the Ebola virus and to develop methods to prevent its spread both in humans and wildlife (as well as to understand and prevent the effects of other diseases on great apes) the following objectives need to be addressed:

- anticipate Ebola outbreaks and populations at risk in order to provide better support to areas that could be affected by the virus;
- establish monitoring teams to determine the existence and progression of the Ebola virus and other serious infectious disease agents in the forest, and their impact on wildlife (affected species, mortality rates, resistance, natural barriers, etc.);
- establish response plans to alert appropriate people to reports of the presence of Ebola hemorrhagic fever and other diseases;
- improve knowledge of the Ebola virus and its ecology (reservoir, mode of transmission between and within species, strains, immunity);
- evaluate ways of reducing the effect of Ebola and other infectious diseases on great apes using techniques such as vaccination programs, separation of reservoir species and affected species in time and space, development of approaches to management of metapopulations (i.e., that are connected by dispersal across the landscape), and other strategies for preventive medicine and improved hygiene practices; and
- using Ebola hemorrhagic fever as an example of disease risk, improve local community education and awareness campaigns, with a view to reducing human contact with and the hunting of great apes.

William Karesh and Patricia Reed

POLITICS AND CONFLICT

Wars and civil conflict have greatly affected a number of great ape range states since 1990, and have had a significant impact on local ape populations (see Map 13.1). In West Africa, major conflicts have occurred in Guinea-Bissau, Sierra Leone, Liberia, and Côte d'Ivoire; in Central Africa, in DRC, Congo, Angola, and CAR; in East Africa, in Sudan, northern Uganda, Burundi, and Rwanda; and in Southeast Asia, in Indonesia, particularly in northern Sumatra where Acehnese separatists have been battling the Indonesian government for many years. Armed conflicts increase the availability of guns, displace people from their homes and farms, and reduce agricultural production; all of these factors can increase hunting levels and illegal logging as people struggle to survive.[36, 37] Much of the money for international conservation comes from bilateral grants, funds that may be frozen when security deteriorates. This can close conservation projects and cause the loss of experienced project staff (e.g. as happened in the Virungas in the mid-1990s). Attempts have often been made, however, to maintain at least

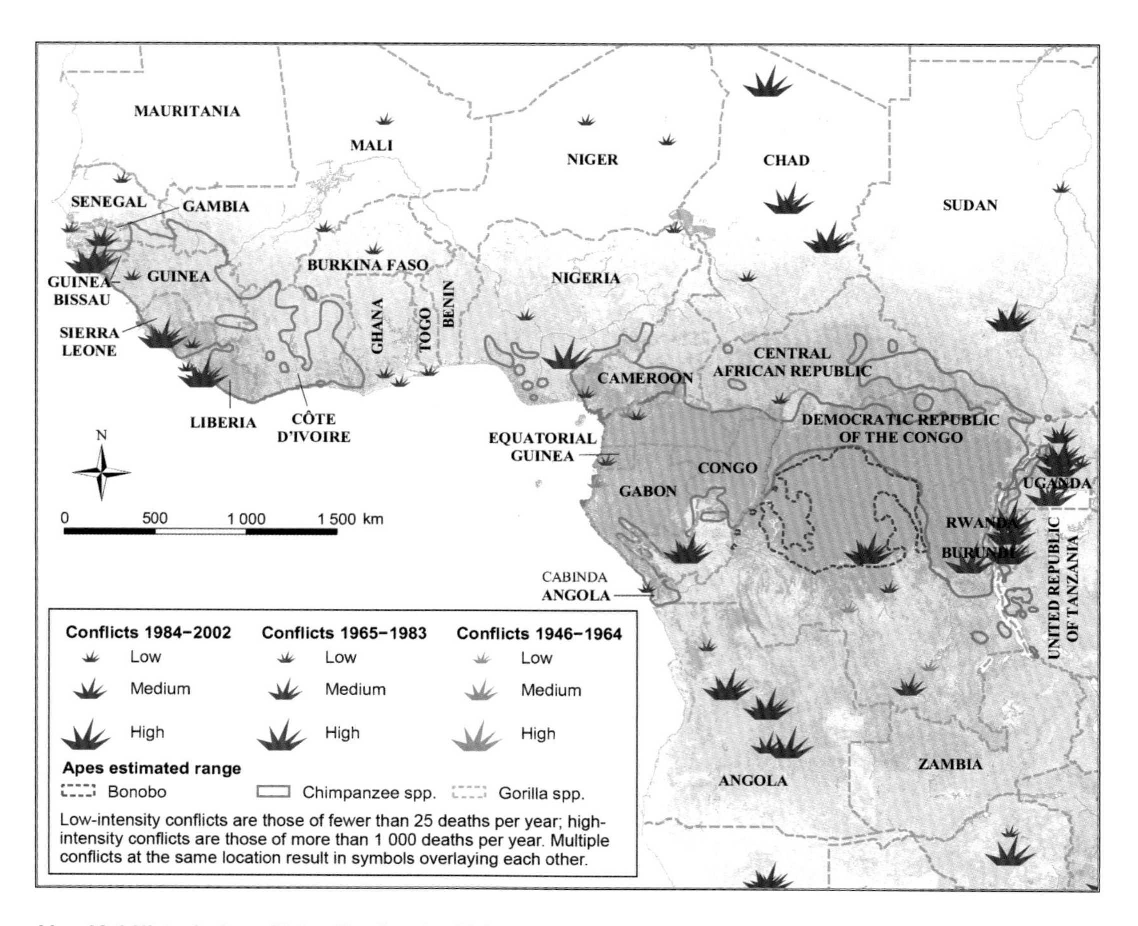

Map 13.1 Historical conflicts affecting the African great ape range states

some activities at great ape sites during periods of conflict.[84, 124, 143]

War kills great apes as well as people, and armed conflict and political unrest have taken their toll on both the eastern lowland and mountain gorilla. Refugees, displaced as a result of conflict, can often put massive pressure on forests and gorilla habitat by uncontrolled harvesting of wood for fuel and of animals for bushmeat. During the war in Rwanda, three of the four refugee camps in North Kivu, DRC, were located in or near to the Virunga National Park. Subsequent conflict in DRC led to looting and destruction of the park infrastructure, and to the deaths of several workers at the Karisoke Research Center in Volcanoes National Park, as well as of many mountain gorillas.[83, 84.]

Conflict diverts revenues to arms and away from social investment in conservation or poverty relief, and deters responsible private investment. Some less scrupulous companies take advantage of the chaos to extract valuable resources, sometimes participating with local military and political actors in maintaining a conflict.[151] Military engagement impoverishes nations and may tempt governments to encourage the extractive industries as a means of liquidating natural resources for financial gain.[37] This is disastrous for great apes and their habitats, as well as for people (especially the poor), although the stifling of economic activity, depopulation, and chaos brought by conflict can occasionally relieve short-term pressures on the environment by, for example, deterring logging companies.

The long-term impacts on great apes of the recent wars in Central Africa are unclear. The return of peaceful conditions, however, is likely to lead to increasing trade in most areas, which may allow more traffic in bushmeat. The purpose of hunting would then be expected to shift from subsistence to profit, but would keep ape populations under pressure. Peace may bring other dangers, including an expansion of industrial-scale logging, mining, and forest clearance for farming, all promoting access and the spread of hunting to new areas.

Less violent political change can also impact ape conservation adversely. For several years after the fall of the Suharto government in 1998, conservation in Indonesia virtually collapsed, and deforestation increased markedly, even in protected areas.[71, 79, 134] Governments rarely have sufficient political will, commitment, or capacity to undertake effective conservation work, which is seldom seen as a high priority. This is especially true in war-torn states. Sudan, for instance, is beginning to emerge from a 30 year conflict that has killed or displaced millions of people, and the country is expected to be engaged in national reconciliation and rebuilding for the next few years, although the conflict and ethnic cleansing in the western province of Darfur that erupted in 2004 may disrupt these plans. Conservation of that country's remnant chimpanzee population is not likely to be viewed as a priority in these circumstances, and Sudan has not yet opted to join the Great Apes Survival Project.

Politicians in many developing countries are beginning to address the impact on natural resources of mushrooming human populations. The great ape range states contain countries with some of the largest (e.g. Nigeria, Indonesia) or densest (e.g. Rwanda) populations on Earth, and some with the shortest life expectancies (e.g. Côte d'Ivoire) and the lowest gross domestic product (e.g. Burundi). Poverty is rife in many of these countries, which places additional burdens on governments and the management of natural resources. Ultimately, the fate of the great apes is linked to the future of the people in their range states, and particularly their capacity to stabilize their populations and meet their social and economic development needs. These solutions will need to be based on wise use of natural resources and respect for other species if great ape populations are to survive.

DISEASE

As closely related animals, humans and great apes can infect one another with a wide range of diseases and parasites. Being far more mobile and in contact with a global spectrum of pathogens, humans are much more likely to introduce illness to small, isolated populations of apes than the other way round. The potential for transmission in the opposite direction is however a strong argument against the consumption of primate meat.

Disease transmission also risks undermining one of the few success stories of great ape conservation – the use of great apes in tourism.[25] Close contact with a group of habituated gorillas or orangutans is an experience for which tourists will pay considerable amounts; tourists are, however, travelers and are therefore most likely to expose apes to new pathogens. A number of cases of cross infection are known among the mountain gorillas of the Virungas, including an outbreak in 1988 suspected to have been of measles or a similar morbillivirus, which killed six habituated females;[166] an epidemic of bronchopneumonia in

William Karesh

A conservation team collecting samples for Ebola diagnostics in Congo.

Elizabeth A. Williamson

Goldminers in Nyungwe National Park, Rwanda.

1990, which affected 26 of the 35 gorillas in a group that had been exposed to tourists, killing two gorillas;[97] and a debilitating skin disease that affected all four members of a gorilla group which had been habituated for tourism, and killed one of them. In this last case, however, the infection was contracted from exposure to local people rather than to tourists.[82]

The continued daily exposure of apes to large numbers of people and their diseases is considered a major potential threat to great ape tourism operations,[23] although this has not yet jeopardized the survival of whole populations of great apes. In contrast, Ebola hemorrhagic fever, one of the most virulent viral diseases known, causes death in 50–90 percent of all afflicted humans and apparently produces an even higher mortality rate among gorillas.[74] This virus was first identified in 1976 after outbreaks in northern DRC and southern Sudan.[162] It is transmitted by direct contact with the blood, body fluids, and tissues of infected mammals. Ebola infections in people have been linked to direct contact with gorillas, chimpanzees, monkeys, forest antelopes, and porcupines found dead in the rain forest (see Box 13.5).

MINING AND OIL

The extraction of minerals and oil causes habitat loss and degradation in a limited area; apes are mainly affected by the associated infrastructure development. Roads and pipeline routes allow hunters to access previously little exploited forests. Oilfield workers and miners frequently consume bushmeat, and may sell it on to supplement their incomes. Major mines and oilfield developments at least have the virtue that they tend to be managed by large companies which, if they wanted to, have the capacity to regulate hunting and wildlife trade in their areas of operation.

The coltan boom in DRC provides an example of a poorly controlled process that caused great damage to great ape populations.[68, 129] Coltan is a black alluvial ore of columbium and tantalum, which is found in riverine deposits in Central Africa. After the tantalum has been extracted, it is used in electronic capacitors, particularly those found in miniaturized equipment such as mobile telephones, laptop computers, and games consoles. Coltan panning, like gold panning, is often undertaken by individual freelance artisanal miners.

Between 1998 and 2003, the war in DRC is

Figure 13.1 Impacts of infrastructure development in tropical Africa

This map is based on GLOBIO 2 analyses for the year 2000. Black indicates a likely high loss in species diversity, red a medium–high loss, and yellow a low–medium loss. Green areas experience low impact of infrastructure development.

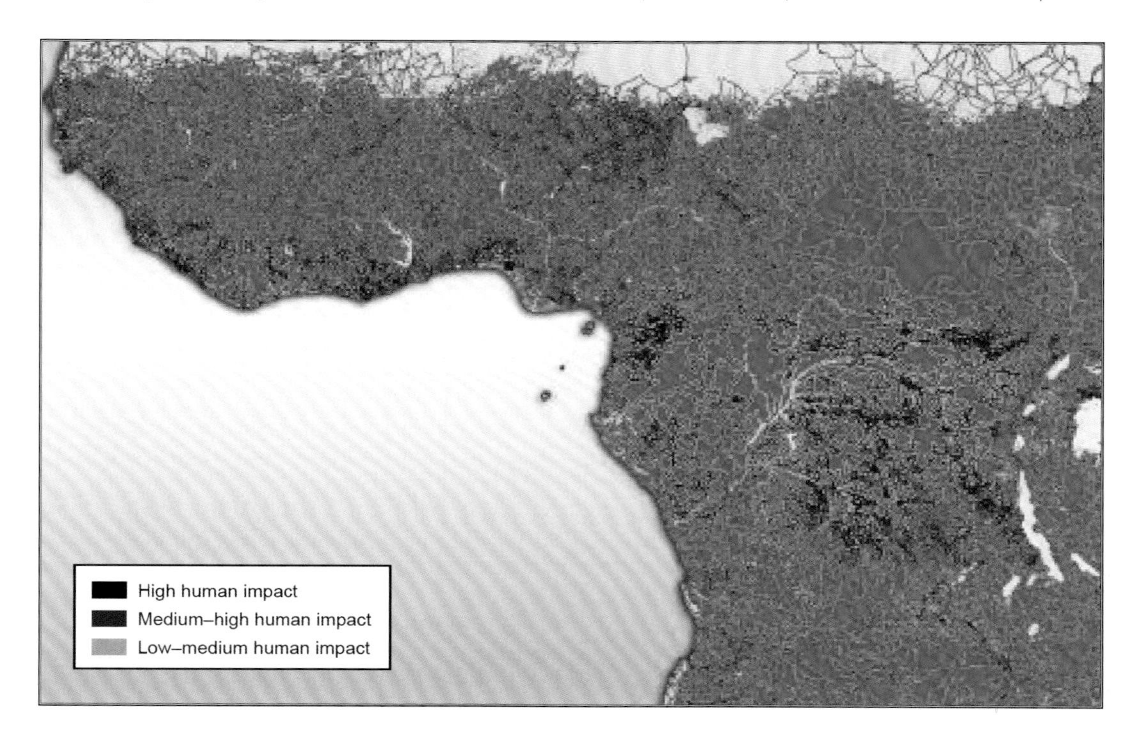

believed to have killed some 3 million people, either directly or indirectly from the effects of displacement. This coincided with a global boom in the coltan market in 2000, followed by a slump in 2001. In DRC, the boom led to a rush of miners in those poorly protected national parks that hold the mineral. Kahuzi-Biega National Park was occupied by about 10 000 people, and Okapi Wildlife Reserve by some 3 000. These included independent workers, their families, and others working as forced labor. By December 2000, 300 professional hunters armed with automatic rifles were working in Kahuzi-Biega National Park, feeding the miners on local wildlife. It is suspected that most of the then population of 8 000 eastern lowland gorillas in the park were eaten, along with all the elephants, and many other animals and birds. A series of reports from the United Nations Security Council and nongovernmental organizations brought the situation to global attention, concluding that the war itself was fuelled by revenue from coltan mining. Various companies and the armies of neighboring countries were noted to have participated in or benefited from the exploitation of DRC's natural resources.[151, 152]

Electronics companies were taken unawares by public dismay at the reported contribution of popular consumer products to both the war and the wildlife slaughter in DRC. Many have since declared that they will avoid purchasing Congolese coltan.[68] There are still exports from DRC, however, and the supply chain is not well audited so it is difficult to assess where this material ends up. There are now moves to legitimize and control the production and trade of coltan in DRC under a mining code developed by the DRC government and the World Bank. During the coltan boom years, the eastern lowland gorilla population may have collapsed from around 17 000 to fewer than 4 000 animals, but insecurity in the region has not yet allowed surveys to confirm this decline.

THREATS AND POSSIBLE FUTURES

Multiplying threats

Great apes are often confronted simultaneously by multiple threats. Assessing their combined effect can be challenging, as these threats often interact with each other as well as with the apes themselves. The simplest way to assess the level of threat posed by a combination of pressures is to

Figure 13.2 Infrastructure development projected for the year 2030

This image reflects GLOBIO 2 scenarios. Black indicates a likely high loss in species diversity, red a medium–high loss and yellow a low–medium loss. Green areas are likely to experience low impact of infrastructure development.

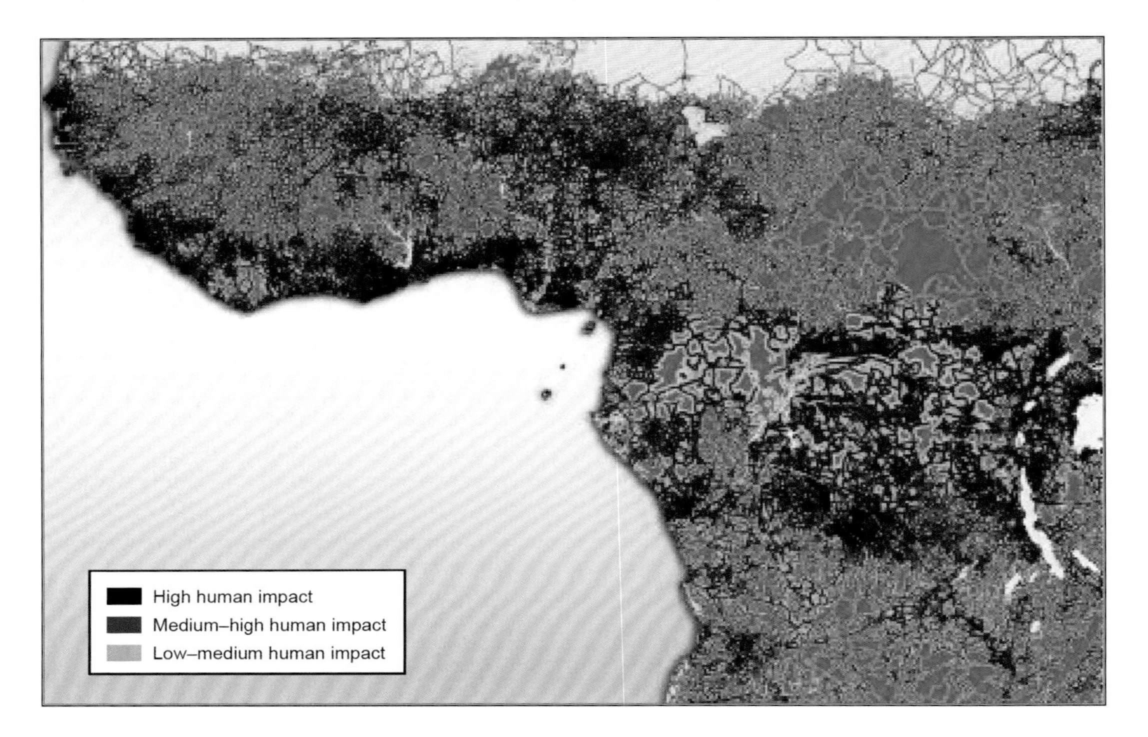

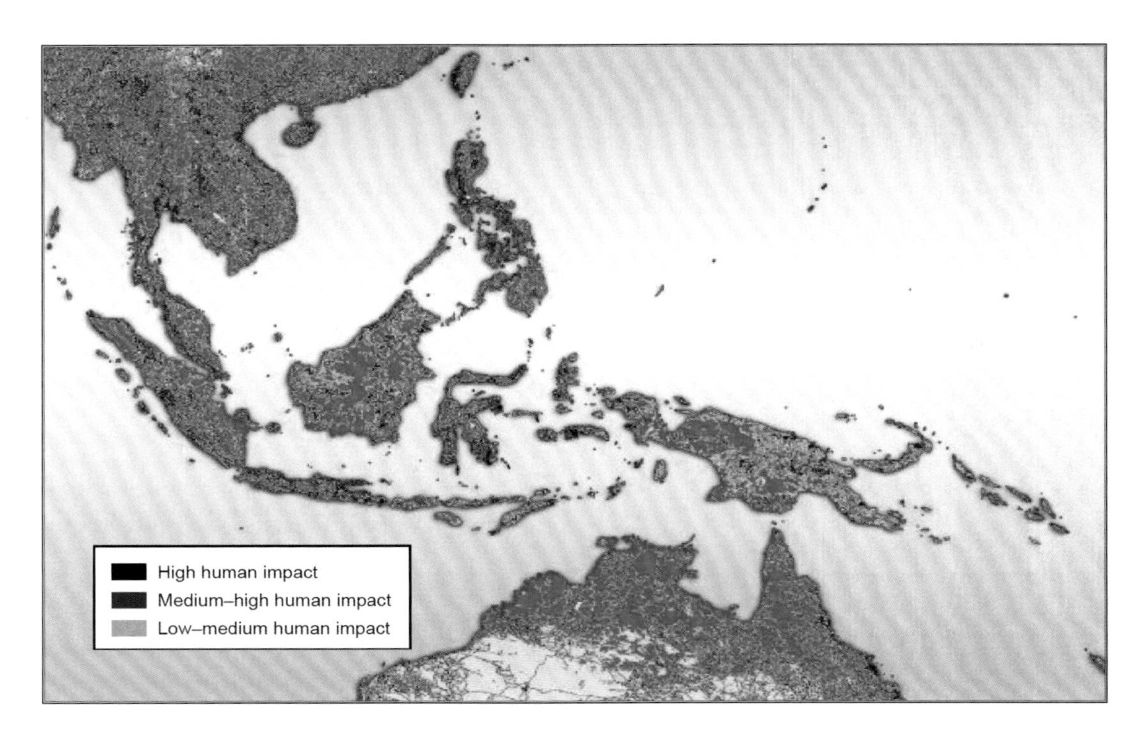

Figure 13.3 Impacts of infrastructure development in Southeast Asia
This map is based on GLOBIO 2 analyses for the year 2000. Black indicates a likely high loss in species diversity, red a medium–high loss, and yellow a low–medium loss. Green areas experience low impact of infrastructure development.

add together the pressures at each location. This can be done by overlaying digital maps of the various pressures over the entire range of a species, using geographic information system (GIS) software. This allows the number of serious pressures at a location to be assessed or ranked, and a combined score to be calculated.

One study employed a ranking system from 1 (least threat) to 4 (greatest threat) to compare the current situation for various pressures in three parts of CAR. Logging, mining, hunting, agriculture, and human presence were calibrated and the results were summed. The combined threat levels for the two sectors of the Dzanga-Ndoki National Park, Dzanga and Ndoki, and the Dzangha Sangha area were determined to be 1.8, 1.5, and 2.4 respectively.[15] Roads were considered one of the most important threats in all three areas, with moderate levels of logging and mining also posing a threat in the Dzangha Sangha area.

Infrastructure development

Roads play a central role in the loss of great apes and of tropical moist forest biodiversity in general. They provide access to mining and logging companies, fragment habitat areas, and facilitate both the transport of bushmeat and access by poachers and settlers.[2, 50, 90, 93, 134, 164]

Outside the few reserves with effective law enforcement, the only places where apes are relatively safe are in very remote areas, swamps, or places where there is powerful local support for protection, such as in the Lac Télé/Likouala-aux-Herbes Community Reserve in northern Congo.[126] Habitat loss mainly takes place through agricultural expansion and burning along road corridors, logging for timber and pulp, and around mining operations.[30] These roads, together with the extractive industries, result in 'boom towns' without sufficient local food supply; this leads to increased demand for bushmeat, which represents a significant income-generating opportunity for families with the lowest incomes.[16, 33, 104] The relationship between roads, extractive industries, and increased bushmeat trade has been confirmed by numerous studies.[19, 106, 117, 142, 148, 164]

The mapping of road networks, settlements, and mining operations can help to indicate areas of probable habitat loss and the degree of exposure of great apes to hunting. On this assumption, the

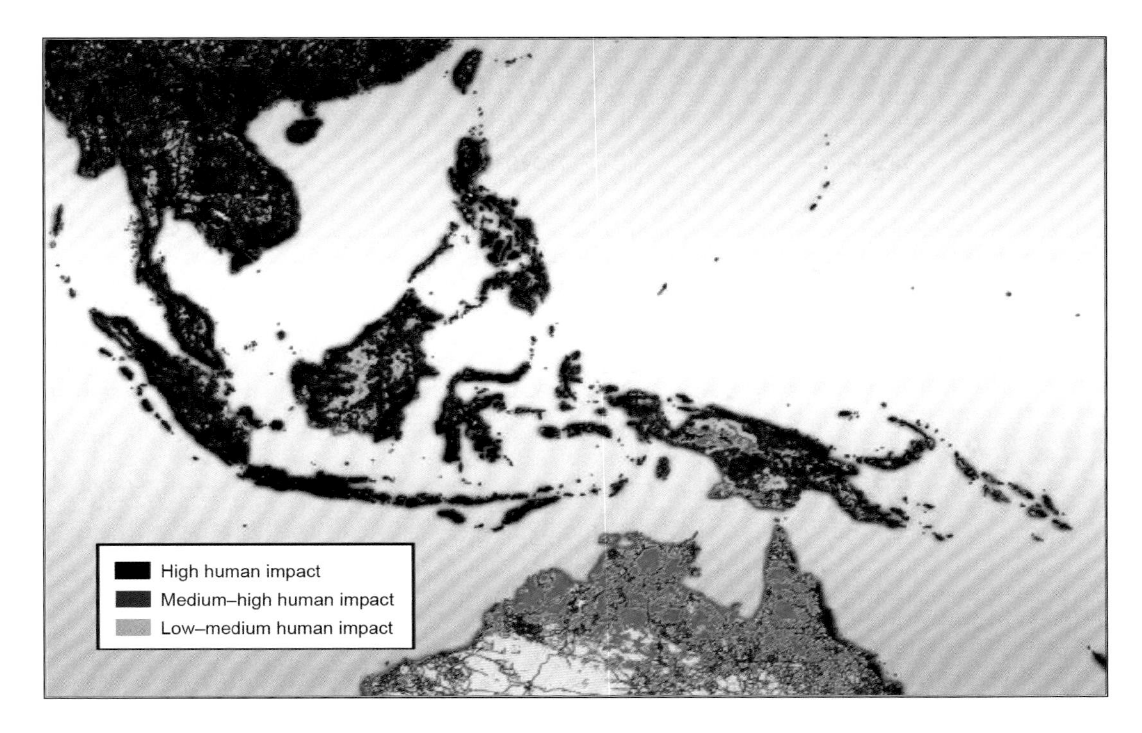

Figure 13.4 Infrastructure development projected for the year 2030
This image reflects GLOBIO 2 scenarios. Black indicates a likely high loss in species diversity, red a medium–high loss, and yellow a low–medium loss. Green areas are likely to experience low impact of infrastructure development.

GLOBIO computer model was developed for the United Nations Environment Programme (UNEP) to help assess and map the environmental impact of human development.[150] This is used to estimate the area of land with reduced biodiversity and abundance of living organisms following infrastructure development. It can also be used to predict the impacts of proposed developments by helping to visualize the zones around roads, major trails, settlements, dams, etc., where there is likely to be a reduction in the abundance of wildlife.[113]

Four zones of impact on biodiversity are defined using this tool:

- high impact (i.e. the area within which more than 50 percent of all recorded species decline by more than 50 percent);
- medium–high impact (i.e. the area within which 25–50 percent of all recorded species decline by more than 50 percent);
- low–medium impact (i.e. the area within which 1–25 percent of all recorded species decline by more than 50 percent); and
- low impact (areas falling beyond the above zones).

The high-impact zone denotes a belt 1 km wide surrounding roads and towns; such an area is generally heavily used by people, and is characterized by logging, farmland, settlements, and very few great apes. Medium- to high-impact zones lie 1–3 km from roads and settlements. This is a typical operating radius for log-skidding (dragging logs to their initial destination), and is where logging is usually most intensive; great apes are known to decline dramatically in areas subjected to intense logging, largely as a result of hunting.[30, 48, 99] Gorillas, chimpanzees, and orangutans have all been shown to use food resources within logged forests, providing that they are not being hunted.[92, 148] Low- to medium-impact zones occur 3–10 km from roads; these relatively intact forests are often subject to heavy hunting pressure. Areas free from hunting and logging generally contain much higher ape population densities.[13, 30, 48, 159] Due to the need to carry loads of meat through the forest, very few hunters will move beyond 10 km from roads; the most intensively hunted zone is typically within 3–8 km of roads in logged areas.

It is possible to simulate future changes in the distribution of these impact zones by using simple

Table 13.6 Projected great ape population declines, 2004 *Red List*

Taxon	Timeframe	Projected rate of decline
Eastern lowland gorilla (*Gorilla beringei graueri*)	2000 + 3 generations	50 percent[a] [20]
Central chimpanzee (*Pan troglodytes troglodytes*)	2000 + 3 generations	50 percent[22]
Eastern chimpanzee (*P. t. schweinfurthii*)	2000–2060	50 percent[b] [21]

a Decline is projected based on recent major losses in the Highlands sector of the Kahuzi-Biega National Park.

b Decline is projected based on habitat loss in East Africa and instability in DRC leading to continuing hunting pressure.

assumptions about the growth and dispersion of infrastructure, for example that it will expand fastest in areas with large human populations, close to natural resource concentrations, near coasts, and around existing infrastructure. This analysis has led to the following conclusions.[113] Less than 30 percent of the habitat of each of the African great apes is currently classified as under low impact from the indirect effects of infrastructure development (Figure 13.1). The future annual rate of degradation of such habitat was projected to exceed 2 percent per year, with 10 percent or less of their habitat remaining in the low-impact category by 2030 (Figure 13.2). These results suggest that great ape habitats will decline rapidly in coming years if current trends continue. Meanwhile, less than 36 percent of orangutan habitat is currently classified as subject to low impact from the indirect effects of infrastructure development (see Figure 13.3). Future scenarios suggest that the annual loss of such habitat will be about 5 percent per year, with less than 1 percent remaining in the low-impact category by 2030 (see Figure 13.4). These figures are consistent with published estimates that 47 percent of the orangutan habitat within protected areas will be lost by 2010.[98]

THE ROAD AHEAD

Some great ape taxa are clearly more threatened than others, but all have experienced or are expected to experience sufficiently worrying declines to merit inclusion as Endangered or Critically Endangered on the 2004 IUCN *Red List*.[77] The threats to great ape survival are difficult to tackle in isolation. Although some excellent models of conservation exist and there is a substantial body of research on the pressures and their impacts, it will take a huge effort to ensure the survival of the more endangered ape species and subspecies.

Protected areas form the core of the conservation strategies of many countries. Good management is crucial for the ape populations living within them. Some argue that protected areas are the best targets for conservation investment as they are well defined units that already exist in national law. Most great apes live outside protected areas, however, and a more inclusive strategy is needed to ensure their survival in the broader rural landscape. Long-term solutions rely on a change in attitudes among the populations and governments of great ape range states. This shift in thinking is required to limit hunting and land-use change within great ape habitats, and to ensure that conservation and development resources are available and appropriately co-managed. Conservation is the theme of the remainder of this volume.

FURTHER READING

Brashares, J.S., Arcese, P., Sam, M.K., Coppolillo, P.B., Sinclair, A.R.E., Balmford, A. (2004) Bushmeat hunting, wildlife declines and fish supply in West Africa. *Science* **306**: 1180–1183.

Formenty, P., Boesch, C., Wyers, M., Steiner, C., Donati, F., Dind, F., Walker, F., Guenno, B.L. (1999) Ebola virus outbreak among wild chimpanzees living in a rain forest of Côte d'Ivoire. *Journal of Infectious Diseases* **179** (Suppl 1): S120–S126.

Galdikas, B.M.F., Briggs, N.E., Sheeran, L.K., Shapiro, G.L., Goodall, J., eds (2001) *All Apes Great and Small, vol.1: African Apes.* Kluwer Academic/Plenum Publishers, Boston, Dordrecht, London, Moscow, New York.

Harcourt, A.H. (1996) Is the gorilla a threatened species? How should we judge? *Biological Conservation* **75** (2): 165–176.

Harcourt, A.H., Parks, S.A. (2003) Threatened primate taxa experience high human densities: adding an index of threat to the IUCN Red List criteria. *Biological Conservation* **109** (1): 137–149.

Hayes, K., Burge, R. (2003) *Coltan Mining in the Democratic Republic of Congo: How Tantalum-using Industries Can Commit to the Reconstruction of the DRC.* Fauna and Flora International, Cambridge, UK. http://www.gesi.org/docs/FFI%20Coltan%20report.pdf. Accessed March 18 2004.

Huijbregts, B., De Wachter, P., Obiang, L.S.N., Akou, M.E. (2003) Ebola and the decline of gorilla *Gorilla gorilla* and chimpanzee *Pan troglodytes* populations in the Minkébé Forest, north-eastern Gabon. *Oryx* **37**: 437–443.

IUCN (2004) *2004 IUCN Red List of Threatened Species.* IUCN–The World Conservation Union, Gland, Switzerland. http://www.redlist.org. Accessed February 28 2005.

Kalpers, J., Williamson, E.A., Robbins, M.M., McNeilage, A., Nzamurambaho, A., Lola, N., Mugiri, G. (2003) Gorillas in the crossfire: population dynamics of the Virunga mountain gorillas over the past three decades. *Oryx* **37** (3): 326–337.

Leroy, E.M., Rouquet, P., Formenty, P. (2004) Multiple Ebola virus transmission events and rapid decline of Central African wildlife. *Science* **303**: 387–390.

Nellemann, C., Newton, A. (2002) *The Great Apes – The Road Ahead.* UNEP/GRID Arendal, GRASP, UNEP-WCMC. http://www.unep-wcmc.org/resources/publications/GrASP/GRASP_5_complete.pdf. Accessed November 28 2004.

Peterson, D. (2003) *Eating Apes.* California Studies in Food and Culture, 6. University of California Press.

Redmond, I. (2001) *Coltan Boom, Gorilla Bust.* A report for the Dian Fossey Gorilla Fund and Born Free Foundation. http://www.bornfree.org.uk/coltan/coltan.pdf. Accessed March 18 2004.

Robertson, J.M.Y., van Schaik, C.P. (2001) Causal factors underlying the dramatic decline of the Sumatran orangutan. *Oryx* **35**: 26–38.

Thibault, M., Blaney, S. (2003) The oil industry as an underlying factor in the bushmeat crisis in Central Africa. *Conservation Biology* **17** (6): 1807–1813.

UNDP (2004) *Human Development Report 2004.* http://hdr.undp.org/reports/global/2004/. Accessed November 2004.

Walsh, P., Abernethy, K.A., Bermejo, M., Beyers, R., De Wachter, P., Akou, M.E., Huijbregts, B., Mambounga, D.I., Toham, A.K., Kilbourn, A.M., Lahm, S.A., Latour, S., Maisels, F., Mbina, C., Mihindou, Y., Obiang, S.N., Effa, E.N., Starkey, M.P., Telfer, P., Thibault, M., Tutin, C.E.G., White, L.J.T., Wilkie, D.S. (2003) Catastrophic ape decline in western equatorial Africa. *Nature* **422**: 611–614.

Wilkie, D., Shaw, E., Rotberg, F., Morelli, G., Auzel, P. (2000) Roads, development and conservation in the Congo Basin. *Conservation Biology* **14**: 1614–1622.

MAP SOURCES

Map 13.1 Conflict data are based on the following sources:

Buhaug, H., Gates, S. (2002) The geography of civil war. *Journal of Peace Research* **39** (4): 417–433.

Gleditsch, N.P., Wallensteen, P., Eriksson, M., Sollenberg, M., Strand, H. (2002) Armed conflict 1946–2000: a new dataset. *Journal of Peace Research* **39** (5): 615–637.

For protected area and other data, see 'Using the maps'.

ACKNOWLEDGMENTS

Many thanks to Thomas Butynski (Conservation International), Alexander Harcourt (University of California, Davis), Ian Redmond (Ape Alliance/GRASP), and David Woodruff (University of California, San Diego) for their valuable comments on the draft of this chapter. Thanks also to Tonya Lander, Nigel Varty, Sarah Ferris, and Valerie Kapos (all of UNEP-WCMC) for research into the literature supporting this chapter.

AUTHORS

Lera Miles, UNEP World Conservation Monitoring Centre

Julian Caldecott, UNEP World Conservation Monitoring Centre

Christian Nellemann, UNEP/GRID Arendal

Box 13.1 Alexander H. Harcourt, Department of Anthropology, University of California, Davis

Box 13.2 Lilian Pintea, Center for Primate Studies, Jane Goodall Institute, University of Minnesota

Box 13.3 Julian Caldecott, UNEP World Conservation Monitoring Centre

Box 13.4 Noëlle Kümpel, Zoological Society of London, and Imperial College London

Box 13.5 William Karesh and Patricia Reed, Field Veterinary Program and Wildlife Conservation Society

CHAPTER 14

Conservation measures in play

NIGEL VARTY, SARAH FERRISS, BRYAN CARROLL, AND JULIAN CALDECOTT

The emerging threats to tropical moist forests in general and great apes in particular have not gone unopposed by people and governments around the world. A host of measures have been discussed and implemented with increasing urgency and with the investment of ever greater resources. These measures have included intergovernmental agreements of various kinds that aim to encourage conservation planning, to regulate trade, and to promote transfrontier cooperation in the management of protected areas and wildlife populations. International nongovernmental organizations (NGOs), national NGOs, bilateral and multilateral donors, and range-state governments have all cooperated on interventions that aim to establish protected areas, improve their management, and engage with people living in and around them. The goal is to encourage and enable improvements in living conditions and compliance with laws that protect great apes and their habitats. These stakeholder groups exert various kinds of influence that support great ape conservation, as discussed in this chapter.

Collaboration among concerned groups has been tentatively extended to include private corporations whose investments in the timber, mining, energy, and infrastructure sectors have all impacted great apes in the past and are likely to do so in the future. Tourism ventures have been established to promote the attractiveness of great apes to nature-oriented visitors for the benefit of conservation. Sanctuaries for great apes confiscated from hunters and traders have multiplied. These are being used skillfully for public education and, in places, to support the reintroduction of great apes to wild habitats where they are likely to be safe. Where empowered to do so, some communities in great ape habitat areas have set aside their own lands as forest refuges in the hope of having a share in revenues from tourism, as well as deriving environmental benefits from the protected forests.

Meanwhile, the scientific community has steadily added to our knowledge of the great apes and their ecosystems, making it easier to communicate the unique attributes of these animals, deepening the interest and excitement felt by all those who value them, and providing guidance to decision makers on how and where best to make investments in great ape conservation. This chapter explores these varied initiatives, in the process telling much of the story of human efforts to achieve a sustainable relationship with the tropical moist forests of the Old World, from West Africa to Borneo.

PARTNERSHIPS

Widespread public interest in great apes created a responsive environment for raising alarm over their deteriorating conservation status. Great ape populations began to decline seriously in the 1960s, as soon as projects involving industrial-scale logging, infrastructure, and plantation development entered their habitats. By the mid-1980s, the fate of tropical rain forests became symbolized by a few charismatic species, including the mountain gorilla, thanks to the influence of primatologists and the media. The murder of Dian Fossey in 1985 highlighted at the international level the devastating impact that poaching was having on the already small and threatened population of mountain gorillas. Populations of all the great apes were under pressure from deforestation, farming, and hunting; armed conflict aggravated the situation in many areas. There was a growing sense of worldwide public urgency, and an increasing willingness to pay for conservation action.

Starting in the early 1970s with the support of WWF–The Global Conservation Organization (formerly the World Wildlife Fund, and still called this in North America) and the Wildlife Conservation Society (WCS, formerly the New York Zoological Society) for orangutan-rehabilitation work at Bohorok and field research at Ketambe (both in the Leuser Ecosystem in Sumatra), numerous international NGOs soon turned their attention to conserving the great apes and their habitats. These included the Jane Goodall Institutes (established in 1977), the Dian Fossey Gorilla Fund International (established in 1978), and the Orangutan Foundation International (established in 1986).

Other well known international organizations became involved, including Fauna and Flora International (FFI, formerly the Fauna and Flora Preservation Society), the African Wildlife Foundation (AWF), Conservation International (CI), and the International Gorilla Conservation Programme (IGCP – involving AWF, FFI, and WWF). Captive apes had often been transported in cramped and inhumane conditions, and then housed in small cages in zoos or used in medical research; a growing animal welfare movement therefore emerged. International organizations such as the International Primate Protection League (IPPL), the Born Free Foundation, and the International Fund for Animal Welfare (IFAW) campaigned for better treatment as well as for the conservation of great apes.

Recognizing that no one institution can solve all problems, partnerships have formed between and among NGOs and government agencies. These have included the International Gorilla Conservation Programme, the Ape Alliance, and the Great Apes Survival Project (GRASP) partnership. GRASP encompasses the governments of the great ape range states, the secretariats of several international conventions, and most of the NGOs concerned with the study, survival, and welfare of great apes. Two United Nations agencies (the United Nations Environment Programme, UNEP and the United Nations Educational, Scientific and Cultural Organization, UNESCO) lead GRASP. They have provided a joint secretariat since December 2003. Such a partnership might be expected to last for a decade or more, during which time the fates of several great ape taxa are likely to be determined.

GRASP was discussed in 2000, founded in 2001, and launched as a Type II Partnership (a voluntary, non-binding agreement to fulfill conservation commitments) at the World Summit on Sustainable Development in 2002. It held the first intergovernmental meeting on great apes at UNESCO in 2003, extending its membership during 2003–2004 to cover all great ape range states (except Sudan) and all key donor countries. Close governmental links allow GRASP to operate at the highest political levels. The partnership aims to provide a framework into which all the individual

Orangutan Foundation

Sanctuaries such as the Orangutan Foundation International's Care Centre in Central Kalimantan form part of the network for public awareness and education.

SOCP

Research staff at the Ketambe Research Station in Sumatra. Ketambe is currently funded by the Sumatran Orangutan Conservation Programme and the Leuser International Foundation.

Martha M. Robbins

Monitoring is one of the essential aspects of conservation that GRASP is working to establish through national great ape survival plans. Here, field assistants are at work in Bwindi Impenetrable National Park.

conservation efforts of governments, wildlife departments, academics, NGOs, UN agencies, and others can be integrated to ensure maximum efficiency, effective communication, and successful mobilization and targeting of resources. GRASP recognizes the autonomy and independence of existing initiatives, but seeks to create synergy among them.

GRASP initially appointed three UN Special Envoys to help raise resources and recognition of the plight of the great apes: Russell Mittermeier, Jane Goodall, and Toshisada Nishida. They, and Special Advisor Richard Leakey, became GRASP Patrons in 2003. As part of the GRASP program, a number of technical missions, seminars, and workshops have been carried out to help establish national great ape survival plans. These identify the current status and recent trends of each great ape population and of their remaining habitat; existing national policy, legislation, and conservation programs; the level of law enforcement; and the impact on ape conservation of extractive industries such as logging, mining, and oil exploration. The plans then set out recommendations for improving existing conservation measures. Each plan aims to give cohesion to the existing work of many agencies, organizations, and individuals to enable resources to be targeted more effectively, to identify areas that are currently neglected, and to improve opportunities for funding. It is envisaged that the plans will be integrated with other relevant processes and documents that relate to national biodiversity conservation and development planning. Such initiatives include national biodiversity strategies and action plans, poverty reduction strategies, and development plans. A series of workshops on national great ape survival plans are being held in the great ape range states, the aim of each being to lay the groundwork for development of each country's national plan. GRASP also participated in the West African Chimpanzee Regional Workshop in Abidjan, Côte d'Ivoire, held by IUCN–The World Conservation Union; the national workshop in Guinea; and the orangutan Population and Habitat Viability Assessment Workshop in Jakarta, Indonesia; it is working to have their recommendations adopted by the relevant governments.

There are a number of other partnerships and partnership-based projects that aim to promote conservation within several great ape range states, including those that engage professional conservationists in common action across national frontiers. Among the most prominent are:

- The Congo Basin Forest Partnership, which involves Cameroon, the Central African Republic (CAR), the Democratic Republic of the Congo (DRC), Equatorial Guinea, Gabon, and Congo, in dialog with other nations from outside Africa, international NGOs, and businesses.[36] It aims to promote economic development, poverty alleviation, improved governance, and natural resource conservation by supporting a network of national parks, protected areas, and well managed forestry concessions. A related goal is to channel assistance to communities that depend on the conservation of the forest and wildlife resources of 11 key landscape areas in the participating range states.
- The Brazzaville Process, which began in 1996 with the Conference on Central African Moist Forest Ecosystems (CEFDHAC), and involves Burundi, Cameroon, CAR, Congo, DRC, Equatorial Guinea, Gabon, Rwanda, and São Tomé and Principe. It aims to facilitate collaboration for the conservation and sustainable use of Central African moist forest ecosystems. Its secretariat is at the IUCN Regional Office for Central Africa, in Yaoundé, Cameroon. Projects have been undertaken on a wide range of topics, including conflict resolution for forest ecosystem management, sustainable use of forest concessions, timber taxes and conces-

sion fees, forest laws and policies, and critical sites for biodiversity conservation.

- A Central African World Heritage Forest Initiative (CAWHFI), which is being developed by an alliance of UNESCO and the Food and Agriculture Organization of the United Nations (FAO), regional governments, international conservation NGOs, and official aid programs. It aims to respond to the increasing threat of illegal hunting and the unregulated trade in bushmeat by promoting and supporting the building of management regimes in transfrontier clusters of outstanding forest protected areas in Central Africa, that will satisfy standards appropriate to the status of World Heritage Sites. Three such transfrontier zones have been identified for the first phase: Gamba-Conkouati between Gabon and Congo; Odzala-Minkébé-Dja-Boumba-Nki between Congo, Gabon, and Cameroon; and the Trinationale de la Sangha between CAR, Cameroon, and Congo.

NATIONAL AND INTERNATIONAL LAW

National legislation and enforcement

Great apes are protected by national law in every country that they inhabit, but enforcement has been poor to nonexistent in many range states. Even in areas designated for conservation, poaching, illegal logging, and mining all have direct impacts on great ape populations. Typically, neither enforcement nor educational resources are sufficient to ensure that conservation legislation is understood, respected, and abided by. In addition, the punishment specified in the legislation is often a fine that is smaller than the financial benefit that the person would gain from committing the crime.

Various measures to control the trade in great ape bushmeat have been suggested and are being addressed on behalf of the major conservation NGOs by the Bushmeat Crisis Task Force, as well as through a number of intergovernmental initiatives.[10, 23, 24, 26, 88, 157, 163] These ideas include investing in law enforcement and increased fines; increasing capacity for the management of protected areas; taxing the sale of legal bushmeat; promoting cheap and sustainable alternative sources of protein for urban consumers and rural subsistence hunters; developing alternative incomes for commercial traders and hunters; encouraging logging and oil companies to control illegal hunting, transport, and consumption of bushmeat at their concessions; and linking of aid and debt relief to verifiable measures of conservation performance.

SOCP

An illegally held orangutan is confiscated by staff of the Sumatran Orangutan Conservation Programme and the Indonesian Conservation Department.

None of these measures is easy or cheap to implement, and action on the ground has so far been insufficient to address the problems fully. Some national governments have taken effective action against poaching of great apes. In Burundi and Uganda, for instance (once-flourishing centers for illegal traffic in chimpanzees), authorities have clamped down on this trade, and orphaned chimpanzees are now rarely seen openly for sale. In most countries, however, agencies involved in wildlife conservation – including the police and customs – lack sufficient capacity, training, and resources to undertake effective enforcement; they are sometimes also insufficiently immune to corruption. As a result, partnerships have developed between government agencies and international NGOs; these provide both financial and training resources.

International donors such as the US Agency for International Development (USAID) and the World Bank increasingly require rigorous, transparent, and independently reviewed environmental impact assessments (EIAs) prior to any major development initiatives. These are seen as a mechanism to encourage more effective environmental protection in developing countries. EIA legislation is however poorly developed and/or policed in many great ape range states. Where such assessments are conducted, they tend to focus on site-specific impacts rather than the wider socioeconomic effects that

Volker Sommer

This young chimpanzee in Ibadan Zoo, Nigeria, was captured near the Cameroon border. Many orphans destined for export or internal trade are confiscated and placed in sanctuaries; some eventually return to the wild.

might lead, for example, to an increase in bushmeat hunting, trade, and consumption. There is a need to improve the regulations on the content and implementation of EIAs in or around ape habitats, but proper enforcement is required for such measures to have any value. For an EIA to be useful, it needs to be commissioned with a serious intent and the mitigation and harm-avoidance measures that it specifies must then be implemented. Plans to open up industrial logging in much of DRC following the end of the civil war, and backed by the World Bank, suggest that there is still a long way to go before it ceases to be acceptable to liquidate a nation's forest estate and biodiversity resources in the name of national development.[143]

International conventions and compliance

Membership of international conventions

A number of international conventions that address different aspects of biodiversity conservation have been agreed and are in force. Of particular relevance to great apes are the Convention on Biological Diversity (CBD), the Convention on International Trade in Endangered Species of Wild Fauna and Flora (CITES), the World Heritage Convention (WHC), the Convention on the Conservation of Migratory Species of Wild Animals (CMS), and a number of regional conventions. The extent to which the range states are parties to these conventions is summarized in Table 14.1.

Convention on Biological Diversity

The CBD came into force on December 29 1993. It establishes three main goals: the conservation of biological diversity, the sustainable use of its components, and the fair and equitable sharing of the benefits arising from their use. Although it does not list species or places of particular conservation concern, the themes, principles, and activities of the CBD are very relevant to great ape conservation. It encourages parties to the conventions to find ways to deal with biodiversity concerns during development planning, to promote transfrontier cooperation, and to involve indigenous peoples and local communities in ecosystem management. Thematic programs covering the biodiversity of inland waters, forests, and mountains address conservation of the habitat of all the great apes. An example is the case study report produced for the CBD on the impact and management of forest logging in the Dja Biosphere Reserve in Cameroon, an area that is home to central chimpanzees and western lowland gorillas.[17, 18]

Convention on International Trade in Endangered Species of Wild Fauna and Flora

Awareness of the threat that unsustainable international trade posed to many animal and plant species led to CITES, which was signed in 1973 and came into force in 1975. By regulating the international trade in endangered species (as listed in its appendices), CITES aims to ensure that international trade in specimens of wild animals and plants does not threaten their survival. Parties to CITES are expected to implement the convention through their national legislation. CITES accords varying degrees of protection to approximately 33 000 species of animals and plants, whether they are traded alive or as "readily recognizable parts or derivatives" such as dried or smoked meat, or as souvenirs, trophies, tusks, or timber. All the great apes are threatened by hunting for food, pets, and curios; although much of the resulting trade is domestic, much also crosses international borders. High levels of chimpanzee trade occurred during the 1950s and 1960s; at least 300 infant chimpanzees were reported to have been exported to Europe during 1950–1956 alone, with about 100 deaths in transit.[5] Much of this trade was driven by demand from the market for live chimpanzees for use in biomedical research.[97]

All six species of great apes are listed in Appendix I of CITES, which means that international trade in them is permitted only in exceptional circumstances, and never primarily for commercial purposes. A permit to trade any of the great apes may be issued only if the movement of the specimen is not detrimental to the survival of the species, as in the exchange of breeding individuals between reputable zoos. The relevant country's designated scientific authority must be satisfied that the proposed recipient is suitably equipped to house and care for any traded live animal or plant species that has been listed in Appendix I.

All great ape range states apart from Angola are parties to CITES, and so must report their imports and exports of CITES-listed species. Total international trade in live specimens of all species of great ape reported between 1975 and 2003 involved 146 *Gorilla* of several subspecies (71 of which were captive bred), 1 284 *Pan troglodytes* (831 of which were captive bred), 30 *P. paniscus* (12 of which were captive bred), and 324 members of *Pongo* species (249 of which were captive bred). Much of the trade in live apes is between zoos and sanctuaries, for captive-breeding programs, for reintroduction into the wild, and for scientific purposes. Some of this trade (19-23 percent) was accounted for by great apes that are part of circuses that regularly move across international borders, most of whom were bred in captivity. It must be noted that these trade data sometimes over-estimate the volume of the trade as, in some instances, an animal that has been moved between countries on more than one occasion will generate multiple records over the course of a year.

As well as live individuals, trade can include body parts, blood, hair, or other specimens for scientific use. Of the specimens reported in international trade for biomedical purposes, the majority involved samples of blood, hair, skin etc.; a total of 57 live chimpanzees were, however, reported to have been traded between 1975 and 2003 for this purpose. The source (wild or captive bred) of 37 of these animals was not reported; the others were from captive sources, or animals obtained prior to the listing of the species on CITES, or from a range state that was not yet party to CITES. Some trade in the bodies of great apes has also been reported. The bodies of five gorillas were exported as hunting trophies from the CAR to the United States of America and France during 1995. These may have been trophy animals obtained prior to the listing of the species on CITES.

Most trade in wild apes is either between African nations or from great ape range states to Europe, North America, and Asia. A systematic evaluation of the accuracy of CITES reporting in relation to the overall great ape trade has not been undertaken, although widespread lack of CITES compliance has been documented in the African range states,[26] and orangutans have been traded under false papers from Indonesia in the recent past.[94]

In addition to the regulation of trade in endangered species, CITES also addresses a number of issues that are pertinent to the conservation of species affected by trade. One such issue is the bushmeat trade; this is considered by the parties to CITES to be an issue of both trade and wildlife

Table 14.1 Great ape range states: parties to international conventions

Range state	CBD[a]	CITES[b]	CMS[c]	WHC[d]	ACC[e]
Angola	yes[f]	no	no	yes	no
Burundi	yes	yes	no	yes	signed
Cameroon	yes	yes	yes	yes	yes
Central African Republic	yes	yes	no	yes	yes
Congo	yes	yes	yes	yes	yes
Côte d'Ivoire	yes	yes	yes	yes	yes
Democratic Republic of the Congo	yes	yes	yes	yes	yes
Equatorial Guinea	yes	yes	no	no	no
Gabon	yes	yes	no	yes	no
Ghana	yes	yes	yes	yes	yes
Guinea	yes	yes	yes	yes	signed
Guinea-Bissau	yes	yes	yes	no	no
Indonesia	yes	yes	no	yes	–
Liberia	yes	yes	yes	yes	yes
Malaysia	yes	yes	no	yes	–
Mali	yes	yes	yes	yes	yes
Nigeria	yes	yes	yes	yes	yes
Rwanda	yes	yes	yes	yes	yes
Senegal	yes	yes	yes	yes	yes
Sierra Leone	yes	yes	no	yes	signed
Sudan	yes	yes	no	yes	yes
United Republic of Tanzania	yes	yes	yes	yes	yes
Uganda	yes	yes	yes	yes	yes

a CBD, Convention on Biological Diversity.
b CITES, Convention on International Trade in Endangered Species of Wild Fauna and Flora.
c CMS, Convention on the Conservation of Migratory Species of Wild Animals.
d WHC, World Heritage Convention.
e ACC, African Convention on the Conservation of Nature and Natural Resources.
f yes: the country is a party to the convention;
signed: the country has signed but is not yet a full party;
no: the country has not signed.

management. Where cross-border trade in bushmeat occurs, it is often unsustainable and illegal. Consequently, in 2000, the parties to CITES set up a Bushmeat Working Group, composed of interested range and donor states. It aims to examine issues raised by the trade in bushmeat, to identify solutions that can willingly be implemented by range states, and to promote awareness and action to achieve better and more sustainable management of the bushmeat trade. The group's initial work was on a case-study area comprising Cameroon, CAR, Congo, DRC, Equatorial Guinea, and Gabon. CITES adopted a resolution in 2004 to report on progress in conservation and trade in great apes, and to work with GRASP and the CBD Secretariat.

World Heritage Convention

Many populations of great apes live in areas that are highly distinctive, species rich, or particularly noteworthy for other reasons. The importance of some of these has been recognized by the WHC, which aims to define and conserve the world's cultural and natural heritage. The WHC was adopted by the General Conference of UNESCO in 1972 and has compiled the World Heritage List, comprising sites or areas nominated according to specified criteria.

Areas nominated under the convention's Criterion (iv) should contain "the most important and significant natural habitats for *in-situ* conservation of biological diversity, including those containing threatened species of outstanding universal value from the point of view of science or conservation." If a World Heritage Site is threatened it may be placed on the List of World Heritage in Danger, which is intended to draw the matter to the world's attention.

The forest edge at Bwindi Impenetrable National Park, one of the national parks that has World Heritage status.

Gordon Miller/IRF

The WHC has been signed by more than 175 states. Parties to the convention must report on the condition of the sites within their borders, on measures taken to conserve them, and on efforts to raise public awareness about them. Of the 23 great ape range states, eight have World Heritage Sites that harbor populations of great apes (see Table 14.2).

In July 2004, the Gunung Leuser National Park (see Box 11.2) in Sumatra was designated part of a 'Cluster Mountain' World Heritage Site; this includes two other parks to the south (Bukit Barisan Selatan and Kerinci Seblat, neither containing orangutans).[134, 153] This initiative derived from an earlier proposal from the Indonesian government to request that UNESCO award World Heritage Site status to the whole Leuser Ecosystem of about 26 000 km^2, rather than to only the 8 900 km^2 of the Gunung Leuser National Park.[135] The rationale for this was that most orangutans in the Leuser Ecosystem (an estimated 3 573 out of 5 598 individuals) live outside the national park, so an excessive focus on the park alone would risk major losses of both habitat and orangutans. Cluster Mountain World Heritage Sites are typically located in high mountain ranges (such as the Himalayas, Andes, or Alps), where the focus of conservation is on the high-altitude habitats. Not all conservationists considered this designation to be appropriate for use in Sumatra, where the greatest biodiversity occurs in, and is totally dependent on, lowland forests.

World Heritage Sites are also protected by the highest category of national designation, normally as a 'national park'; many of those in which great apes live have, however, been placed on the List of World Heritage in Danger. In particular, sites in DRC that are home to eastern gorillas, chimpanzees, and bonobos have been seriously affected by armed conflict and its consequences. The World Heritage Committee has therefore undertaken to provide support and assistance to DRC in cooperation with IUCN and other institutions, such as the World Bank and the United Nations Foundation.

It has been proposed that a new designation of World Heritage Species be created, and that all the great apes be given this status.[61] This idea enjoys strong, although not universal, support among conservationists.[135] If this proposal were accepted,

Table 14.2 World Heritage Sites containing great ape populations

Range state	World Heritage Site	Year inscribed	Species or subspecies	World Heritage in Danger
Cameroon	Dja Faunal Reserve	1987	western lowland gorilla, chimpanzee	
Côte d'Ivoire	Taï National Park	1982	chimpanzee	
	Comoé National Park	1983	chimpanzee	
Democratic Republic of the Congo	Virunga National Park	1979	mountain gorilla, chimpanzee	√
	Kahuzi-Biega National Park	1980	eastern lowland gorilla, chimpanzee	√
	Garamba National Park	1980	chimpanzee	√
	Salonga National Park	1984	bonobo	√
	Okapi Faunal Reserve	1996	chimpanzee	√
Guinea and Côte d'Ivoire	Mount Nimba Strict Nature Reserve	1981	chimpanzee	√
Indonesia	Tropical Rainforest Heritage of Sumatra	2004	Sumatran orangutan	
Malaysia	Kinabalu Park	2000	Bornean orangutan	
Senegal	Niokolo-Koba National Park	1981	chimpanzee	
Uganda	Bwindi Impenetrable National Park	1994	mountain gorilla, chimpanzee	
	Rwenzori Mountains National Park	1994	chimpanzee	

new World Heritage Sites would then be designated specifically for the protection of viable populations of these species. A new Protocol to the World Heritage Convention would be needed to establish this new category.

Convention on the Conservation of Migratory Species

The CMS aims to conserve terrestrial, marine, and avian migratory species in all national jurisdictions that they visit or through which they pass. Appendix I of the CMS lists migratory species that, according to the best scientific evidence available, are endangered. Although not migratory in the usual sense, mountain gorillas regularly cross international borders and, because they are of particular conservation concern, the subspecies is listed in Appendix I of the CMS. Since 1997, the mountain gorilla has also been designated for 'Concerted Action' on the basis of the conflicts in DRC and the problems facing its habitat. This designation requires parties to undertake particular actions to help conserve the subspecies and its habitat. The Scientific Council of the CMS has also noted the opportunities for cooperation with other conventions in protecting the habitat of mountain gorillas in DRC.

All three of the mountain gorilla range states, (DRC, Rwanda, and Uganda) are parties to CMS. Rwanda was the most recent to join, in June 2005. The Convention Secretariat has agreed to promote a possible agreement, under Article IV of the CMS, to cover the conservation of the shared mountain gorilla population of the three countries. This agreement could also cover key World Heritage and other protected sites.[74]

In 2004, the Terrestrial Mammals Working Group of the CMS proposed that the special designation of mountain gorillas be extended to all gorillas.

Compliance with international agreements

The success of all these conventions depends on the ability of the parties to implement them effectively, including compliance with the conventions' reporting requirements. These commitments demand the allocation of trained staff and resources to cover operating costs; when other government priorities are taken into account this is often beyond the means of great ape range states. No comprehensive assessment of the effectiveness of these conventions has yet been carried out, but the CBD has adopted a measurable target.[35] This is the '2010

Gordon Miller/IRF

A billboard at the Equator crossing near Lake Mburo National Park, Uganda, warns of the dangers of corruption.

target', by which the parties committed themselves to achieve by 2010 a significant reduction of the current rate of biodiversity loss at the global, regional, and national level.

Regional agreements and activities

Regional agreements

All African apes are listed under 'full protection' Class A of the African Convention on the Conservation of Nature and Natural Resources (ACC). This entered into force in 1969, and is binding on the vast majority of Africa's 21 great ape range states (Table 14.1).[113] It is designed to validate and encourage conservation and the wise use of natural resources by individual countries and by countries working together. Its significance is mainly as a policy guide and a basis for developing specific measures to conserve resources such as great apes that are important "from an economic, nutritional, scientific, educational, cultural and aesthetic point of view," in the words of the convention.

The Forest Law Enforcement and Governance (FLEG) process is an initiative organized by donors such as the World Bank, and involves the participation of governments, NGOs, and civil society. It focuses on combating the threat posed to forests by illegal logging and trade, corruption, and poaching. Ministerial and other meetings have been organized in Africa (AFLEG) and Asia (FLEG Asia), to examine potential partnerships between producers and consumers, the private and public sector, and donors, that can help address illegal forest exploitation. These meetings have led to ministerial declarations and action plans, an example being the October 2003 Yaoundé Ministerial Declaration on Africa Forest Law Enforcement and Governance (AFLEG).[145]

The Yaoundé Forest Summit of 1999 brought together leaders from the Congo Basin countries to address the growing threats to the forests of the region and to look at ways to help ensure their integrity and survival. The principal outcome was the 'Yaoundé Declaration' of intent to promote transfrontier cooperation for biodiversity conservation in and around the Congo Basin, involving Cameroon, CAR, Chad, Congo, DRC, Equatorial Guinea, and Gabon. It established firm political commitments, such as the creation of new forest protected areas, developed plans to combat illegal logging and poaching of wildlife, and broadened the application of sustainable forest-management strategies. This paved the way for both action and new partnerships, such as the accord between Cameroon, Congo, and CAR in 2000, that established collaborative management of over 28 000 km^2 of forest with harmonized forestry policies, in the newly created Trinationale de la Sangha conservation area. To date, a total of 34 000 km^2 of new protected areas have either already been gazetted or are in the process of being so in the Congo Basin; efforts have been made to strengthen the sustainable management of existing protected areas, which amount to a total area of 135 000 km^2.[146]

The African Ministerial Conference on the Environment, held in Maputo in 2003, adopted the Action Plan of the Environmental Initiative (APEI) of the New Partnership for African Development (NEPAD). This includes the African Protected Areas Initiative (APAI), which is an Africa-led initiative to mobilize African institutions and expertise to enhance the role of protected areas as tools for safeguarding biodiversity, sustaining ecosystem processes, and contributing to livelihoods and sustainable development across the continent. It aims to encourage and enable the development of conservation areas, the building of the capacity to both implement the CBD and to manage biodiversity and knowledge about it, and the networking of African experts as well as institutions.

Regional and species action plans

A number of regional action plans, which involve several range states, and species action plans have been developed to guide and organize conservation of great apes. Notable recent examples from the Primate Specialist Group of the IUCN Species Survival Commission (SSC) include:

- Status survey and conservation action plan for African primates;[109]
- Status survey and conservation action plan for West African chimpanzees;[88, 89]
- Population and habitat viability assessment (PHVA) for mountain gorillas;[70, 138, 160]
- PHVA for eastern chimpanzees in Uganda;[50] and
- PHVA for orangutans.[135, 148]

A number of other species-specific plans have also been developed, including two on bonobos,[44, 147] and one on chimpanzees in Uganda.[117] These documents are sets of recommendations; prior to the involvement of GRASP, few appear to have been incorporated into national planning documents such as national biodiversity strategies and action plans.

The IUCN's West African Chimpanzee Action Plan reviews existing information on the status of the two subspecies that occur in West Africa (up to the border between Nigeria and Cameroon), *Pan troglodytes verus* and *P. t. vellerosus*.[87] It comprises a set of national profiles and a regional assessment of threats and recommendations for action. It assesses logging, agriculture, crop raiding, bushmeat hunting, primate sanctuaries, rehabilitation experiences, and the threat from infectious diseases. The plan was based on a workshop held in 2002 in Abidjan, Côte d'Ivoire, attended by 72 scientists, conservationists, policy makers, protected area managers, and donors from 15 countries.[89]

PHVAs are facilitated by the IUCN's Conservation Breeding Specialist Group, but involve local stakeholders and various organizations with an interest in the conservation of target species. The process uses mathematical models within a participatory workshop framework to produce a strategic recovery plan for a threatened species and its habitat. The process uses data on demography, genetics, and ecology, as well as estimates of threats such as current and predicted land-use patterns; it is explicitly designed to broaden stakeholder involvement and enhance knowledge sharing.[91, 161] A PHVA workshop that focused on orangutans was held in Jakarta, Indonesia, in January 2004.

NONGOVERNMENTAL ORGANIZATIONS

A large number of international and national NGOs, working in partnership with national authorities and local NGOs, have been involved in developing and implementing conservation programs for the great apes. Many of these are now partners in GRASP (see Annex), and their activities are summarized in this volume, especially in the country profiles in Chapters 16 and 17. Their activities can only be touched on here to give an idea of the breadth of their involvement.

The international NGOs active in Africa include the International Gorilla Conservation Programme (IGCP) partnership (as discussed above in the section entitled 'Partnerships'), Conservation International (CI), Fauna and Flora International (FFI), the Wildlife Conservation Society (WCS), and WWF–The Global Conservation Organization.

WCS has a very active program in Central and East Africa, including great ape conservation projects in Congo, Gabon, DRC, Uganda, and Rwanda.

CI has a major program in West Africa, a region it has identified as a 'hotspot' for biodiversity.[14] It also secured core funding and provided technical and logistical support for the development of the IUCN's West African Chimpanzee Action Plan. IGCP's goal is the conservation of mountain gorillas and their habitats in DRC, Uganda, and Rwanda, and to increase cooperation between their protected area authorities. Despite civil unrest and other setbacks, the project has achieved considerable success in helping the authorities protect mountain gorillas in the Virungas since the 1980s.

As well as its role in IGCP, WWF has projects across Africa, including support for forest conservation projects in Cameroon, Côte d'Ivoire, Gabon, and CAR.

A rangers' truck sports the WWF logo. WWF was one of the first NGOs to become involved in great ape conservation issues.

Gordon Miller/IRF

Other groups active in the field include Les Amis des Bonobos du Congo, Berggorilla und Regenwald Direkthilfe, the Bonobo Conservation Initiative, Born Free Foundation, Cameroon Wildlife Aid Fund, Dian Fossey Gorilla Fund Europe, Dian Fossey Gorilla Fund International, Frankfurt Zoological Society, the Jane Goodall Institutes, Pan African Sanctuary Alliance, the Wild Chimpanzee Foundation, and the Zoological Society of London.

In Southeast Asia, much NGO effort is invested in field conservation and orangutan rehabilitation. In Malaysian Borneo, Hutan runs an innovative community conservation and research project along the Kinabatangan River, Sabah. In Indonesian Borneo, the Orangutan Foundation International funds patrols in Tanjung Puting National Park, rehabilitates and releases orphan orangutans in Lamandau Strict Nature Reserve, and supports research into conservation and forest restoration. The Borneo Orangutan Survival Foundation rehabilitates and releases orphans in Balikpapan and other parts of Kalimantan, and is involved in proposals to protect the Mawas area, 5 000 km^2 of peat-swamp forest inhabited by orangutans. The Sumatran Orangutan Conservation Programme (SOCP) has established a similar release program in Bukit Tiga Puluh National Park in Jambi province. It is also the only organization that is currently carrying out field research in Ketambe, keeping the research site open; it conducts most of the survey and monitoring work concerning the status and distribution of the wild orangutan population in the area. Two related networks exist, linking workers on both islands: the Orangutan Network, which links research scientists involved in conservation;[112] and the Orangutan Conservation Forum, proposed by the 2004 PHVA workshop, which aims to coordinate education and communication efforts, including advising on the preparation of the national great ape survival plans for Indonesia and Malaysia.[135]

The Leuser Ecosystem (see Box 11.2) is a key habitat for Sumatran orangutans. It includes the Gunung Leuser National Park (8 900 km^2) and extensive areas of protection and production forests in the provinces of North Sumatra and Nanggroe Aceh Darussalam, amounting to about 26 000 km^2 in total. The whole area is the subject of the long-term Leuser Development Programme (LDP), a partnership project between the government of Indonesia and the European Union, on which the EU has spent around US$39 million and the Indonesian government about US$7.5 million since 1996. The Indonesian government has assigned management of the whole area to an NGO created for this purpose, the Leuser International Foundation (LIF), which has these rights for nine years, extendable to 30. The LDP is now supporting LIF in managing the Leuser Ecosystem in its first nine year phase.[95] On the other hand, in early 2004 the Indonesian president Megawati Sukarnoputri gave the initial go ahead for the Ladia Galaska road project, which was planned to pass through the Gunung Leuser National Park (but see Box 11.2).

In Sabah, the Kinabatangan Orangutan Conservation Project (KOCP) was established in 1998 to secure one of the largest Malaysian populations, in the Kinabatangan floodplain of eastern Sabah. The KOCP is supported by WWF-Malaysia and works closely with the Sabah Wildlife Department, local communities, and other stakeholders to find ways to conserve wild orangutans in multiple-use forests. The KOCP has identified an orangutan population of several thousand individuals in the Kalabakan area, between the Maliau Basin and Danum Valley Conservation Areas.[79] This was assessed as being perhaps the population of Bornean orangutans with the best long-term viability, provided that logging (both legal and illegal) could be brought under control and associated forest fires prevented. The area had a narrow escape in 2002, when a Malaysian–Chinese agreement to convert 2 414 km^2 of natural forest into *Acacia mangium* pulpwood plantation was terminated by the Sabah government.[106] The existence of this agreement, however, had already been used as an excuse for clearfelling (the removal of all the trees, with a view to subsequent planting) from extensive areas of forest.

PROTECTED AREAS

Overview

In terms of their habitat needs, the interests of great apes and humans often come into conflict; on the whole, apes need natural forests free of disturbance while humans rely on farmland, plantations, or forests that are intensively managed for production purposes. Areas of great ape habitat must therefore be set aside with permanent legal and actual protection if ape populations are to survive in the wild. The same analysis applies to many other tropical moist forest organisms, so protected areas are the chief means of conserving biodiversity in this type of ecosystem.[119] National parks and other areas where the forest may not be damaged and wildlife may not be hunted are fundamental to the conservation of great apes. All great ape range states have protected area systems, typically covering 5–15 percent of their national territory, although relatively few of these areas are important for great apes and some, in reality, exist only on paper.

It is particularly challenging for the smaller protected areas to maintain a healthy ape population. Small populations of great apes are highly susceptible to extinction due to random catastrophes (such as infection by Ebola virus, or exposure to forest fires), reduced genetic variability, poaching, and other forms of human disturbance. Moreover, the ability to transfer between communities is essential for the viability of great ape populations; one that is composed of small isolated groups has only a precarious future. Population densities of great apes are typically only 0.3–1.0 individual per square kilometer; large areas of protected habitat are therefore needed to support populations that are likely to be viable in the long term. Such populations should generally be of at least 1 000 individuals living in an area of several hundred to several thousand square kilometers.[100]

Michael Huffman

Martha M. Robbins

Top: Rubondo Island, United Republic of Tanzania, where 17 chimpanzees were released into the wild in the 1960s, growing by 2002 to a self-sustaining population of 40. Above: Deforestation up to the boundary of Bwindi Impenetrable National Park, Uganda.

Great apes in existing protected areas

The Bornean orangutan, like any widespread species, includes within its range both protected and unprotected forests. Indonesian national parks containing orangutan populations include Kutai, Tanjung Puting, Gunung Palung, and Betung Kerihun. Most Sumatran orangutans are found in the Leuser Ecosystem, which is under active management but still subject to illegal logging. In Malaysian Borneo, the most significant population in Sarawak is found in the Lanjak-Entimau Wildlife Sanctuary (which now incorporates Batang Ai National Park and is connected to Betung Kerihun). In Malaysia, a wildlife sanctuary is more strictly protected under state legislation than a national park. Likewise, orangutans occur in several areas in Sabah that are strongly protected under state legislation, including the Danum Valley Conservation Area.

Much of the existing population of the eastern lowland gorillas is within formally protected areas (although the status of this population is uncertain due to reported heavy poaching, and insecurity preventing surveys), and all mountain gorillas in the Virungas and Bwindi are within national parks. The position for the eastern chimpanzee is less clear because extensive areas of the eastern Congo Basin have yet to be surveyed. Similarly, no precise estimate exists for protected area coverage for the central chimpanzee or western lowland gorilla. There are 26 protected areas within the western chimpanzee's geographical range (covering 6.6 percent of its distribution), and three in the range of the Nigeria-Cameroon chimpanzee with similar coverage.[87] Finally, bonobos occur in only two officially protected areas in DRC, one very large (Salonga National Park, 36 560 km^2) and the other much smaller (Luo Reserve for Scientific Research, 358 km^2).

The effectiveness of protected areas

As with law enforcement, many existing management programs for protected areas in Africa and Southeast Asia are heavily funded by NGOs and external donors and would not be able to function without this support. Many of the national parks and reserves in great ape range states are critically understaffed and underfunded, and are little more than 'paper parks'.[149] Indeed, by 2000, declining great ape populations had been reported in almost all of 24 protected areas where great ape research programs had been conducted.[100] The collapse of eastern lowland gorilla populations since the late 1990s has occurred despite the large protected areas overlapping their range. On the other hand, even 'paper parks' generally fare better in the long term than areas that have no protected status at all, as people often respect their boundaries.[29, 72] They also deter certain kinds of investments, such as major infrastructure projects by publicly accountable donor agencies, which are bound by law to commission environmental impact assessments, and by public opinion to take some notice of them.

It has long been accepted that protected areas must be perceived as beneficial by local people as well as by governments if they are to survive and be maintained. Park administrations therefore usually try to reach out to nearby communities to explain the benefits of conservation, and there is increasing pressure for protected areas to generate revenues that can be shared between local communities and governments. In some cases, it may be possible to meet both requirements through income from tourism, research communities, filming fees (from documentary makers), and local community development projects, supplemented by external donor support; this is not, however, feasible for all areas in the Congo Basin.[19, 164] For Central Africa in particular, it is not realistic to expect national governments to bear all the costs of maintaining functional protected areas. In Cameroon, for example, the opportunity cost (i.e. the revenues not obtained from using a resource in other ways, sustainably or otherwise, because it is being protected) to the government of not logging a tropical forest has been estimated at about US$15 000 per km^2 annually;[162] in this sense, the cost of setting aside a 2 700 km^2 protected area would exceed US$40 million per year. Governments will naturally argue that the international community should contribute at least the value of the global environmental benefits generated by such an investment.

This is the policy basis for the Global Environment Facility (GEF), which finances the 'incremental' cost of measures aimed at biodiversity conservation. This incremental cost is the difference in price between what a responsible government should invest in the interests of its own people (e.g. to secure water catchment services and tourism revenues), and what the world community would like to have done (e.g. to secure a globally important ecosystem, or the carbon-storage function of forests). Part of the challenge is that conservation is a slow, long-term business, requiring recurrent investment in perpetuity, while governmental and donor interest in any given conservation initiative operates within shorter timeframes. To help solve this, the GEF has often endowed trust funds and trust-like mechanisms, equipping them to invest a one-time-only grant with the intention that the return on capital will underwrite a share of conservation expenditure indefinitely.[62] This method was used in the Mgahinga-Bwindi Impenetrable Forest Conservation Trust, centered on two national parks in southwest Uganda where mountain gorillas occur.[139] This trust was established in 1995 with initial funds from the GEF; it provides a sustainable source of funding for park management, research, and community conservation projects;[118] globally, such examples are rare.

Ian Redmond/UNESCO

Rangers in Volcanoes National Park, Rwanda, destroying snares set by poachers.

TRANSFRONTIER PROTECTED AREAS

Many great ape populations extend across national frontiers; in places, these coincide with protected (or otherwise managed) habitat areas that have been established on either side of a border. In principle, these provide opportunities for reserve managers in neighboring countries to cooperate to improve great ape protection in a combined area. One of the best examples is the cluster of protected mountain gorilla habitat in the Virungas, in the management of which the authorities of Rwanda, Uganda, and the DRC cooperate closely. The Yaoundé Ministerial Declaration of 1999 was intended to promote transfrontier cooperation for biodiversity conservation in and around the Congo Basin. Among other things, so far it has led to:

- recognition and endorsement of the creation of Trinationale de la Sangha, which has a core area of around 7 300 km^2, in Cameroon (Lobéké National Park), CAR (Dzanga-Ndoki National Park), and Congo (Nouabalé–Ndoki National Park);
- the creation of two new forest national parks in Cameroon (Campo Ma'an and Mbam et Djerem);
- the establishment of a transborder conservation initiative between Congo (Bambama-Lékana National Park) and Gabon (Batéké Plateau National Park);
- the founding in 2002 of the Congo Basin Forest Partnership, which seeks to build upon many existing unilateral, bilateral, and multilateral programs in the region, including collaboration between governmental and NGO stakeholders.

A program funded by the European Union (Appui à la Gestion Intégrée des Ressources Naturelles des bassins du Haut Niger et de la Haute Gambie, AGIR) is working in West Africa to establish transboundary conservation agreements between Guinea and neighboring countries with contiguous forest, much of which is chimpanzee habitat. It includes Niokola-Koba National Park (Senegal); Badiar National Park (Guinea); Mount Nimba Strict Nature Reserve (Guinea–Côte d'Ivoire); the proposed Bafing-Falémé Protected Area (Guinea–Mali); and the proposed Guinea-Bissau–Guinea Protected Area.

Since 1995, there have been efforts to establish a transfrontier biodiversity conservation area that incorporates the Lanjak-Entimau Wildlife Sanctuary in Sarawak and the Betung Kerihun National Park in West Kalimantan, Indonesia, a complex with a combined total area of 11 000 km^2. This initiative is being coordinated by the International Tropical Timber Organization (ITTO) and the Forest Department in Sarawak, with WWF-Indonesia and the Park Management Unit of Betung Kerihun National Park in Indonesia.[80]

Gordon Miller/IRF

Mgahinga Gorilla National Park in Uganda is adjacent to Virunga National Park in the Democratic Republic of the Congo and to the Volcanoes National Park in Rwanda.

MANAGEMENT OUTSIDE PROTECTED AREAS

Logging

Most great apes live outside existing or planned protected areas,[149] so protection and management of great ape habitat in the wider environment is crucial to their survival. We should therefore consider how the survival of great apes can be reconciled with logging, which is one of the major influences on their habitats outside protected areas. To do this, it is necessary to understand more fully the impact of timber harvesting on great ape behavior and ecology.

For gorillas, selective logging may improve the availability of easily digestible herbaceous and similar plant foods in their environment. For these animals, the threat is less the alteration to the forest ecosystem caused by logging, but rather the greatly increased hunting pressure brought about by forestry workers living off the land, and the improved access to commercial hunters and transport opportunities provided by logging roads and vehicles. The situation for orangutans is less

Ian Redmond

Logging in Gabon.

clear, as there is evidence both of serious impacts and remarkable adaptation associated with logging in various parts of Borneo and Sumatra. In areas where hunting is not a serious threat, the survival of orangutans in selectively logged forest appears to be determined by the quality of the residual stand in terms of fruit productivity. Even where there are many fruit trees, however, logged forest can and does burn; fire has a devastating impact on orangutans. Chimpanzees are also capable of great adaptation to ecological conditions in logged forest, but there are few data on the response of bonobos to logging on its own. Their greater consumption of herbaceous vegetation suggests that they might be reasonably successful in disturbed forest. As with gorillas, it is hunting and the fragmentation of forest areas by logging roads and settlements that are more likely to destroy chimpanzee and bonobo populations.

In principle, therefore, the selective extraction of timber at moderate intensities from a forest need not cause the complete loss of great ape populations. The impact could be reduced if logging companies took steps such as:

- suppressing hunting on their concessions;
- blocking access to logging areas;
- destroying roads, canals (e.g. in Bornean swamp forests), and bridges after operations are completed;
- leaving patches of unlogged forest of sufficient size to preserve a moist microclimate and taking other steps to prevent forest fires; and
- taking much greater care in the logging process to preserve fruit trees.

The latter would include:

- taking a detailed inventory prior to tree felling;
- marking of fruit trees;
- ensuring directional felling of target trees;
- the use of aerial log extraction, rather than skidding, to reduce damage to trees that have not been chosen for felling; and
- the avoidance of post-felling silvicultural treatments such as the cutting of climbers.

All these measures require greater investment by the logging company and may reduce profit margins; there are, however, potential benefits from being able to demonstrate to consumers that wood is produced in a 'great ape friendly' way.

Many great ape populations, particularly in Central Africa, live in forests managed for timber production; the timber industry is a key component in the economic and social development of most range states. A positive engagement with logging companies is therefore seen as crucial to the future protection of great apes, and some companies have indeed taken an active interest in adopting a 'code of conduct' (as proposed by the Ape Alliance in 1998) to reduce the impact of their activities on wildlife. A few initiatives have worked directly with logging companies to reduce hunting. One example promoted as a successful model is a joint project in Congo, between the Wildlife Conservation Society, the Ministry of Forests and Environment, and Congolaise Industrielle des Bois, a logging company. This partnership seeks to control hunting and trade in bushmeat, especially of great apes, in forestry concessions near Nouabalé-Ndoki National Park,[13] although some critics have pointed to the lack of independent evaluation or monitoring. Other businesses have also tried to control bushmeat hunting in their areas of influence. In Cameroon, for instance, the oil company Esso tried to police and prevent trade in bushmeat during the construction of the Cameroon–Chad pipeline during 2001–2003.[52, 53]

Media exposure of the links between logging and commercial bushmeat hunting has also been shown to be effective in changing business attitudes. The Inter-African Forest Industries Association (IFIA) represents 14 logging companies that together held concessions on 140 000 km^2 of forest and employed 20 000 people in the late 1990s. It has committed its members to take action to prevent commercial hunting.[149] The World Bank has also established a working group of chief executive officers of European timber companies, including IFIA. This aims to promote better forest protection and management (including protected areas), to enable investment in local forest industries, and to reduce the impact of the bushmeat trade on vulnerable species. The group is working to develop cooperation between private companies, NGOs, and the World Bank, in order to foster sustainable and integrated use of forest resources in tropical Africa, with an emphasis on the Congo Basin and West Africa. It is also seeking to promote a 'code of conduct' for forestry companies in Africa,[15] but NGO observers and even the World Bank have been frustrated at the slow progress and lack of concrete results.

Environmentally responsible logging operations and schemes to track timber through a 'chain of custody' from producer to consumer, like that developed by the Forest Stewardship Council (FSC), have been introduced in recent years. These are promoted by NGOs (such as WWF, Greenpeace, and Friends of the Earth) as a mechanism for tackling illegal logging and developing sustainable forest management. The Tropical Forest Trust works with its corporate sector members to increase the area of certified tropical forest. Certification is developing slowly, however, in Africa and Southeast Asia. There are two FSC-certified sites in Uganda, one in Malaysia, and one in Sumatra.[152] Existing certification schemes generally rate poorly, however, in addressing wildlife management issues.

Mining

A comparable approach has been adopted by international conservation groups to the certification of mined materials, and to advocacy surrounding their use. For example, international advocacy campaigns have had some effect in stopping companies from purchasing coltan from Central Africa.[73]

As discussed in Chapter 13, coltan is an ore from which tantalum (a heavy metallic element) is extracted; it is used in certain surgical and aerospace alloys and for making heat-resistant capacitors for mobile phones and other electronic products. High prices led to a coltan rush in 2000, with thousands of artisanal miners entering Kahuzi-Biega National Park and the Okapi Faunal Reserve in eastern DRC. The resulting demand for bushmeat had a devastating impact on gorilla and other wildlife populations.[122] Until 2004, much of Kahuzi-Biega National Park was not controlled by the park authorities, but was occupied by coltan miners and militia.[73] Coltan was also being used to finance various armies in DRC during the war,[155] so was described as 'conflict coltan', further strengthening the grounds for a boycott. The aim of the campaign to boycott coltan was, in effect, to lower the value of DRC coltan to the electronics companies that ultimately were buying it.

A different approach was adopted by the Dian Fossey Gorilla Fund Europe in the 'Durban Process', which brought together all the stakeholders and sought investment in a system of environmentally and socially responsible mining outside protected areas, that would still allow poor artisanal miners to benefit from this valuable resource.[46] New partnerships between business and conservation groups are now emerging. Fauna and Flora International (FFI), for instance, is working with the UN Global Compact, a UN-sponsored network, to promote responsible corporate citizenship, to develop markets for ethically sourced coltan as an investment for peace in DRC.[73] The aim is to persuade large western companies that use coltan (such as Motorola, Sony, Hewlett Packard, and Nokia) to buy the mineral only from environmentally and socially responsible operations in DRC.

Forest restoration

Great ape populations can be fragmented among patches of unconnected habitat, within which they

Ian Redmond

Coltan (tantalum ore), confiscated by wardens at Kahuzi-Biega National Park, Democratic Republic of the Congo.

Akihiro Hirata

A tree nursery in Guinea, part of a program to create a 'Green Corridor' between Bossou forest and Mount Nimba in order to link two chimpanzee populations.

are vulnerable to chance factors. Forest restoration aiming to reconnect such patches or increase their size has been little promoted in the great ape range states. The few examples of successful projects include the two FSC-certified forests in Uganda, both restoration projects financed with carbon-storage funds. One is located within the Kibale National Park and aims to increase the area of chimpanzee habitat.[37, 133] In Kinabatangan, Sabah, efforts at forest and wetland restoration are underway, countering the ongoing loss of forest,[166] but the rate of forest clearance for palm oil plantations by far exceeds the rate of reforestation.

Connecting forest protected areas and other habitat blocks by establishing and protecting 'corridors' of forest has been proposed by a number of conservationists[51, 117] A population and habitat viability assessment (PHVA) on the eastern chimpanzee in Uganda estimated that the extinction risk to some of Uganda's chimpanzee populations is reduced by as much as 55 percent when populations are linked by corridors.[50] If options to maintain large populations are limited, ensuring the opportunity for individuals to disperse is critical. Highly fragmented populations of great apes that might benefit from the creation of corridors include the Cross River gorilla in Nigeria,[51] and populations of Sumatran orangutans in the Leuser Ecosystem, Indonesia.[135] On a broader spatial scale, the West African chimpanzee population would benefit from habitat restoration and the protection of existing corridors.[86] The chimpanzees at Bossou (Guinea) have survived in an isolated forest patch, and are now likely to benefit from a corridor project to link Bossou with the central Mount Nimba forest area.[63, 86]

Local communities

The loss of food sources following forest degradation and clearance has led some great ape groups to forage in agricultural areas, bringing them into conflict with farmers. Chimpanzees are most commonly documented as crop raiders, and their impacts on crops are greatest when fruit is scarce in the forest.[78, 104, 105] Orangutans also often raid fruit crops adjacent to the forest, and may be killed; this is becoming an important issue in parts of Sumatra. In southern Burundi, this issue has been tackled through a project that employed a small group of well trained people to protect villagers from raiding chimpanzees; they monitored the locations of chimpanzee groups, and drove them away from villages and crops.[67] Gorillas have also been implicated in crop raiding, for example outside Bwindi Impenetrable National Park in Uganda,[92] where managers have devoted significant attention to finding ways to limit the park's impacts on local communities.

Most conflicts of interest between people and wildlife are resolvable through the payment of sufficient compensation for loss, and people can be quite tolerant of damage caused by wildlife if they stand to benefit from wildlife revenues. Various arrangements for negotiating and paying compensation for loss, injury, and death have been tested since the beginning of wildlife conservation in the early years of the last century; benefit-sharing arrangements are increasingly coming to be expected by both the public and the conservationists.

Matters become more complex when a community is required to give up its right to use a nature reserve in any way; an opportunity cost is then being imposed that must also be offset by some other set of benefits, if it is not to cause friction. Diverse options are available for achieving sustainable agreements. These range from easement contracts to compensate for certain lost rights (e.g. hunting, logging, or farming) while allowing for continued use of the forest for collecting medicinal plants and other items of little conservation concern, through to outright land purchase. All require funding, however, which is why international donor support is often so welcome, and why benefit-sharing schemes (based, for example, on gorilla tourism) are so attractive as a self-financing alternative to grant aid. Nevertheless, it is important to recognize that people who give up land rights so that the rest of society can have a nature reserve have earned the right to be paid fairly. Participation in the protection of a natural landscape that produces clean water, flood control, soil protection, carbon sequestration, and great ape benefits deserves appropriate remuneration.

INTEGRATING APPROACHES

It has increasingly been recognized that biodiversity conservation and sustainable development are inextricably linked. Consequently, addressing the most immediate concerns of local people living around protected areas, such as improving health care and access to education, has become a major feature of conservation policy.[4, 103, 136] Two common approaches that attempt to link the conservation of natural resources and the development needs of local people are integrated conservation and development projects (ICDPs) and community-based conservation. These both aim to alleviate poverty and make rural livelihoods more sustainable, especially in and near protected areas.[1, 29, 69, 108, 158, 159] Such projects typically promote activities such as:

- beekeeping;
- agroforestry, i.e. growing trees along with other crops for building poles, firewood, charcoal, silage, etc.;
- tree nurseries, for species bearing fruit and other foodstuffs, fire wood, and palm oil;
- sustainable farming, for cash and subsistence crops;
- ecotourism, where appropriate;
- environmental education;
- family planning;
- clean water;
- low-fuel stoves;
- health care; and
- microcredit loans.

If hunting is allowed, it is typically managed through licensing procedures, harvest quotas, monitoring of animal off-takes, and enforcement. ICDPs with objectives that include the protection of great ape populations include those for:

- the Oban and Okwangwo Divisions of Cross River National Park in Nigeria;[29, 30, 32]

Iroko Foundation

African giant snails for sale on the roadside in the Niger Delta.

- the Korup National Park and Dja Faunal Reserve in Cameroon;
- the long-running Lake Tanganyika Catchment Reforestation and Education (TACARE) project founded by the Jane Goodall Institutes in western United Republic of Tanzania (and others founded by the same institution, such as the Mengamé Reserve project in Cameroon);
- the project at Nyalama in Guinea[93], which is funded by the European Union and the US Agency for International Development; and
- the Leuser Development Programme in northern Sumatra.

At its heart, an ICDP is a transaction between various stakeholder groups, with the aim of meeting all their needs, typically including those of:

- official donor agencies to relieve poverty;
- participating international NGOs to conserve biodiversity;
- local NGOs and community groups to achieve greater self-sufficiency, autonomy, and collective wellbeing; and
- national governments to deliver on political and international commitments.

Fulfilling all these agendas in a balanced and harmonious way is extremely difficult, particularly in view of the different scales of time and space by which each group judges success. Donors want to design and implement a project of limited duration and expense, with a favorable mid-term and post-project evaluation. National governments want significant benefits delivered quickly to electorates. Local people want to feel more secure, empowered, and hopeful about their long-term future. Conservation NGOs often want a protected area with clear and stable boundaries, that is safe from both hunting and harvesting, and from foreseeable threats that cannot be resisted by well trained and well equipped enforcement staff. Some variations occur around these hoped-for outcomes, but it is a rare ICDP that satisfies everyone. The most successful tend to be those that are slow acting, forum based, and that create genuine partnerships of common interest and understanding between international NGOs, local NGOs, and community groups – with the agendas of both external government and donor agencies sometimes being given secondary importance. In other words, to save a local ecosystem and the great apes within it, local transactions are needed; these must be continuous, ongoing, and cumulative in effect, as is discussed more fully in Chapter 15.

MANAGING THE BUSHMEAT CRISIS

People who live in hunting cultures (such as those in the non-Muslim areas of Borneo, West Africa, and Central Africa) are used to consuming meat from many vertebrate species. They may define their identity partly in terms of their hunting prowess and dietary diversity and, for some ethnic groups, through certain taboos. The fact that a wild animal can be trapped for free is also often appreciated in a rural setting, as is the fact that wild-caught meat may be available in the local market at prices lower than are charged for farmed meat.[54] The 'bushmeat crisis' in West and Central Africa emerges as a result of these preferences being expressed by a growing population with improved transport infrastructure and access to remote areas full of wildlife. This trade is serviced by energetic hunting and trading fraternities; like specialists in any other commodity, these are often clan based and possess secretive commercial contacts over large areas. To the extent that the bushmeat trade is illegal (as it is for great apes), these groups may be thought of as criminal gangs that run a lucrative and shady business. That said, weaknesses in both rural conservation education and law enforcement mean that not all great ape hunters realize that their activities are illegal.

Early attempts to curb this trade were founded on the belief that people consume bushmeat simply because they need protein, and sell bushmeat just because they need cash. In the context of integrated conservation and development projects (ICDPs), the idea arose that the provision of alternative sources of both protein and cash would allow effective enforcement programs in the prevention of this trade. 'Protein projects' have therefore been developed to provide a nutritional supplement or an alternative to hunted wildlife. Rural families have been provided with animals to raise for meat (e.g. rabbits, cattle, or chickens), or support for keeping quick-breeding and culturally acceptable species. Examples of these include cane rats (*Thryonomys swinderianus*) and African giant snails (*Archachatina marginata*). The Développement d'Alternatives au Braconnage en Afrique Centrale (DABAC) project (known in English as Development of Poaching Alternatives in Central Africa) operates in the great ape range states of Gabon, Cameroon, and Congo.[10, 41] It is now becoming clearer, however, that a regional bushmeat issue cannot only be addressed in such ways, and must vigorously target:

- sources of supply through, for instance, hunting bans in logging concessions and the closure of disused access roads;
- market demand through information and education campaigns delivered through schools and all media, to replace the concept of

Jobogo Mirindi/Virunga NP

Rangers and a poached gorilla in the Virunga National Park, Democratic Republic of the Congo.

bushmeat as virtuous with that of it being dirty, primitive, and frequently illegal; and
- trading links through the treatment of the traffickers of illegal bushmeat like any other criminal gang.

In parts of Ghana, the consumption of bushmeat declined dramatically after an intense and pervasive awareness campaign.[111] In the Nigerian states of Cross River and Akwa Ibom, occupational hunting and poaching has become a 'dirty man's job', leading to a drastic decrease, over the last decade, in the number of full-time hunters.[51] These examples offer important clues to how a permanent reduction in hunting pressure might be achieved, through an integrated process that brings a whole range of influences to focus on achieving cultural change, including explanation and outreach, prohibition and enforcement, community mobilization, and investment for sustainable livelihoods.

GREAT APE TOURISM

Where nature-oriented tourism contributes in a sustainable, benign way to support the wellbeing of local people, ecosystems, and wildlife, it is known as ecotourism. Numerous operations that provide great ape tourism claim to meet this strict definition, but there is often a question about how much local communities benefit. Great ape tourism generates income from entry fees, permits, and tracking fees; during 1985–1998, the annual income from gorilla tourism in individual protected areas ranged from US$60 000 to over US$500 000[3, 164] (see Table 14.3). Revenues to individual parks from gorilla tourism are some of the highest in the world,[3, 68] and demand is sustained despite price increases. Great ape tourism is well developed for mountain gorillas in all three range states; for chimpanzees, it is available in Tanzania, Uganda, and (before the war) in DRC. It is also provided for the viewing of partially rehabilitated, confiscated orangutans in Malaysia (and, increasingly, in Indonesia). Tourism in many other range states is considered to be one potentially sustainable way to generate revenue that can be channeled into conservation, and that will encourage community support for conserving great apes as well as other charismatic species and their habitats. The Convention on Biological Diversity has produced guidelines and case studies to promote the sustainable planning and management of tourism activities in vulnerable ecosystems and habitats of major importance for biological diversity.

Table 14.3 Gorilla tourism revenues in East and Central Africa[164]

Site	Period	Visitors/year	Revenue/ year (US$)
Volcanoes National Park, Rwanda	1985–1989	5 800	525 000
Virunga National Park, DRC	1986–1990	2 800	250 000
Kahuzi-Biega National Park, DRC	1988–1991	2 000	200 000
Bwindi Impenetrable National Park, Uganda	1994–1996	2 800	450 000
Mgahinga Gorilla National Park, Uganda	1995–1996	1 200	60 000

In some countries, such as Uganda, gorilla-based tourism is seen as a means of alleviating poverty and as an ideal conservation tool; it is able to attract sufficient visitors to merit significant investment.[9, 96, 165] In 2004, both Uganda and Rwanda increased the price of an individual gorilla-tracking permit from US$250 to US$350 for one hour spent with a family of mountain gorillas, making it the most expensive wildlife-viewing experience in the world.[123] The income, however, is usually subject to the conflicting demands of national and subnational authorities, and is also vulnerable to changes in conservation costs. In Uganda, the distribution of income from gorilla tourism at Mgahinga Gorilla National Park has varied significantly according to national policy. The proportion distributed to local communities has been consistently small;[2] there has been conflict both within the Uganda Wildlife Authority (UWA) over the distribution of ecotourism income among parks, and between the UWA and the Ugandan government. Central government has sometimes viewed the income from tourism as a justification for reducing the budget allocation to the UWA.[3, 9]

While ecotourism may benefit local communities by bringing improved road access or employment opportunities, these are often not recognized as compensating for the perceived costs to the local communities imposed by conservation.[3] The potential for ape-based tourism to secure benefits for local communities can be further limited by factors such as war and unrest that cause tourism income to fluctuate.[26, 27]

Nevertheless, hundreds of people live off the Bwindi Impenetrable National Park in Uganda, where foreign tourists trek to view gorillas and where local people work as rangers, guides, and camping staff, or sell food, crafts, and entertain-

Stuart Chape

Ecotourists traveling by boat through the Kinabatangan area in Malaysia.

ment to the tourists (see Box 8.4). In the Buhoma Valley just outside Bwindi, many new local businesses now offer goods and services to visitors.

Other ecotourism opportunities in Africa include the chance to view:

- western lowland gorillas in Dzanga-Ndoki National Park in CAR[19, 107] and Lopé National Park in Gabon;[86]
- chimpanzee tourism in Kibali National Park in Uganda;[115] and
- the famous mountain gorillas in the Volcanoes National Park in Rwanda.[84]

In Southeast Asia, ecotourism has grown up around several orangutan rehabilitation centers in Malaysia and Indonesia, but lack of controls and poor planning has led to serious criticism of the practice[132, 137] (see Box 14.1). Tanjung Puting National Park in Central Kalimantan and Bohorok in northern Sumatra are unusual in that tourists can observe wild orangutans deep in the forests. At Tanjung Puting there are also free-living rehabilitants who spend time around Camp Leakey, the long-term research site. There are plans for ecotourism centers at Sungai Wain and at Kutai National Park in East Kalimantan.

Another way to look at great ape tourism is to ask whether it has actually contributed to saving forests, quite apart from its benefits to park administrations and local communities. It is possible that the answer to this may be yes, as the forests around Bohorok in northern Sumatra and also those near Ketambe have generally suffered far less from illegal logging than have other nearby areas. This effect presumably arises thanks to both the greater presence of outside observers where the tourism occurs, and the benefits of tourism that might be valued by those who would otherwise be logging or hunting in the area.

Tourism cannot protect all populations of great apes. Analyses of tourist revenues have shown that it is highly unlikely that the costs of managing protected areas in the Congo Basin can ever be generated fully in this way.[164] Mountain gorillas live in small habitat islands that can be both easily accessed by tourists and intensively protected by relatively small numbers of rangers. They may be able to survive with the support of ecotourism, but this mechanism is unlikely to support the conservation of eastern and western lowland gorillas and other great apes that occur over wide ranges and in difficult-to-access areas.[19, 117] If carefully developed as part of a range of activities, however, it could contribute toward the costs of conservation and bring infrastructure development that could benefit both conservation work and local communities.

Tourist spending generally contributes to the profits on investments by international hotel chains, airlines, and tour operators, as well as on imported goods and services used in the tourism industry. Tourism is a fickle business; few ecotourists travel to areas at risk from war or civil strife, or where there are severe health problems. The turbulent 1990s in the African Great Lakes region, however, demonstrated that gorilla tourism is remarkably durable. There are always at least some enthusiasts prepared to accept certain risks in the hope of a meaningful encounter with gorillas. The temporary collapse of the gorilla-watching tours to Rwanda and DRC during the 1990s due to the civil wars and genocide there led to an increased demand to view gorillas in Uganda; as soon as the fighting stopped, gorilla tourism began to pick up again in both Rwanda and DRC. The world now contains many wealthy people who are prepared to pay to encounter semi-wild great apes; they will keep coming, as long as the apes are there.

Even well planned ecotourism has potential costs, however, in particular the risk of disease transmission between humans and apes[27, 75] (see Chapter 8). Great ape tourism can also be seen as merely another form of exploitation for entertainment and commercial gain; there is an ongoing debate over whether it is morally acceptable to subject any wild ape to any practice (including habituation and tourism) that intrudes and exposes that ape to potential or actual harm.[26]

CONSERVATION *EX SITU*

Overview

The principles of zoo-based conservation are set out in the *World Zoo Conservation Strategy*,[39] which is currently being revised. Publication of the *World Zoo and Aquarium Conservation Strategy*[64] is expected in 2005. Both documents emphasize:

- the need for zoos to maintain populations of animals (to minimize loss of genetic diversity and maximize retention of natural behaviors) in conditions that promote good animal welfare and allow expression of natural behaviors;
- the ambassadorial and educational role of such populations;
- that populations of captive animals are valuable in furthering knowledge of the biology of species; and
- that such populations should link with conservation activity in the wild, whether through breeding for reintroduction or raising funds for conservation *in situ*, either by the holding establishment or by other organizations (it is increasingly clear that the latter role is much more important than the former for most species held in zoos).

Management of captive populations

The management of captive populations of great apes is carried out through regional collaborative breeding programs managed by the American Zoo and Aquarium Association (AZA), the European Association of Zoos and Aquaria (EAZA), or the Australasian Regional Association of Zoological Parks and Aquaria (ARAZPA). All use studbooks as their basic management tool (see Table 14.4).

All species other than chimpanzees are managed as separate species or subspecies. The AZA manages chimpanzees of all four subspecies as one population, while the EAZA (which has a relatively large number of *Pan troglodytes verus* in its region), manages that subspecies as a separate population from chimpanzees of unknown origin or mixed subspecific status. Movements of animals between different collections typically take place on the recommendation of the designated species coordinator, who takes account of a number of factors in making recommendations. These include genetic factors such as avoidance of inbreeding or founder contribution to the population, and social factors such as the age or social history of the animals concerned. The species coordinator also identifies animal-management issues that need to be addressed or researched. These might include investigation into causes of death or illness, or research into reproduction or social biology.

Table 14.4 Captive-breeding programs for great apes

Taxon	Authority[a]	Studbook
Chimpanzee (*Pan troglodytes* spp.)	AZA, ARAZPA	regional
Western chimpanzee (*Pan troglodytes verus*)	EAZA	regional
Bonobo (*Pan paniscus*)	EAZA, AZA	international, regional
Western gorilla (*Gorilla gorilla gorilla*)	EAZA, AZA, ARAZPA	international, regional
Eastern lowland gorilla (*Gorilla beringei graueri*)		international[b]
Bornean orangutan (*Pongo pygmaeus*)	EAZA, AZA, ARAZPA	international, regional
Sumatran orangutan (*Pongo abelii*)	EAZA, AZA, ARAZPA	international

a ARAZPA, Australasian Regional Association of Zoological Parks and Aquaria; AZA, American Zoo and Aquarium Association; EAZA, European Association of Zoos and Aquaria.
b Only a small number of old eastern lowland gorillas are present in zoos, and an active program for this subspecies is not possible.

David W. Liggett (www.daveliggett.com)

Captive great ape breeding programs are taking place in many zoos, including at the Columbus Zoo and Aquarium, which houses the bonobo pictured here. She was born in 2000.

Box 14.1 ORANGUTAN TOURISM

Orangutan tourism has operated continuously since the early 1970s. From the outset, it focused on rehabilitant orangutans, i.e. orangutans captured for the illegal wildlife trade as infants, then confiscated from illegal captivity and rehabilitated to forest life. Four rehabilitation sites have been heavily involved: Sepilok (Sabah), Tanjung Puting (Central Kalimantan), Bohorok (northern Sumatra), and Semenggoh (Sarawak). All had conservation education and fundraising as secondary aims, and promoted tourism in the expectation that it would generate benefits to both.[12, 128] This proved highly attractive to tourists, largely because it is far easier to view rehabilitants than wild orangutans. Wild orangutans are semisolitary, elusive, and typically stay high in the forest canopy; rehabilitants are habituated to humans, comfortable near the ground, and visit accessible feeding sites daily, and on schedule.

When rehabilitation projects and rehabilitant-focused tourism were launched, knowledge of orangutan readaptation was relatively limited and little thought was given to the negative impacts of tourism. These became evident when experts began assessing orangutan rehabilitation in the late 1970s. Two sites were already experiencing heavy tourist usage by then: Bohorok attracted up to 5 000 visitors annually; Sepilok drew up to 17 000.[12] Tourism-related problems that surfaced included excessive rehabilitant–human contact and undermining of feralization of the ex-captives.[58, 128] Perhaps more seriously, tourists and staff at rehabilitation projects were shown to be sources of diseases (such as tuberculosis, hepatitis B, and poliomyelitis) that could pass to rehabilitants, and then into the wild population.[58, 126]

Most experts therefore recommended change. Some argued that tourism could be, and had been, managed effectively to reap both economic and educational benefits.[11, 58] Others argued that visitor control and the benefits promised were rarely realized, and any benefits simply did not offset the costs to the readaptation and health of the orangutans.[99, 128] One recommendation for change was to restrict tourist-rehabilitant contact.[126, 128] Efforts were made to discontinue tourism at some sites, but this proved difficult because the sites and local businesses had become dependent on this source of revenue.[11] In other words, the rehabilitation of orangutans was coming to be driven by economic rather than conservation interests, with rehabilitants being encouraged to stay around for display rather than to resume independent forest-based life.

Tourism at orangutan-rehabilitation sites soared after the early 1980s in such an uncontrolled manner that it greatly intensified the costs to ex-captives. Many of the problems identified in the 1970s persisted into the 1990s and some worsened, often because the initial recommendations for change from experts had not been adopted.[125] By the turn of the century, a disturbing picture of orangutan-focused tourism was emerging, with little evidence to support claims of economic or educational benefits, and growing worries about adverse consequences to both health and behavior.

One of the few systematic studies on orangutan tourism was a case study at Tanjung Puting focused on tourist–orangutan interactions and the educational rationale.[131, 132] The expectations and behavior of the tourists were not tempered by the minimal educational programming provided. Many of them interpreted young ex-captives to be much like human infants and

Education and the 'ambassadorial' role

Great apes are a focus of attention for zoo visitors. Interpretation through a variety of media conveys messages on the endangered status of apes and the reasons for it. These media include static information boards, interactive computer-based presentations, short films, and oral presentations given by keepers, volunteers, or education personnel. In 2001, EAZA ran a year-long campaign on the issue of bushmeat, with the aim of raising awareness and funds, and organizing a petition to African leaders and the European Parliament to demand better control of the illegal bushmeat trade. The campaign information boards were seen by millions of people across Europe; it raised about US$50 000 and collected 1.9 million signatures on a petition to the European Parliament. The latter resulted directly in a report on the bushmeat trade being adopted by the European Parliament in January 2004. This report recognized the bushmeat trade generally as a livelihood- and poverty-related issue. It also recognized that illegal trade is a major threat to the

sought every chance to hold and cuddle them, either unaware or unwilling to acknowledge the health hazards this created. Others insisted that rehabilitants were not 'real' and sought out wild orangutans, potentially increasing pressure on wilder areas. A later survey of other rehabilitation sites confirmed that tourists were rarely informed of rehabilitation issues or of how they should behave with ex-captives.[125]

We know of no systematic studies on the economic impacts of orangutan-focused tourism on local communities. Available information suggests patterns similar to those typical of wildlife tourism elsewhere: most economic benefits are captured by large (external) businesses; few benefits reach local communities. International operators may, for example, advertise high-priced ecotourism experiences at orangutan rehabilitation sites (costing e.g. US$3 600 per person for a 12 day visit, in addition to international airfare). Some money from associated work reaches local people, but local salaries and costs in Indonesia and Malaysia are very low by international standards, so much of the income tends to remain outside local or even national hands. Local businesses (e.g., hotels, restaurants, guides, transport, shops) have flourished around some sites but in such an uncontrolled and poorly managed fashion that they have exacerbated health and behavioral problems for the rehabilitants, and aggravated the deterioration of their habitats.[125]

Recognizing the extent of these problems and the lack of evidence of benefits, the community of orangutan specialists officially recommended that tourism no longer be allowed in areas with rehabilitant orangutans that have been reintroduced to forest life, or where the orangutans are eligible for reintroduction.[137] Tourism focused on wild orangutans was considered acceptable, but only with very careful controls.

There has been, as yet, very little systematic research on orangutan-focused tourism. Research would be most welcome on impacts of tourism on the success of ex-captive orangutan rehabilitation; disease transmission from tourists to ex-captives and from ex-captives to wild orangutans; economic impacts of orangutan tourism on local communities; and educational impacts on both the tourists and the local people.

Anne E. Russon and Constance L. Russell

A Bornean orangutan at the Sepilok Rehabilitation Centre in Sabah, Malaysia.

Elaine Marshall

African great apes, and recommended that measures to help address it be supported by the European Union through the European Commission.

Links to conservation *in situ*

An increasing number of zoos contribute to conservation in the field by supporting projects in range states. The support provided may be financial, gifts in kind, or the provision of technical expertise. Projects so supported include pilot ecotourism projects; research into the bushmeat trade; research into conservation status; support of sanctuaries and reintroduction projects; and support for both rangers and the management of protected areas. The involvement of zoos varies too; some raise funds for particular projects from their own visitors, while others provide grants from central funds. A growing number become actively involved in the supported projects.

Organizations such as the Zoological Society of London, Wildlife Conservation Society, Antwerp Zoological Society, and Frankfurt Zoological Society

Ian Redmond

Orphaned chimpanzees confiscated by authorities and housed in a makeshift sanctuary, Lwiro, Democratic Republic of the Congo.

have been involved in field conservation for many years. Others, such as Bristol Zoo Gardens, have more recently become involved directly in field projects. The pilot EAZA conservation database (which contains data from around 20 percent of EAZA zoos) shows that at least 22 zoos support great ape field conservation projects directly. At least 18 projects benefit from that support, which amounted to over US$530 000 in value between 1999 and the end of 2002.

Finally, zoos have successfully reintroduced captive-bred individuals of various species into the wild. So far the only attempt to do so for great apes is the project initiated by the late John Aspinall, in which Kent-born gorillas from Howletts Zoo were sent to Gabon to join a rehabilitation program for orphans of the bushmeat trade. In this case, both individuals died but a second attempt is underway.[42, 123] Captive-bred great apes are likely to have particular difficulty in the wild before they are able to assimilate culturally transmitted skills such as foraging, parenting, and interacting with other ape groups. There has been more success with wild-born but orphaned apes, as described below.

SANCTUARIES, REHABILITATION, AND RELEASE

Overview

Sanctuaries have arisen largely on an *ad hoc* basis as a crisis-management measure and are seen by some as being of limited conservation value. Laws against keeping or trading live animals, however, can only be enforced if the confiscating authority has a suitable facility in which to place the animals in question. The alternative of euthanasia is deeply unattractive for endangered species in general, and for great apes in particular. Government-approved sanctuaries have therefore become an accepted solution to the dilemma of what to do with great apes that have been confiscated or discarded as pets, performers, or research subjects. In addition, they can provide valuable opportunities for conservation education and for raising public concern and awareness for the plight of wild ape populations.

The number of great apes held in sanctuaries has increased in recent years,[114] probably reflecting growth in logging, habitat destruction, and the bushmeat crisis in Africa, as well as deforestation and forest fires in Indonesia and Malaysia. These institutions, of which there are now over 50 worldwide, hold animals that are either donated to them or are confiscated by the local wildlife authorities. Sanctuaries holding apes are, for the most part, in range states or in states close to the natural range, but there are a small number outside of Africa, Indonesia, and Malaysia that hold former laboratory animals or animals that were confiscated in Europe that have not yet been repatriated to range states. Sanctuaries in range states have a number of roles in great ape conservation:

- in caring for individual animals, they fulfill a role in animal welfare;
- by providing a place where confiscated animals can be held, they help the authorities implement robust confiscation policies that are an important part of the enforcement of wildlife law and the protection of great apes;
- by providing animals for reintroduction, they may be able to help restore wild populations; and
- by exposing local people and visitors to great apes, they can have an important educational impact, helping people to appreciate the value of the species concerned and to understand the conservation problems affecting their wild counterparts.

Animals in sanctuaries often arrive in poor health, having frequently experienced inadequate housing, inappropriate social groups, or solitary confinement, and a poor diet. Indeed, at least four out of five gorilla orphans die before reaching expert care.[121] This presents challenges in management that do not usually arise for zoo animals, and results in a strong focus on animal health and welfare at sanctuaries.

Some sanctuaries have been instrumental in establishing protected areas, an example being the Afi Mountain Wildlife Sanctuary in Cross River state, Nigeria. This was prompted by the efforts of the NGO Pandrillus and the Drill Rehabilitation and Breeding Centre.[34] At least one sanctuary, the Tchimpounga Sanctuary in Congo, is developing a long-term program centered on community-based conservation principles, including a health dispensary to serve local communities.[82] Financing and long-term planning for conservation of captive populations at sanctuaries have been difficult to achieve and there is a need for general guidelines on the establishment of great ape sanctuaries. These should address liaison with host governments, local communities and authorities, site location, long-term sustainability, management practices, animal management, and health issues.[114]

African sanctuaries

The first great ape sanctuary in Africa was established in 1969 in the Gambia by Stella Brewer.[25] The Chimpanzee Rehabilitation Trust Gambia continues to protect and monitor a naturally reproducing population of 63 chimpanzees on three islands in the River Gambia National Park.[34, 40] Many other primate and wildlife sanctuaries have also been established throughout sub-Saharan Africa; in 2000, these formed a network, the Pan African Sanctuary Alliance (PASA). It encourages liaison between the sanctuaries, their sponsors, and conservation organizations, and lobbies more effectively than each could on its own for a conservation approach that includes rehabilitation and reintroduction.[45, 56]

PASA's 2003 Workshop Report lists 20 sanctuaries, 19 of which are PASA members, and 18 of which hold great apes.[130] The apes held in these 18 sanctuaries in June 2003 included 632 chimpanzees (of unknown subspecies), 67 gorillas (believed to be mostly western lowland gorillas), and 27 bonobos. The large proportion of chimpanzees reflects their greater ability to survive ill treatment and poor captive conditions; it may also indicate higher hunting rates or their greater geographical range (or both).

Confiscated great apes typically come to sanctuaries as infants or juveniles. They are mainly part of a local pet trade, which almost always emerges as a side effect of bushmeat hunting (see Box 14.2). Young animals that survive an attack by hunters are often considered potentially more valuable as an item for sale than for consumption, provided they can be kept alive long enough to be sold. The number of all great apes in sanctuaries is likely to continue to increase as some range states adopt more robust policies on the confiscation of illegally held animals. This policy shift has certainly been seen in Cameroon, where recent court cases have resulted in substantial fines and prison sentences for those found in possession of illegally held apes.

Rondang Siregar

Orphaned Bornean orangutans at the Wanariset Rehabilitation Centre near Balikpapan in East Kalimantan, Indonesia.

Southeast Asian sanctuaries

An estimated 600 formerly captive orangutans were under the care of rehabilitation centers in Malaysia and Indonesia in 2002.[129] There are three centers in Malaysia (including one in Peninsular Malaysia, where orangutans do not occur in the wild); four in Indonesia, with three in Borneo and one in Sumatra. Bohorok, an earlier center at Gunung Leuser National Park in Sumatra, officially ceased to take in new orangutans from 1995. Several centers are tourist attractions, as well as simply caring for and releasing captive orangutans (see Box 14.1). Many young orangutans have, in the past, entered a thriving local pet trade, with large numbers being exported illegally to Taiwan and Thailand. More robust application of wildlife laws in the range states and in Taiwan has resulted in increasing numbers of illegally held captive animals being moved to sanctuaries for care and, if possible, for rehabilitation and release.

In Indonesia and Malaysia, orangutan rehabilitation centers have formed a network and held a series of workshops to exchange experiences and develop improved practices. At the 2002 Orangutan Conservation and Reintroduction Workshop, it was

Box 14.2 SENDJE, AN ORPHANED CHIMPANZEE

Thousands of tons of bushmeat are consumed every year in Central Africa, and much of it comes from animals that are considered to be Vulnerable, Endangered, or Critically Endangered under the *Red List* of IUCN–The World Conservation Union. There are many factors that encourage bushmeat consumption, including cultural tradition, increasing human populations, and expanding access to forest areas. There are also many consequences, including the ecological impacts of eliminating seed dispersers, and the outbreak of diseases such as Ebola hemorrhagic fever following contact between humans and wild animals. Another effect is the growing number of orphaned great apes. What follows is the story of one such ape that also sheds light on the role of the expatriate community in Equatorial Guinea.

Equatorial Guinea has a quickly growing, petroleum-fueled economy that has attracted increasing numbers of white expatriates. Many of these are willing to buy orphaned chimpanzees, (even though this is illegal), either because of their 'cuteness' or to prevent them from being eaten by local people. By selling one baby chimpanzee to an expatriate, the average hunter is able to make more money than by selling 20 blue duikers to the bushmeat market; this has become too well understood by the local hunting population. Before the oil boom, great apes were seldom hunted for meat, as the adults are heavy to carry. Whole groups of chimpanzees and gorillas are now killed to provide orphans for sale to expatriates. By buying chimpanzees 'in order to save them', the expatriate market has stimulated great ape hunting and the pet trade.

This is the story of an orphaned chimpanzee in Equatorial Guinea, known as Sendje. She was named after the village closest to the forest from which she came. Once orphaned, she eventually came under the care of a conservation project in Equatorial Guinea; Brigid Barry became her surrogate mother:

I first heard about this baby chimpanzee arriving in the city when I saw the corpses of two adult chimpanzees in the bushmeat market. This is not an unusual sight, as dead chimpanzees were seen in the market a few times a week. What was unusual was that one of the adults was a lactating mother. A local employee of the project then described how a live baby chimp had also arrived in the market and that a taxi driver had taken it off to try to sell it to the expat community.

Sendje with carer. Young chimpanzees need constant physical contact.

Lise Albrechtsen

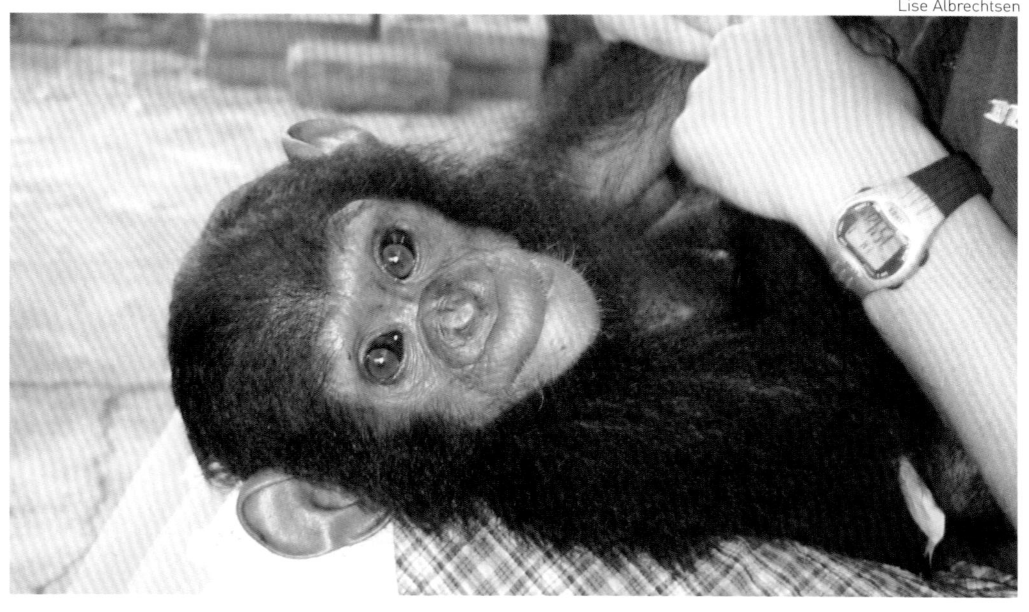

Later that night, a group of drunken Europeans at the local disco bought Sendje for a large sum of money because they felt "sorry for her and thought perhaps they could return her to the forest." The prospects for a lone juvenile chimpanzee in the forest are extremely poor; without her mother, she would begin to starve at once, and would be taken by a predator or scavenger within days. After two days these new owners contacted Brigid, claiming that they no longer wanted Sendje as she was too much of a handful. When she was brought into Brigid's care, Sendje was only semiconscious. This was perhaps due to a gash on her crown, which she might have suffered when her mother was shot, but her condition seemed more likely to be due to her diet over the previous two days of bacon sandwiches and beer. It was estimated that she was only about eight or 10 months old. As chimpanzees drink their mother's milk for up to 18 months, Brigid concentrated on giving her a diet of powdered milk alongside a course of supplementary vitamins and calcium. Within a week she regained the sheen of her coat and became quite active.

Young chimpanzees raised in captivity are even more demanding than human babies. During the night, they must be fed about three times and require up to three diaper changes. Unlike human babies, for the first 18 months of life a chimpanzee does not relinquish physical contact with its mother or another member of its community. For the human 'chimp-sitter' this becomes difficult. Sleeping with a furry creature attached to you through a sweaty tropical night is not pleasant. Even having a shower or bending down to tie one's shoe laces becomes tedious when the activity is performed to a constant accompaniment of ear-piercing screeches and bites. Brigid was unable to leave the house with this chimpanzee, as she did not want any of the locals to think that she was a potential buyer for other captured wildlife.

With the financial help of an employee of an American oil company, two daytime chimp-sitters were employed and a garden was found with plenty of trees for Sendje to climb. At nights, Brigid and one other person took turns at chimp-sitting.

But what of Sendje's future? At two or three years of age, she would become a serious threat, as an adult chimpanzee is much stronger than an adult human. There are currently no animal sanctuaries in Equatorial Guinea, and there was no hope of reintroducing Sendje to the forest. After many searches and much discussion with European and American primate experts, the Sanaga-Yong Chimpanzee Rescue Center in Cameroon agreed to take her. The oil company offered to take her to Cameroon in its private jet, the necessary vaccinations were administered, and CITES export and import papers were processed. Sendje finally made the journey to live among other chimpanzees at the sanctuary.

Lise Albrechtsen

A recently killed adult chimpanzee in the Mondasi bushmeat market in Bata, Equatorial Guinea.

Sendje's situation is not ideal. She will never return to the life she knew before her mother was killed. If she remains in good health, she might live for 50 years in captivity and could give birth there. Let's hope that the future of chimpanzee conservation does not lie behind metal bars. Although Sendje was saved from the cooking pot, the expatriates who bought her at the disco contributed to the growth of the pet trade for orphaned chimps. Over the following two months, Brigid was asked three times whether she would like to buy a chimpanzee baby, offers that were rejected and used as a way to raise awareness of the issue among the expatriate community. It is hoped that, with enough rejections, the hunters will return to thinking that these endangered apes are too heavy to carry out of the forest and that the juveniles are not worth capturing. Another thing to hope for is that the national laws protecting endangered species will eventually be enforced. With these changes, Sendje's sad history and future life in captivity would become among the last such experiences for the chimpanzees of Equatorial Guinea.

Lise Albrechtsen and Brigid Barry

Benoît Goossens/HELP International

A field assistant holds an adult chimpanzee who has been tranquilized for the journey to the release site in Conkouati-Douli National Park, Congo.

Benoît Goossens/HELP International

This female was one of the first chimpanzees to be released at Conkouati-Douli (with four other individuals), in 1996. In 2003, she gave birth to a male infant.

resolved to establish the Orangutan Conservation Forum (OCF). At the 2004 Population and Habitat Viability Assessment Workshop in Jakarta, NGOs pledged over US$25 000 in funds to support the OCF, as its establishment had foundered through lack of resources. The forum is intended to provide policy advice, media awareness, and networking to improve conservation effectiveness.

Reintroduction and translocation

Although sanctuaries provide food, shelter, care, social interactions, and varying degrees of freedom (from spacious cages to electrically fenced enclosures), they are still a form of captivity. Rehabilitation and reintroduction to the wild are therefore often proposed for confiscated apes, and many sanctuaries work toward this goal. Reintroduction is often not possible for a variety of reasons, however, since an animal's capacity to respond to rehabilitation efforts depends on several variables, including its age and the conditions and duration of its prior captivity. Animals with severe behavioral problems or deformities, for example from snare-related injuries, cannot be released safely into the wild.[34]

Reintroduction is rarely a straightforward task. There are often questions over the desirability of mixing apes of different genetic provenance (confiscated apes are usually of unknown origin). There is also potential for disease transmission to wild populations, increasing competition for scarce food resources between reintroduced and native apes, and the possibility of aggression between them.[86] New genetic research and advancing DNA techniques can now be used to analyze genetic makeup. Mitochondrial DNA analysis, for example, showed that a female infant confiscated in DRC was an eastern lowland gorilla from Kahuzi-Biega National Park;[60] in 2004, testing was planned to identify the origins of over 100 orangutans being held illegally at a Thai park.[7] Reintroduction is pointless unless a secure habitat is available where there is little or no hunting.[34, 109] If rehabilitant apes are released near human settlements, their familiarity with humans sometimes leads them to be aggressive, to raid crops, and enter villages.[26]

Any release sites must therefore be chosen with great care. In Indonesia, the national policy is now to allow release at sites only where the local orangutan populations have been lost. As with any reintroduction project, it is critical to ensure that the new individuals will not succumb to the same pressures that destroyed the original population. As reintroductions are costly and time consuming,[150] some argue that the money and effort would be better spent on tackling the factors leading to the creation of orphan apes in the first place.[26]

These same concerns also apply to translocation, a procedure in which animals are captured from the wild, given veterinary care during a period of quarantine, and then released in a new location.[31] It has been suggested that translocation be used as a way to consolidate isolated individuals and small groups of great apes into a single larger population in one area, which would be easier to protect and would have a larger and more diverse gene pool.

One potential candidate for such treatment could be the Cross River gorilla, which is widely scattered among disjunct sites in Nigeria and Cameroon[51] (see Box 7.1). There are so many potential drawbacks to this approach, however, including its high cost and the health risks posed to great apes newly in contact with people and with each other, that such an effort should only be considered as a last resort.

It is clear that techniques and protocols for successful rehabilitation and release of great apes must continue to be developed, and suitable protected sites should be identified for reintroductions, as the number of confiscated animals will continue to grow. Captive breeding of great apes for release into the wild is not a cost-effective conservation strategy at present. For it to be effective, not only must animals breed successfully in captivity, but the many problems of reintroduction must also be solved. Many believe that the main value of captive breeding of great apes is not as a tool for conserving wild populations, but as a means to satisfy demands for biomedical and scientific research and for the apes' educational value in zoos,[109] to the extent that these uses are deemed ethical.[26]

Rehabilitation and release of African great apes has been carried out at a number of sites; these involve mostly chimpanzees that have been released onto islands in rivers, and that generally require continued supplementary feeding. An exception, however, is the release by the Frankfurt Zoological Society of 17 wild-born chimpanzees on the 240 km^2 Rubondo Island in Lake Victoria, United Republic of Tanzania, in the late 1960s. The chimpanzees were from West Africa originally and had spent between three and a half months and nine years in captivity in Europe.[124] These 'problem' animals were largely left to their own devices, but had given rise by 2004 to a self-sufficient breeding population of more than 40 individuals living in three groups.[22, 77, 101]

There has been one successful release of chimpanzees into mainland forest, with 25 of 34 released animals surviving, three having been confirmed to have died, and the fate of six being unknown[55] (see Box 4.5). The first equivalent program for gorillas started in 1994 in Congo; poor initial survival rates later improved.[42, 43] This was followed by reintroduction of gorillas in Gabon in 1998, and of chimpanzees in Congo from 1996.

Rehabilitation and release is an inherently attractive prospect, of interest to both the media and to politicians, so it remains firmly on the agenda in many places. It has been carried out for orangutans since the mid-1960s, and five projects currently have release programs at a number of sites. Some have continued supplying food to released animals, while others do not provision, in order to encourage independence. Minimum criteria for both behavior and health have been recommended for individual great apes prior to release, and criteria have also been proposed for the suitability of release sites.[129] Provided that animals are judged to be suitable for release, survival rates are good.[134]

PUBLIC EDUCATION

Beside the fundamental reform of governance, investment, trade, national economic development, and poverty eradication, conservation education is arguably the most important tool for saving the great apes. Without awareness and understanding among stakeholders, other measures are likely to be ineffective. Most great ape conservation programs therefore have an educational component. All of the African range states have significant weaknesses in the areas of education and literacy,[87] so approaches intended to increase public awareness of conservation issues must be designed to take account of constraints within the target societies and educational systems.

Various techniques have been used, based for instance on the use of comic books and theater. An example of the latter is a play developed by the Wild Chimpanzee Foundation to raise awareness in towns and villages near forests with chimpanzee

Ian Singleton/SOCP

Orangutans released at the reintroduction site adjacent to Bukit Tiga Puluh National Park in Jambi province are subsequently monitored – and sometimes join in.

Michael Huffman

Two WaTongwe research assistants, second generation members of the Kyoto University Mahale research team, and a Tanzanian PhD candidate working together in the field.

populations. Performances, given in French by the theater troupe Ymako Teatri, have reached over 8 000 people in 17 communities around the Taï National Park in Côte d'Ivoire. The play uses actors accompanied by live music to portray chimpanzee behavior, the effects of hunting on chimpanzees, and the connections between animal and human communities. It emphasizes human connections to chimpanzees through totems and ancestry, depicting a conflict that arises between families whose totem is the chimpanzee and a hunter who shoots and kills a chimpanzee. The play can be adapted to local customs to suit different audiences but conveys the message that people should not kill and eat apes, because they are closely related to humans.

Conservation education projects are frequently an important part of the work of ape sanctuaries,[56] where animals serve as a focal point for explaining the purpose and importance of wildlife conservation. Urban Africans, Indonesians, and Malaysians seldom get a chance to see wild animals in their natural habitat, but visits to sanctuaries provide stimulating opportunities, especially for classes of schoolchildren. Examples include:

- the Chimpanzee Rehabilitation Trust, at the River Gambia National Park, which has helped to eliminate the illegal trade in young chimpanzees in Gambia;
- the education program of the Lola ya Bonobo Sanctuary close to Kinshasa in DRC, which has the largest group of captive bonobos in the world and reaches up to 10 000 students each year;
- the Uganda Wildlife Education Centre and Ngamba Island Chimpanzee Sanctuary, which play a similar role in Uganda;
- the Drill Rehabilitation and Breeding Centre in southeastern Nigeria, which is managed by the NGO Pandrillus and has exposed many thousands of Nigerians to drills and chimpanzees, both at its former location in urban Calabar and more recently in the Afi Mountains; and
- the Limbé Wildlife Centre, Cameroon, which is visited by more than 30 000 people annually and whose outreach program visited more than 100 schools and 11 000 students in 2000.[34]

The Pan African Sanctuary Alliance (PASA) is also working with the Bushmeat Crisis Task Force to develop tools to improve education and public awareness, at both local and national levels. They hold workshops to bring educators from different sanctuaries together to share experiences and develop materials.

CONSERVATION-ORIENTED RESEARCH

Field research has become intimately linked to conservation efforts, as primatologists have witnessed the decline of the great apes and the destruction of their habitats, and have become strong advocates for conservation.[38] Research projects bring in foreign exchange, provide some employment for local people, and focus attention on the area of study. If the site concerned is already protected, the research project and its staff can be a source of practical support and knowledge to those responsible. If it is not protected, the results of the study or survey may persuade local authorities that it should be.

Examples of long-term research programs that have been highly influential in favor of great ape conservation include those at Karisoke Research Center in the Volcanoes National Park, Rwanda,[116] Gombe National Park in Tanzania,[65, 66] Bossou in Guinea,[102, 167] the Taï National Park in Côte d'Ivoire,[20] Tanjung Puting National Park in Indonesian Borneo,[59] and Ketambe in the Gunung Leuser National Park, Sumatra.[127] These and others are described in the relevant species chapters and country profiles of this volume. Many recent research projects seek to involve local communities in conservation activities, and frequently include conservation education and ecotourism elements (e.g. the collaboration between WWF and the Ba'Aka people of Dzanga-Ndoki National Park in CAR).[71]

COMMUNITY INITIATIVES

Communities may declare protected areas where they have the legal authority to do so and sufficient awareness of the advantages to be gained by doing so. The Lossi area, some 50 km southwest of Odzala National Park in Congo, was placed under such protection after the community was sensitized by researchers attracted by the high densities of western lowland gorillas found there. The community became aware of the potential of gorilla-based tourism.[16] A pilot tourism project was developed, gorilla groups habituated, and monitored from 1994 onwards; hopes were high that benefits would be obtained by local communities. The Lossi Gorilla Sanctuary was however hit by an outbreak of Ebola hemorrhagic fever in 2002, in which the gorilla population was devastated, and two habituated groups were wiped out.[157]

The Mbe Mountains Community Forest was established in the early 1990s by the people of the area to link the Okwangwo Division of Cross River National Park and the Afi River Forest Reserve in Cross River state, Nigeria.[110] This decision was informed by the educational work of Pandrillus in the area, by the wish to help protect Cross River gorillas, and by the general sense of wishing to participate in the process of conservation and sustainable development prompted by the investment from governmental, international NGO, and donor agency sources in and around Cross River National Park.[29]

Decentralization in Indonesia following the fall of the authoritarian Suharto regime in 1998 has often been blamed for an increase in illegal logging throughout the country,[81] as local officials became able to award logging concessions without reference to Jakarta. In one respect, however, the changes are beginning to prove beneficial. If local government leaders decide that the interests of their people are best served by conservation, then they have the authority to take proactive steps in a way that was impossible previously. Thus, the regency (*kabupaten*) of Mandailing Natal in North Sumatra, under pressure from 30 000 local people who had petitioned the Resident (Bupati), proposed the creation of the Batang Gadis National Park, an area of 1 080 km^2. The park was endorsed by central government and inaugurated by President Megawati in May 2004.[143] This new 'bottom-up' national park has been hailed by Conservation International (CI) as a new model for conservation cooperation between the central and regency level of Indonesian society.[141] Apart from preventing disasters, local people expect the new park to encourage cooperation with NGOs such as CI, which has negotiated the sale of organic coffee from the surrounding area through the international Starbucks chain. A similar community-driven initiative is leading the protection of the Sungai Wain forest near Balikpapan, East Kalimantan.[134]

These local conservation initiatives exist in the context of a global advance in local-scale involvement in forest management. Throughout Latin America, for example, local governments have become involved in planting trees, fighting fires, zoning, managing parks, granting permits, and charging fines; hundreds of municipalities have their own offices and commissions working on forestry and the environment.[57] Likewise, in the Philippines, implementation of the 1991 Local Government Code has authorized municipal governments to set areas of up to 50 km^2 aside as community forests to safeguard water catchment and other environmental services.[28] They are increasingly doing this in partnership with local and international NGOs and community groups. These decentralized arrangements are paralleled in Africa through the African Forest Law Enforcement and Governance process (see earlier section on Regional activities and agreements, this chapter). They give indigenous peoples, small farmers, foresters, and local environ-

SOCP

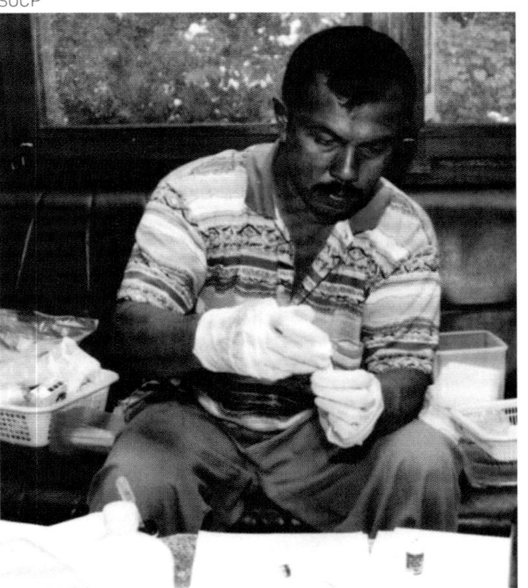

Fecal samples, collected from wild orangutans for parasitology and DNA analysis, being processed at the Sumatran Orangutan Conservation Programme's Ketambe Research Station, Indonesia.

Julia Lloyd

Staff at Kibale National Park, Uganda, have recently received vital equipment for work in the field.

mental groups new opportunities to participate and to deter unwelcome outside interests. It is expected that the next generation of great ape conservation projects will take full advantage of this redistribution of authority over the environment.

FINANCING GREAT APE CONSERVATION

Over the years, significant amounts of money have been committed to great ape conservation or to projects that are likely to benefit great apes directly or indirectly. The NGO community has raised and spent several tens of millions of US dollars on great apes since the late 1970s; more recently governments and official donors have joined in. The US Great Ape Conservation Act of 2000, for example, committed about US$1 million each year to support a program of small grants (in the US$30 000 range). The US government plans to spend up to US$53 million during 2003–2005 on conservation and sustainable resource use through the Congo Basin Forest Partnership, and this is likely to be at least matched by other donors and partner governments. Of the US contribution, US$12 million per year for three years will be disbursed through the Central African Regional Program for the Environment (CARPE), a program begun in 1995 by the US Agency for International Development with a mandate to assess the environment in nine countries in the region.

Meanwhile, the United Nations Foundation has contributed US$3.3 million through the Central African World Heritage Forest Initiative, matched by NGO partners in 2003–2007.[154] The foundation also contributed substantial funds in 2001 to support the five World Heritage Sites in DRC during that country's debilitating civil war.

Since 1992, the European Commission has committed some US$50 million to the program Conservation and Rational Use of Forest Ecosystems in Central Africa (ECOFAC). This has as a priority the involvement in its activities of forest-dwelling peoples.[49] ECOFAC is engaged directly with the conservation of the western gorilla and chimpanzee by, for example, supporting gorilla-based tourism, biodiversity assessments, and primate censuses in Odzala National Park in Congo and Lopé National Park in Gabon, and the administration of the Ngotto Classified Forest in CAR.

Since the early 1990s, the European Commission has also supported the Leuser Development Programme in the range of the Sumatran orangutan, various projects to encourage reform of the Indonesian timber industry, and the suppression of illegal logging (among the most serious threats to orangutans in Indonesia).

Finally, the Global Environment Facility (GEF) has invested many millions of dollars in the protection of great ape habitat areas, typically in matched-funding arrangements with bilateral and other multilateral donors and lenders, and with range-state governments. Examples of relevant GEF grants include US$4.43 million to the Bwindi Impenetrable National Park and Mgahinga Gorilla National Park Conservation Project in Uganda, and US$10.46 million to the Conservation of Transboundary Biodiversity in the Minkébé–Odzala–Dja Interzone Project in Gabon, Congo, and Cameroon.

With all this money being spent, it might be surprising to see how little long-term security for great apes has been achieved. It is less surprising, however, when one considers the scale, pace, and momentum of the human exploitation of tropical forests: the vast private investments in infrastructure, mining, and logging; and the almost unimaginable diversity of actions and actors, side effects, unexpected events (from wildfires to coups), and cascades of destructive consequences associated with these processes. In comparison with these hundreds of billions of dollars and hundreds of millions of people, a few tens of millions of dollars for forest conservation will have to be spent very wisely indeed if they are to make much difference. Wisdom is increasing with experience, however, and there is the hope that the next US$25 million spent on great ape conservation will have much more impact than the last. It is in this context that in 2003 Klaus Toepfer, the Executive Director of the United Nations Environment Programme, set a fundraising challenge to GRASP by declaring that an additional US$25 million was urgently needed to finance adequately the efforts needed to lift the threat of imminent extinction from the great apes.[151]

FURTHER READING

Beck, B.B., Stoinski, T.S., Hutchins, M., Maple, T.L., Norton, B., Rowan, A., Stevens, E.F., Arluke, A., eds (2001) *Great Apes and Humans: The Ethics of Coexistence.* Smithsonian Institution Press, Washington, DC.

Brookfield Zoo (2001) *The Apes: Challenges for the 21st Century.* Conference proceedings. Chicago Zoological Society, Brookfield, Illinois. http://www.brookfieldzoo.org/content0.asp?pageID=773. Accessed June 13 2005.

Brown, D. (1998) *Participatory Biodiversity Conservation: Rethinking the Strategy in the Low Tourist Potential Areas of Tropical Africa.* Natural Resource Perspectives **33**. Overseas Development Institute. http://www.odi.org.uk/nrp/33.html. Accessed June 14 2005.

Caldecott, J. (1996) *Designing Conservation Projects: People and Biodiversity in Endangered Tropical Environments.* Cambridge University Press, Cambridge, UK.

Cowlishaw, G., Dunbar, R. (2000) *Primate Conservation Biology.* Chicago University Press, Chicago.

Dian Fossey Gorilla Fund (2003) *The Durban Process: Report of the Meeting.* Dian Fossey Gorilla Fund and Independent Projects Trust. http://www.durbanprocess.net. Accessed June 16 2004.

ECOFAC (2004) Conservation et utilisation rationnelle des ECOsystèmes Forestiers d'Afrique Centrale. http://www.ecofac.org/. Accessed July 23 2004.

Hacker, J.E., Cowlishaw, G., Williams, P.H. (1998) Patterns of African primate diversity and their evaluation for the selection of conservation areas. *Biological Conservation* **84** (3): 251–262.

Kormos, R., Boesch, C., Bakarr, M.I., Butynski, T.M., eds (2003) *West African Chimpanzees: Status Survey and Conservation Action Plan.* IUCN/SSC Primate Specialist Group. IUCN, Gland, Switzerland.

Leuser Development Programme (n.d.) The Leuser Development Programme. http://www.eu-ldp.co.id/. Accessed June 16 2004.

Price, S.V. (2003) *War and Tropical Forests. Conservation in Areas of Armed Conflict.* Haworth Press, Binghamton, New York.

Rao, M., van Schaik, C.P. (1997) The behavioral ecology of Sumatran orangutans in logged and unlogged forest. *Tropical Biodiversity* **4** (2): 173–185.

Rijksen, H.D. (1982) How to save the mysterious 'man of the forest'? In: de Boer, L.E.M., ed., *The Orang Utan: Its Biology and Conservation.* Dr W. Junk Publishers, The Hague. pp. 317–341.

Russell, C.L. (1995) The social construction of orangutans: an ecotourist experience. *Society and Animals* **3**: 151–170. http://www.psyeta.org/sa/sa3.2/russell.html. Accessed June 16 2004.

Wallis, J., Lee, D.R. (1999) Primate conservation: the prevention of disease transmission. *International Journal of Primatology* **20** (6): 803–826.

Wilkie, D.S., Carpenter, J.F., Zhang, Q.F. (2001) The under-financing of protected areas in the Congo Basin: so many parks and so little willingness-to-pay. *Biodiversity and Conservation* **10** (5): 691–709.

ACKNOWLEDGMENTS

Thanks to Ian Redmond (Ape Alliance/GRASP) for review; to Rebecca Kormos (Conservation International) and Eugène Rutagarama (African Wildlife Foundation) for comments on material that was merged into this chapter; and to Helen Corrigan and John Caldwell (both of UNEP-WCMC) for CITES data.

AUTHORS

Nigel Varty, UNEP World Conservation Monitoring Centre
Sarah Ferriss, UNEP World Conservation Monitoring Centre
Bryan Carroll, Bristol Zoo Gardens
Julian Caldecott, UNEP World Conservation Monitoring Centre
Box 14.1 Anne E. Russon, York University, and Constance L. Russell, Lakehead University, Canada
Box 14.2 Lise Albrechtsen, University of Oxford, and Brigid Barry, Tropical Biology Association

CHAPTER 15

Lessons learned and the path ahead

JULIAN CALDECOTT

In theory at least, a great ape population could survive indefinitely within a landscape made up of large patches of food-rich forest connected by broad corridors of forest to allow dispersal, interspersed with farmlands and prosperous villages inhabited by people who neither hunt apes nor set snares, who receive good health care, and who do not mind their crops being raided. The establishment of nature reserves, ecological restoration, enrichment planting, education, compensation, and various other forms of public, private, or community investment might create such an ideal landscape. Putting such a vision into effect is challenging, however, not least because most great ape range states face significant socioeconomic challenges. This can make it difficult for them to undertake the organized, long-term social investments that successful conservation seems to require.

Elizabeth A. Williamson

Tree planting around Nyungwe National Park, Rwanda.

For most of the great ape range states, this type of investment requires external support. The cost of this proposed 'help' is often monetarized and expressed in the form of programs made up of projects. The financing of these projects has become the chief priority of an industry comprising government departments, donor agencies, nongovernmental organizations (NGOs), and consulting firms. This industry is a large one, but not large enough to counter the overall effects of the human assault on tropical moist forests. It is fair to ask whether we could do better.

A wide range of techniques have been developed by conservationists to target aspects of the great ape survival crisis. In the process, some have been tested and abandoned. The key lesson learned is that conservation programs need to be adapted to their circumstances and, generally, have to involve a long-term commitment of appropriate resources; what is almost always needed is an irreversible shift in the way in which people perceive and relate to each other and to their environment. A sustainable conservation project educates and empowers, encourages and enables people to live better lives (in their own terms), causes values to shift, and involves effective partnerships. Following these principles, perhaps we should be examining their potential application in the context of the range state societies with regard to their peoples' history and culture, hopes and fears, attitudes to forests and great apes, and expectations placed on government.

While doing so, we should be aware that there are many stakeholder groups involved in any conservation enterprise that targets great apes, and that these groups have different perspectives and interests (whether conscious or not). Broadly, these stakeholder groups are:

- educated middle classes of all countries, who donate to conservation NGOs and apply political pressure to their governments, thereby directly or indirectly providing the backbone of international financing for conservation;
- range state governments, struggling to meet their international obligations, to sustain a degree of national legitimacy and public support, and eager for as much external funding as possible (on the best terms possible), whether in the form of private investment or official donor assistance;
- national conservation and forest-management agencies and their staff, all more or less duty-bound to implement legislation concerning the management of wildlife, protected areas, and forests, while usually also starved of resources with which to do so;
- private companies and government agencies with interests in agriculture, logging, oil and gas extraction, mining of hard-rock or alluvial minerals, and in the building of roads and other infrastructure in the vicinity of great ape habitat areas; and
- local people, members of communities who live in rural areas in and around the forests where great apes live, often poor in monetary terms, isolated from urban decision-making processes and poorly educated (in cosmopolitan terms), yet typically strong in group identity, perceived values, and in an awareness of what they want to achieve in life.

The values and aspirations of any of the above stakeholders may – but do not always – conflict with the interests of the final key stakeholder group:

- great apes, whose natural history has been explained in this volume, with their specific needs for an intact and suitable environment free of hunting, disease, and disturbance if they are to breed and raise their young successfully and, thereby, to survive in the wild.

Ian Singleton/SOCP

Conservation awareness at Medan International School in Sumatra, Indonesia.

Conservation is achieved by some or all of the human stakeholder groups working together in partnerships to clarify and, if necessary, to adjust their roles, rights, responsibilities, and relationships. In this approach, the great apes are somewhat objectified as 'animals of concern', but this is not to forget their inherent value as intelligent beings and as our close biological relatives. They are however victims, with a collective worldwide population equivalent to, say, Stuttgart in Germany. It falls, in particular, to conservation NGOs and scientists to understand and represent the interests of the great apes in dialogue with the other human stakeholders.

WORKING WITH EDUCATION

People who value wildlife care enough to justify or to change their own behavior, or at least to feel guilty about their negative effects on wildlife. A successful process of great ape conservation must validate and promote the notion that these animals are sufficiently valuable to refrain from killing, eating, or abusing them, or from destroying their habitat. This will have a practical effect if it guides behavior among stakeholders that are able to influence events on the ground.

Education can encourage and enable people to

develop the mental tools with which to understand not only their environments and communities, but also their personal potential to achieve adequate and sustainable livelihoods. If the value and plight of great apes are an integral part of educational processes, then people are likely to become much better able to appreciate them. If social, economic, and technological circumstances are right, this appreciation may lead people to act in favor of great ape survival and against the factors that threaten it.

Efforts toward conservation may be ineffective when they adopt a single tactic in isolation, such as offering alternatives to bushmeat without also enforcing hunting or access bans, or offering livelihood support without also providing environmental education and enforcement. Conservation is likely to be more effective if it delivers a balanced program that includes:

- advisory materials on laws concerning wildlife and protected areas – these should be developed with local people, trialed in appropriate languages (to ensure clarity of meaning), and widely distributed;
- educational materials in a variety of formats (community radio, posters, magazines, etc.) that encourage understanding of ecology and a change in the perception of hunting and bushmeat consumption;
- programs that encourage and enable local people to analyze their own environments and plan their own development accordingly, while actively defending their interests against those of outsiders;
- clearly marked boundaries for conservation areas;
- explanation and enforcement of hunting and access rules;
- forums for dialog and conflict resolution among stakeholders;
- assistance in developing sustainable livelihoods;
- managed expectations that encourage people to solve their own problems through community organization, self reliance, and assertiveness in seeking outside resources.

As such a program is delivered, community benefits arising from the conservation process will become more easily apprehended by all participants. Such benefits can include any or all of those listed below. The precise mix of priorities in each place, and how they are presented in an educational context, will vary according to ecological and social circumstances.

- **Providing harvests within the reserve**. Not all protected areas set aside to preserve great ape populations need exclude all access and use. 'Extractive' reserves, with harvesting regimes organized with local people, can provide a renewable supply of materials such as medicinal plants, thatching, structural wood, climbing palms, and a range of foodstuffs, spices, dyes, drinks, etc. These either meet household needs directly, or provide opportunities for the processing of raw materials and the sale of products such as foods or handicrafts.
- **Providing harvests outside the reserve**. Even if a protected area is closed to human access and use, it may have a buffer zone in which harvesting could take place for uses such as those listed above. Both the protected area and the buffer zone will act as breeding and feeding grounds for mobile wildlife species that may serve the surrounding areas as legal sources of bushmeat, dispersers of economically useful plants, crop pollinators, and pest control.
- **Providing local ecological services**. Protected forests can safeguard local communities from the ill effects of drought, flooding, fire, soil erosion, landslides, and the drying out of aquifers, wells, and springs.
- **Resisting global warming**. Forests sequester gaseous carbon dioxide. Globally, increasing atmospheric carbon dioxide levels contribute significantly to climate change through an enhanced greenhouse effect. There are emerging mechanisms for the global community to help meet the cost of forest-based carbon storage, which are already forming the basis of activities such as the restoration of forest at Kibale National Park, Uganda. There may be considerable scope for proposals by community stewardship groups.
- **Protecting genetic resources**. Many forest species contain chemicals which are important in the ethnobiological traditions of local people, and which might be developed as commercial products via bioprospecting. Local people may benefit from these discoveries, provided that adequate contractual provision is made for access and benefit sharing on

mutually agreed, fair, and equitable terms (including the appropriate recognition of intellectual property rights).

- **Providing resources for education and tourism.** Even a small number of recreational visits or school parties each year can make a big difference to a community's economic prosperity, both directly from the sale of goods (e.g. refreshments, handicrafts) and services (e.g. guidance, accommodation), and indirectly by raising the profile of the area outside the realm of conventional decision makers.
- **Local opportunities.** Research, management, tourism, education, training, and other activities in and around a reserve create many different kinds of jobs, from which local people can benefit as their training and experience increase. These activities will also generate new information (making ecosystems more interesting to local people), while encouraging them to enroll as students at educational institutions, thereby helping to transfer new skills and technology to local people.
- **Providing a clean and beautiful environment.** Local people benefit from access to natural scenery and often a lack of crowding and pollution, all of which can help provide a healthy and relaxed lifestyle. Nature reserves are special places; people living near them are stewards of unique resources, and have an important role that is increasingly being recognized.
- **Preserving traditional values.** Cultures are distinctive mixtures of traditions, languages, technologies, beliefs, and art forms, all of which are rooted in the ecosystems where their people live. A nature reserve can help to preserve those links, keeping alive valuable local ideas and ways of doing things.
- **Preserving options.** People who live within a viable ecosystem close to a nature reserve often have a broader range of options for development than others. This is because they can use their resources wisely to determine their own future, without risking the loss of their means of survival and prosperity.
- **Improving tenure security.** A nature reserve can be used by local people to obtain legal rights to occupy or use nearby areas. This may mean agreeing to respect permanent reserve boundaries, but secure tenure outside will often give more benefits than are lost by doing so. These can include clearer rights of inheritance, and better local ability to resist unwelcome immigration or theft of resources by outsiders.

This approach is fundamentally 'pro-livelihood' rather than simply 'anti-bushmeat'. The basic model aims to clarify and refine the particular elements of the relationship between educators and communities, and the content of community environmental and livelihood discoveries, that have most impact on bushmeat consumption and on forest protection. Improved knowledge could then be used to formulate action guides, tools, and priorities for extending the process adaptively to other locations in other range states.

If local communities can be helped to meet their protein requirements through domesticated or rapidly maturing plant and animal species, the need for bushmeat is reduced.

Gordon Miller/IRF

Gordon Miller/IRF

Box 15.1 GREAT APES, CONSUMERS, AND THE MEDIA

Over the last 25 years, many millions of pounds have been raised to help protect mountain gorillas in Rwanda – and have been spent on paying the guards' wages, supplying uniforms and equipment, and operating educational programs; millions more have come to the country from visitors who are willing to pay to see the gorillas, thus persuading Government to give its full backing to conserving them. There is no question that the publicity brought to them by television has been crucial in bringing these things about.

David Attenborough

John Sparks/naturepl.com

This is perhaps one of the most famous pictures from a television sequence of two dissimilar primates together. But what has that fame achieved? Are mountain gorillas better off thanks to David Attenborough? Have the animal stars actually benefited from that publicity, from *Life on Earth* onwards? There's no question the world knows more about gorillas than it did and, arguably, has acted in response.

As a result of this media attention, all the great apes should, in theory, be better off. Their actual conditions are discussed in depth elsewhere in this volume, but two issues are particularly worth considering here.

Many television stations now broadcast high-quality wildlife programs, but these often make no reference to conservation. Yet the eastern gorillas of the Democratic Republic of the Congo (DRC) have been devastated by the mining of coltan, a mineral found in riverbeds in the rain forest and consumed by mobile phone and computer hardware companies servicing the global TV-viewing public. South-east Asian forests are also being converted to palm oil plantations, and their wildlife destroyed. If we eat processed foods, use shampoo, employ a builder who uses Indonesian plywood, we are (in theory and probably in practice) contributing to the deaths of orangutans – perhaps even while also donating money to help conserve them.

On television, commercials for palm oil pro-

WORKING WITH LIVELIHOODS

Two key lessons for conservation have emerged from efforts to improve livelihoods:

- increasing wealth among local people can give them better tools with which to degrade their environment, while also potentially attracting outsiders who will then do the same; and
- if improvements in the livelihoods of local people are based on their obtaining a share of the wealth generated by wildlife, they will resist attempts by outsiders to destroy it.

From the first lesson emerges an abundance of measures built into the family of social investments known as integrated conservation and development projects (ICDPs). In this approach, overtly development-oriented and conservation-oriented efforts are deployed in a symmetrical way in the project area. The development activities aim to boost livelihoods in the support or buffer zone around a protected area, using such measures as agroforestry, aquaculture, microlivestock rearing, and community forestry. The conservation activities focus on demarcating and patrolling the boundary of the protected area, training and equipping enforcement staff, leafleting surrounding villages to let them know what the rules are, and reaching out to local people to explain ecology and the national and global importance of the biodiversity sheltering in the nearby protected area.

If properly done, in a coordinated, respectful way, with adequate resources and a realistic time-frame, this approach can work quite well. Problems arise, however, when ICDPs achieve much more 'development' than 'conservation', when the links between the aims are not adequately explained, when the project cycle leads to a discontinuity in

ducts like shampoo, chocolate, ice cream, or cosmetics are inserted within escapist productions about orangutans apparently living in an endless rain-forest paradise. The combined expenditure on advertising and publicity for palm oil products far exceeds the money available for conserving orangutans, and also far outweighs any effect that conservation broadcasts may have. Perhaps those multinationals should devote some of their profits to great ape conservation? Innocently, or with culpable negligence, we are connected with international corporations (sometimes as shareholders), Indonesian politics, and global markets. Individually we may be powerless to change the status quo, but perhaps *en masse* we could tip the balance.

As Attenborough convincingly reminds us, television shows us the wonder of these primates and their habitats. Could the media directly repay and sustain its source out there in the wild? The producers of 'reality' conservation programs that reveal the conservation crisis have a dilemma: not every viewer will rush home to learn about the bushmeat trade and the killing of apes to feed humans. Those that do sit through such an important but unattractive subject as shown in *The Ape Hunters* on BBC4 (UK) are often already aware of the issues; they are likely to be few in number and unlikely to be watching a mainstream TV channel. The newspapers and radio regularly carry serious environmental reports; given the relative audience shares, however, attractive entertaining moving pictures have more impact on the popular impression of the situation than disturbing information about bushmeat. Documentaries will probably continue to show and to 'celebrate' the great apes, filming in lovely Gombe National Park, while carefully avoiding the issues of its continuing degradation and endangerment. If Gombe, its chimpanzees, and Jane Goodall's work are reduced to nothing over the next 10 years, viewers will then turn off their televisions and turn on the program makers to accuse them of misleading, even lying about the true situation. Television should use its pictures and potential to help save these on-screen winners but potential real-world losers.

One way forward is to take film production into communities, making relevant films for local audiences that would help them to explore and value their own wildlife, and make a tangible impact on attitudes on the ground. Important progress was made in early 2005 when the first batch of 11 award-winning programs on the great apes, donated to the Great Apes Film Initiative by BBC Worldwide and Granada International, were taken to Congo Brazzaville for local showings there and in Cameroon. Filmmakers for Conservation (FFC) were crucial in this initiative, and their plan is to spread such films more widely in great ape range states.

Richard Brock

external support, when rule breaking by villagers is met by harsh treatment, leading to a breakdown in trust, or when the success of the project attracts outsiders to the area. It is a rare ICDP that does not suffer from one or more of these problems.

From the second lesson emerges a family of conservation activities, inspired by the Communal Areas Management Programme for Indigenous Resources (CAMPFIRE) in Zimbabwe in the 1980s.[2,3] This community-based natural resource management (CBNRM) method has been adapted to many locations ever since, including community fish sanctuaries in the Philippines, reef-guarding communities and scuba ecotourists in Indonesia, bioprospecting programs in Costa Rica, and great ape tourism in Uganda and elsewhere. The concept is that if wildlife resources are redefined as an asset of local people (whether completely, or shared with other stakeholders), and if revenue streams from their use are distributed fairly according to the ownership, then the whole incentive structure will automatically change, along with values and behaviors. Villagers who, one year, would cooperate with gangs of elephant poachers would, the next, be reporting the poachers to the police and digging waterholes for the elephants (as happened in Zimbabwe).

For this approach to work well, an acceptable distribution of ownership and revenue must be negotiated and implemented in a transparent way; revenue streams and their distribution must also offset the marginal value to an individual of cheating the system. Even small amounts of revenue per person can be enough to change attitudes, if they are regular and come to be expected; future earnings can then be factored into decisions, such as whether or not to set a snare in a particular location. Problems arise when there are disputes

Alastair McNeilage

Gordon Miller/IRF

Census training (top) and community conservation awareness (above) in Uganda.

over the proportion of the earnings going to external stakeholders, for example when gorilla tourism is used to finance parks elsewhere,[1] or when the source of revenues dries up, as occurred as a result of armed conflict in the Democratic Republic of the Congo (DRC). Even so, once a revenue stream from a wildlife resource has been experienced, attitudes to that resource change; as in DRC, local people may still consider wildlife worth keeping in the hope that revenues from it will, one day, be restored.

The main challenge to implementing either type of livelihood-based conservation strategy lies in achieving just and durable arrangements, in creating a long-term social contract in the minds of all concerned. For integrated conservation and development projects, the costs imposed on local people by having a nearby protected area are offset by the benefits offered in the form of livelihood support by (or on behalf of) the conservation agency concerned. In systems like CBNRM, the costs imposed by tolerating wildlife are offset by earnings from wild species, mediated by some publicly accountable institution such as an NGO or trust fund. This trade-off must be clear, transparent, consistent, and trustworthy, and would be fatally undermined by corruption, brutality, or tribal favoritism (all factors with which range state villagers are deeply familiar). In both kinds of enterprise, local people must have sufficient authority to negotiate freely enough to protect and advance their own interests. There must be willingness on the part of government to allow them this authority; this frequently requires a degree of decentralization or other local empowerment.

STRATEGIC CONSERVATION PRIORITIES

Expert workshops, population and habitat viability assessments (PHVAs), national great ape survival plans, and the GRASP strategy have all contributed to defining a long list of urgent actions that need to occur in the range states if great apes are to survive in the wild, as summarized in the country profiles that follow. These interventions would involve some combination of the measures described above and in Chapter 14, adapted to the precise circumstances of the target area and population, and based on an understanding of the challenges reviewed in Chapter 13. In general terms, though, the strategic priorities can be summarized as follows:

- **Protected area management**. Rehabilitate, demarcate, and manage the protected areas that already exist, aiming for high standards of professional and institutional competence, with resource allocation, capacity building, and technology transfer as required in each case.
- **Surveys and gap analyses**. Use conservation science (including the latest global positioning systems, geographic information systems, and remote sensing) to establish where great apes are, and to identify sites for new protected areas that may be needed.
- **Environmental education**. Build local support for and participation in livelihood systems that require ecosystem conservation and are be-

nign to great apes, while also encouraging the use of sanctuaries, zoos, tourism operations, and protected areas as educational resources for range state populations.

- **Suppression of the great ape bushmeat trade**. Close down sources (through enforcement, good management of protected areas, alternative-livelihood provision, and private-sector cooperation), demand (through public education and alternatives), and distribution mechanisms (through targeted enforcement and cooperation across national borders).
- **Transfrontier cooperation**. Reward and replicate the initiatives that have already occurred in and around the Congo Basin, and implement policy level agreements.
- **Private-sector cooperation**. Encourage extractive industries to adopt approaches friendly to great apes through partnerships, certification, investigative journalism, and consumer and shareholder demand; work together with retailers, traders, customs departments, and civil society groups in importing countries.
- **Community initiatives**. Understand local motivations to conserve and find ways for local people to have the freedom to pursue the conservation of forests and biodiversity, in their own interests.
- **Other activities**. No such list is complete without also noting the need to evaluate every other resourceful idea that might have emerged already or be about to emerge, such as a new breakthrough in controlling the Ebola virus, or a new approach to promoting forests as sacred places.

In doing all these things, there must be a strong emphasis throughout on minimum cost, maximum local self-sufficiency, and effective cost recovery. The global community will also need to invest significantly in these measures if they are to succeed. This investment need not be wholly financial; self-funded volunteers, scientists, and journalists can help a great deal. A range of inexpensive incentives can be used by overseas governments to boost conservation initiatives. Sovereign debt relief or removal of trade barriers would free up or increase range state resources.

Nevertheless, large amounts of cash will be needed every year for the foreseeable future to pay for wages, equipment, training, and running costs of national conservation agencies and management teams working in protected areas. This will enable them to demarcate and patrol boundaries; to sustain outreach programs affecting millions of people; to survey, study, and report on the populations and habitats being conserved; and to manage the resulting knowledge effectively, putting in place early-warning and feedback mechanisms.

Money will also be needed to finance various forms of compensation for local communities, to capitalize their enterprises through grants or microcredit, and to buy logging concessions or easements for conservation purposes. Those who give up other options to participate in conservation so that the world can have nature reserves and great ape populations deserve to be rewarded in ways that they themselves appreciate. There are formidable challenges to obtaining political consent in the wealthier parts of the world for the scale of taxation needed to fund all this, and to organizing and endowing sustainable financing mechanisms needed to deliver meaningful change (locally, transparently, and in perpetuity). This demands inspired leadership and great creativity.

The last is at a particular premium, for tropical biodiversity problems are urgent and global willingness to solve them through taxation is limited. There is therefore a need for interim measures that will make things happen in new ways. Nations could 'adopt' great ape species, for example, by giving themselves the task of doing whatever is needed to ensure their survival. With sufficiently persuasive advocates, it is possible to imagine an animal-loving people like the British or Japanese adopting the bonobo as a national mascot. Meanwhile, private companies could 'adopt' World Heritage Sites or Biosphere Reserves that contain great apes, underwriting their management costs and drawing on a permanent source of knowledge about the natural history of 'their' area to sustain and enrich their public image.

In the final analysis, the limiting factor is not whether the peoples of the world care about great apes, rain forests, and the health of planetary ecosystems; many millions clearly do, and this number has nowhere to go but up. The crucial issue is the lack of practical means by which we can express our care in ways that will make a real difference. This is an organizational challenge, and one that human institutions (whether public, private, governmental, or nongovernmental) are uniquely equipped to meet.

Tamar Ron

A poster for raising conservation awareness in Cabinda, Angola.

CONCLUSIONS

Conservation activities do not operate in a vacuum, but are influenced and frequently limited by a variety of external situations, events, and demands. These include war and civil conflict, political unrest, unrestrained corporate exploitation of nature, human population increase, the effects of natural disasters and climatic anomalies, other government sector policies and plans, economic recession, and the crippling poverty in many range states, as well as the attitudes, behaviors, and beliefs of local people.

Since the mid-1970s, the great apes have attracted increasing interest and concern in the international community. Efforts to conserve them have taken many forms and have involved many actors in many different locations. Intergovernmental activities have often been led by UN agencies or by the UN-sponsored multilateral environmental agreements, such as the Convention on Biological Diversity and World Heritage Convention. The resulting national biodiversity strategies and action plans, World Heritage Sites, and transfrontier conservation agreements have helped to create an enabling environment for great ape conservation, and have undoubtedly helped preserve certain areas of great ape habitat.

Meanwhile, national governments, aided and encouraged by multilateral and bilateral official aid programs and by local and international NGOs, have set aside many more protected areas than are designated as globally significant under the multilateral environmental agreements. These measures are insufficient, however. This is partly because so many protected areas are 'paper parks', without the resources to defend themselves against mounting threats, and partly because so many great ape populations and their habitats occur entirely outside protected areas and are vulnerable to hunting. Wider and deeper-acting initiatives that address the underlying causes of habitat destruction are also required, but are seldom forthcoming. The great apes are, therefore, still in decline.

Many conservationists now believe that great ape conservation requires social mobilization and a public willingness to invest in sustainable development that will enable local peoples to improve their circumstances in ways that are compatible with great ape survival in the wild. Yet the range states have many problems and few resources, especially in Africa. These include widespread severe poverty and economic and political conditions (including, in several cases, armed conflict) that do not encourage long-term planning and investment. Meanwhile, governments face severe temptation to liquidate timber and mineral resources in order to service debt, or to meet current obligations for political expenditure. This can be aggravated by pressure to accept investments by multinational corporations that wish to extract these resources.

Widespread habitat destruction has often been the inevitable result, complicated for the great apes by hunting pressure facilitated by easier access to formerly remote areas. International charities and official donors interested in sustainable development and biodiversity (or great ape) conservation are able to deploy resources that are paltry and influence that is trivial compared with what is available to the private sector in alliance with governments. It is nevertheless possible that a combination of conditions being attached to major public investments, pressure on corporations from consumer groups and investigative journalists, and rewards for companies that cooperate with certification schemes and agree to adopt values such as those of the UN Global Compact, may result in more sustainable use of natural resources.

Meanwhile, much can be achieved through partnership-based field projects, building on the lessons learned from the educational and other mechanisms described above. These all involve complex processes interacting with environments that are themselves complex and dynamic. In these circumstances, the best approach is to adopt a number of key principles that can be applied consistently regardless of the details, allowing the process to adapt to the environment where it is being nurtured. These key principles are outlined below.

1. **For an educational project:**

   - maintain respectful, trust-building forms of dialog at all times;
   - recognize that local people are the actors and that outside assistance is for them to use in their own interests; and
   - seek ways to turn talk and ideas into action as swiftly as possible, to encourage confidence and to build 'action competence'.

2. **For programs such as integrated conservation and development projects:**

   - seek to achieve symmetry between overtly development-oriented and conservation-oriented efforts in the project area;
   - explain clearly and consistently the 'social contract' involved in efforts to conserve biodiversity, and promote improved but sustainable livelihoods;
   - be alert to the temporary nature of project interventions and build into them sustainable financing mechanisms and other means to promote continuity;
   - avoid harsh enforcement action, except where consensus has been built locally around the need for this; and
   - find ways for local residents to cooperate in excluding unwelcome immigration to or extraction of resources from project areas.

3. **For projects using community-based natural resource management:**

   - build consensus around an ownership model for wildlife resources that includes local people and has clear consequences for the division of benefits flowing from their use, so that communities receive sufficient benefits to offset the opportunity costs incurred;
   - maintain consistency in the formula used to distribute benefits and transparency in the distribution itself; and
   - ensure that benefits are distributed in forms that are acceptable to participating communities.

Chapters 16 and 17 review the circumstances of great apes in every country where they occur in the wild. They reveal how the principles explained in this chapter have been derived, and also how difficult it will be to apply them effectively unless they are accompanied by humility, respect, persistence, and a willingness to learn and adapt.

FURTHER READING

Adams, W.M., Infield, M. (2003) Who is on the gorilla's payroll? Claims on tourist revenue from a Ugandan national park. *World Development* **31** (1): 177–190.

Cowlishaw, G., Dunbar, R. (2000) *Primate Conservation Biology.* Chicago University Press, Chicago.

Lutz, E., Caldecott, J.O. (1996) *Decentralization and Biodiversity Conservation.* World Bank, Washington, DC.

Martin, R.B. (1986) *Communal Areas Management Programme for Indigenous Resources (CAMPFIRE).* Department of National Parks and Wildlife Management, Harare.

AUTHORS

Julian Caldecott, UNEP World Conservation Monitoring Centre
Box 15.1 Richard Brock, Living Planet Productions

Where are the great apes and whose job is it to save them?

IAN REDMOND

African great apes inhabit the now fragmented belt of tropical moist forest and woodland stretching from Senegal in the west to Tanzania in the east. The orangutans of Asia are now restricted to parts of Sumatra and Borneo.

Despite the detailed maps in this volume, there are many uncertainties about where great apes are or were, until recently, found. Opinions even differ as to how many countries still have surviving populations. Their exact distribution is

Map 16a Great ape distribution in Africa

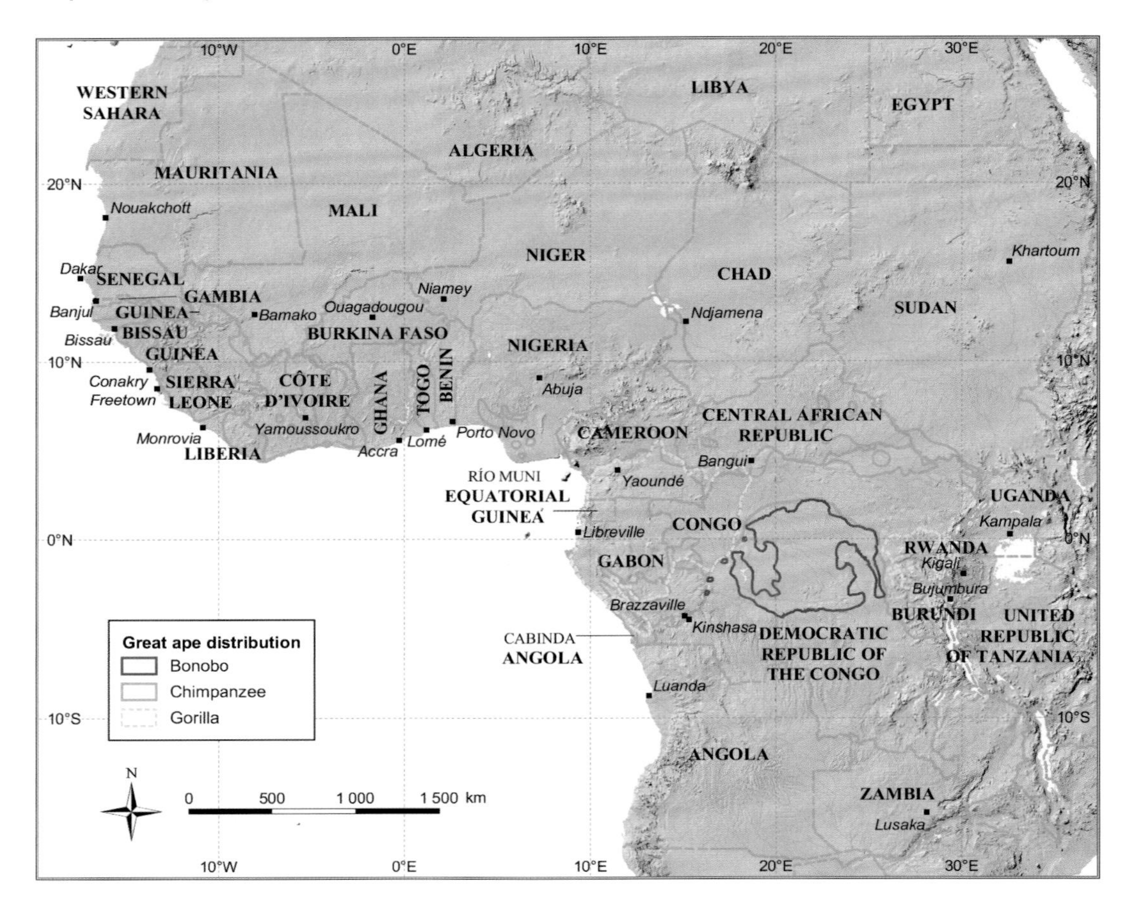

poorly documented in most cases. Even for those countries where they are reported to be extinct, it is seldom possible to say when, or even if, the last individual has been killed, because individual apes may live for more than 50 years. Moreover, where a species has been lost but its habitat survives and is contiguous with occupied great ape habitat across a national border, the possibility of natural re-colonization exists. If the reasons for local extinction can be removed, deliberate reintroduction is also possible – especially where captive individuals, such as those confiscated from illegal animal traders, are in need of a release site that meets the guidelines of IUCN–The World Conservation Union. Thus, the number of great ape range states is not constant, and can go up as well as down.

Twenty-three countries are known to have great apes now (Table 16, Map 16a, Map 16b), and in five more they are reported to have been present within living memory (Table 16).

Chimpanzees may once have inhabited the forests and woodlands of 26 African countries, but are now probably limited to 21. The two gorilla species are found in a total of 10 countries and bonobos in only one.

In Southeast Asia, Bornean orangutans are found in parts of two countries, and Sumatran orangutans restricted to Indonesia. In some countries, such as Rwanda and Uganda, great apes are found mainly in national parks and are the basis of a profitable, carefully regulated tourist industry, whereas in others, such as Indonesia and Cameroon, more apes live outside protected areas than within them, and numbers are spiralling downwards. In several countries numbers are down to 100–200 individuals, and for some it is too late.

Confirmation of disputed historical reports is difficult, and even settling questions of current distribution faces the perennial problem of limited funds for surveys. Some of the intriguing questions awaiting a definitive answer include:

Map 16b Great ape distribution in Southeast Asia

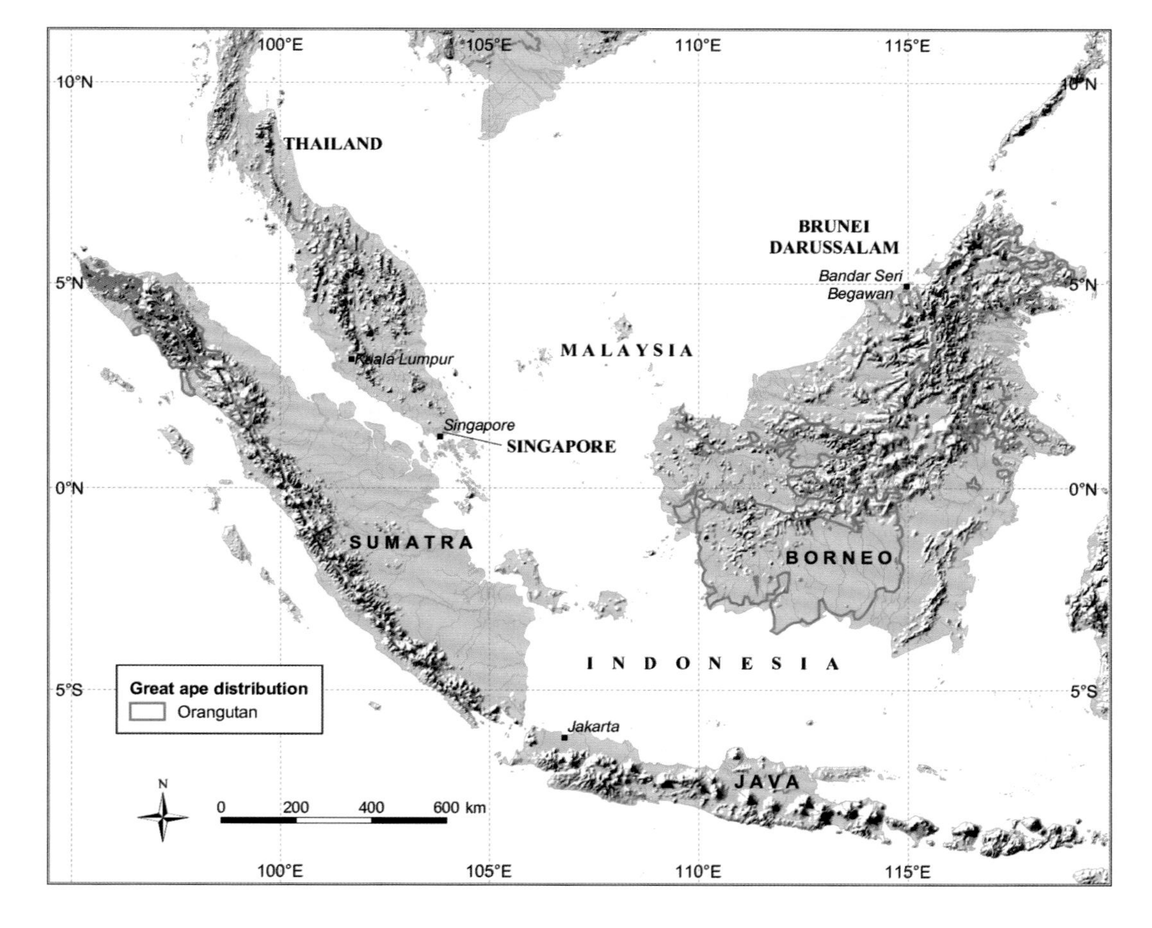

- Did chimpanzees ever live in Gambia? Old hunters say they used to see chimpanzees[2,4] but some authors dispute this.[1]
- Have resident chimpanzees been extirpated in Burkina Faso? There are persistent reports of at least seasonal movements from across the Côte d'Ivoire border[3] and some say resident chimpanzees survive.[6] Bance Soumayila,[7] a biodiversity expert from Ouagadougou, is convinced that chimpanzees are still present in Burkina Faso on the border with Côte d'Ivoire, near Comoe National Park.
- A recent report of chimpanzees near Bunkpurugu on the Ghana–Togo border raises a question mark over their extinction from Togo. This village is nearly 200 km north-northwest of the existing queried record on the map in the Ghana country profile, so Togo now has two possible populations.
- Are western lowland gorillas still found in the Bas-Fleuve region of the Democratic Republic of the Congo? The forests are contiguous with those of Angola's Cabinda province and Congo, where both western lowland gorillas and

Table 16 Countries hosting great apes

Genus:	Chimpanzee (*Pan*)				
Species:	Bonobo	Chimpanzee			
Subspecies:		western	Nigeria-Cameroon	central	eastern
WEST AFRICA					
Benin		EXTINCT			
Burkina Faso		EXTINCT?			
Côte d'Ivoire		x			
Gambia		EXTINCT			
Ghana		x			
Guinea		x			
Guinea-Bissau		x			
Liberia		x			
Mali		x			
Senegal		x			
Sierra Leone		x			
Togo		EXTINCT?			
CENTRAL AFRICA					
Angola				x	
Cameroon			x	x	
Central African Republic				x	
Congo				x	
Dem. Republic of the Congo	x			x?	x
Equatorial Guinea				x	
Gabon				x	
Nigeria		?	x		
EAST AFRICA					
Burundi					x
Rwanda					x
Sudan					x
Uganda					x
United Republic of Tanzania					x
Zambia					EXTINCT
SOUTHEAST ASIA					
Indonesia					
Malaysia					

central chimpanzees are still found in the Mayombe Mountains.

- Did chimpanzees ever occur naturally in Zambia? One reliable observer reports having heard chimpanzees in the forests on the southernmost shores of Lake Tanganyika, in what is now Zambia, back in the 1960s,[5] but the species is not usually listed in books on Zambian wildlife.
- Did orangutans once live in Brunei? The current patchy distribution of Bornean orangutans is partly a reflection of human hunting practices, but are there ecological reasons which would have prevented them living all across north Borneo in the past (see chapter 10)?

WHO WILL SAVE THE GREAT APES?

A glance through the popular literature from the past 40 years shows that people have been trying to save great apes for decades. Bookshelves are filled with the inspiring stories of heroic primatologists, struggling to save their study animals and habitats. Conservation organizations have launched appeals to fund projects. Park guards put their lives

Gorilla (*Gorilla*)				**Orangutan (*Pongo*)**			
western		eastern		Bornean			Sumatran
Cross River	western lowland	eastern lowland	mountain	northwest	northeast	central	
	x						
x	x						
	x						
	x						
	EXTINCT ?	x	x				
	x						
	x						
x							
			x				
			x				
				x		x	x
				x	x		

on the line to protect great apes. Educators have worked to inspire appreciation of their many values. Nevertheless, year after year, the area of viable habitat shrinks and the number of great apes dwindles. This is not the result of a concerted effort to eliminate them. It is more a case of collective negligence.

Great apes are being driven to extinction because people in the developed world, the 'haves', are not taking the trouble to source raw materials from ape habitats carefully enough. Timber and rattan, gold and tantalum, palm oil and rubber are among the raw materials that come from ape habitats, from the ground beneath them or from plantations that have replaced them. The hard reality is that the relatively small sums available for conservation are no match for the massive economic pressures to exploit or destroy ape habitats. And the 'haves' can hardly point the finger of blame at the 'have nots' – people attempting to lift themselves out of poverty by responding to those same economic pressures (or opportunities, depending on your perspective) – whilst ignoring their own role in the equation.

If the responsibility for the recent dramatic decline in great apes is shared by the global economic community, it would seem unjust for the job of halting that decline to fall solely to the governments and people of the range states – those countries where the great apes live. The United Nations defines 'least developed countries' as those with a per capita income of under US$800 per annum – and this includes 16 of the 23 countries that still have natural populations of great apes (also see Table 13.5).

To help countries work together, the Great Apes Survival Project (GRASP) was launched by the United Nations Environment Programme (UNEP) in 2001 and, joined by UNESCO, was registered as a Type II Partnership in 2002 at the World Summit on Sustainable Development in Johannesburg. This is a new kind of partnership, bringing together governments, UN bodies, nongovernmental organizations (NGOs), civil society, and private sector interests, all of whom share a common goal: in this case, to ensure the survival of great apes across their natural range. The challenge for the GRASP partnership is how to weave the many conservation activities, existing and newly identified, into a coherent global strategy to halt the decline in great ape numbers, and then – most importantly – how to find the funds to implement it.

NATIONAL GREAT APE SURVIVAL PLANS

Conservation action can be taken at many levels – from international laws to state-run protected areas, through community initiatives, NGO projects, private sector activities, and, of course, the actions of individuals. Each is important and necessary. Usually, major nongovernmental actions require the consent of the relevant authorities. Legally, therefore, the final responsibility for saving the great apes rests with the governments of those countries where they live. But, in the long run, governments can only make and enforce laws with the support of the people who elect them. People will only give that support if they understand the purpose behind the laws, and feel the government is helping them to improve their standard of living. Thus enforcement efforts must be complemented by education, and by sustainable development initiatives that are compatible with the survival of great apes and their habitat. All these activities require funds and trained personnel, but most of the countries with great apes lack both. Long-term conservation success is like a three-legged stool, with three equally indispensable legs: political will, support of the people, and sufficient, sustainable finance. GRASP takes the view that the survival of great apes is of concern to the whole of humanity, and has called upon the wealthier nations to help fund the prioritized, budgeted actions identified in each region.

To initiate GRASP, UNEP invited the governments of range states to designate a focal point in government and begin developing a national plan of action. It was requested that these plans should be developed in consultation with all those with an interest in great apes and their habitat, and be adopted as government policy. This task is complicated by the fact that responsibility for great apes and their habitat is often divided between several government departments – forestry, environment, tourism, national parks, rural development, etc. Each country organizes these responsibilities differently and, when a new administration takes over, the arrangements may change. In addition to this shifting mix of relevant government departments, GRASP has sought to involve local and international NGOs and academics, local communities living in or adjacent to ape habitat, and private sector interests, mainly in tourism and the extractive industries.

Helping 23 nations to develop a coherent cross-sectoral policy to ensure that great apes survive has proved quite a challenge, and yielded a range of results. Some countries quickly set up formal national committees, while others took a less structured approach; several sought funds for national workshops, and one or two had yet to respond by May 2005. Where the response was most energetic, the structure of the national great ape survival plan (NGASP) soon took shape, covering five areas:

(i) Where do great apes occur in the country and what is the current pattern of land use in and around their habitat? This immediately identifies the main stakeholders.
(ii) What are the threats to great apes in each area, and who is involved? The threats may well vary from region to region within a country, and must be understood if effective action to counter them is to be taken.
(iii) What is the status of current conservation efforts, including legislation, law enforcement, NGO projects, and local traditions – everything that protects apes and their habitat from destruction or overexploitation.
(iv) If great ape populations are still declining despite the activities listed in (iii) – and they are in virtually every location where data have been collected – what more do the best available experts think should be done to halt that decline and who is best placed to do it?
(v) Having identified the causes of the decline in each area, and what actions need to be taken by whom, the final question is who will finance it?

These plans must be concrete enough to have measurable indicators of success (or failure) but flexible enough to adapt to new challenges, such as rising prices of commodities from ape habitats causing 'gold-rush' scenarios (e.g. coltan, palm oil), emerging diseases (e.g. Ebola), inbreeding in newly fragmented or reduced populations, and, perhaps the biggest challenge of all, climate change and the ensuing disruption to weather patterns and vegetation zones.

The reality in many range states, however, is that even the answer to question (i) is not known in any detail. In such cases, the first recommendation would be to find the funds to carry out surveys of remote regions, filling in the information gaps identified in the following country profiles. Only on the basis of good, up-to-date knowledge can effective plans be developed and implemented. As the profiles illustrate, progress towards the goal of stable ape populations at natural densities, in well managed, sustainably utilized forests, is – to say the least – patchy. But GRASP has provided a global framework for all those working towards this end. By working as a partnership within a broadly agreed global strategy, each partner brings different strengths to the task. It is hoped that the GRASP whole will therefore be greater than the sum of the individual parts and, moreover, that each partner will gain from being part of the whole rather than working in isolation.

As with any long-term endeavor, however, it may be decades before we can judge whether GRASP has succeeded or failed. The bottom line for conservation is the area of habitat left and the number of animals living in that habitat, fulfilling the ecological role they evolved to perform. We can but hope that in future editions of this atlas, the country profiles will show that the 20th century decline has been halted, and viable populations of all the great apes have survived for the benefit of everyone.

Ian Redmond
Chief Consultant, GRASP

FURTHER READING

Bailey, N.D., Eves, H.E., Stefan, A., Stein, J.T., eds (2001) *BCTF Collaborative Action Planning Meeting Proceedings.* Bushmeat Crisis Task Force, Silver Spring, Maryland. http://www.bushmeat.org/may2001.htm.

Beck, B.B., Stoinski, T.S., Hutchins, M., Maple, T.L., Norton, B., Rowan, A., Stevens, E.F., Arluke, A., eds (2001) *Great Apes and Humans: The Ethics of Coexistence.* Smithsonian Institution Press, Washington, DC.

CITES BWG/IUCN, Ly, U., Bello, Y. (2003) *Study on Wildlife Legislation and Policies in Central African Countries.* http://www.cites.org/common/prog/bushmeat/rep_legislation.pdf.

FAO (2003) *State of the World's Forests.* http://www.fao.org/forestry/site/sofo/en. Accessed May 26 2004.

Fishpool, L.D.C., Evans, M.I., eds (2001) *Important Bird Areas in Africa and Associated Islands: Priority Sites for Conservation.* Pisces Publications and BirdLife International, Cambridge and Newbury, UK.

Forests Monitor (2001) *Sold Down the River. The Need to Control Transnational Forestry Operations: a European Case Study.* http://www.forestsmonitor.org/reports/solddownriver/title.htm. Accessed June 16 2004.

IUCN (1992) *Protected Areas of the World: A Review of National Systems, vol. 3: Afrotropical.* IUCN, Gland, Switzerland and Cambridge, UK.

Kormos, R., Boesch, C., Bakarr, M.I., Butynski, T.M., eds (2003) *West African Chimpanzees: Status Survey and Conservation Action Plan.* IUCN/SSC Primate Specialist Group. IUCN, Gland, Switzerland.

Lee, P.C., Thornback, J., Bennett, E.L. (1988) *Threatened Primates of Africa. The IUCN Red Data Book.* IUCN, Gland, Switzerland and Cambridge, UK.

Oates, J.F. (1996) *African Primates: Status Survey and Conservation Action Plan (Revised Edition).* IUCN/SSC Primate Specialist Group. IUCN, Gland, Switzerland.

UNDP (2004) *Human Development Report 2004.* http://hdr.undp.org/reports/global/2004/. Accessed November 28 2004.

World Bank Group (2005) Countries and Regions. http://www.worldbank.org. Accessed June 17 2005.

World Commission on Protected Areas (2004) World Database on Protected Areas. UNEP-WCMC. http://sea.unep-wcmc.org/wdbpa/index.htm. Accessed June 17 2005.

CHAPTER 16

Africa

REPUBLIC OF ANGOLA

GEMMA SMITH

BACKGROUND AND ECONOMY

The Republic of Angola is bordered by Namibia to the south, Zambia to the east, the Democratic Republic of the Congo (DRC) to the north and northeast, and the Atlantic Ocean to the west. Extending over a total land area of 1 246 700 km^2 with a coastline of approximately 1 670 km, Angola is divided into 18 administrative provinces. One of them, Cabinda province or the Cabinda enclave, lies to the north of the Congo River and is separated from the rest of Angola by about 30 km of DRC territory. It is therefore the most northerly part of Angola, and is bounded by DRC, Congo, and the Atlantic Ocean. This is the only part of Angola in which great apes occur.

The history of Angola is intimately bound up with a long Portuguese colonial presence. This was extended far beyond the period in which other European countries dissolved their empires in the 1940s, 1950s, and 1960s, because Portugal was being run by an ultraconservative military clique until the 'Carnation Revolution' of 1974, which introduced democracy. Before then, the priority of the Portuguese government had been to maintain a military hold over its possessions, which also included Mozambique, Guinea-Bissau, and East Timor, rather than to negotiate independence and national development of its former colonies. The abrupt collapse of this system in Portugal propelled all four territories into independence, at which point armed nationalist groups, which had previously been encouraged by the USSR and/or China and were of a leftist political complexion, sought to seize power. In the prevailing Cold War atmosphere of the time, this provoked the USA to encourage military intervention by its allies, South Africa in Mozambique and Angola, military factions in Guinea-Bissau, and Indonesia in East Timor. Terrible civil war and genocide ensued in all four countries, before stable governance was eventually attained.

Angola is thus recovering from more than three decades of warfare, first with the Portuguese, and then a civil war between the National Union for the Total Independence of Angola (União Nacional para a Independência Total de Angola, UNITA) and the Angolan government. This conflict was lengthy due to the involvement of Cold War interests, with USSR-backed Cuban assistance to the Angolan government and US-backed South African assistance to UNITA. Major oil companies were also involved, such as in the notorious 'Angolagate' scandal of 2000, in which the international press reported that arms-for-oil deals and corruption had thrived among corporate and political institutions in both France and Angola.[15] The end of the Cold War and the introduction of inclusive democracy in South Africa altered these conditions, and the war ended in April 2002, following the death of the long-term UNITA leader Jonas Savimbi.[7, 20] Fighting has ended in all areas of the country except in Cabinda province.[23]

There are growing signs of economic recovery in Angola, but major problems remain with virtually every element of infrastructure and government service throughout the country, including communications, roads, and basic education and health services. Angola exported oil worth US$3–5 billion in 2003, about 87 percent of state revenue, but around 82 percent of the population continue to live in extreme poverty, 42 percent of Angolan children aged five or less are underweight, one child now dies of preventable diseases and malnutrition every three minutes (i.e. 480 every day), and life expectancy is 44 years. Even to the extent that state revenues are known, somewhere between

Map 16.1 Great ape distribution in Angola

Data sources are provided at the end of this country profile

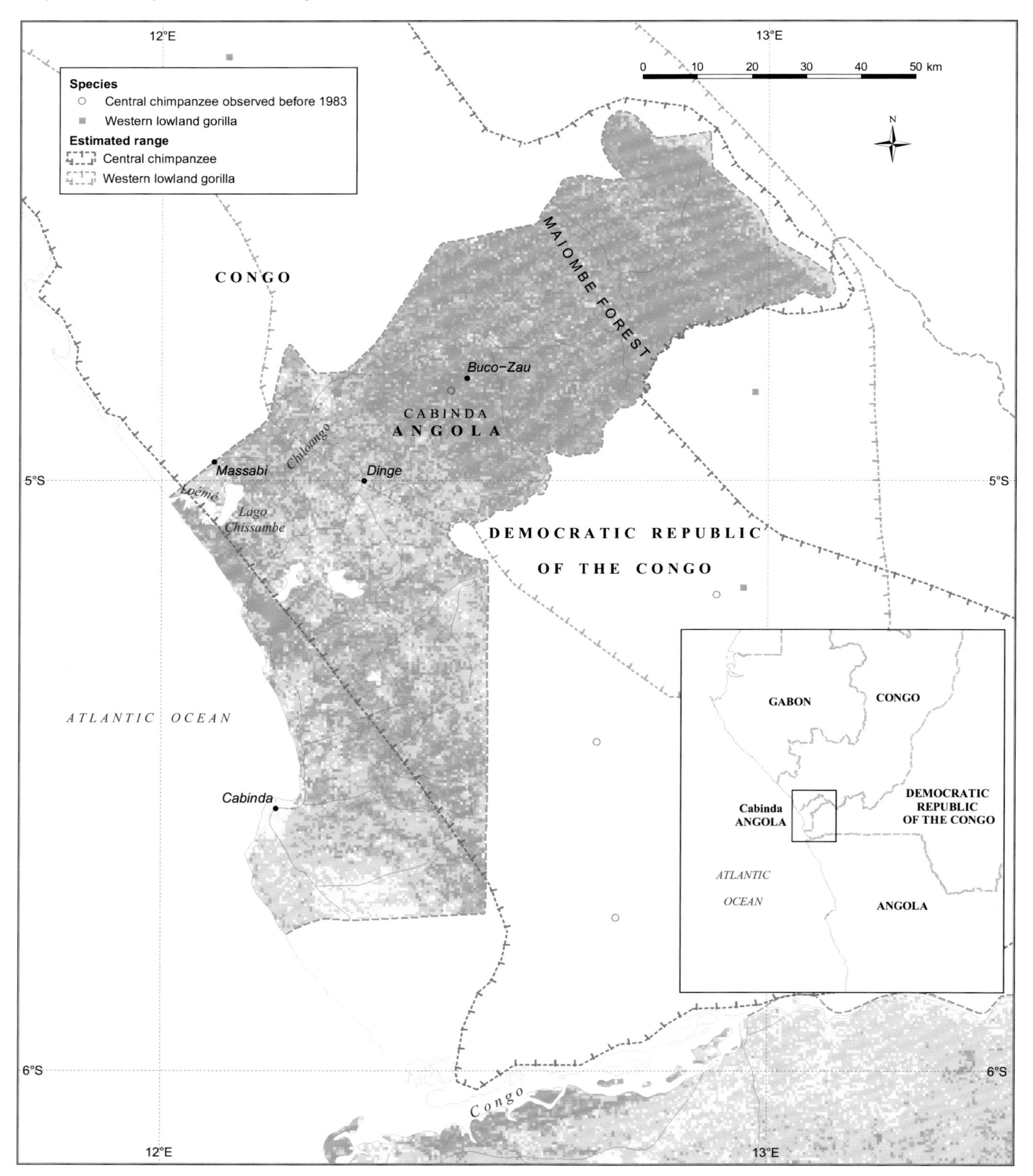

US$1.3 and US$1.7 billion are missing from the state coffers.[15]

Recent population estimates are difficult to obtain due to the conflict, which killed at least half a million people and displaced around another 4 million. Since the ceasefire in 2002, a large number of internally displaced persons have returned to their places of origin, particularly in rural areas, as have many refugees who had settled in neighboring countries.[7] In 2003, the country's total population was estimated as 13.5 million people, growing at a rate of 3 percent per year.[24] Adult literacy was 42 percent in 1998.[3]

After Nigeria, Angola is the second largest oil producer in sub-Saharan Africa, with most of its crude oil production located offshore in the shallow waters of Cabinda province. Cabinda produces more than half of Angola's oil and accounts for nearly all of its foreign exchange earnings.[6]

With stakes so high, it is not surprising that there are still political tensions in Cabinda. The Front for the Liberation of the Enclave of Cabinda (FLEC), a separatist group, is demanding a greater share of oil revenue for the population of the province. The Angolan government has ruled out complete independence for the province, but in 2002 announced its willingness to open talks with separatist groups, with a view to agreeing some measure of autonomy. Military operations in Cabinda since then, however, suggest that peace is still distant, with increasing numbers of civilians being killed by both sides.[1, 23] Cabinda has a population of about 300 000 people, with another 20 000 or more now living in refugee camps in DRC and Congo.

DISTRIBUTION OF GREAT APES

The western lowland gorilla (*Gorilla gorilla gorilla*) and the central chimpanzee (*Pan troglodytes troglodytes*) both occur in Cabinda, at the southwestern edge of their range. No definitive population figures exist, although residents have reported sightings of both throughout the Maiombe (Mayombe) forest,[12] where chimpanzees are believed to be widespread.[16, 19] On the basis of potentially suitable habitat, it was estimated in 1988 that at least 200–500 chimpanzees could occur in Cabinda, but there may be more.[13] Gorillas have been thought to be rare there since the 1970s.[4]

Cabinda lies within the Congo Basin and is the part of Angola that is most rich in different species. Mean annual rainfall is 850 mm along the coast and 1 300 mm (spread over at least 10 months of the year) inland. Mangrove communities fringe the lower Congo River and its tributaries, and there are substantial areas of permanent swamp forest as well as lakes in the coastal plain. Forests characteristic of the Guinea–Congo biome occur in the interior. These are tall and semideciduous, with a canopy up to 50 m in height, and are dominated by tree genera such as *Gilletiodendron*, *Librevillea*, *Julbernardia*, and *Tetraberlinia* (all Leguminosae-Caesalpinioideae). The understory is sparse, and in many parts of Cabinda (particularly in the south), it has been replaced with low-density coffee plants.[5, 8] The highest annual rainfall is recorded in the Maiombe forest, which supports Cabinda's populations of great apes, and extends across 2 000 km^2 of the mountainous northeastern interior and into DRC and Congo. This area is the southern margin of tropical moist forest and great ape distribution in western Africa.[16]

THREATS

Angola had one of the richest yet least known wildlife resources in Africa, but it has been seriously affected by conflict and Cabinda is no exception to this. Field studies have been deterred by conflict, but local information has led the United Nations Development Programme (UNDP) to conclude that hunting for bushmeat for both subsistence and commercial purposes, the pet trade, and to protect crops is probably widespread. In November 2003 a live infant chimpanzee was seen for sale in Massamba, with the hunter reporting that an adult male and female had been killed.[17] The use of apes as bushmeat is not traditional in Cabinda, but is now occurring, mainly under the influence of ready markets among soldiers and in the neighboring DRC and Congo.[17]

Meanwhile, the Maiombe forest continues to be subjected to a high rate of degradation, mainly due to heavy logging and poaching, for subsistence and commercial purposes.[17] These observations suggest that the Angolan populations of both great ape species may be declining, and their distribution contracting.

LEGISLATION AND CONSERVATION ACTION

International agreements

Angola ratified the World Heritage Convention in 1991 (but as yet has no World Heritage Sites), the UN Convention to Combat Desertification in 1997, the Convention on Biological Diversity in 1998 (but has not yet completed a national report on implementation of the convention, or a national biodiversity strategy and action plan), and the UN Framework Convention on Climate Change in 2000. Angola is not a party to the Convention on International Trade in Endangered Species of Wild Fauna and Flora (CITES), the Convention on Migratory Species, the Convention on Wetlands of International Importance (Ramsar), or the African Convention on the Conservation of Nature and Natural Resources. Angola is, however, party to the Memorandum of Understanding concerning Conservation Measures for Marine Turtles of the Atlantic Coast of Africa, the Treaty of the Southern African Development Community, and the Protocol on Shared Watercourse Systems in the Southern African Development Community.

National legislation

Wildlife legislation in Angola dates from 1911, with the establishment of the *Fundo de Caça*, a fund into which monies from hunting licenses were deposited.[9] The first hunting regulations were approved in 1929, and the creation of national parks and reserves was mentioned in the 1936 *Regulamento*. Legislation for the conservation of soil, fauna, and flora continued to develop, and in 1955, it was consolidated through Decree No. 40040. This legislation formed the basis of the Hunting Regulations (*Regulamento de Caça*), first published in 1957, and amended frequently since. Legislation prohibiting the export of live animals, including monkeys and parrots, was introduced in the early 1990s.[5, 10]

Article 24 of the Angolan Constitution invests the state with responsibilities for environmental protection. Since 1998, all biodiversity conservation and protected area management has been governed by the Basic Law of Environment (*Lei de Bases do Ambiente*, No. 5/98). This moved responsibility for biodiversity from the Forestry Development Institute (IFD), within the Ministry of Agriculture and Rural Development, to the Ministry of Environment (as of 2000 the Ministry of Fisheries and Environment and, since 2003, the Ministry of Urban Affairs and Environment).[19] After this history, the division of responsibilities is not yet very clear in practice. The Forestry Development Institute remains in overall charge of Angola's forest sector, however. It is represented in all 18 administrative provinces of the country. The National Directorate of Agriculture and Forest (DNAF) also shares some forest responsibilities in relation to policy formulation and guidance. The provincial government of Cabinda has emergency powers to prohibit trade in wildlife or associated derivatives from Cabinda within Angola or across its borders.[16] Despite these various efforts, it is reported that wildlife protection laws are scarcely enforced either inside or outside protected areas, and illegal hunting, harvesting, and settlements inside protected areas occur regularly. Endangered wildlife and products, including infant chimpanzees and other primates, African grey parrots, bushmeat, ivory, etc., are sold fairly openly in markets in Luanda and throughout the country.[19]

Protected areas

Angola has a long-established system of protected areas. The first national park, Parque Nacional de Caça do Iona (Iona NP), was established by Regulation No. 2421 of October 2 1937. Decree No. 40040 of January 20 1955 provided the first comprehensive nature conservation legislation for the country, covering all aspects of conservation and use of game, and providing for the establishment of national parks, reserves, and controlled hunting areas. It also established a Nature Conservation Council, which laid out regulations governing national parks. Further refinement of protected area legislation over recent years has led to the definition of additional protected area categories, which now comprise national park (outstanding sites where public access is allowed), strict nature reserve (for total protection), partial reserve (for licensed extraction only), regional nature park (for nature protection), and special reserve (for protection of particular species).[5, 9]

National parks are defined by Article 13 of the Hunting Regulations as an area subject to the direction and control of public authorities, reserved for protection, conservation, and propagation of wild animal life and indigenous vegetation, and furthermore for the conservation of objectives of esthetic, geological, prehistoric, archeological, or other scientific interest, for the benefit and enjoyment of the public.[8] Strict nature reserves, in contrast, were intended to offer total protection to wild fauna and flora (Article 14). Hunting, killing, or capture of animals, or the collection of plants other than for scientific or management purposes, authorized by the director general, is prohibited in partial reserves (Article 15).

Within special reserves (Article 16) the killing

of certain species is prohibited, in order to support their conservation.[8]

Finally, the *Diploma Legislativo* 88/72 defines a regional nature park as "an area reserved for the protection and conservation of nature, in which hunting, fishing, and the collection of, or destruction of, wild animals and plants, and the execution of industrial, commercial, or agricultural activities is prohibited or conditioned."[8]

Huntley and Matos (1994)[10] concluded that the almost continuous civil war in the country since 1974–1975 has had significant impact on Angolan protected areas and conservation efforts, particularly relating to populations of large mammals, even in large protected areas. Many protected areas lack wardens; poaching, settlement incursion, the cultivation of crops, and the illegal collection of timber and firewood have impacted those sites near human population centers.

The Maiombe forest in Cabinda is critical for great ape conservation in Angola, yet is virtually unprotected in law or practice. The only designated conservation area is the Cacongo Forest Reserve, which was established in 1930 for forestry purposes.[16] This was originally 650 km^2 in area, but more than half was excised in 1962, when another forest reserve (Alto Maiombe) was degazetted. BirdLife International[2] has identified a 400 km^2 part of Maiombe as an important bird area (IBA), centered on 04° 40'S 12° 30'E. It is located on the watershed of the Loémé and Chiloango Rivers, north northeast of Buco-Zau, and has the greatest number of species in Angola that are restricted to the Guinea–Congo forest biome.[5] The avifauna of the area is virtually unstudied, yet its designation as an IBA by the international conservation community can only help to strengthen the case for its full protection.

United Nations support

UNDP is supporting the Angolan government in seeking ways of sustainably protecting the environment and managing natural biological resources.[21] Key projects are described below.

- **Development of a national biodiversity strategy and action plan.** Through this project, which is funded by the Global Environment Facility (GEF) and UNDP and started in November 2004, it is hoped that the status of the biodiversity of Angola will be defined, the pressures to which it is exposed documented, and priority actions to ensure its conservation and sustainable use identified.
- **Improved Environmental Planning and Conservation of Biological Diversity in Angola (project ANG/02/005).** Implemented in 2002–2004 by UNDP with support from the Norwegian Agency for Development Cooperation (NORAD), this project aims to strengthen national capacity to protect the environment and manage natural biological resources, especially in planning, monitoring, evaluating, and reporting on the state of the nation's environment and its implementation of international environmental conventions. NORAD had some initial work, from 2000, at the request of the Ministry of Fisheries and Environment[16]. The project was justified in terms of improving the long-term life quality of the population of Angola, through strategically planned sustainable management of natural resources. Key outputs of the project include a finalized national environmental action plan; the development and implementation of key community-based components of a national strategy for the conservation of biological diversity; launching a comprehensive study on the state of Angola's environment; establishing an environmental database; and enhancing the capacity of environmental NGOs to undertake effective community-based conservation initiatives, environmental advocacy, and education.
- Following the UNDP/NORAD project, the Cabindan provincial government has undertaken to resource the further development of this work. Plans for the study and conservation of the Maiombe forest and its ape

Tamar Ron

The Friends of Nature club of Ganda-Cango, Angola. Members make a commitment not to eat bushmeat.

populations are being developed. Initiated through legislative developments at national and provincial level, the proposal includes the designation of a new protected area and a separate sanctuary for the rehabilitation of orphaned wildlife including chimpanzees and gorillas. It is based on the involvement of local communities in developing alternatives to the currently unsustainable use of forest species, and education and public awareness campaigns. At the request of the Ministry of Fisheries and Environment, UNDP helped to coordinate an initiative to establish a regional task force for the joint protection of the Maiombe forest and its flora and fauna by the three countries that share it (Angola, DRC, and Congo). Mainly based on the participation of local communities, the initiative has been approved by all three governments.

Other conservation efforts in the Maiombe area include an awareness and consultation process with resident communities and, within the armed forces, the development of voluntary 'friends of nature' clubs (*amigos da natureza*), whose members have agreed not to eat bushmeat except in extreme survival situations.[18] The association of oil operators in Cabinda, headed by ChevronTexaco, established the CABGOC Protocol with the Angolan government in 2002, and provides modest support to the Maiombe forest area.

FUTURE CONSERVATION STRATEGIES

There is virtually no enforcement of environmental law in Angola, so the overwhelming priority for great ape conservation must be building fundamental capacity for enforcement and education at governmental, nongovernmental organization, and community level throughout Cabinda, and especially in and around the Maiombe forest. Six main areas of environmental concern also need to be addressed at the national level in parallel with efforts in Cabinda. These are deforestation, soil impoverishment, erosion, and desertification;[7] the dependence of poor households on traditional energy sources;[11] the depletion of fish stocks; pollution by the petroleum industry; the loss of biological diversity; and poor environmental sanitation.[22] Civil conflict has resulted in the breakdown of the protected area system established prior to independence (covering 6.5 percent of the national territory), and there is a risk that some unique ecosystems will be lost.[22] While wartime conditions have also provided some protection for both plant and animal species, due to rural depopulation, Angola's especially rich crop plant genetic resources could be threatened by the large-scale introduction of imported commercial seeds during the postconflict agricultural rehabilitation process.[14]

The priority needs of great apes include a thorough census of Cabinda to identify potentially viable populations. Education, enforcement, public awareness, and alternative livelihoods are the basic requirements for the sustainable protection of biodiversity in Cabinda. Effective law enforcement is essential to halt commercial hunting, which may be much more destructive than subsistence hunting in this area, which has no strong ape bushmeat tradition. Hunting for subsistence can be resolved mainly through the combination of developing alternatives and awareness campaigns.

There are several immediate priorities,[21] including the further development of an extensive awareness campaign among soldiers, police, resident communities, and others within the forest, to ensure their cooperation in the protection of the forest and its biodiversity. The devastated socio-economic status of local communities, however, means that an education and awareness campaign is not likely to be sufficient on its own, but must be accompanied by measures to encourage and enable local people to achieve alternative and sustainable livelihoods. Such efforts are within the framework of the objectives of the Cabindan provincial government, with the additional aspects of assuming social responsibility, improving relationships with communities, and visibility.

From August 2005, Cabinda's Department of Agriculture, Fisheries, and Environment will employ the biodiversity advisor originally funded by UNDP/NORAD. The department is also working with a provincial environmental nongovernmental organization, Gremio ABC. Priorities include the recruitment of a law enforcement unit of rangers (*fiscais*), mainly from resident communities, to provide some alternative livelihoods, and a feeling of ownership. There is an urgent need to train a first group of rangers that could then take part in establishing the law enforcement unit and in training others. In 2003, it was agreed that the budget allocated under the UNDP/NORAD project for the training of rangers would be used to contribute to a first national course for rangers by the Southern African Wildlife College, in Kissama NP, including places for six trainees from Cabinda.

Finally, a transfrontier conservation initiative is needed for the protection of the Maiombe forest, in response to the harvesting and cross-border smuggling of wildlife, timber, and associated derivatives. This would need to involve all three countries that share the forest: Angola, Congo, and DRC. UNDP has produced a proposal for the development of such an initiative, and it has been distributed to the international scientific and donor community.[16]

FURTHER READING

FAO (2001) *Forestry Outlook Studies for Africa (FOSA) – Angola Country Study.* FOSA Working Paper – FOSA/WP/02. FAO, Rome. ftp://ftp.fao.org/docrep/fao/003/X6772E/X6772E00.pdf. Accessed May 19 2005.

Huntley, B.J., ed. (1994) *Botanical Diversity in Southern Africa.* National Botanical Institute, Pretoria.

IUCN (1992) *Angola: Environmental Status Quo Assessment Report – Executive Summary.* IUCN Regional Office for Southern Africa, Harare.

IUCN (1992) Peace in Angola. *IUCN Bulletin* 23.

Ministry of Fisheries and Environment (2002) *Report to the Committee for the Review of the Implementation of the Convention to Combat Desertification.* Luanda. http://www.unccd.int/cop/reports/africa/national/2002/angola-eng.pdf. Accessed May 19 2005.

MAP DATA SOURCES

Map 16.1 Great apes data are based on the following source:

Butynski, T.M. (2001) Africa's great apes. In: Beck, B.B., Stoinski, T.S., Hutchins, M., Maple, T.L., Norton, B., Rowan, A., Stevens, E.F., Arluke, A., eds, *Great Apes and Humans: The Ethics of Coexistence.* Smithsonian Institution Press, Washington, DC. pp. 3–56.

With additional data by personal communication from Ron, T. (2003).

For protected area and other data, see 'Using the maps'.

ACKNOWLEDGMENTS

Many thanks to Tamar Ron (UNDP) for her valuable comments on the draft of this section, and for information about the Maiombe forest.

AUTHOR

Gemma Smith, UNEP World Conservation Monitoring Centre

REPUBLIC OF BURUNDI

GEMMA SMITH

BACKGROUND AND ECONOMY

Located in central Africa, the Republic of Burundi is a small, steep country that rises from the eastern shore of Lake Tanganyika. The land area is 25 650 km^2, or 27 834 km^2 including Burundi's section of the lake. Elevations only fall below 1 000 m along the lake's shores (773 m). A mountain range, the Congo–Nile Ridge, that reaches 2 670 m, runs roughly north–south along the western boundary, providing the highest land in the country. Most of the country comprises flat plateaus and rolling hills. It is bordered by Rwanda to the north, by the United Republic of Tanzania to the south and east, and by the Democratic Republic of the Congo (DRC) to the west.[2, 11] Burundi's population was estimated at 6.2 million in 2004,[2] with up to 300 people per square kilometer in some areas.[1, 2, 13] The population is growing at an annual rate of 1.9 percent.[21] It includes three main ethnic groups: the Hutu (over 83 percent), the Tutsi (less than 15 percent), and the Twa (1 percent).[18] Approximately 52 percent of inhabitants over the age of 15 are literate.[2]

Burundi's economic situation has varied in

Map 16.2 Chimpanzee distribution in Burundi

Data sources are provided at the end of this country profile

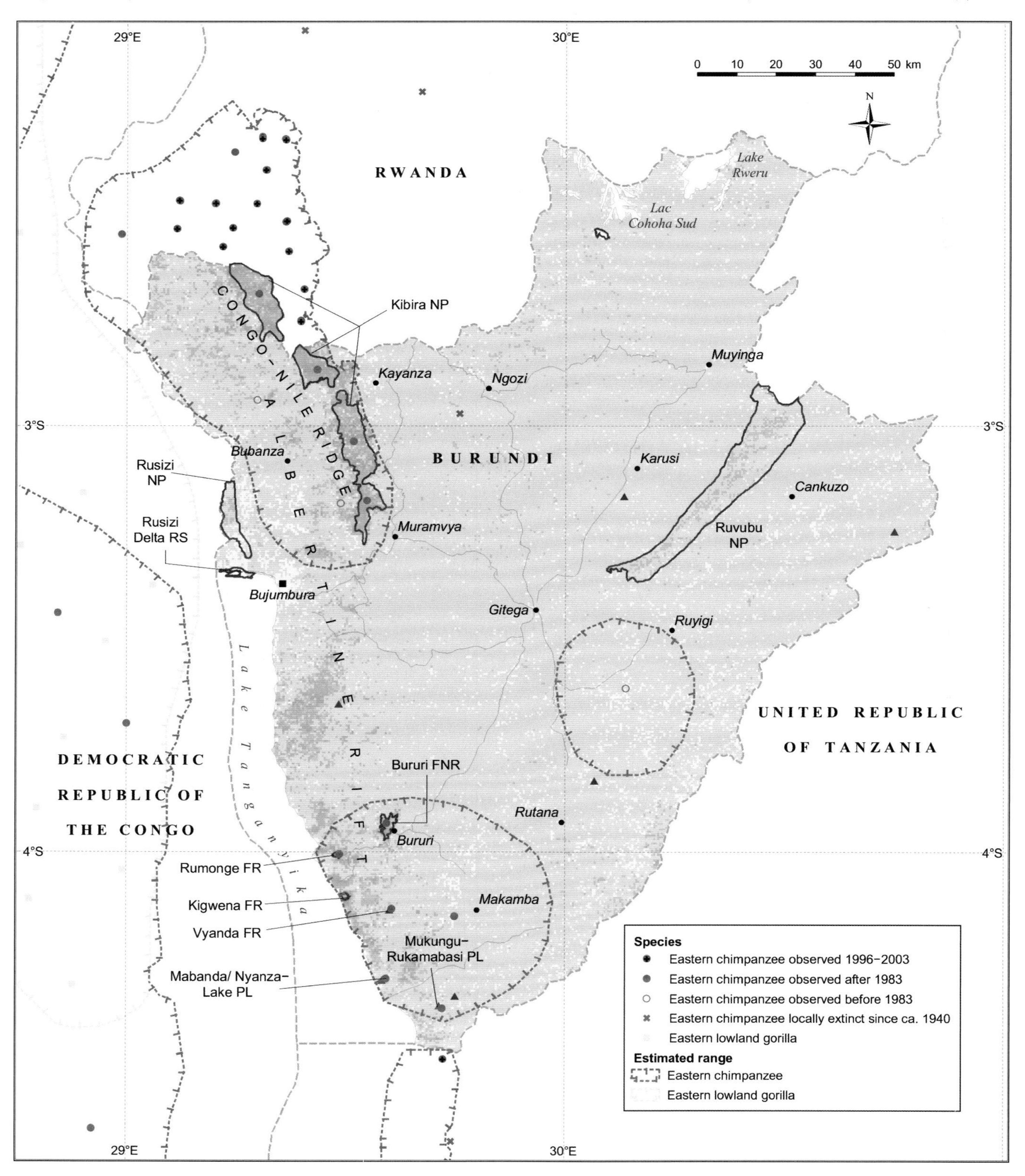

recent years due to political disturbances and conflict. In 2002, the gross domestic product (GDP) was US$719 million and the gross national income (GNI) per person was US$110.[21] The economy is based on subsistence agriculture,[13, 17] with 90 percent of the population engaged in farming.[1] Coffee is Burundi's main export, while principal food crops are cassava, bananas, maize, sorghum, and beans. Livestock are abundant and there is heavy grazing pressure in many locations, leading to damaging soil erosion in this high rainfall area.[13]

The varied topography, soil, and climate support a large number of vegetation types.[11] Most of the natural vegetation is a mosaic of East African evergreen bushland and secondary wooded grassland with abundant *Acacia* trees (Leguminosae-Mimosaceae), with large areas of afromontane vegetation to the west. Miombo woodland dominated by *Brachystegia* and *Julbernardia* (both Leguminosae-Caesalpinioideae) exists along the southeast border, with small patches of transitional rain forest in the northwest. Much of the natural vegetation has been degraded by grazing and farming. Prior to the conflicts of the 1990s, the natural and planted forests of Burundi together extended over 2 000 km², about 8 percent of the country.[7] In 2000, however, only an estimated 940 km² remained under forest, 3.7 percent of total land area. The forests are under intense pressure from legal and illegal logging, leading to rapid deforestation.[6]

Burundi gained independence from Belgium in July 1962. Since then it has been plagued by tension between the Tutsi minority, who had traditionally ruled the country, and the Hutu majority, and Burundi has been the scene of one of the most intractable conflicts in Africa. The first democratic elections occurred in 1993, when Burundians chose their first Hutu head of state, Melchior Ndadaye, and a parliament dominated by the Hutu Front for Democracy in Burundi (Frodebu). Within months Ndadaye had been assassinated, setting the scene for years of Hutu-Tutsi violence in which an estimated 300 000 people have been killed. After 10 years of conflict, in April 2003, Domitien Ndayizeye, a Hutu, succeeded Pierre Buyoya, a Tutsi, as head of Burundi's three year transitional powersharing government.

DISTRIBUTION OF GREAT APES

The eastern chimpanzee (*Pan troglodytes schweinfurthii*) is the only great ape to occur in Burundi. Relatively little is known about Burundi's chimpanzees, but 300–400 individuals are thought to exist there. Surveys in 1987 found two small populations: 200–250 in Kibira National Park (NP) and 30–50 in the Rumonge and Bururi reserves.[13, 14] In 1992, based on surveys by Trenchard,[16] further estimates were made for populations in the Vyanda Forest Reserve and the Mabanda/Nyanza-Lake and Mukungu-Rukamabasi Protected Landscapes (see Table 16.1).

Most of Burundi's chimpanzees live in Kibira NP, which covers approximately 403 km² at an altitude of 1 600–2 666 m in the northwest of the country.[5, 8] It stretches along the north–south mountain range of the Congo–Nile Ridge, and is contiguous with Nyungwe forest in Rwanda, forming a montane forest block of approximately 1 300 km². Chimpanzees are found in each major part of the park.[3] Several other primate species are also present, including the Endangered species L'Hoest's guenon (*Cercopithecus l'hoesti*). Key tree species include *Newtonia buchananii*, *Albizia gummifera* (both Leguminosae), and *Entandrophragma excelsum* (Meliaceae).[13] A herbaceous layer and trees occur on rocky soils even at 2 300 m altitude, while bamboo species occur at 1 900–2 300 m. Some trees grow as high as 50 m. It is estimated that about 16 percent of the park comprises primary evergreen forest, mostly montane.[13] Since the crisis of 1993, however, much damage has occurred through uncontrolled destruction by armed gangs, bush fires, illegal timber harvesting, agricultural encroachment, goldpanning, poaching, collection of medicinal plants, and charcoal production.[7]

Bururi Forest Nature Reserve is located in southwest Burundi, to the west of the town of Bururi, with an altitudinal range of 1 900–2 307 m and an

Table 16.1 Estimated chimpanzee populations in Burundi, 1992[8, 10]

Park, reserve	Vegetation	Area (km²)	Number of chimpanzees
Kibira NP	Afromontane forest	403	200–250
Bururi Forest Nature Reserve	Afromontane forest	33	30–50
Rumonge Forest Reserve	Miombo woodland	6	10–15
Vyanda Forest Reserve	Miombo woodland	45	50–68
Mabanda/Nyanza-Lake and Mukungu-Rukamabasi Protected Landscapes	Gallery forest	85	15–20
Total			305–403

The impact of gold mining at Mabayi, near the Rwandan border.

Geoffroy Citegetse

area of 33 km^2. It contains about 16 km^2 of semi-evergreen forest, and receives an average annual rainfall of 1 200–2 400 mm. Despite its small size, 93 tree species have been recorded here, with species of *Myrianthus* (Cecropiaceae) and *Strombosia* (Olacaceae) being dominant, and those of *Newtonia* (Leguminosae), *Tabernaemontana* (Apocynaceae), and *Entandrophragma* being common.[13]

Vyanda Forest Reserve is situated nearby, and comprises miombo woodland habitat, as well as a Zambezian flora, including drier savanna vegetation from East Africa and lowland evergreen forest common in DRC.[3] Rumonge Forest Reserve is also situated in western Burundi, at an altitude of about 850 m, and has an area of some 6 km^2. It is predominantly composed of uniform *Brachystegia* forest. Other primate species present include olive baboons (*Papio anubis*) and vervet monkeys (*Chlorocebus aethiops*).

The protected landscapes of Mabanda/Nyanza-Lake and Mukungu-Rukamabasi are located in the south of the country at an altitude ranging between 900 and 1 600 m, and total approximately 85 km^2 in area. Natural vegetation covers about 37 km^2, and is made up of open forests dominated by *Brachystegia*, wooded savannas, grassland savannas, submontane gallery forests, and low grassland. A study of the fauna of these protected areas remains to be made, but olive baboons and aardvarks (*Orycteropus afer*) are found here.[3]

THREATS

Chimpanzees are vulnerable to hunting, logging, and forest clearance, so they are assumed to be in decline, but the rate of loss is unknown. The effects of warfare and conflict have further exacerbated the impacts of habitat loss.[14] The montane forests of Kibira NP suffered much land conversion and forest degradation during the 1990s.[7] Habitat fragmentation particularly affects the chimpanzees of Rumonge, Vyanda, and Mabanda/Nyanza-Lake, which now move from one small forest block to another. Along the way, they come into close contact with humans and are sometimes killed.[8]

LEGISLATION AND CONSERVATION ACTION

International agreements

Burundi signed the Convention on Biological Diversity in 1992; it came into force in 1993 and was ratified in 1997. Two national reports on implementation of the convention, and a national biodiversity strategy and action plan have been developed.[1, 9] Both the UN Convention to Combat Desertification and the UN Framework Convention on Climate Change have been signed and ratified (both in 1997). Burundi is also a party to the Convention on International Trade in Endangered Species of Wild Fauna and Flora (CITES), acceding to the convention in 1988, and it ratified the Convention on Wetlands of International Importance (Ramsar) in 2002.

Burundi has ratified the World Heritage Convention (May 19 1982), and participates in UNESCO's Man and Biosphere (MAB) Programme, although it currently has no designated MAB sites or properties inscribed on the World Heritage List. It is not party to the Convention on Migratory Species, despite the presence of several species listed by the convention within its territory. Other regional biodiversity treaties and agreements which Burundi has signed include the African Convention on the Conservation of Nature and Natural Resources.

National legislation

The Institut National pour l'Environnement et la Conservation de la Nature (INECN), created in 1980, is responsible for the management of national parks, nature reserves, and natural monuments. Part of the Ministère de l'Aménagement du Territoire, de l'Environnement et Tourisme (Ministry of Land Planning, Environment, and Tourism), it is also responsible for organizing and implementing biodiversity assessments and conservation activities, for organizing public conservation education activities, for making proposals for new protected areas, and for ensuring that tourist sites are used sustainably.[13] INECN is also the CITES management and scientific authority for Burundi.

There is no national law that specifically protects chimpanzees, although export of primates is illegal.[14] As chimpanzees are thought to be restricted to protected areas, their current populations are widely assumed to be sufficiently protected from hunting by the protected area legislation. The 1971 regulation on hunting and protection measures for certain animal species includes 12 articles relating to hunting rights, the special permits required for hunting, methods of hunting prohibited, and the listing of protected species. Other relevant national conservation legislation is outlined below.[4, 8, 13]

- **Law No. 1/06 (1980)** provides for the establishment of national parks and nature reserves, within which hunting and habitation of protected areas are prohibited, as is land exploitation within 1 km of the boundaries.
- **Law No. 1/02 (1985)**, the Forest Code, which provides for establishing protection forests, forest reserves, and reforestation areas, is administrated by the Forest Service, and aims to protect soils and conserve animal and plant species under threat within forest reserves and protection forests.
- **Law No. 10 (2000)**, the Environment Code, is Burundi's key legislation to allow the management and protection of the environment as an integral part of Burundi's National Environment Strategy. In 163 articles, the code addresses administrative responsibilities, environmental impact assessment procedures, protection and development of natural resources including biodiversity, the human environment and cultural heritage, control of all forms of pollution, and penal provisions.
- **Decree No. 100/007 (2000)** delineates the boundaries of Kibira NP, Rusizi NP and four natural forest reserves (Bururi, Rumonge, Vyanda, and Kigwena), and defines their management objectives and the mode of protection and conservation of the flora within the sites. The protected landscapes were also identified here, but the precise boundaries had not yet been delineated.[3]

Protected areas

Burundi is one of the few African countries that did not have an established protected area system during colonial times, and it was not until 1980 that formal protected area legislation came into existence here.

Currently, four types of national protected area are designated in Burundi: national parks (*parcs nationaux*), natural reserves (*réserves naturelles*), natural monuments (*monuments naturels*), and protected landscapes (*paysages protégés*).[11, 13, 19] Fourteen national protected areas have been designated: national parks (two), natural reserves (seven), natural monuments (one), and protected landscapes (four). Together they extend over 1 277 km^2 and cover 4.6 percent of the total land area.[8, 13, 19]

To date Burundi has designated one international protected area, a Ramsar Site designated in 2002, and covering 10 km^2 of the shore of Lake Tanganyika.

Conservation projects

In 1992, INECN in collaboration with the Jane Goodall Institute (JGI) initiated the Kibira Chimpanzees Project – but this was derailed by the war. In 2002, WWF–The Global Conservation Organization started to support work in Kibira NP, aiming to avert encroachment and destruction of the area as the country returned to peace. This partnership work is carried out through the WWF ecoregional project for the Albertine Rift Forests, and is being undertaken with Burundi's parks authority and two local nongovernmental organizations (NGOs): the Association Burundaise pour la Protection des Oiseaux (Burundi's BirdLife affiliate, ABO) and the Organisation pour la Défense de l'Environnement au Burundi (ODEB). The project is funded by the MacArthur Foundation. It provides training support to park authorities to improve natural resource management, and aims to empower and involve local communities in conservation activities, while exploring and promoting alternative sources of income or benefits.[22] In recent times, ABO and other environmental organizations and media in Burundi have been organizing awareness campaigns to develop an understanding among forest users of the economic and ecological benefits that the forest offers.[7]

The project supports and reinforces the efforts of l'Association d'Encadrement, de Production et de Vulgarisation (AEPV-DUFASHANYE) to help local populations affected by war. It does this by promoting activities to generate income, supporting forestry and small plantations around the parks, and by raising awareness among the people and the administrative and military authorities about the Parks for Peace concept, in order to transform this park into a site that will be respected by everyone, even during periods of conflict.[7]

IUCN–The World Conservation Union and the Burundian government have completed an orientation phase of the Parks for Peace project, which includes the Kibira NP in Burundi, as well as Virunga NP in DRC and the Volcanoes NP in Rwanda. The initiative aims to resolve conflicts and impacts upon protected areas through enhancing partnerships and dialogs at national and local levels, as well as providing training and support for survey work. The aim is to reduce conflicts and the threat of poaching, logging, mining, and other destructive land uses in protected areas.The Wildlife Conservation Society (WCS) is also working in the region through its Albertine Rift Programme. [20]

There are no sanctuaries or reintroduction centers in Burundi, primarily due to the relatively unstable political situation, which, in 1994, prompted the Jane Goodall Institute to request permission from the Burundian and Kenyan governments to relocate the 20 chimpanzees from its rehabilitation center in Bujumbura, Burundi, to Kenya. These orphans had mostly originated in DRC.[12, 15]

FUTURE CONSERVATION STRATEGIES

The status of national parks and forest reserves in Burundi urgently needs assessment, especially regarding the increasing isolation and resulting vulnerability of the eastern chimpanzee to extinction. The establishment of additional protected areas may be required. Improved environmental governance is essential. Although new environmental and biodiversity legislation has been developed, additional capacity to implement this is likely to be required, as well as tight controls on encroachment into protected areas, poaching, and timber harvesting.

As well as undertaking more general biodiversity assessments, there is a need to undertake surveys within protected areas where chimpanzees have been historically recorded, to determine presence and numbers.

It is essential to continue raising awareness about the environment and its current degradation. A few national organizations are already doing this, but further efforts are needed. The constructive efforts of international NGOs (such as WWF and WCS) should continue and their efforts and their effectiveness should be monitored.

FURTHER READING

Habonimana, A. (2003) *The Magnificent Kibira Park Turned into a Land of Devastation.* Submitted to the UN Convention on Desertification. www.unccd.int/publicinfo/localcommunities/burundi-eng.pdf. Accessed September 27 2004.

Powzyk, J. (1988) *Tracking Wild Chimpanzees in Kibira National Park.* Lothrop, Lee and Shepard, New York.

MAP DATA SOURCES

Map 16.2 Great apes data are based on the following source:

Butynski, T.M. (2001) Africa's great apes. In: Beck, B.B., Stoinski, T.S., Hutchins, M., Maple, T.L., Norton, B., Rowan, A., Stevens, E.F., Arluke, A., eds, *Great Apes and Humans: The Ethics of Coexistence.* Smithsonian Institution Press, Washington, DC. pp. 3–56.

With additional data from:

INECN (1992) *Projet Chimpanzées de la Kibira – Jane Goodall Chimpanzees, Conservation and Research Project.* Proposal submitted to the Jane Goodall Institute.

MINITERE (2004) *Rwanda's National Great Apes Survival Plan 2004–2009.* Final draft. Ministry of Lands, Environment, Forestry, Water, and Natural Resources, Republic of Rwanda.

For protected area and other data, see 'Using the maps'.

ACKNOWLEDGMENTS

Many thanks to Geoffroy Citegetse (ABO), Adrien Habonimana (ABO), Jean Rushemeza (INECN), and an anonymous reviewer for their valuable comments on the draft of this section. Thanks also to Brigid Barry (Tropical Biology Association) for editorial assistance.

AUTHOR

Gemma Smith, UNEP World Conservation Monitoring Centre

REPUBLIC OF CAMEROON

PATRICE TAAH NGALLA, LERA MILES, AND JULIAN CALDECOTT

BACKGROUND AND ECONOMY

The Republic of Cameroon is located in Central Africa on the Gulf of Guinea, with Nigeria to the west, Chad to the north, the Central African Republic (CAR) to the east, and Congo, Gabon, and Equatorial Guinea to the south. It covers approximately 475 440 km^2, with forested lowlands in the south, a coastal plain in the southwest, moist savanna woodlands in the highlands of the northwest and central provinces, a seasonal marshland around Lake Chad in the extreme northwest (now somewhat desiccated), and the Mount Cameroon range of forested mountains in the southwest.

Cameroon had an estimated population of 15.7 million people in 2003, which was growing at about 2 percent per year.[15] Approximately half live in urban areas such as Douala and Yaoundé, with most of the rural population residing in the south and central regions of the country. In 2002, the gross national income (GNI) per person was US$560, and the gross domestic product (GDP) for the country was US$9.4 billion,[66] 9 percent of which derived from timber products.[16] The national economy has a strong dependence on oil, timber, and cocoa exports; the major national industries are the production of timber and textiles.

The present country was formed in 1961 through the merger of the former French Cameroon with the English-speaking southwest, which was then administered by the UK (and had been part of the larger German colony of *Kamerun* until the First World War). Compared to other countries in the region, it has experienced relative stability (apart from a border dispute with Nigeria over the Bakassi Peninsula, ongoing since 1992), allowing the steady development of infrastructure, agriculture, and the oil and timber industries.

DISTRIBUTION OF GREAT APES

Overview

The western lowland gorilla (*Gorilla gorilla gorilla*), Cross River gorilla (*G. g. diehli*), central chimpanzee (*Pan troglodytes troglodytes*), and Nigeria-Cameroon chimpanzee (*P. t. vellerosus*) all occur in Cameroon. This diversity is a result of the biogeographically unique, transitional nature of the Cross-Sanaga area of southwest Cameroon and southeast Nigeria. This region contains the Nigeria-Cameroon chimpanzee and Cross River gorilla, as well as other restricted-range primates such as the drill (*Mandrillus leucophaeus*). Western lowland gorillas and central chimpanzees are widespread to the south of the Sanaga River (Map 16.3a).

Western lowland gorillas

The most recent estimate is that there are 15 000 western lowland gorillas in Cameroon.[54] These include some of the 10 000 gorillas thought to live in a triangle of protected areas on the common frontiers of Cameroon (Lac Lobéké National Park (NP)), CAR (Dzangha-Ndoki NP), and Congo (Nouabalé-Ndoki NP), collectively called the Trinationale de la Sangha. The core area is around 7 300 km^2, with buffer zones of 21 000 km^2.[2, 56] Other recorded locations include the Campo-Ma'an NP, Forest Management Unit (Unité Forestière d'Aménagement) 003, Mengamé Gorilla Sanctuary, Dja Wildlife Reserve (also known as Dja Faunal Reserve, and designated as a Biosphere Reserve and a World Heritage Site), the Abong-Mbang Forest Reserve, Deng Deng Forest Reserve, a small pocket of forest outside Nanga Eboko, and the Southeast Forest Technical Operational Unit (Unité Technique Opérationnelle), which covers most of the eastern province of Cameroon, from the Lomié–Batouri axis to the border with CAR and Congo.[40, 27] Gorillas are thought to be rare and possibly extinct in the Douala-Edéa Wildlife Reserve.[14] A 2002–2003 survey in the Mengamé Gorilla Sanctuary found a greater population of apes than expected – 1 200 western lowland gorillas and 200 central chimpanzees. The animals were found at a higher density further from human villages and settlements.[19, 34]

Cross River gorillas

In Cameroon, the Cross River gorilla is confined to the Takamanda Forest Reserve, the Mone Forest Reserve, the Mbulu Hills Community Forest, and other highland areas of forest to the east and south of these,[40, 46] as well as the nearby forests in Nigeria. These are the northernmost and westernmost gorilla populations, separated from the range of the western lowland gorilla by some 200 km.[12] The range over Cameroon and Nigeria is limited to at most 200 km^2 of largely unconnected forest fragments, spread over a total area of at least 2 500 km^2.[12, 45] There are

Map 16.3a Chimpanzee distribution in Cameroon

Data sources are provided at the end of this country profile

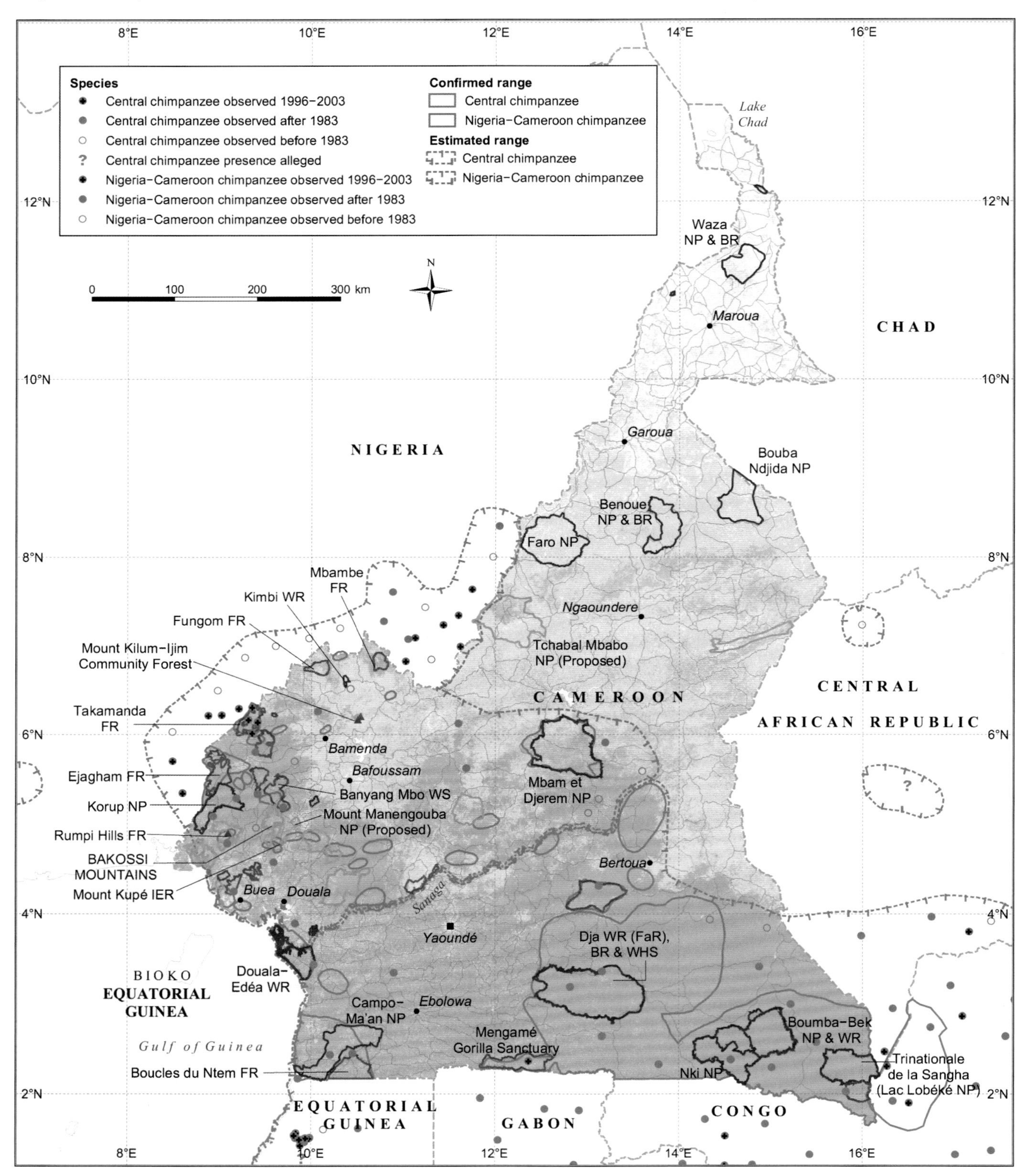

known to be some 205–250 weaned Cross River gorillas, of which about 150 live in Cameroon.[46] It is suspected that the total population size is slightly larger, and further surveys are underway. More details can be found in Chapter 7 of this volume.

There is also a small population of gorillas north of the Sanaga River in the Ebo forest (a proposed national park), which appears to be in danger of extinction.[17] The taxonomic status of this population is unknown; measurements of the only skull available, a male, give ambiguous results. Genetic analysis of fresh fecal samples, or freshly shed hairs (though these typically yield lower levels of DNA), may help to resolve the taxonomic status of these Ebo gorillas.[41]

Nigeria-Cameroon chimpanzees

The Nigeria-Cameroon chimpanzee occurs in the same forests as the Cross River gorilla, and also has a wider distribution within southwest Cameroon.[13] In addition to the areas mentioned above, its range includes Korup NP, Banyang Mbo Wildlife Sanctuary, Rumpi Hills Forest Reserve (also a proposed wildlife reserve), the Bakossi Mountains Wildlife Reserve (proposed), Mount Kupé and Mount Manengouba NP (proposed), and some areas to the north of Takamanda Forest Reserve.[40, 42, 46] Population data are limited, but a 2005 workshop in Brazzaville estimated that there are around 3 380 Nigeria-Cameroon chimpanzees in Cameroon. The largest populations are thought to be at Korup, Takamanda, and Ebo-Ndokbou.

Central chimpanzees

Central chimpanzees are distributed in Cameroon over much the same geographical range as the western lowland gorilla, i.e. the southern moist forests. In 2001, there were estimated to be about 35 000 chimpanzees in total in the country, which, based on the numbers above, would imply that there are at least 30 000 central chimpanzees.[55] Protected areas containing central chimpanzees include Campo-Ma'an NP, Lac Lobéké NP, Dja Wildlife Reserve, Douala-Edéa Wildlife Reserve, Boumba-Bek NP and Wildlife Reserve, Nki NP, and Mengamé Gorilla Sanctuary.[19, 55] Their population densities appear to vary with habitat type, and are particularly high in the Dja area (see Table 16.2).

THREATS

The principal threats to the great apes of Cameroon are hunting and logging.[55] Approximately 238 580 km^2, or half the country's land area, was forested in 2000, with an estimated decrease of 2 220 km^2 per year.[23] Large areas have been lost to shifting cultivation, plantation development, and forest fires, or else degraded by fuelwood collection and commercial selective logging. In the southern province of Cameroon, for example, the cumulative number of logging concessions from 1959 to 1999 covered 76 percent of the total forest area.[24, 32] In addition, some companies have been shown to fell trees illegally over much larger areas than are granted in their concessions.

Logging causes fragmentation of forest areas, opens up access for hunters who use wire snares and fire arms,[48] and increases local demand for bushmeat. Logging routes also serve as conduits for the meat to be transported into market towns.[16] Bushmeat offers both an income to hunters and an affordable source of animal protein to rural and urban people, being cheaper than beef in many areas.[55] Hunting and encroachment are recognized threats for many of Cameroon's reserves.[40] There are indications that Cross River gorilla numbers have declined significantly as a result of illegal hunting.[52] Chimpanzees were wiped out in the Kilum-Ijim forest in northwest Cameroon in 1987–1998.[38] It has been estimated that about 44 gorillas[36] and over 50 chimpanzees[5] are killed annually in the Dja Wildlife Reserve. Dja's wildlife is also under pressure from logging, with active concessions surrounding the reserve, although technically this is a protected buffer zone[5, 7, 8, 11]

Other large-scale development in forest areas also increases hunting intensity and the local market for bushmeat. During the construction of

Table 16.2 Central chimpanzee population density estimates in Cameroon

Reserve	Weaned individuals per km^2
Dja Wildlife Reserve	1.2 (0.81-1.77)[18] 0.7 (0.6-0.9)[57] 0.8 (0.6-1.0)[62]
Outside northern periphery of Dja Reserve	1.1[18]
Dja mangrove forest	0.61[63]
Lac Lobéké NP	0.14[63] 0.17[60]
Mengamé Gorilla Sanctuary	0.18[19, 34, 35]
Ntibonkeuh	0.64[34]
Campo-Ma'an NP	0.63-0.78[39]
Boumba-Bek NP and Wildlife Reserve, Nki NP	0.3 (0.2-0.4)[4]

Map 16.3b Gorilla distribution in Cameroon

Data sources are provided at the end of this country profile

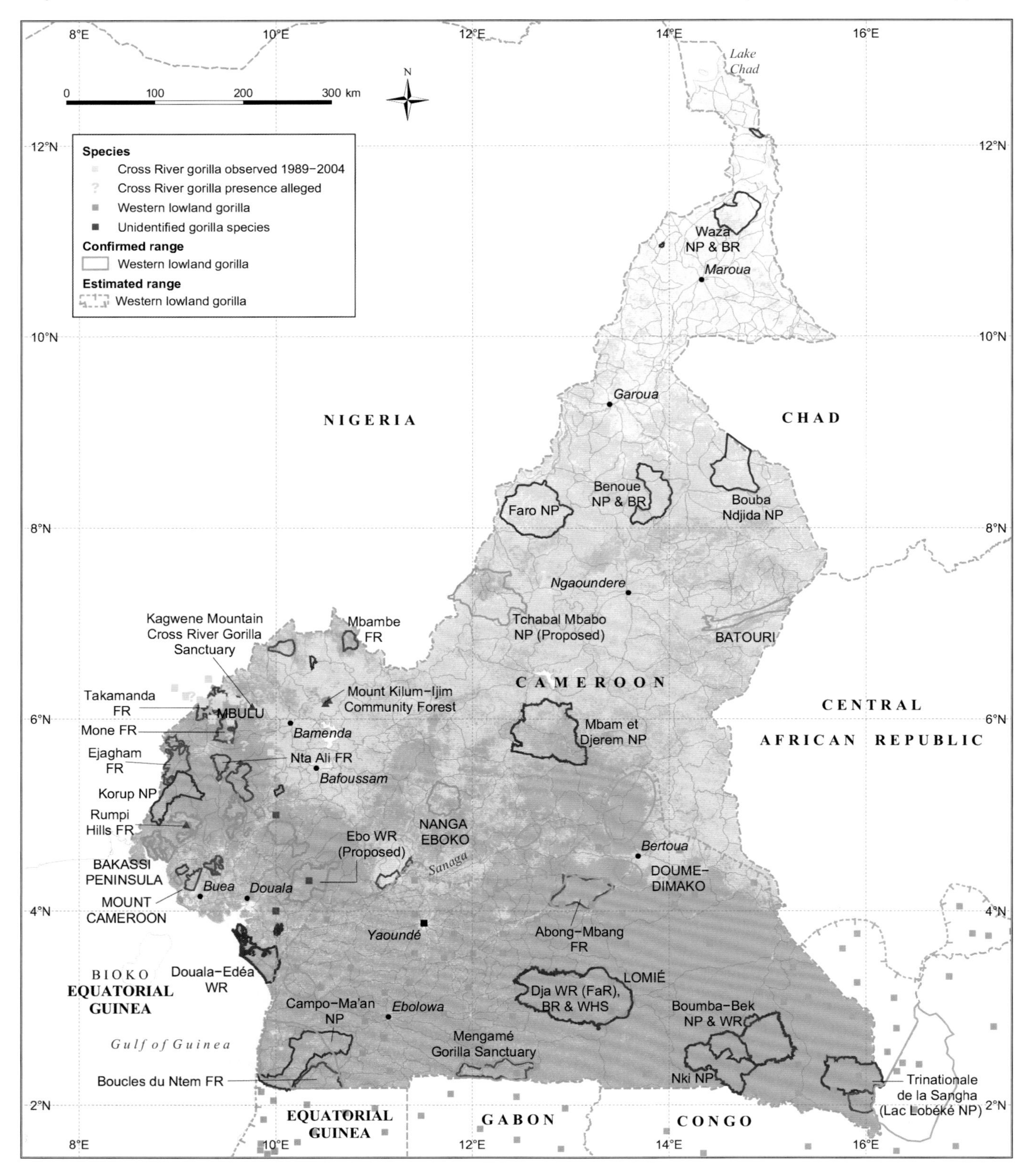

the Cameroon–Chad pipeline in 2001–2003, for example, the oil company Esso reported on several cases of bushmeat purchase by its workers, which it discovered in the course of attempting to enforce stronger environmental standards.[20, 21]

The net result of all these factors is an inferred steady decline among great ape populations in Cameroon.

LEGISLATION AND CONSERVATION ACTION

National legislation

Law No. 94/01 (1994) sets out the country's forestry, wildlife, and fishery regulations, and lists gorillas and chimpanzees as Category A species, which are fully protected against hunting, capture, or sale, in whole or in part. Protected areas such as national parks and wildlife reserves may be established under the auspices of the Direction de la Faune et des Aires Protégées (DFAP) of the Ministry of Environment and Forestry (MINEF), which is also responsible for the protection of the country's biodiversity in general. Article 7 of the *Document des Normes* calls for a protected buffer zone around each protected area, to shield it from hunting and other activities that might damage forest health and biodiversity. Taxes and permit requirements have also been imposed on hunters with the aim of reducing indiscriminate hunting in protected areas.[36]

Many of the country's forest reserves were set up by the British colonial administration. The aim was to protect watersheds, restrict agricultural expansion, and to conserve areas for future timber exploitation. Designation as a forest reserve does not therefore automatically offer protection from future logging concessions. Article 11 of the *Document des Normes* prohibits forest development activities in forest reserves as well as conservation areas and their buffer zones,[11] but the designation may be removed and the area auctioned for timber.[27]

The same legislation provides for six types of logging license to be issued, together with some provision for community forests.[16, 24] Companies that break the terms of these licenses can be fined and disqualified from bidding for further licenses. Under the supervision of the Ministry of Environment and Forestry, the Agence Nationale d'Appui au Développement Forestier (ANAFOR) is responsible for forest inventory and management, promotion of the use of timber species, soil protection, desertification control, and forest regeneration.[16] Resources with which to monitor and enforce the implementation of Law No. 94/01 are, however, scarce.

The Ministry has plans for a wildlife revenue enhancement program, which would secure tax revenues from wildlife-based income in a similar way to the existing Forest Revenue Enhancement Program.[9]

Protected areas

Formally established protected areas that contain great ape populations in Cameroon include the following national parks, forest reserves, and wildlife reserves.[30, 40, 55, 59]

- **Nigeria-Cameroon chimpanzees and Cross River gorillas:** Takamanda Forest Reserve (the Ministry of Environment and Forestry has recently proposed increasing the protection status of this area),[50] Mone Forest Reserve.
- **Nigeria-Cameroon chimpanzees:** Mbam et Djerem NP, Korup NP, Ejagham Forest Reserve, Banyang Mbo Wildlife Sanctuary, Fungom Forest Reserve, Rumpi Hills Forest Reserve, and others.
- **Central chimpanzees and western lowland gorillas:** Lac Lobéké NP, Campo-Ma'an NP, Dja Wildlife Reserve/Biosphere Reserve, Nki NP, Mengamé Gorilla Sanctuary, and Boumba-Bek NP and Wildlife Reserve.
- **Central chimpanzees:** Douala-Edéa Wildlife Reserve.

Of these areas, Dja Wildlife Reserve is the largest at 6 236 km^2. Several reserves fall close to international boundaries, which means that territorial boundary conflicts with neighboring countries can influence conservation outcomes. An example is the Nigeria–Cameroon boundary disagreement over the Bakassi Peninsula. This has hampered cooperation between the two countries in managing the several protected areas (e.g. Korup NP, Cross River NP) in the transfrontier range of the Cross River gorilla and Nigeria-Cameroon chimpanzee.

International support

The government of Cameroon has received significant support from the international conservation community, both nongovernmental and governmental, for the protection of great apes and the biodiversity of their habitats. Some of the highlights are listed here.

- The Korup Forest Project in and around Korup NP began in the mid-1980s. It was among the

Lake Beme, a crater lake in the Bakossi Mountains, Cameroon.

Bethan Morgan

first major integrated conservation and development projects invested in by the UK government in partnership with nongovernmental organizations (NGOs) (initially the Earthlife Foundation, and then later WWF–The Global Conservation Organization). A series of such investments has been made in Korup NP and the neighboring Cross River NP (Oban Division) in Nigeria by the UK Department for International Development, the European Commission, and others.

- The Israel-based NGO, The Last Great Ape (LAGA), was formed in the early 1990s to monitor, track down, and ensure prosecution of poachers.[37] In 2001, the first recorded person was imprisoned after having been caught trying to sell a baby chimpanzee.[44]
- The European Union program, Conservation and Rational Use of Forest Ecosystems in Central Africa (ECOFAC), has been active in the Dja Wildlife Reserve.[24] For example, global positioning systems (GPS) were used to locate and map paths, snares, and camps used by hunters in the reserve.[6, 10]
- The Wildlife Conservation Society (WCS) has been working in Cameroon since 1998, integrating field research, education, and outreach, as well as support and capacity building of local government agencies.[26, 50] Within the Cameroon-Nigeria border region, WCS supports Cross River gorilla research and conservation, together with biological surveys in both countries. Technical support is provided to the Mbam et Djerem NP and the Banyang Mbo Wildlife Sanctuary, with financing from multiple donors including the Dutch Directorate General for International Cooperation (DGIS) and the Foundation for Environment and Development in Cameroon (FEDEC). WCS also partners the Cameroon Rail Company (CAMRAIL) in a program to control the illegal transportation of bushmeat.
- In 1999, the government of Cameroon announced the launch of a new trust fund to help finance the effective management of protected forest areas, and an initial donation of US$500 000 was made by WWF.[67] The UK, International, and US branches of WWF lead projects in various protected areas, while WWF-Cameroon has additional research, policy, and education projects.
- The department of Conservation and Research for Endangered Species (CRES) of the Zoological Society of San Diego has been conducting surveys on large mammals in Cameroon since 2000. It has also assisted Cameroon with gazetting several new protected areas in Bakossiland, including the proposed Bakossi Mountains Wildlife Reserve.[42]
- In 2002, the Jane Goodall Institute (JGI) agreed to establish a community-based conservation and wildlife research program in the then newly declared Mengamé Gorilla Sanctuary (1 150 km^2), located on the border of Cameroon and Gabon.[19]
- Within and outside the protected areas, the UK-based Global Witness is working with the Ministry of Environment and Forestry as an independent observer to help improve governance and transparency in the forest sector, concentrating on the issue of illegal logging.[32]
- Collaborative agreements have been signed between Cameroon and BirdLife International and between Cameroon and the US-based World Resources Institute (WRI), aimed at effective conservation and monitoring of forest resources and biodiversity.
- The UK-based Living Earth, Bristol Zoo Gardens, and Fauna and Flora International (FFI) have all sponsored research or education activities.[64]
- The US-based Bushmeat Project works to involve hunters in forest and fauna protection in eastern Cameroon.[49]

The government of Cameroon has also entered into agreements with the governments of neighboring and nearby countries, including the following:

- Nigeria and Cameroon have established an agreement to protect the Cross River gorilla. A collaborative partnership has also been formed between the relevant government departments, WCS, Fauna and Flora International, the German overseas development agency GTZ, and the Nigerian Conservation Foundation (NCF).
- Cameroon, CAR, Congo, Chad, Equatorial Guinea, and Gabon all signed the 1999 Yaoundé Declaration (see Chapter 14). It outlines plans to create new cross-border protected forest areas in the Congo Basin.
- The Conference of Central African Moist Forest Ecosystems (CEFDHAC), coordinated by IUCN–The World Conservation Union in Yaoundé, was later nominated to lead an intergovernmental process based on the Yaoundé Declaration.[1, 16] The initiative includes:
  - the endorsement of the existing 7 300 km^2 trinational network of protected areas between Cameroon, CAR, and Congo as the Trinationale de la Sangha:[67] the Cameroon sector is Lac Lobéké NP, which supports western lowland gorillas and central chimpanzees;[67]
  - the creation of two new national parks in Cameroon: Campo Ma'an and Mbam et Djerem; these were established in compensation for the biodiversity impacts of the Chad–Cameroon pipeline scheme, through an offsite environmental enhancement program.[65]
- Other emerging transfrontier initiatives with Cameroonian involvement include the Central African World Heritage Forest Initiative, the Congo Basin Project, and the Congo Basin Forest Partnership. Protected areas involved include Dja, Mengamé, and Campo Ma'an.[16]

Research and education

Various educational and research institutes exist in Cameroon. In particular, the School for Training of Wildlife Specialists at Garoua, supervised by the Ministry of Environment and Forestry, is the principal school of its type in Francophone Africa.[43] The country has six universities, of which at least the University of Dschang includes research related to wildlife conservation within its forestry program. The Research Institute for Agriculture and Development (IRAD, formerly the Institute for Zootechnical Research) is based in Nkolbisson near Yaoundé, under the Ministry of Scientific Research. Research projects carried out by foreign nationals include the Durrell Wildlife Conservation Trust's investigation into bushmeat supply and demand in the Cross River gorilla region.

Public education and awareness is a component of most current conservation and development projects and a major theme of the national NGOs. Local communities are encouraged to protect biodiversity, both for conservation and symbolic reasons – apes and other mammals are important in some traditional ceremonies. The enforcement of a traditional hunting ban to protect the sacred Mount Kupé, for example, led to a significant decrease in hunting from mid-1994.[61] Similarly, chiefs in the Ma'an community are now making an effort to put an end to poaching, and within the Cross River gorilla range, local communities have agreed to protect ape populations.[33] Formal awareness projects include:

- In Defense of Animals – Africa (IDA-Africa) has built an educational center and is launching a national radio campaign.
- The Cross River Gorilla Project, led by WCS, incorporates a local education component consisting of slide shows, posters, and leaflets focused on the conservation of the Cross River gorilla and other endangered wildlife.[50]
- The Club des Amis de la Nature is a loose association of groups located mainly in schools and universities.[26]
- The Great Apes Project (Projet Grands Singes) working in the Dja area has been involved in capacity building in higher education by European and Cameroonian students, who assist in training and research in topics related to ape protection.
- The Bushmeat Project in south Cameroon has provided material for use in schools and training workshops.

Conservation projects

- The Sanaga–Yong Chimpanzee Rescue Center, run by In Defense of Animals – Africa, opened in August 1999 in Central province. It was originally dedicated to the rescue and rehabilitation of adult chimpanzees rather than younger orphans, but young chimpanzees have more recently been accepted. It held 39 chimpanzees in September 2004.[47]
- The Cameroon Wildlife Aid Fund (CWAF) works with the Ministry of Environment and Forestry to care for the animals at Yaoundé

A drill at the Limbé Wildlife Centre.

Ian Redmond

Zoo at Mvog-Betsi, and their involvement has led to great improvements in animal welfare and education at the zoo. CWAF also works in the Mefou NP, where the Michael Leo Rion Sanctuary was opened in 2001 with the support of the US-based Gorilla Foundation. This sanctuary cares for and rehabilitates bushmeat orphans. Animals are released into a restricted area. The UK's Bristol Zoo Gardens also provides support to CWAF.

- The Limbé Wildlife Centre is active in rehabilitating captured great apes and other species. In January 2004, it was caring for 36 chimpanzees (mostly Nigeria-Cameroon subspecies), 11 western lowland gorillas, and one Cross River gorilla.[28] It receives financial support from the Arcus Foundation.[3]
- In 2004, the NGO Pandrillus organized the repatriation of two confiscated western lowland gorillas from Nigeria to Cameroon – the first time the two countries had cooperated to resolve an instance of illegal cross-border trade in a species listed by the Convention on International Trade in Endangered Species of Wild Fauna and Flora (CITES).[29]

FUTURE CONSERVATION STRATEGIES

Protected areas

Following the Second International Workshop and Conference on the Conservation of the Cross River Gorilla held in Limbé, Cameroon in 2003,[51, 53] the governments of Cameroon and Nigeria acknowledged the need to protect Cross River gorilla habitat. As mentioned in the country profile for Nigeria, this would entail the establishment of a transfrontier protected area for the Takamanda–Okwangwo complex, in particular by upgrading the protection status of the Takamanda Forest Reserve. As part of the development of a land-use plan for the Takamanda–Mone–Mbulu area, the WCS and the Ministry of Environment and Forestry are currently working on the creation of a protected area on Kagwene Mountain, in the forests of Mbulu and Njikwa. Further recommendations include the urgent need to strengthen protection and law enforcement measures for all Cross River gorilla populations.

Three areas have been identified by the Ministry as priorities for protection in the Congolian lowland evergreen forest of the extreme southeast: Boumba-Bek NP and Wildlife Reserve, Lac Lobéké NP, and Nki NP. Better demarcation of existing protected area boundaries might help to discourage illegal logging.

The National Great Ape Survival Plan (NGASP) Workshop held in Cameroon suggested that Ebo, Makombe, Mbulu, Mbargue, Kupé, and Bakossi forests were priorities for great ape conservation.[40] Protection forests are also proposed in the western mountain region near Mount Cameroon: Etinde, Mabeta–Moliwe, Kilum Mountain (Mount Oku), and Bakossiland areas. In 2002, the Kupé chiefs voted in favor of the proposition that an integrated ecological reserve be declared in the Mount Kupé area.[2]

Logging concessions

It has been recommended that logging companies be requested to produce great-ape-sensitive management plans for their concessions. This would include controlling illegal hunting and financing law enforcement, as well as providing protein alternatives to workers and local communities affected by logging.[40]

Capacity building

For Cross River gorillas, it has been recommended that a management committee be established, and that research and conservation capacity is built in government departments, universities, and NGOs.[53] In general, reserves and parks suffer from a chronic lack of staff, equipment, and infrastructure. Only the Korup NP has both a management plan and chief warden. Action is needed to strengthen the Ministry of Environment and Forestry's ability to enforce forestry legislation, including hunting laws.[31, 32, 40, 55] In particular, an increase in the number and training of ministry wildlife monitoring staff has been widely recommended.[16, 40]

Research

Further, better-coordinated research and monitoring is needed on ape distributions and populations throughout Cameroon.[40, 55] In particular, basic research into the ecology, distribution, and population biology of the Cross River gorillas should be expanded.[53] A study of the Ebo gorilla population to the south is underway.[41]

Education and community development

Information and education campaigns are needed on a large scale to inform Cameroonians of the endangered status of the great apes and to tell them about wildlife protection laws. Training workshops for community members and groups should also be organized, to cover conservation issues and the use of other protein sources as alternatives to bushmeat.[22, 40] Support should be given to developing alternative methods of generating income for people currently engaged in hunting.[55]

FURTHER READING

Bikié, H., Collomb, J-G., Djomo, L., Minnemeyer, S., Ngoufo, R., Nguiffo, S. (2000) *An Overview of Logging in Cameroon.* Global Forest Watch. http://www.globalforestwatch.org/common/cameroon/english/report.pdf. Accessed June 12 2005.

Comiskey, J.A., Sunderland, T.C.H., Sunderland-Groves, J.L., eds (2003) *Takamanda: The Biodiversity of an African Rainforest.* Smithsonian Institution, Washington, DC.

Dupain, J., Guislain, P., Nguenang, G.M., De Vleeschouwer, K., Van Elsacker, L. (2004) High chimpanzee and gorilla densities in a non-protected area on the northern periphery of the Dja Faunal Reserve, Cameroon. *Oryx* **38** (2): 209–216.

Global Witness (2005) *Forest Law Enforcement in Cameroon: Third Summary Report of the Independent Observer December 2003–June 2005.* http://www.globalwitness.org/reports/show.php/en.00072.html. Accessed June 12 2005.

Gonder, M.K., Oates, J.F., Disotell, T.R., Forstner, M.R., Morales, J.C., Melnick, D.J. (1997) A new West African chimpanzee subspecies? *Nature* **388**: 337.

Matthews, A., Matthews, A. (2004) Survey of gorillas (*Gorilla gorilla gorilla*) and chimpanzees (*Pan troglodytes troglodytes*) in southwestern Cameroon. *Primates* **45**: 15–24.

MINEF (2003) *Cameroon Action Plan for the Survival of Great Apes and Endangered Primates.* Workshop report, Mfou, Cameroon, March 18–20. Cameroon Ministry of the Environment and Forestry (MINEF).

Sunderland-Groves, J.L., Jaff, B., eds (2004) *Developing a Conservation Strategy for the Cross River Gorilla.* Proceedings of the 2nd International Workshop and Conference on the Cross River Gorilla. Wildlife Conservation Society.

MAP DATA SOURCES

Maps 16.3a and b Great apes data are based on the following source:

Butynski, T.M. (2001) Africa's great apes. In: Beck, B.B., Stoinski, T.S., Hutchins, M., Maple, T.L., Norton, B., Rowan, A., Stevens, E.F., Arluke, A., eds, *Great Apes and Humans: The Ethics of Coexistence.* Smithsonian Institution Press, Washington, DC. pp. 3–56.

With additional data by personal communication from Bergl, R. and Sunderland-Groves, J.L. (2005) and from the following sources:

Dowsett-Lemaire, F., Dowsett, R.J. (2001) A new population of gorillas *Gorilla gorilla* and other endangered primates in western Cameroon. *African Primates* **5**: 3–7.

Halford, T., Ekodeck, H., Sock, B., Dame, M., Auzel, P. (2003) *Statut des populations de gorilles (*Gorilla gorilla gorilla*) et de chimpanzés (*Pan troglodytes troglodytes*) dans le Sanctuaire à Gorilles de Mengamé, Province du Sud, Cameroun: densité, distribution, pressions et conservation.* MINEF and Jane Goodall Institute, Yaoundé. http://www.janegoodall.net/news/assets/RapportGrandsingesMengamefinal2003.pdf. Accessed August 16 2004.

Morgan, B.J. (2004) The gorillas of the Ebo forest, Cameroon. *Gorilla Journal* **28**: 12–14. http://www.berggorilla.de/english/gjournal/texte/28ebo.html. Accessed November 24 2004.

Sunderland-Groves, J.L., Maisels, F., Ekinde, A. (2003) Surveys of the Cross River gorilla and chimpanzee populations in Takamanda Forest Reserve, Cameroon. In: Comiskey, J.A., Sunderland, T.C.H., Sunderland-Groves, J.L., eds, *Takamanda: The Biodiversity of an African Rainforest.* Smithsonian Institution, Washington, DC. pp. 129–140. http://nationalzoo.si.edu/ConservationAndScience/MAB/researchprojects/appliedconservation/westafrica/Takamandabook/Chapter_9.pdf. Accessed July 13 2004.

Various authors (2003) Draft map for working purposes prepared at GRASP meeting, Cameroon, March 18–20 2003.

For protected area and other data, see 'Using the maps'.

ACKNOWLEDGMENTS

Many thanks to Jean LaGarde Betti (University of Dschang), John Ngong Fonweban (University of Dschang), Roger Fotso (Wildlife Conservation Society), Elizabeth Gadsby (Pandrillus), Bethan Morgan (Zoological Society of San Diego), John F. Oates (Hunter College, City University of New York), and Jacqueline Sunderland-Groves (Wildlife Conservation Society) for their valuable comments on the draft of this section. Thanks also to Brigid Barry (Tropical Biology Association) for editorial assistance.

AUTHORS

Patrice Taah Ngalla, Limbé Botanical and Zoological Gardens
Lera Miles, UNEP World Conservation Monitoring Centre
Julian Caldecott, UNEP World Conservation Monitoring Centre

CENTRAL AFRICAN REPUBLIC

NIGEL VARTY

BACKGROUND AND ECONOMY

The Central African Republic (CAR) is a landlocked country, covering 622 984 km^2. Barthélémy Boganda led the country to independence, but was killed in 1959 shortly before secession from France. In 1960, David Dacko became the first independent president of CAR. Elections were held in 1964, in which Dacko was the only candidate, representing the only party.[22] Deteriorating economic circumstances led to a 1966 coup d'état led by Colonel Jean-Bédel Bokassa, who later obtained virtually absolute power; in 1972 he declared himself president for life. In 1977, Bokassa was crowned as emperor of a renamed country, the Central African Empire. Opposition to his erratic rule led to riots and massacres in 1979, followed by a French-backed coup that reinstalled Dacko as president and allowed the republic to be restored. Popular unrest, strikes, army mutinies, foreign interventions, and coups d'état occurred periodically during the 1980s, 1990s, and into the 2000s, as the country struggled to develop democratic institutions; the current head of state, François Bozize himself came to power through a coup in 2003. The constitution was then suspended, but following a general election held in early 2005 Bozize was returned as president.

CAR is among the least developed countries in the world, with a Human Development Index ranked 169th out of 177 countries in 2004,[43] and a life expectancy at birth of only 40 years.[43] The economic situation has deteriorated in recent years due to severe political disturbances, combined with violence and looting. In 2002, the gross domestic product (GDP) was US$1.1 billion and the gross national income (GNI) per person was less than US$300.[48] The major sources of national income are mining (mostly for diamonds) in the southwest, center, and north of the country, and the sale of timber from logging, mainly in the moist forests of the southwest. The population was estimated to be about 3.8 million in 2001,[22] with a growth rate of 1.6 percent.[15] Most of the people are concentrated in the cities and rural areas of the south and west, leaving the north and east largely unpopulated.

The Aka people (also known as Ba'Aka and Bayaka) number about 20 000 in CAR and inhabit the southwestern rain forests and savannas. Like the other forest-dwelling hunter–gatherer peoples of Central Africa (the BaKa of Cameroon and Gabon, the Twa and the Mbuti of the Democratic Republic of the Congo (DRC), all of whom are sometimes called 'Pygmies'), they are often subordinate to their Bantu neighbors. Their homelands are now being logged, hunted, and settled by outsiders, leading to an erosion of their livelihoods and cultural integrity, including the loss of traditional hunting skills, techniques, and technologies. The latter include driving animals such as duiker into nets, and hunting with spear and bow-and-arrow.

Forests and woodland savannas of more than 10 percent canopy cover were estimated to cover about 229 000 km^2 in 2000, or 32 percent of the land area.[19] Most of CAR is too dry for closed canopy rain forest, which only ever occurred in a narrow band along the southern edge of the country, amounting to about 8 percent of the land area.[17] It is now restricted to the southwest corner, between Gamboula and the capital city Bangui, and the southeastern Bangassou region.[17] The Bangassou forests are patchy and degraded,[2] and isolated from other rain forests by farmland, which is concentrated in the south, even though it only accounts for about 3 percent of land use in CAR overall.[15]

DISTRIBUTION OF GREAT APES

The western lowland gorilla (*Gorilla gorilla gorilla*) and the central and eastern subspecies of the chimpanzee (*Pan troglodytes troglodytes* and *P.t. schweinfurthii*) are all found in CAR. The country is at the northeastern edge of the range of the western lowland gorilla and central chimpanzee. Neither gorillas nor chimpanzees have been censused at a national level,[11] but the Dzanga-Sangha area has been surveyed more thoroughly than most other sites in CAR.[2, 40] Gorillas are restricted to the forests of the southwest. They occur in both the Dzanga and Ndoki sectors of Dzanga-Ndoki National Park (NP) (1 222 km^2), in the Dzanga-Sangha Dense Forest Special Reserve (3 359 km^2),[3, 4] and in the Ngotto forest, which was declared an integral reserve in 1996 and is a proposed national park,[10] and there are several records of gorillas from the surrounding area.

The central chimpanzee also occurs in the forested southwest, but not abundantly so.[13] A density of only 0.16 individuals per square kilometer was reported in the Dzanga-Ndoki NP in 1996–1997,[3] and 0.44 individuals per square kilometer in the Ngotto Classified Forest in 1998–1999.[10] Central chimpanzees have also been reported at one location in the northwest.[12] The eastern chimpanzee is reported from the forests of the southeast, which have never been properly surveyed for the species,[12, 28] and again, there is one recorded locality in the north. Eastern chimpanzees were previously known in central parts of the country, but their current distribution is poorly known and there are few recent records.

A population of about 40 chimpanzees of an as yet undetermined subspecies (presumably eastern) has also been reported from the island of Nabolongo.[1, 29, 30] The island is to the south of Bangassou forest in the Mboumou River, which forms part of the southern border of CAR with DRC. There no longer seem to be any in Bangassou forest itself,[47] and the island population continues to be hunted by Congolese.[46]

Numbers

Based on a mean density of 0.25 individuals per square kilometer occurring over the area of forest in a 1985 map,[5] it was estimated that 9 000 gorillas may be present over 36 000 km^2.[24] Based on nest counts over 783 km^2 of transects through different habitat types in 1984–1985, extrapolated over a 1967 vegetation map modified to take account of later human habitation, it was estimated that there were between 4 806 and 7 830 gorillas in the southernmost 6 000 km^2 of forest alone.[14] This is a more positive picture than that given by the 1980 estimate of 500 gorillas.[27]

An estimate published in 1987 gives a national population of chimpanzees at 800–1 000, again based on the area of suitable habitat.[42] In a 2003 review, however, the estimate was 800–1 000 for the central subspecies alone, as well as an unknown number for the eastern subspecies.[12] The difference in both cases reflects an improvement in knowledge rather than an actual increase in numbers.

THREATS

An estimated 300 km^2 of forest (about 0.1 percent of the total) are cleared in CAR each year.[19] Logging concessions cover 50–75 percent of the remaining forests.[36] In the rain-forested southwest, 86 percent of forests had been allocated to concessions by 2000.[23] High transport costs mean that logging in CAR tends to be highly selective, but the operation of these concessions inevitably opens up previously inaccessible areas to hunting. Both gorillas and chimpanzees are vulnerable to hunting, logging, and forest clearance, so they are assumed to be in decline, although the rate of loss is unknown.

There is little evidence that gorillas are hunted systematically in CAR,[33, 39] but chimpanzees are a known prey, albeit at an unknown level. The bushmeat trade is considered the greatest threat to wildlife conservation in CAR, even inside the pro-

Map 16.4 Great ape distribution in the Central African Republic

Data sources are provided at the end of this country profile

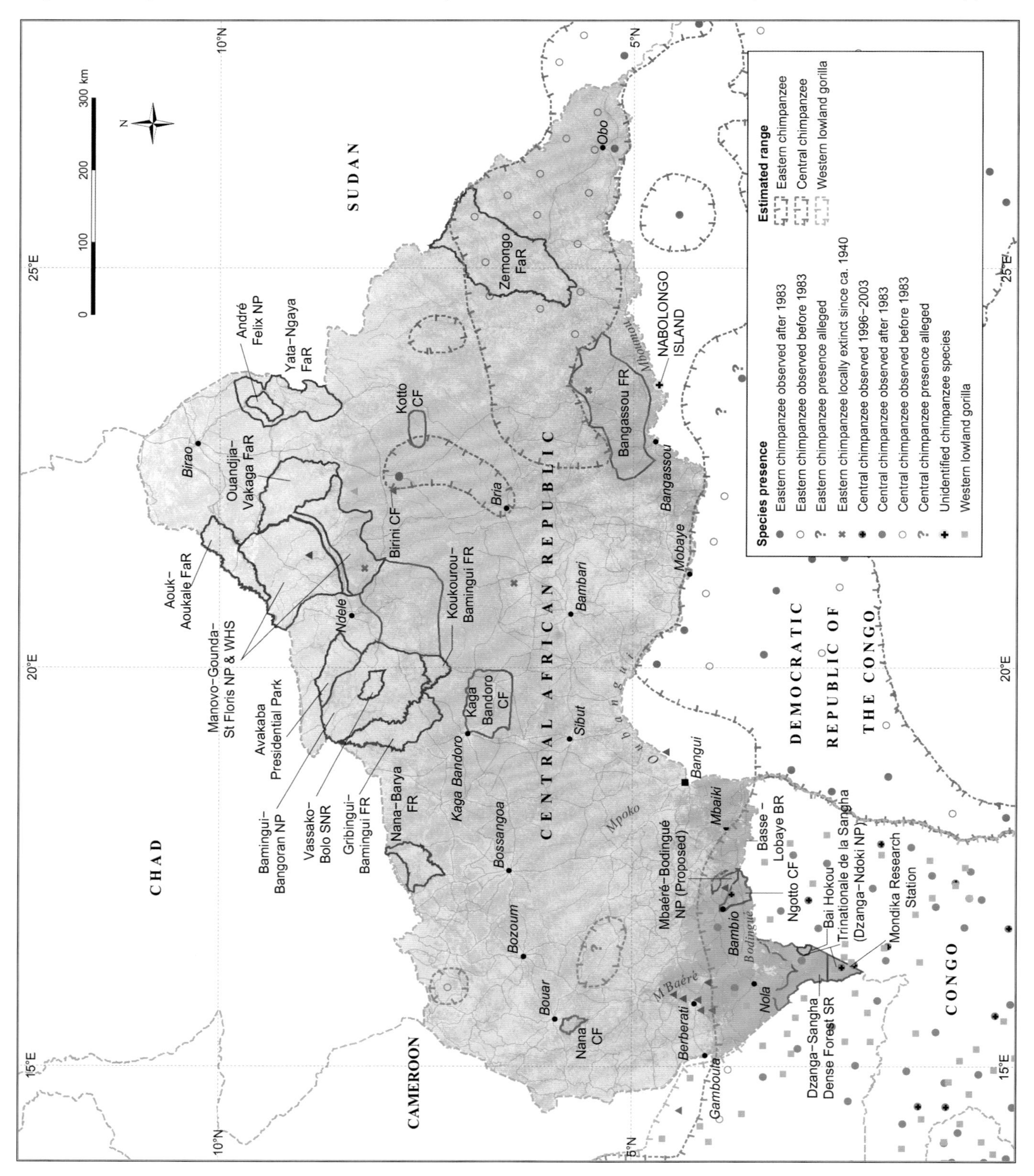

tected area system. In forest areas, bushmeat is less expensive than chicken, goat, or *kinin* caterpillars (an important food source for forest-dwelling Aka people).[16] Firearms have been cheap in recent years,[32] but most local hunters in the forest use nets and cable snares.[2, 34] Capture and injury by snares placed for other animals is probably the most serious hunting threat to great apes in CAR.[2] A study in the Dzanga-Sangha Special Reserve found that at any one time approximately 60 people were using wire snares in an area of about 1 000 km^2.[34]

Hunting with firearms, including automatic assault rifles, is more common in the north and east, where it is often undertaken by well organized gangs of poachers from Sudan.[17] Hunting pressure was reported to be very high in the Ngotto forest in the 1990s, particularly near the town of Bambio,[21] but to have later decreased somewhat.[8] The Aka and Bofi people living in the reserve appear to focus on small game, and often hunt with nets designed for this purpose.[26] The Bangassou forest is heavily hunted throughout and unless this changes soon, it will become an 'empty' forest.[30, 37] Hunting in the east and southeast of the country, including the buffer zone of the Zemongo Faunal Reserve, has also been intense.

LEGISLATION AND CONSERVATION ACTION

National legislation

The Ministry of the Environment, Waters, Forests, Hunting, and Fishing is responsible for wildlife conservation and the use of natural resources in CAR. These are governed by Ordinance No. 84.045 of July 27 1984, which deals with the protection of wildlife, and Law No. 90.003 of June 9 1990, which pertains to the Central African Forestry Code. Customary hunting is authorized throughout the territory of CAR, with the exception of integral reserves and national parks.[31] Great apes are listed in Category A as 'completely protected' under Ordinance No. 84.045.

International agreements

CAR ratified or acceded to the Convention on Biological Diversity in 1995, the Convention on International Trade in Endangered Species of Wild Fauna and Flora (CITES) in 1980, the UN Convention to Combat Desertification in 1996, the African Convention on the Conservation of Nature and Natural Resources in 1969, and the World Heritage Convention in 1980 (under which one site has been designated, the Manovo-Gounda-St Floris NP). The country has also designated two Biosphere Reserves under UNESCO's Man and Biosphere (MAB) Programme, one of which is Basse-Lobaye near the Ngotto forest. Preliminary steps for the ratification of the Convention on Wetlands of International Importance (Ramsar) have been conducted but final ratification was delayed due to political instability. The Mbaéré-Bodingué area in the Ngotto forest was proposed as the first national Ramsar Site.[7]

CAR collaborates with Congo and Cameroon in managing a three-way transfrontier reserve known as Trinationale de la Sangha (TNS). The Dzanga-Ndoki NP is contiguous with the Lac Lobéké NP of Cameroon and the Nouabalé-Ndoki NP in Congo, which together form the 7 300 km^2 core of the TNS conservation area. Among other achievements, the tripartite TNS agreement, signed in December 2000, has resulted in joint ranger patrols and exchange of intelligence, leading to some successful antipoaching missions.

Protected areas

There are three principal categories of protected area in CAR:[17; 44] strict nature reserves (one site), national parks (five sites), and faunal reserves (eight sites). There is also one special reserve (Dzanga-Sangha Special Reserve) and one private reserve (Avakaba Presidential Park). While the protected area system includes almost 11 percent of the land area, only 32 percent of the protected areas are thought to be adequately managed, and three are completely unmanaged.[4] The latter include the Zemongo Faunal Reserve, where the eastern chimpanzee is still believed to occur. About 50 areas, covering around 1 percent of the national territory, have been gazetted as forest reserves ('classified' or 'gazetted' forests), intended for the sustainable production of forest products.[2, 44] Few of these are actively managed.[2]

The Dzanga-Ndoki NP in the extreme southwest of the country covers 1 220 km^2 and is divided into the Dzanga (495 km^2) and Ndoki (725 km^2) sectors, which are joined to the 3 359 km^2 Dzanga-Sangha Special Reserve. The topography of the Dzanga sector is relatively flat, and it comprises a patchwork of primary and secondary forest, with much herbaceous undergrowth.[3] Parts of the park were selectively logged prior to 1982.[3, 38] The climate here is characterized by a dry season of three months (December–February) and a long rainy season, with a drier period in June–July.[3] Gorilla

A logging operation in the Central African Republic.

Ian Redmond

densities of 1.6 individuals per square kilometer have been recorded within the park; that is 10 times the density of chimpanzees.[3] Full protection is accorded to forests within the national park, while traditional and safari hunting, agroforestry development, and commercial logging have occurred in the special reserve.[13, 34] In the early 1990s, a scheme was developed to share revenues from tourism in the reserve and park, with 40 percent going to local communities and 50 percent to reserve administration for salaries, upkeep, and maintenance.[31] Since 2001, the reopening of a logging concession within the reserve has led to a huge influx of outsiders, and the reopening of a network of roads throughout the area. There has been a resultant upsurge in commercial bushmeat hunting and ivory poaching, in both the special reserve and the national park.[30]

Gorillas and chimpanzees are also reported from Ngotto forest,[9] part of which may become the Mbaéré-Bodingué NP.[10] Ngotto, covering about 10 000 km^2, is the second largest moist forest area in CAR after Dzanga-Sangha. It is a dense semi-deciduous forest with large tracts of *Raphia* (Palmae) swamp forest along the M'Baéré and Bodingué Rivers. Situated in the northern part of the Guineo-Congolian forest block, and bordering the southern limits of the Sudanian wooded savanna, the area has high species richness, which is enhanced by its position astride the three major biogeographical zones of Central, West, and East Africa. More than 115 species of mammal (including 13 primates) and more than 320 species of bird are found in the Ngotto forest,[18] which is regarded as an area of national importance for the conservation of at least four monkey species.[10] Estimated gorilla densities are lower at Ngotto (0.34–0.40 weaned gorillas per km^2) than in the Dzanga-Sangha region, probably due to the lesser abundance of herbaceous vegetation there.[8] The opposite pattern is seen in chimpanzees. Based on these densities, an estimated population of 295–350 weaned gorillas and 380 weaned chimpanzees occurs in the proposed Mbaéré-Bodingué NP (872 km^2).[6]

Conservation and field projects

Studies on western lowland gorillas in CAR began in the mid-1980s, when extensive surveys (e.g. Fay 1989[20]) revealed high densities of animals in the Dzanga-Sangha forest. Research has since been conducted into nesting, feeding and foraging, and ranging behavior of gorillas and the impact of human activities on the gorillas (e.g. Remis 1997, 2000[38, 39]). Much of this has been taking place at three study areas within the Dzanga-Ndoki NP, and has included successful efforts to habituate gorillas for tourism by teams including CAR nationals. Ape research has had a fairly constant presence at Dzanga-Sangha since the early 1990s, despite economic and political fluctuations and the consequent instability of the developing tourism sector. Environmental education activities have also been undertaken in the area. Gorilla research, in particular, has contributed significant development opportunities for local, national, and international participants.[25] Research into chimpanzee ecology and behavior has been more limited in CAR.[8] A feasibility study on habituation of apes to ecotourism in the Ngotto forest was carried out in 2000–2001, but found that densities were too low and the hunting pressure too high for success.[41]

A number of international organizations have supported conservation projects at the Dzanga-Ndoki NP and Dzanga-Sangha Special Reserve, including WWF–The Global Conservation Organization, the Wildlife Conservation Society (WCS), the World Bank, the US Agency for International Development (USAID), and the German overseas development agency GTZ. WWF supports the administration of the complex as part of an integrated conservation and development project. Activities are focused on protected area management, rural development, tourism, sustainable forestry management, and applied ecological and social research, and have included recruitment and training of new wildlife guards, tourist guides, and trackers, and development of an antipoaching strategy for the area. There are also two gorilla habituation sites in the Dzanga-Sangha area – Mondika Research Station and Bai Hokou. The latter has been developed for a successful WWF tourism program, which includes extensive health monitoring of gorillas.[2, 46]

The CAR component of the European Union program, Conservation and Rational Use of Forest Ecosystems in Central Africa (ECOFAC), supported the administration of the Ngotto forest until its withdrawal in 2003.[46] One of the objectives of the project was to set up a pilot sustainable logging operation, based on a management plan and a set of general operating conditions established by ECOFAC and the government. An area of 872 km^2, situated within a triangle of forest between the Bodingué and M'Baéré Rivers that supports gorillas and central chimpanzees, has been set aside for total protection and is patrolled by a team of 'ecoguards' recruited and trained by the project.[8]

The Bangassou forest, where the eastern chimpanzee has been recorded, was the subject of a US$3.5 million community conservation initiative financed by the Global Environment Facility (GEF). There is little law enforcement by the local authorities at present, but it is planned that the next phase of the GEF project will provide for the recruitment and deployment of paid ecoguards.[30, 46] No sanctuaries for captive great apes exist in CAR.[2]

FUTURE CONSERVATION STRATEGIES

As yet there is no conservation action plan for the great apes in CAR, nor are great apes known to be highlighted in any national environmental plans. Recommendations have been made by a number of authors, however, regarding specific areas where gorillas and chimpanzees occur, and regarding the region as a whole for great ape conservation measures.

CAR will need continued financial as well as technical assistance to deal with ape conservation and protected area management. There is a real need for capacity building, regional collaboration, political commitment, and the development of sustainable long-term funding mechanisms.[4] Further information on ape population sizes and distribution in CAR would greatly assist in conservation planning.[35]

Some authors recommend that most ape conservation investment in western equatorial Africa should be focused on formally protected areas, where great ape populations are likely to be viable in the long run, with an immediate large investment in law enforcement to tackle uncontrolled poaching and, over the longer term, building of national capacity to conduct all aspects of protected area management.[45]

One study assessed the conservation potential of the protected areas of CAR based on a variety of measures including threats from logging, mining, hunting, grazing, farming, villages and roads, biodiversity significance, integrity of the area, and the effectiveness of its management. The Dzanga-Ndoki NP and the Dzanga-Sangha Special Reserve had the highest ratings, making them, with the high density of gorillas they support, a priority for continuing great ape conservation and research efforts.[4, 35] Potentially adding to the set of protected areas under consideration, the ECOFAC program has recommended that the Mbaéré-Bodingué area in Ngotto forest be given national park status.[10, 18] It has also been suggested that logging companies leave an undisturbed strip of forest connecting this area with Dzanga-Ndoki.[10] A new effort to conserve the Bangassou forest to protect chimpanzees has also been suggested;[35] it is to be hoped that the GEF project is instrumental in achieving this.

Increased wildlife law enforcement is frequently seen as a priority over the country as a whole (e.g. Blom *et al.* 2004[4]). It has also been suggested that logging and mining companies should be compelled to control illegal bushmeat hunting, transport, and consumption on their concessions. Meanwhile, funding for efforts to control hunting and the Ebola virus should be generated in developed countries. Regional governments could then be offered economic incentives for ape protection, linking aid and debt relief to verifiable measures of conservation performance.[45]

FURTHER READING

Blom, A., Almasi, A., Heitkönig, I.M.A., *et al.* (2001) A survey of the apes in the Dzanga-Ndoki National Park, Central African Republic: a comparison between the census and survey methods of estimating the gorilla (*Gorilla gorilla gorilla*) and chimpanzee (*Pan troglodytes*) nest group density. *African Journal of Ecology* **39**: 98–105.

Blom, A., Yamindou, J., Prins, H.H.T. (2004) Status of the protected areas of the Central African Republic. *Biological Conservation* **118** (4): 479–487.

Carroll, R.W. (1988) Relative density, range extension, and conservation potential of the lowland gorilla (*Gorilla gorilla gorilla*) in the Dzanga-Sangha region of southwestern Central African Republic. *Mammalia* **52** (3): 309–323.

Fay, J.M. (1989) Partial completion of a census of lowland gorilla (*Gorilla g. gorilla*) in the Central African Republic. *Mammalia* **53**: 203–215.

Remis, M.J. (1997) Ranging and grouping patterns of a western lowland gorilla group at Bai Hokou, Central African Republic. *American Journal of Primatology* **33**: 111–133.

Remis, M.J. (2000) Preliminary assessment of the impacts of human activities on gorillas *Gorilla gorilla gorilla* and other wildlife at Dzanga-Sangha Reserve, Central African Republic. *Oryx* **34** (1): 56–65.

MAP DATA SOURCES

Map 16.4 Great apes data are based on the following source:

Butynski, T.M. (2001) Africa's great apes. In: Beck, B.B., Stoinski, T.S., Hutchins, M., Maple, T.L., Norton, B., Rowan, A., Stevens, E.F., Arluke, A., eds, *Great Apes and Humans: The Ethics of Coexistence*. Smithsonian Institution Press, Washington, DC. pp. 3–56.

With additional data by personal communication from Maisels, F. (2004) and from the following sources:

Blom, A., Almasi, A., Heitkönig, I.M.A., Kpanou, J-B., Prins, H.H.T. (2001) A survey of the apes in the Dzanga-Ndoki National Park, Central African Republic: a comparison between the census and survey methods of estimating the gorilla (*Gorilla gorilla gorilla*) and chimpanzee (*Pan troglodytes*) nest group density. *African Journal of Ecology* **39**: 98–105.

Brugière, D., Sakom, D. (2001) Population density and nesting behaviour of lowland gorillas (*Gorilla gorilla gorilla*) in the Ngotto Forest, Central African Republic. *Journal of Zoology* **255**: 251–259.

Brugière, D., Sakom, D., Gautier-Hion, A. (2005) The conservation significance of the proposed Mbaéré-Bodingué National Park, Central African Republic, with special emphasis on its primate community. *Biodiversity and Conservation* **14** (2): 505–522.

For protected area and other data, see 'Using the maps'.

ACKNOWLEDGMENTS

Many thanks to Allard Blom (WWF-US), David Brugière (SECA-BRLi Consulting Company), Fiona Maisels (Wildlife Conservation Society), and Elizabeth A. Williamson (University of Stirling) for their valuable comments on the draft of this section.

AUTHOR

Nigel Varty, UNEP World Conservation Monitoring Centre

REPUBLIC OF THE CONGO

NIGEL VARTY

BACKGROUND AND ECONOMY

The Republic of the Congo is a Central African country with a land area of 342 000 km^2. Its population was estimated at 2.95 million in 2003 with a growth rate of 1.5 percent.[7] Its gross domestic product (GDP) for 2002 was estimated at US$3.0 billion with a gross national income (GNI) of US$720 per person.[47] Congo is one of Africa's main petroleum producers, with significant potential for further offshore development. Petroleum (mostly offshore) and timber were Congo's two largest sources of foreign exchange in 2002, while agriculture accounted for only 10 percent of GDP.[7] More than half of the population lives in the southern cities of Brazzaville, Pointe-Noire, and Loubomo. Much of rural Congo has fewer than four inhabitants per square kilometer.

Congo is one of the most densely forested countries on the African continent. Forests cover 220 600 km^2, or about 64.6 percent of the land area, including some 830 km^2 of plantations.[14] In 2003, permanent crops and arable land were estimated to account for only 0.6 percent of the total land cover.[7] The main forest areas are the Mayombe (Maiombe) and Chaillu massifs in the southwest (which are southern extensions of the Lower Guinea forest block), and the Sangha and Likouala regions in the north (which are part of the Congo Basin).

Independence from the French was ratified in 1960 when the former Middle Congo became the Republic of the Congo. Until 1990, the dominant political system was based on an interpretation of Marxism, but this was abandoned, and a more market-oriented government was elected in 1992.[7] Following the outbreak of civil war in 1997, former Marxist President Sassou-Nguesso was elected, but a period of civil unrest followed until the beginning of this century. Negotiations eventually led to a new draft constitution, which was approved by referendum in 2002,[17] leading to a national election won by Sassou-Nguesso, and subsequently allowing more stable conditions to prevail. Most people displaced by the war have returned, malnutrition and death rates have dropped, the rail link between Brazzaville and Pointe-Noire has reopened, and the security situation is slowly improving.[17] In March 2003, the main rebel group signed a new commitment to peace with the government, and its disarmament and reintegration into society has commenced.

DISTRIBUTION OF GREAT APES

The western lowland gorilla (*Gorilla gorilla gorilla*) and central chimpanzee (*Pan troglodytes troglodytes*) occur in Congo. The Congo and Oubangui Rivers form the eastern edge of the range of both the western lowland gorilla and central chimpanzee.

A large proportion of the world's western lowland gorillas are thought to live in Congo. They occur in the densely forested northern region[15] and, to a lesser extent, central parts of Congo, and in forested areas in the southwest of the country. Based on 1990 habitat maps and censuses in the north, the national population has been estimated at 34 000–44 000 gorillas,[5, 15, 20] but it is feared to have decreased markedly by the early part of the 21st century. While the 1990 figure is substantially larger than the 1980 estimate of 1 000–3 000,[23] the actual population trend has been negative.[41]

Up to 30 000 of these gorillas are found in the Odzala-Koukoua National Park (NP). The highest known density of lowland gorillas (11.3/km^2) is found in the park's extensive Marantaceae forests. Over a five month period in 1994 and 1995, 427 nests were recorded along line transects.[2] The park was expanded in 2001 to absorb the adjacent Lékoli-Pandaka Faunal Reserve and M'boko Hunting Reserve. The resulting protected area covers 13 456 km^2, and includes the Maya Nord salt clearing and its surrounding Marantaceae forests, a study area estimated to support 500 gorillas.[27] Social stability, high birth rate, and apparent low infant mortality have indicated that this subpopulation was doing well.[27] By September 2004, however, there were fears of a devastating Ebola epidemic in the park (see below).

Benoît Goossens/HELP International

The River Louvandzi, bordering the chimpanzee release site in the Conkouati area.

High densities of gorillas have also been reported in the vast Likouala swamp area of north central Congo between the Oubangui and Sangha Rivers, and both east and west of the Likouala aux Herbes River.[3, 15] They also occur in the 4 190 km^2 Nouabalé-Ndoki NP, and the surrounding forests.[38] The Oubangui River is thought to be the most easterly limit to the range of the species, unless the populations in southwest Democratic Republic of the Congo (DRC) are not extinct after all.[16] In the past, gorillas have also been reported to be numerous in the Kouilou Basin part of the Mayombe forest block,[12] but these are considered to be vulnerable because of logging and associated hunting,[2] and it has been suggested that their fate depends on the success of the protected areas in which they occur.[20]

Central chimpanzees occur in the north and southwest of Congo along the border with Gabon but are less widespread in the central region.[6, 26] In 1991, it was thought that there were only 3 000–5 000 chimpanzees in Congo,[39] but since then, the estimates have been revised, with a 2003 figure of around 10 000 chimpanzees.[6, 25] The highest recorded density of this species in Central Africa has been reported in Odzala-Koukoua NP (2.2/km^2).[2] Chimpanzees are not uncommon in the Likouala swamp forests, and densities of 0.1–1.3/km^2 have been recorded in the Lac Télé/Likouala-aux-Herbes Community Reserve.[16, 34] In the Motaba River area in

Map 16.5 Great ape distribution in Congo

Data sources are provided at the end of this country profile

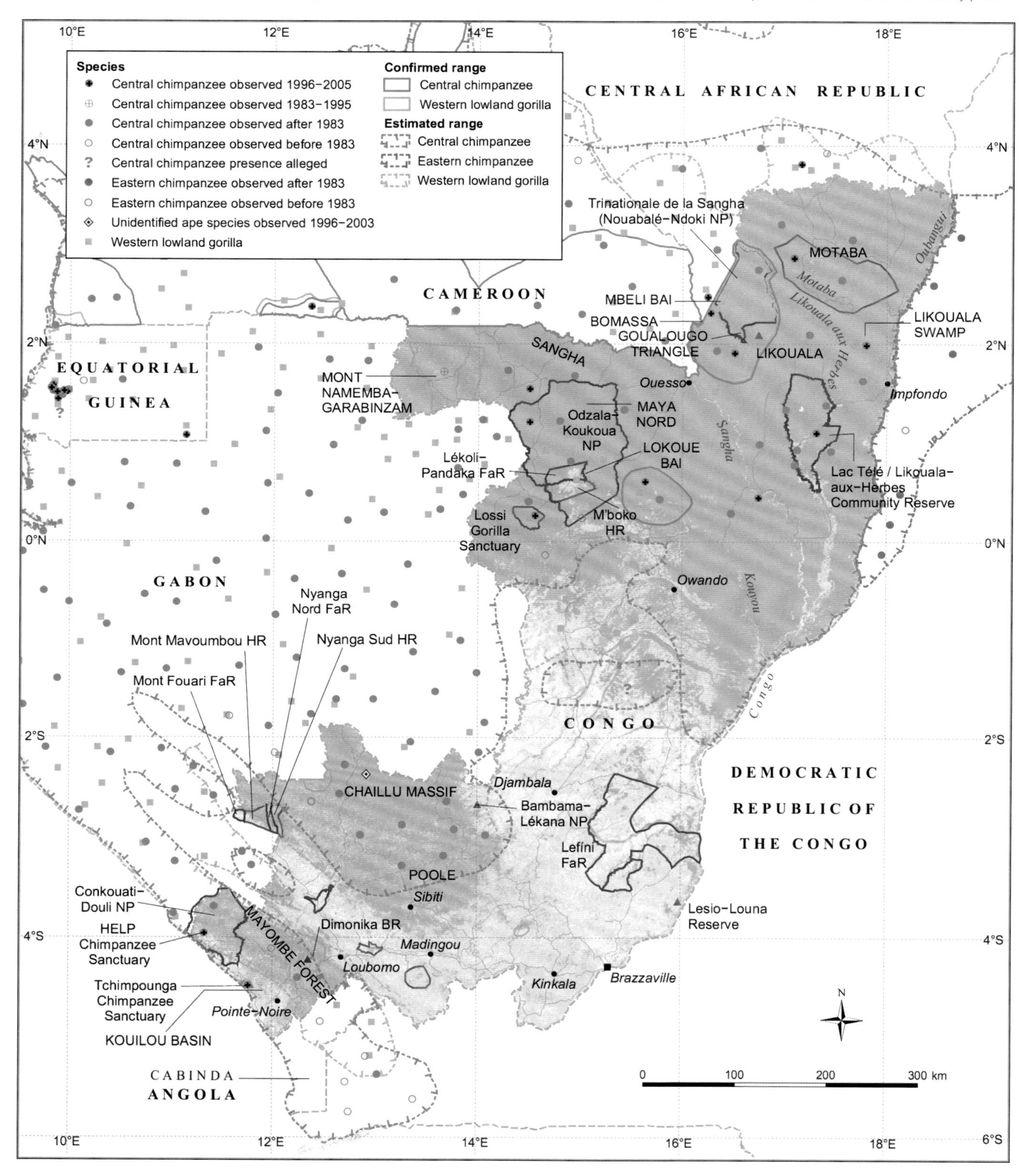

northeast Congo, a density of 0.3/km^2 was found.[24] The density in southwest Congo was found to be lower than that in the Nouabalé-Ndoki NP in the north, but about the same as in Equatorial Guinea and in Gabon.[21]

THREATS

Both gorilla and chimpanzee populations in Congo are threatened by human conflict, the expansion of logging, and associated hunting.[26, 45] The Ebola virus is a major threat in the border region with Gabon and a gorilla population was wiped out at Lossi, in a community-protected forest some 50 km southwest of Odzala-Koukoua NP. [41] The population in the park itself was feared to be under threat in 2004.[19]

The effects of civil war

In the civil war and its aftermath, an unknown number of great apes were killed and conservation efforts were hindered as a result of the overall disintegration of law and order. The Brazzaville zoo was raided for meat to feed hungry people, and has since closed. The John Aspinall Foundation (JAF) and the Jane Goodall Institute (JGI) airlifted the gorillas and chimpanzees from the zoo in 1997. Most of the fighting occurred in the south of the country, however, with little conflict reaching the more remote northern forests where apes are most abundant. The general breakdown of infrastructure that occurred during the war would have further reduced travel to those forest regions that had been previously more accessible. This is likely to have helped protect much of the country's wildlife during this period.

Hunting

In the Motaba region of northeast Congo in 1992, 40 percent of Aka (Ba'Aka or Bayaka) men were recorded as eating chimpanzee or gorilla meat, and hunting of apes occurred in every part of the area.[24] Hunters killed an estimated 0.01 gorillas and 0.02 chimpanzees per square kilometer per year, which represented 5 percent and 7 percent of their populations respectively. This was higher than the replacement rate, and so considered unsustainable.[4, 24] These numbers represented local consumption only; market trade was unquantified. Hunting, including with the use of automatic weapons, is recognized as a major management problem in the Lac Télé/Likouala-aux-Herbes Community Reserve.[44] In the 1980s, staff at the Brazzaville gorilla orphanage estimated that 400–600 gorillas were killed each year in northern Congo, although this may have been an underestimate.[35, 36] Many of the gorilla parts and orphans brought to the city were also believed to come from the southwestern Mayombe region.[15]

There has been an increased demand for bushmeat in towns and cities in Congo and neighboring countries as urban incomes have increased. Figures from the markets in Brazzaville indicate that gorilla and chimpanzee carcasses account for about 2 percent of the total number of animal carcasses, and 2.23 percent of the total weight of meat for sale.[4] Gorillas and chimpanzees are also frequently killed or maimed by traps and snares intended for other forest animals such as antelope.[4]

Some local traditions involve great apes. The Kwélé, Kota, Mboko, and Djem ethnic groups in north Congo eat gorilla meat as part of a circumcision ritual for young men.[18] In some other parts of Congo, such as the Odzala and Ndoki regions, powdered chimpanzee and gorilla bones and hair are traditionally added to the bath water of babies to improve strength and health.[28] Conversely, chimpanzees in and around Conkouati-Douli NP in the Poole region are protected by local taboos against eating their meat; Vili tribal tradition states that chimpanzees are reincarnations of people. Continued commercial logging in Conkouati-Douli NP, and the consequent substantial immigration, however, has diluted these indigenous beliefs with those of cultures that do not consider chimpanzee meat taboo.

Disease

Ebola hemorrhagic fever now rivals hunting as a threat to apes in Congo.[41] The Ebola virus has occurred sporadically in Congo and Gabon since the late 1990s, and continues to spread. The disease has claimed over 140 human lives, probably as a result of the victims eating meat from infected primates. In late 2002, an outbreak of Ebola was reported in the north of Congo on the border with Gabon; in some areas more than 90 percent of the populations of gorilla and chimpanzee were killed during this single episode. The epidemic appears to have killed all but seven of a study population of 143 gorillas in the Lossi Gorilla Sanctuary, for example. An Ebola outbreak has not yet been confirmed in the nearby Odzala-Koukoua NP,[8, 22, 41] but by September 2004, an 80 percent decline in gorilla sightings in the park's Lokoue Bai sparked fears that Ebola had arrived.[19]

Other diseases reported in apes in Congo

Joanna Setchell & Benoît Goossens/HELP International

A newly released chimpanzee at Conkouati lagoon takes a last look back.

include yaws, a debilitating disease closely related to syphilis, with 24 from a sample of 420 gorillas being infected.[27]

Habitat loss

It is estimated that 170 km^2 of forest was cleared annually in Congo from 1990 to 2000, representing about 0.1 percent of the total per year.[14] Timber is also selectively exploited wherever dryland forest is accessible. Selective logging can improve the habitat for gorillas, which can often find good food supplies in secondary forest, but in reality, the associated hunting reduces gorilla numbers.[15] Some of the forests of the north have remained unlogged so far because of poor road infrastructure, but this is changing and today many of the remaining pristine forest management units have been assigned to logging companies.[11] In Congo, roads established and maintained by logging concessions intensify bushmeat hunting by providing hunters with greater access to relatively unexploited populations of forest wildlife, lowering the cost of transporting bushmeat to market and, in many cases, offering a local market among logging company workers.[46]

LEGISLATION AND CONSERVATION ACTION

International legislation

Congo has ratified or acceded to the Convention on Biological Diversity (1993), the Convention on International Trade in Endangered Species of Wild Fauna and Flora (CITES) (1983), the UN Convention to Combat Desertification (1999), the African Convention on the Conservation of Nature and Natural Resources (1981), and the World Heritage Convention (1987). The country also participates in UNESCO's Man and Biosphere (MAB) Programme under which two Biosphere Reserves have been designated: Odzala-Koukoua NP and Dimonika in the Mayombe mountains.

National legislation

The main laws dealing with wildlife conservation and use are Law 48/83, which protects gorillas and chimpanzees,[26, 38] Law 49/83 of April 21 1983, and Decree 85/879 of July 6 1985. Law 16/2000 of November 20 2000 establishes the Forestry Code, which aims to achieve sustainable management of forest ecosystems in Congo. The Ministry of Forest Economy and the Environment (MFEE) is responsible for wildlife conservation and regulating use, including management of protected areas. The Congolese government has recently banned the production of shotgun cartridges.[5]

Congo is part of the Congo Basin Forest Partnership (CBFP), an international initiative including governments, nongovernmental organizations (NGOs), and the private sector, which is "committed to conserving the forests of the Congo Basin while promoting sustainable economic and social development in the region."

Protected areas

There are three main categories of protected area in Congo: national parks (four sites), faunal reserves (six sites), and hunting reserves (four sites).[11] There are also four faunal sanctuaries and a number of other reserves.[42] The existing protected area network covers an estimated 11 percent of the country's land area. The Congolese government has recently launched an evaluation of the country's protected areas, with an emphasis on assessing gaps between parks, with support from the Wildlife Conservation Society (WCS), and aims to create a national service for protected area and wildlife management.[43]

Western lowland gorillas and central chimpanzees are found in all three national parks. The Nouabalé-Ndoki NP covers an area of 4 193 km^2.[38] It contains about 2 percent of all Congo's forests. Most of the periphery to the west and south has been logged once, at a tree intensity of about 100/km^2, and active logging concessions largely surround the park.[31, 38] Human impact on the park and in the Bomassa buffer zone is considered minimal.[11] WCS administers the park in collaboration with the MFEE.

In 2001, the Odzala-Koukoua NP was expanded to include the Lékoli–Pandaka Faunal Reserve and M'boko Hunting Reserve (both to the south of the park) and an enormous area of forest to the north, west and east, previously assigned for timber exploitation. The park now covers 13 456 km^2 on the northwestern edge of the Congo River watershed, with altitudes from 300 to 600 m.[2, 42] In the absence of human influence, gorilla density tends to be higher where there is more herbaceous vegetation, such as Marantaceae species (see Chapter 7). Over 90 percent of the forest of Odzala is fairly open-canopy Marantaceae forest with dense undergrowth. The northern area of rain forest contains many clearings (swampy *bais*) that are particularly attractive to large mammals, including gorillas. Odzala has the highest recorded densities of western lowland gorillas (mean 5.4 /km^2) and chimpanzees (mean 2.2/km^2) in Central Africa.[2] Wherever gorillas and elephants coexist, the latter would be expected to improve gorilla habitats, through seed dispersal as well as by opening up the forest and encouraging the growth of herbaceous vegetation.[30, 38]

Apart from the period during the civil war from 1997 through 1999 when conservation activities were reduced, the Odzala-Koukoua NP has been administered since 1992 with the support of the European Union program, Conservation and Rational Use of Forest Ecosystems in Central Africa (ECOFAC), which has brought about a significant decrease in poaching. The Odzala-Koukoua NP was declared a Biosphere Reserve in 1977.

Conkouati-Douli NP is situated on the Atlantic coast in the southwest of Congo and covers 5 045 km^2.[42] Gorillas and chimpanzees are said to be common in the northern part of the park. The terrestrial part of the park includes several active logging concessions and many villages in an 'ecodevelopment zone'. Hunting has proved difficult to control, since the Conkouati area contributes significantly to the bushmeat market of Pointe-Noire, and this has severely affected large mammal populations. The park was managed by IUCN–The World Conservation Union for five years until 1999, since which time it has been managed by WCS, in collaboration with the MFEE.

The Lac Télé/Likouala-aux-Herbes Community Reserve in the Likouala swamp region supports both gorillas and chimpanzees. WCS is becoming involved in the management of this reserve, designated as a Ramsar Site in 1998 and owned by the local communities. Finally, the Lossi Gorilla Sanctuary is a small reserve effectively created by the local community, who established an ecotourism project here. It is also used for research.

Transfrontier initiatives

Congo is a partner in Trinationale de la Sangha (TNS), a transborder conservation zone in which protected areas in Congo, Cameroon, and the Central African Republic (CAR) are managed in common (see CAR country profile). The core protection zone comprises the national parks of Nouabalé-Ndoki (Congo), Lac Lobéké (Cameroon), and Dzanga-Ndoki (CAR), which together cover about 7 300 km^2. All support important populations of western lowland gorillas and central chimpanzees.

In January 2004, the government of Congo announced the creation of the Bambama-Lékana NP, which contains chimpanzees, as part of a transboundary protected area with Batéké Plateau NP in neighboring Gabon. The government also plans to connect Conkouati-Douli NP to the Mayumba NP in Gabon.

Similarly, there is an emerging proposal for a Mayombe transboundary area, to include parts of Congo, Angola, and DRC.[37] If taken up by national governments, this would be an extensive area including fully protected and limited-use zones. Congo's part of the forest seems to have been particularly affected by logging.

Conservation field projects

Ecotourism, including gorilla watching, is being developed at several sites around Congo, including Odzala-Koukoua NP and Nouabalé-Ndoki NP.

WCS has been involved in a number of conservation projects in Congo, including a project with the MFEE and the Congolaise Industrielle des Bois (CIB), a logging company, around the periphery of Nouabalé-Ndoki NP. A set of agreed guidelines for hunting has been developed, which includes a ban on the hunting of apes and other endangered species and the export of meat from the concession. As a result, commercial bushmeat hunting has been reduced, and it was reported in 1999 that the incidence of gorilla and chimpanzee hunting had dropped by an estimated 90 percent over a two year period.[13] In addition, the company gave up a large part of its concession in an area where there are apes with little fear of humans, presumably because they have not had previous contact with people. The World Bank has become interested in promoting the project as a model for other concessions in Central Africa.[1, 33] In addition, the Congolese government now requires all logging companies operating in northern Congo to pay for 'ecoguards' and wildlife management on their concessions.[1]

Research into the population, ecology, and social behavior of gorillas and/or chimpanzees in Congo has taken place at a number of sites, mostly in the north of the country. These include Mbeli Bai, an open, swampy forest clearing that covers 1.3 km^2, and the Goualougo Triangle, both located in Nouabalé-Ndoki NP;[32] Maya Nord *bai* in the Odzala-Koukoua NP;[27] the Likouala swamp area;[15] and at Lossi Gorilla Sanctuary, 50 km southwest of Odzala-Koukoua NP.[2]

Sanctuaries

There are a number of great ape sanctuaries and reintroduction projects in Congo. In 1986, the John Aspinall Foundation (JAF) started a gorilla orphanage in the grounds of Brazzaville zoo. In 1994, the Lesio-Louna Sanctuary (now gazetted as a reserve) was established adjacent to the existing Lefini Faunal Reserve by the JAF in partnership with the Congolese government.[10] The area was to be used for gorilla rehabilitation. In 2004, the southern sector of the Lefini Reserve was combined with the Lesio-Louna Reserve, creating a protected area of 1 700 km^2. There are currently 23 gorillas in the project's care, 15 of which have been successfully reintroduced. There has been one birth in the reintroduced group.[9]

The Tchimpounga Chimpanzee Sanctuary covers 73 km^2, lies 50 km north of Pointe-Noire, was opened in 1992, and is run by the Jane Goodall Institute (JGI); it is the largest sanctuary of its kind in Africa. It houses at least 115 chimpanzees, with increasing numbers of orphans arriving. In most cases, the arrivals are young chimpanzees confiscated by the Congolese authorities from hunters trying to sell them. The center also engages in local community development work and environmental awareness programs.

A Congolese NGO called Habitat Ecologique et Liberté des Primates (HELP), established in 1991, also cares for young chimpanzees confiscated by the Congolese authorities. Since November 1996, 36 orphan chimpanzees have been released into the forests of Conkouati-Douli NP (previously Conkouati-Douli Faunal Reserve). To monitor the success of the introduction, a team of Congolese field workers tracks the chimpanzees each day. Of 20 animals released, 14–17 survived and readapted to life in the wild.[40] HELP no longer rescues new orphan chimpanzees, and the release of the last suitable candidates was planned for 2003. Post-release monitoring will continue until 2013 to allow assessment of the reintroduction process (see Box 4.5).

FUTURE CONSERVATION STRATEGIES

An international meeting of primatologists and other conservation experts was held in Brazzaville in March 2003 under the GRASP initiative to discuss conservation action for great apes in Congo. A draft national great ape survival plan (NGASP) was produced.[29] The main recommendations from the working groups at the meeting are summarized below.

- Increase effectiveness of surveillance and antipoaching brigades in forest concessions, create a national antipoaching brigade, establish an interministerial committee to help combat poaching, and publicize the law on conservation of wildlife among the army, police, customs, and courts.
- Ensure effective surveillance of protected areas, including better supply of equipment, designate some protected areas as World Heritage Sites, and promote transborder management of great apes.
- Develop a monitoring system for illegal activities concerning great apes and encourage the national networking of environmental information management, with an emphasis on great apes.
- Undertake research into the status, distribution, and biology of great apes in Congo, including their population dynamics and diseases, develop national research into great apes, and establish a database of all scientific studies of great apes.
- Develop a national policy on tourism, promote tourism among departments of government, create incentives to promote local tourism throughout the country, and carry out a feasibility study of ecotourism in areas inhabited by great apes.
- Develop alternative sources of income for rural communities such as agroforestry, fish farming, and beekeeping.
- In all forest concessions in Congo, promote the results of the joint project of the WCS, CIB, and the MFEE.
- Create an autonomous wildlife and protected area management agency.

- Undertake both education and conservation-awareness activities.

Particular concerns over the threat of the Ebola virus to great ape populations in Congo have also been raised and calls have been made for increased research on vaccines and ways to vaccinate people and wild apes.[41] A second workshop was held in Brazzaville in March 2003 on Ebola and preventative healthcare, through the support of the MFEE, the Ministry of Health and Population, and ECOFAC. The workshop shared understanding of the virus and started to formulate a strategic approach to developing preventative measures and further research.

FURTHER READING

Bermejo, M. (1999) Status and conservation of primates in Odzala National Park, Republic of the Congo. *Oryx* **33** (4): 323–331.

Blake, S., Rogers, E., Fay, J.M., Ngangoué, M., Ebéké, G. (1995) Swamp gorillas in northern Congo. *African Journal of Ecology* **33** (3): 285–290.

Bowen-Jones, E., Pendry, S. (1999) The threat to apes and other animals from the bushmeat trade in Africa, and how this threat could be diminished. *Oryx* **33** (3): 233–246.

Fay, J.M., Agnagna, M. (1992) Census of gorillas in northern Republic of Congo. *American Journal of Primatology* **27**: 275–284.

Kano, T., Asato, R. (1994) Hunting pressure on chimpanzees and gorillas in the Mobata River Area, northeastern Congo. *African Study Monographs* **15** (3): 143–162.

Leroy, E.M., Rouquet, P., Formenty, P., Souquière, S., Kilbourne, A., Froment, J.M., Bermejo, M., Smit, S., Karesh, W., Swanepoel, R., Zaki, S.R., Rollin, P.E. (2004) Multiple Ebola virus transmission events and rapid decline of central African wildlife. *Science* **303**: 387–390.

MFEE (2003) *Rapport final de l'atelier sur l'elaboration d'un plan national pour la survie des grandes signes. GRASP: Great Apes Survival Project.* Ministère de l'Economie Forestière et de l'Environnement, Republic of Congo.

Nishihara, T. (1995) Feeding ecology of western lowland gorillas in the Nouabalé-Ndoki National Park, Congo. *Primates* **36** (2): 151–168.

Walsh, P., Abernethy, K.A., Bermejo, M., *et al.* (2003) Catastrophic ape decline in western equatorial Africa. *Nature* **422**: 611–614.

Wilkie, D., Shaw, E., Rotberg, F., Morelli, G., Auzel, P. (2000) Roads, development, and conservation in the Congo Basin. *Conservation Biology* **14** (6): 1614–1622.

MAP DATA SOURCES

Map 16.5 Great apes data are based on the following source:

Butynski, T.M. (2001) Africa's great apes. In: Beck, B.B., Stoinski, T.S., Hutchins, M., Maple, T.L., Norton, B., Rowan, A., Stevens, E.F., Arluke, A., eds, *Great Apes and Humans: The Ethics of Coexistence.* Smithsonian Institution Press, Washington, DC. pp. 3-56.

With additional data by personal communication from Maisels, F. (2004), Poulsen, J. (2005), Ron, T. (2003), Stokes, E. (2003), and from the following sources:

Goossens, B., Setchell, J. (2003) Home free. *BBC Wildlife* **21** (2): 30–35.

Poulsen, J.R., Clark, C.J. (2004) Densities, distribution, and seasonal movements of gorilla and chimpanzees in swamp forest in northern Congo. *International Journal of Primatology* **25** (2): 285–306.

Various authors (2003) Annotated map prepared by UNEP-WCMC and modified at the meeting of the national GRASP team in Brazzaville, Congo on April 2 2003.

For protected area and other data, see 'Using the maps'.

ACKNOWLEDGMENTS

Many thanks to Amos Courage (John Aspinall Foundation), Ian Redmond (Ape Alliance/GRASP), and Emma Stokes (Wildlife Conservation Society) for their valuable comments on the draft of this section.

AUTHOR

Nigel Varty, UNEP World Conservation Monitoring Centre

REPUBLIC OF CÔTE D'IVOIRE

ILKA HERBINGER, CHRISTOPHE BOESCH, ADAMA TONDOSSAMA, AND EDMUND MCMANUS

BACKGROUND AND ECONOMY

The Republic of Côte d'Ivoire is in West Africa, bordered by Liberia and Guinea to the west, Mali and Burkina Faso to the north, Ghana to the east, and the Gulf of Guinea to the south. It has a land area of 318 000 km^2. In 2004, the population was estimated to be 17.3 million, growing at about 2.1 percent per year, and mainly living in the southern coastal region and the main commercial center of Abidjan.[3]

In 1960, Côte d'Ivoire gained independence from France, with Félix Houphouët-Boigny as president and chairperson of the single ruling party. This arrangement lasted until 1990, when a multiparty system was established. In the first contested presidential election, Houphouët-Boigny was re-elected and remained in office until his death in 1993.[5] The economy of Côte d'Ivoire is dominated by agriculture and related services, which together engage about two thirds of the population. It is historically one of the world's largest producers and exporters of coffee, cocoa beans, and palm oil, so its economy is sensitive to international prices for these foodstuffs. These prices were high in the mid-1990s, which, together with other factors such as the discovery of offshore petroleum resources, allowed international debt to be accumulated. Annual growth in gross domestic product (GDP) was about 5 percent in the period 1996–1999.

This period of relative prosperity was brought to an end by a military coup on December 25 1999, leading to instability, counter-coups, and a civil war, which continued into 2003, when peace agreements brokered with the support of the Economic Community of West African States (ECOWAS) and France began to take effect. Although there were further troubles in 2003, the agreed peace process continued and several thousand French and West African troops remained to keep order and facilitate the disarmament, demobilization, and rehabilitation of combatants. The West African contingents were replaced in 2004 by a United Nations peacekeeping force. However, an upsurge in violence in November 2004 slowed down the ongoing peace process, and triggered a UN embargo on arms dealing with either side. The economy contracted each year between 2000 and 2004, and the outlook was poor for 2005.

DISTRIBUTION OF GREAT APES

Only one species of great ape is found in Côte d'Ivoire, the western chimpanzee (*Pan troglodytes verus*). It has historically been widespread throughout the country, with most individuals living in the rain-forest zone (Map 16.6). Surveys in 1986–1988[11] and 1989–1990[14] suggested a total population of 11 000–12 000 individuals, with over half the chimpanzees living in protected areas, including national parks (NPs) and reserves, and the remainder living in poorly protected classified forests or unprotected areas. In 2003, the population size was estimated to be 8 000–12 000 individuals,[8] which according to the figures in the other country profiles, would add up to some 21–36 percent of the total population of western chimpanzees.

Sites that are thought or confirmed to hold important numbers of chimpanzees include the Taï NP, N'Zo Faunal Reserve, Mont Péko NP, Mont Sangbé NP, Mount Nimba Strict Nature Reserve, Comoé NP, Marahoué NP, and the classified forests of Haute Dodo, Cavally-Goin, Haut Sassandra, Haut Bandama, Bossématié, and the nearby Songan-Tamin-Mabi-Yaya complex and Monogaga Classified Forest.[11, 14] The high proportion of the chimpanzee population living within protected areas reflects the fact that most of the country's forests – once the largest in West Africa – have been heavily logged, and deforestation is now widespread outside the protected area system.

It has been estimated that there are around 4 500 chimpanzees in the Taï NP (0.4–1.7/km^2), and up to 1 500 individuals in the nearby N'Zo Faunal Reserve and Haute Dodo and Cavally-Goin Classified Forests.[8, 11, 14] The last published census for Mont Péko NP (1989–1990) estimates a total population of 78,[14] but a later census suggested a significant population of about 320 chimpanzees.[9] There might still be about 400 chimpanzees in the classified forest of Haut Sassandra, which is connected by forest corridors to Mont Péko NP.[11] Mont Sangbé NP might hold at least 55 chimpanzees,[14] or up to 260 as estimated in a more recent survey.[10] Mount Nimba Strict Nature Reserve, which straddles Côte d'Ivoire, Liberia, and Guinea, is thought to have about 59 chimpanzees in the Côte d'Ivoire section.[14] The whole reserve, together with

Map 16.6 Chimpanzee distribution in Côte d'Ivoire

Data sources are provided at the end of this country profile

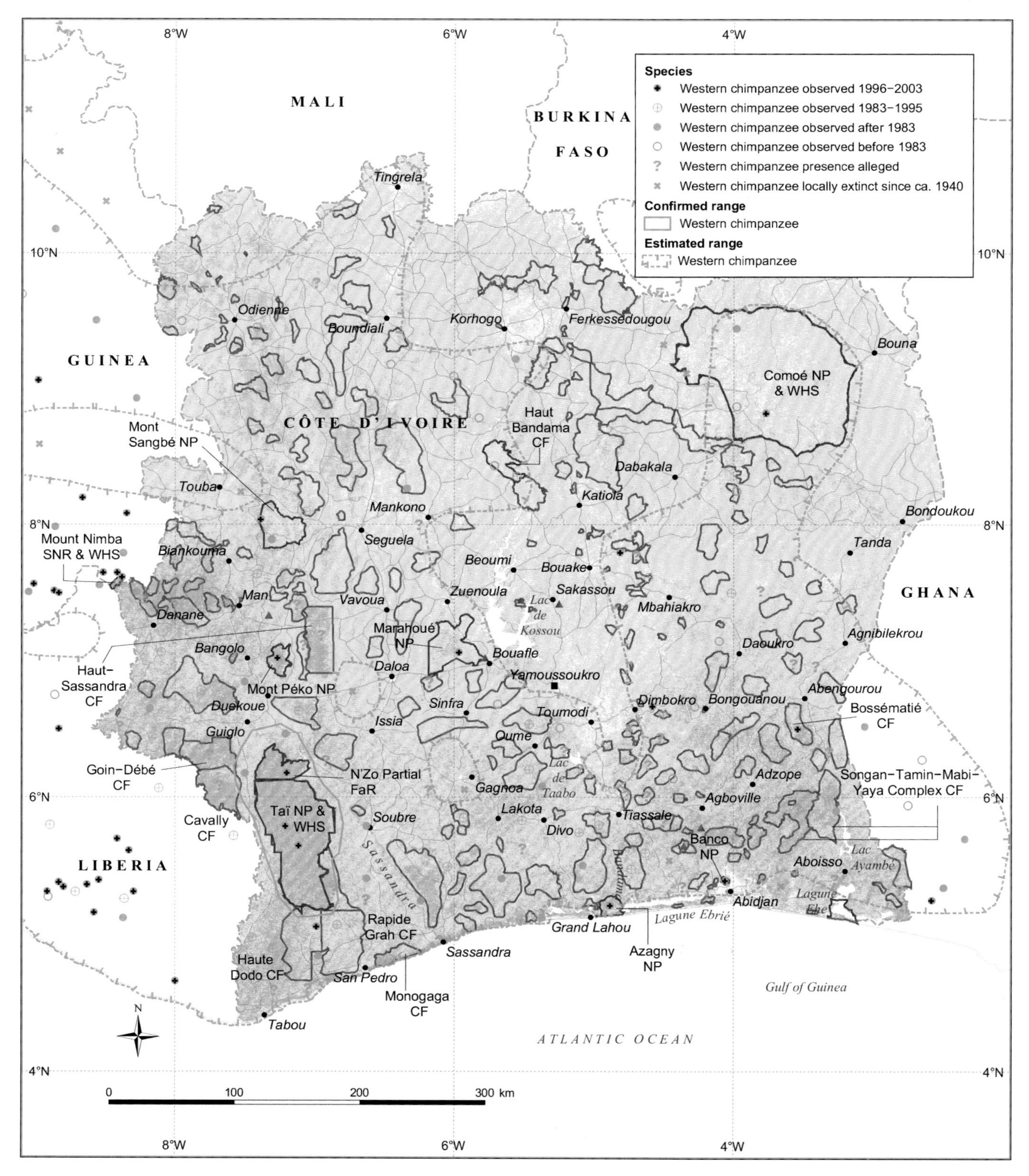

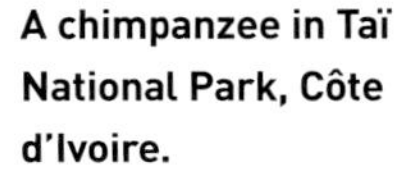

A chimpanzee in Taï National Park, Côte d'Ivoire.

Ilka Herbinger/Wild Chimpanzee Foundation

adjacent classified forests, might contain over 300 chimpanzees. Comoé NP was estimated to have about 470 individuals.[14] Marahoué NP had the highest density of chimpanzees in Côte d'Ivoire, estimated to be 6.4/km^2, with a population of 1 407 individuals (1989–1990).[14] A rapid census in 1998, however, found only one nest and had only four auditory contacts with chimpanzees, confirming their presence, but suggesting that the population had greatly declined.[15] The total population of chimpanzees for Haut Bandama was estimated to be 300.[14] The classified forests of Bossématié and the nearby Songan–Tamin–Mabi–Yaya complex may hold a combined total of 600 individuals.[11] Approximately 100 chimpanzees may still exist in the Monogaga Classified Forest.[11, 14] The future of chimpanzees in Côte d'Ivoire will depend upon the management of these mostly isolated populations.

THREATS

The southern half of Côte d'Ivoire originally supported about 160 000 km^2 of rain forest, but immigration and agricultural expansion resulted in widespread deforestation. By 1966, annual deforestation was estimated at about 5 000 km^2,[12] a figure that was maintained into the 1980s. It later declined to about 2 650 km^2 (3.1 percent) per year from 1990 to 2000. Only 71 170 km^2 of forest remained by the end of that period,[4] with much of it degraded. Chimpanzee numbers have declined even more rapidly, from an estimated 100 000 in the early 1960s to 8 000–12 000 by 2003.[8]

Other threats to chimpanzees in Côte d'Ivoire include hunting, trade in bushmeat, and disease. Chimpanzees are hunted for meat, for ingredients for traditional medicine, and because they raid crops. Chimpanzee meat is sold in urban markets and village restaurants.[2] However, taboos against killing or consuming chimpanzees exist throughout Côte d'Ivoire. Religious taboos exist among Muslims, as Islam forbids the consumption of the meat of primates. Traditional and cultural taboos mostly stem from legends connecting chimpanzees and humans: these tell of humans being transformed into chimpanzees and chimpanzees assisting humans during sicknesses, accidents, births, and times of war. Local traditions could be used to promote the conservation of chimpanzees.

Zoonotic diseases including monkeypox and Ebola are a threat to these small populations, as are diseases introduced to the wild population through interactions with farmers, hunters, researchers, or tourists. In 1994, chimpanzees in Taï NP suffered from an infection of a new subtype of the Ebola virus, and 25 percent of the community under study died.[6] In 1999, an epidemic of acute respiratory disease reduced that community by a further 25 percent.[7] More recently, the bacterial disease anthrax killed at least eight chimpanzees in Taï NP, which was the first time that this illness was known to have affected great apes in a rain-forest setting.[1, 13] This infection was tentatively attributed to the deforestation that has allowed cattle transport routes from Mali and Burkina Faso to pass close to the park, so that the chimpanzees may have caught the disease from passing livestock.

LEGISLATION AND CONSERVATION ACTION

Côte d'Ivoire is a signatory to the Convention on Biological Diversity, the Convention on International Trade in Endangered Species of Wild Fauna and Flora (CITES), the International Tropical Timber Agreement, and the UN Convention to Combat Desertification. In 1995, the government adopted a national strategy for parks and reserves; in 1996, a national environmental action plan was created and is currently being implemented.

Present chimpanzee-related conservation projects include the Ecotourism Project of the Projet Autonome pour la Conservation du Parc National de Taï which offers, among other activities, a guided visit to a habituated chimpanzee community in the southern part of the Taï NP. The Wild Chimpanzee Foundation (WCF) carries out environmental education projects in the vicinity of chimpanzee populations in Côte d'Ivoire, including interactive theater plays, a newsletter, and film presentations. WCF is also involved in implementing a national biomonitoring program, including the development of better methods to estimate the chimpanzee population in Côte d'Ivoire and to identify immediate threats to it. WWF–The Global Conservation Organization has provided financial support to various WCF activities.

The main field research project on the country's chimpanzees is also being carried out in Taï NP, under the supervision of Christophe Boesch, and focuses on various aspects of their behavior, ecology, cognition, and genetics, and on disease transmission between chimpanzees and humans.

FUTURE CONSERVATION STRATEGIES

Some of the priorities for conservation action targeting the chimpanzees of Côte d'Ivoire are outlined here.

- **Protection**. Assign a higher protected area status to the N'Zo Faunal Reserve and Cavally-Goin and Haute Dodo Classified Forests. Create as many protected forest corridors as possible to link the country's fragmented chimpanzee populations. In particular, create corridors to link the chimpanzee populations of Cavally-Goin and Taï NP, and to link these Côte d'Ivoire chimpanzees to the nearby populations in Liberia. Protect and enlarge the two forest corridors that link Mont Péko NP to Haut Sassandra Classified Forest. Improve law enforcement and surveillance efforts.
- **Surveys**. Conduct a survey of the current chimpanzee population, habitat, and bushmeat hunting in the largest forest block in the Guinean belt, which includes Taï NP, N'Zo Faunal Reserve, Cavally-Goin, and Haute Dodo. Repeat a nationwide survey, including Marahoué NP, Comoé NP, Mount Nimba Reserve, and the Haut Sassandra and Songan–Tamin–Mabi–Yaya complex. This will allow the rate of population decline to be estimated.
- **Education.** Education and awareness-raising campaigns in rural and urban areas have been suggested, including a program of regular visits of school classes to protected areas. Banco NP in the center of Abidjan should play a crucial role in such education programs. Ecotourism development could also be initiated in promising sites such as Taï NP, Mont Sangbé NP, or the Monogaga Classified Forest.

FURTHER READING

Chatelain, C., Gautier, L., Spichiger, R.A. (1996) A recent history of forest fragmentation in southwestern Ivory Coast. *Biodiversity and Conservation* **51**: 37–53.

Formenty, P., Boesch, C., Wyers, M., Steiner, C., Donati, F., Dind, F., Walker, F., Le Guenno, B. (1999) Ebola virus outbreak among wild chimpanzees living in a rain forest of Côte d'Ivoire. *Journal of Infectious Diseases* **179** (Suppl 1): S120–S126.

Herbinger, I., Boesch, C., Rothe, H. (2001) Territory characteristics among three neighboring chimpanzee communities in the Taï National Park, Côte d'Ivoire. *International Journal of Primatology* **22** (2): 143–167.

Kormos, R., Boesch, C., Bakarr, M.I., Butynski, T.M., eds (2003) *West African Chimpanzees: Status Survey and Conservation Action Plan.* IUCN/SSC Primate Specialist Group. IUCN, Gland, Switzerland.

Marchesi, P., Marchesi, N., Fruth, B., Boesch, C. (1995) Census and distribution of chimpanzees in Côte d'Ivoire. *Primates* **36**: 591–607.

MAP DATA SOURCES

Map 16.6 Chimpanzee data are based on the following sources:

Butynski, T.M. (2003) The chimpanzee *Pan troglodytes*: taxonomy, distribution, abundance, and conservation status. In: Kormos, R., Boesch, C., Bakarr, M.I., Butynski, T.M., eds, *West African Chimpanzees: Status Survey and Conservation Action Plan.* IUCN/SSC Primate Specialist Group. IUCN, Gland, Switzerland. pp. 5–12.

Herbinger, I., Boesch, C., Tondossama, A. (2003) Côte d'Ivoire. In: Kormos, R., Boesch, C., Bakarr, M.I., Butynski, T.M., eds, *West African Chimpanzees: Status Survey and Conservation Action Plan.* IUCN/SSC Primate Specialist Group. IUCN, Gland, Switzerland. pp. 99–109.

With additional data from:

Fischer, F., Gross, M. (1999) Chimpanzees in the Comoé National Park, Côte d'Ivoire. *Pan Africa News* **6** (2): 19–20.
Hoppe-Dominik, B. (1991) Distribution and status of chimpanzees (*Pan troglodytes verus*) on the Ivory Coast. *Primate Report* **31**: 45–57.
Marchesi, P., Marchesi, N., Fruth, B., Boesch, C. (1995) Census and distribution of chimpanzees in Côte d'Ivoire. *Primates* **36**: 591–607.
For protected area and other data, see 'Using the maps'.

ACKNOWLEDGMENTS
This country study draws extensively on the Côte d'Ivoire chapter in the IUCN/SSC Primate Specialist Group's *West African Chimpanzees: Status Survey and Conservation Action Plan.*[8] Thanks to reviewer Gottfried Hohmann (Max Planck Institute for Evolutionary Anthropology) and to a further reviewer who wishes to remain anonymous.

AUTHORS
Ilka Herbinger, Wild Chimpanzee Foundation
Christophe Boesch, Wild Chimpanzee Foundation
Adama Tondossama, Taï National Park
Edmund McManus, UNEP World Conservation Monitoring Centre

DEMOCRATIC REPUBLIC OF THE CONGO

Nigel Varty

BACKGROUND AND ECONOMY
The Democratic Republic of the Congo (DRC, formerly Zaïre) covers 2 345 410 km^2, making it one of the three largest countries in Africa. It gained independence from Belgium in 1960, and after a time of political instability, an army coup d'état installed General Mobutu as president in 1965. His rule was largely unchallenged until the late 1980s, by which time opposition to the corruption of his regime was growing, and there were army mutinies in 1991 and 1993.[23] The genocide in Rwanda in 1994 further destabilized the situation, as President Mobutu gave sanctuary and support to Hutu refugees from the former Rwandan Army. The Tutsi inhabitants living in the east of the country then allied themselves with other discontented armed groups and launched a successful attack on the Zairean army and the Mobutu regime in 1996. Their leader, Laurent Kabila, was installed as president, but within two years the political situation in eastern DRC began to unravel and Congolese rebels, supported by Rwanda and Uganda, seized major towns there, prompting Zimbabwe, Angola, Namibia, and Chad to intervene. By May 1999, diverse rebel armies controlled large areas in the north and east of the country, and had begun to finance themselves (and their foreign allies) by looting mineral resources such as diamonds, gold, and coltan.[66, 67] President Kabila was assassinated in 2001 and his son, Joseph Kabila, was sworn in as president, promising a program of improved democracy, human rights, and economic liberalization. Active diplomacy by South Africa and the United Nations has since achieved a settlement between the government and the main armed groups, creating a transitional government in 2002 that is still in place, despite repeated coup attempts and episodes of combat in the east of the country.

This recent history is fundamental to the challenge of conserving the great apes in DRC, and to the economic circumstances of the country's 55 million people, a population that is growing at about 2.9 percent annually according to 2003 estimates.[13] The Belgian colonial government created little infrastructure or social capital, although at independence there was a transport system that included strategically interconnecting roads, rivers, and railways.[1] Little more was done by the Mobutu regime after an initial flurry of public works in 1966–1974 (mostly funded by foreign loans), and much that had been created was later destroyed by neglect, underinvestment, and war. The country's abundant natural resources include cobalt, copper, gold, diamonds, zinc, uranium, tin, silver, coal, manganese, tungsten, tantalum, cadmium, petroleum, timber, and untapped hydroelectricity resources, yet in 2002–2003, gross domestic product (GDP) was estimated at US$5.7 billion, gross national income (GNI) per person was less than US$90, or US$0.23 per person per day,[1, 74] and the Human Development Index position of DRC was 168th out of 177 countries ranked globally.[63] Agriculture accounted for 55 percent of GDP, and industry 11 percent.[13] DRC possesses more than half of Africa's tropical closed broad-leaved forests.[73] Forests cover 1 352 070 km^2 (about 59.6 percent) of the country,[21] making DRC one of the two most forested countries in Africa in percentage terms. Permanent crops and arable account for only 3.5 percent of the land cover.[13] Having reviewed the political economy of DRC from 1960 to 2000, analysts at the International Monetary Fund concluded that poor economic policies, bad governance, and war had contributed to the country's economic decline during this 40 year period, but that the right policies are now being put in place to pave the way for a restoration of economic growth.[1]

DISTRIBUTION OF GREAT APES

Both the bonobo (*Pan paniscus*) and eastern lowland gorilla (*Gorilla beringei graueri*) are endemic to DRC. The mountain gorilla (*G. beringei beringei*) and the eastern chimpanzee (*P. troglodytes schweinfurthii*) are also present. It is not known whether the western lowland gorilla (*G. gorilla gorilla*) and the central chimpanzee (*Pan t. troglodytes*) still persist in the Bas-Fleuve region, to the north of the mouth of the Congo River.

The bonobo has a highly fragmented and discontinuous distribution (see Map 16.7a).[61] The area occupied has been estimated at no more than 118 000 km^2.[57] Much of the 500 000 km^2 of potential habitat has not been surveyed, however, which leads to great uncertainty about population numbers.[60] Numbers of bonobos estimated over the last 10 years range from 10 000 to more than 100 000.[10, 19, 50, 60, 61] Bonobos are known from many localities. There are currently conservation and research efforts within the Salonga National Park (NP) and the Lomako forest (a proposed forest reserve), between the Lomami and Lualaba Rivers, along the Kasai-Sankuru River, in the Lukuru Wildlife Research Project field site, at Wamba, Lac Tumba, and most recently in the Kokolopori forest.[19, 60, 68]

The eastern lowland gorilla has a discontinuous distribution east of the Lualaba River and west of the Albertine Rift, and from the northwest corner of Lake Edward in the north to the northwest corner of Lake Tanganyika in the south. It occupied roughly 15 000 km^2 in the early 1990s,[25] down from an estimated 21 000 km^2 in 1959–1960.[52] Most gorillas lived in the Kahuzi-Biega NP, whose 6 000 km^2 is spread over a separate mountain and lowland sector, connected by a forested corridor. The park's population suffered a severe decline in the conflict of the late 1990s. Some sources estimated that only a few thousand gorillas remained by 2001.[45, 49]

The mountain gorilla population is restricted to the southern section of the Virunga NP. It forms part of the larger 'Virunga' population, inseparable from the gorillas in the contiguous Mgahinga Gorilla NP in Uganda and the Volcanoes NP in Rwanda. An estimated 183 mountain gorillas were resident in DRC in 2001[10] – about half the Virunga population. Mount Tshiaberimu in the northern sector of the Virunga NP supports a small population of eastern lowland gorillas, which has been identified as distinctively different from other eastern lowlands, but is not recognized as a subspecies. The range of the distinctive Bwindi population of mountain gorillas, which is largely restricted to Uganda's Bwindi Impenetrable NP, also includes part of DRC. Hence, DRC hosts all known variants of the eastern gorilla.

Eastern chimpanzees occur in forested areas to the north and east of the Congo River. They appear to be widely distributed in the Ituri Forest Reserve[28] and are present in the eastern regions parallel to the Albertine Rift, stretching from the Sudan border in the north to the southern end of Lake Tanganyika.[34] They also occur in the northwest, near the border with Congo.[55] In the Kahuzi-Biega NP, a density of 0.4/km^2 was reported,[26, 30] and in the Itombwe Massif, DRC, a total of 1 100 was estimated in a study area of 1 600 km^2.[42] Their total population has been estimated at 70 000–100 000 individuals.[56] This is a significant upward revision of the 1980 estimate of around 6 200 chimpanzees.[31] However, current estimates are provisional, as little chimpanzee habitat in DRC has ever been surveyed, especially in the north of the country, and the impact of the civil war is unclear. Those areas that have been surveyed need to be revisited.[11]

The western lowland gorilla is probably absent from its former range in the extreme west of DRC, north of the Congo River.[10, 22] It is thought to have become extinct prior to 1980.[31] The central chimp-

Map 16.7a Bonobo and chimpanzee distribution in the Democratic Republic of the Congo

Data sources are provided at the end of this country profile

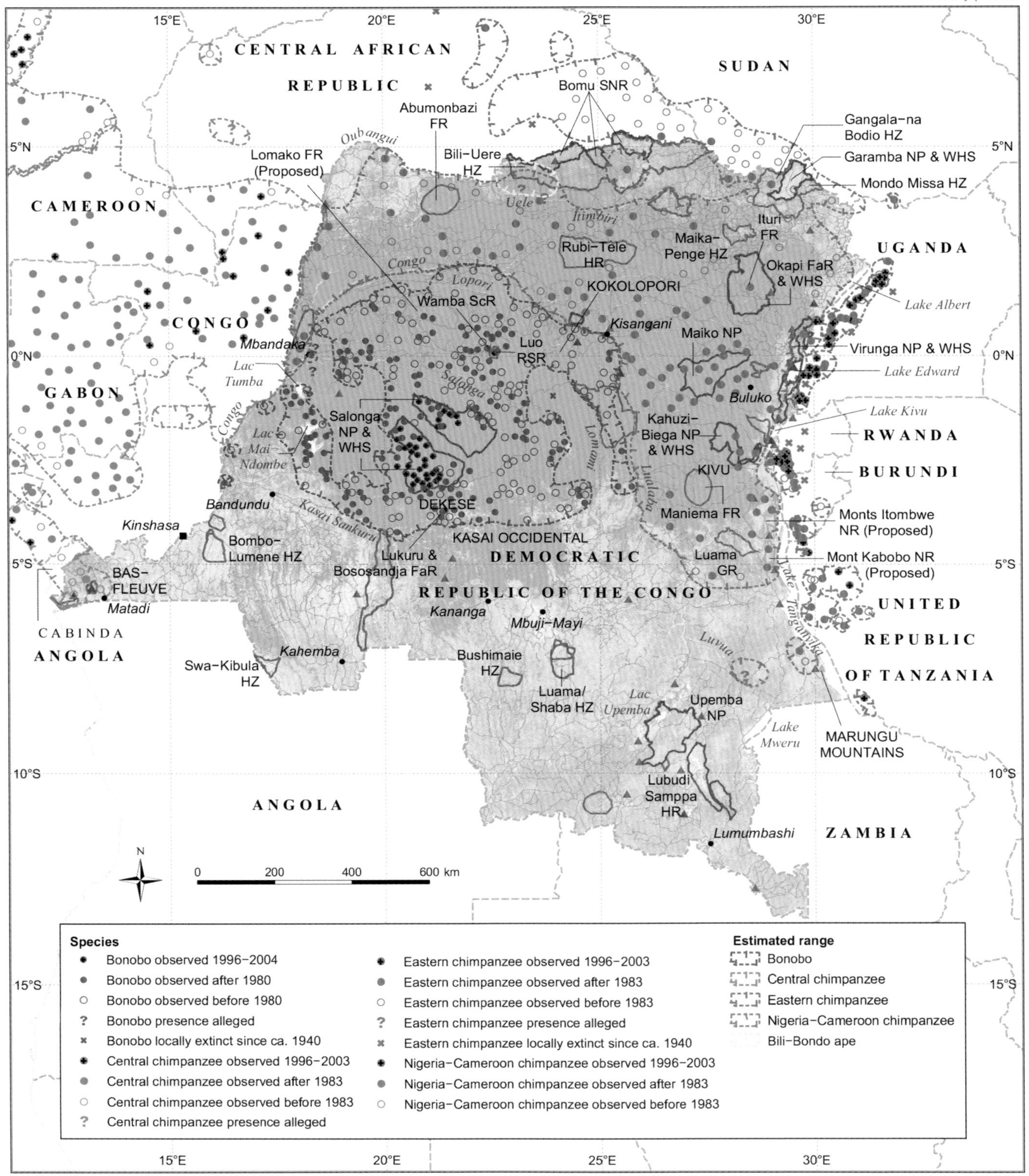

anzee also formerly occurred in the far west of DRC, but there is no information on its present status or distribution.

THREATS

Impacts of the 1996–2002 conflict

With the exception of the mountain gorilla, populations of all the great apes of DRC appear to be decreasing.[7, 10, 26, 46, 53] Although no firm figures are available, the eastern lowland gorilla seems to have been badly affected towards the end of the 20th century. There is particularly little information about the current status of the chimpanzees in DRC.[59]

Before the conflict, eastern DRC had some of the highest human densities in Central Africa, with around 300 individuals per square kilometer in the province of Kivu.[35] The Rwandan genocide and civil war in 1994 led to an influx of nearly 1 million refugees from Rwanda and Burundi.[30] About 860 000 refugees were concentrated in the vicinity of the Virunga NP (7 900 km^2). Some 332 000 had gathered near Kahuzi-Biega NP, an area of 6 000 km^2 originally established to protect DRC's eastern lowland gorilla population.[17, 30] Large areas of forest were cleared,[25, 27, 46] and the huge demand for fuelwood and food led to incursions into protected areas.[16, 46]

In 1996, civil war broke out in DRC. By late 1999, all the national parks were located within or near rebel territory. Many eastern lowland gorillas in Kahuzi-Biega NP and Maiko NP fell victim to the large-scale slaughter of wildlife by fighters and refugees.[49, 68, 75] The high price of coltan (columbium and tantalum) ore between 1998 and 2000 led to a further invasion of Kahuzi-Biega NP and the Okapi Faunal Reserve, by an estimated 10 000 and 3 000 people respectively.[44, 49] DRC is unusual in that there are few fixed coltan mining concessions with secure tenure, and so coltan is generally mined with shovels along rivers.[6] Professional hunters accompanied these miners and their dependents. Local warlords and rebel leaders are believed to have sold much coltan in order to raise money for weapons, used in hunting as well as warfare.[66, 67]

In the mid-1990s, between 12 800 and 21 900 chimpanzees were estimated to live in the 30 530 km^2 area covered by the Maiko NP, Kahuzi-Biega NP, and Okapi Faunal Reserve in eastern DRC.[27] Okapi (13 726 km^2) supported a population of 7 500–12 000 chimpanzees.[27] Even prior to the conflict, commercial hunting and mining was an issue, with four large villages sited within the reserve.[27] The bushmeat consumed in the Okapi reserve during the conflict included the meat of at least three chimpanzees.[20] As figures for the impact of the conflict on the chimpanzee population are not available, it is urgent that fieldwork be carried out to establish their conservation status.

The situation for the eastern lowland gorilla is little clearer. In 1998, there were estimated to be around 16 900 eastern lowland gorillas in existence,[25] but following the war, it is difficult to know how many remain. Population surveys have only been carried out in a few sites[45, 49] (see Status and trends in Chapter 8), with many conservationists therefore being unwilling to provide an overall estimate. It is certain that by 1999, the highland sector (10 percent) of Kahuzi-Biega NP had lost about 50 percent of its gorilla population, including 88 percent of the gorillas habituated for tourism.[75] These habituated animals were particularly easy to shoot. By 2001, the park's gorilla population, estimated to be 15 000 in a 1995 census,[26] had been substantially reduced by hunting for bushmeat.[49] In spring 2004, park officials regained control of much of the park, but estimated that 100 coltan, casserite (tin ore), and gold mining areas were still used.[32] Apart from a few well observed family groups, the current status of the park's gorillas is unknown, so there is a clear need for field surveys to help establish their conservation status.

The Maiko NP (10 830 km^2) constitutes the northern limit of the eastern lowland gorilla in eastern DRC, with an estimated 859 individuals reported present in 1996, along with 4 000–5 000 chimpanzees.[27] There were seven gorilla groups here making up distinct northern, southern, and western populations, but the latter is thought to be extinct.[25] Gorillas are absent from large areas of apparently suitable habitat in the park.[27] Threats include hunting and mining, but there is little timber extraction and the human population is relatively sparse.[27] The impacts of coltan mining here are unknown,[49] so survey work is urgently needed.

The Tayna Gorilla Reserve (700 km^2) was set up between the Maiko and Virunga NPs by two local communities, the Batangi and Bamate Nations. It was officially recognized by DRC as a 'private reserve' in 2002, which simply indicates that it is managed by the community rather than by the state wildlife department. Preliminary surveys suggest that it supports between 225 and 360 eastern lowland gorillas.[36, 70] Patrol teams protect the reserve wildlife from exploitation, and a 5 km buffer

Map 16.7b Gorilla distribution in the Democratic Republic of the Congo

Data sources are provided at the end of this country profile

zone is protected from conversion. The Dian Fossey Gorilla Fund International's Community Conservation Program helped to establish Tayna, and is working with other local communities on seven other proposed reserves in the vicinity. There are tentatively thought to be between 700 and 1 400 gorillas in Tayna and these adjacent areas. The eight communities are working together through the Union of Associations for Gorilla Conservation and Community Development in Eastern DRC (UGADEC), and wish to promote development through conservation rather than through unsustainable use of natural resources.[69] The new reserves include Bakumbule Primate Reserve, Usala Gorilla Reserve, Bakano Forest Reserve, Ngira'Yitu Community Reserve, and Punia Gorilla Reserve. In addition, Initiative Locale pour la Sauvegarde de la Nature (ILSN, Local Initiative to Safeguard Nature) is active over the Masisi territory, and Action Communautaire pour la Protection de la Nature Itombwe Mwenga (ACPN-IM, Itombwe Mwenga Community Action for the Protection of Nature) is involved in the Itombwe area.[69] Itombwe forest was thought to have supported about 1 155 gorillas in 1998,[41] and to have been under moderate hunting pressure at that time.[25] The forest area includes two proposed nature reserves (Monts Itombwe and Mont Kabobo), a forest reserve (Maniema), and a game reserve (Luama).

The rate of habitat loss for the eastern lowland gorilla is probably the highest of any of the gorilla subspecies.[46] It now occupies only about 13 percent of its geographic range, which reflects the extent of the fragmentation of the populations involved.[10] The small Masisi and Mount Tshiaberimu populations are particularly vulnerable.[25] In 1998, it was thought that 16 animals lived in the forests around Mount Tshiaberimu[12] and 28 in the Masisi area.[39] On the boundary of Mount Tshiaberimu, there was extensive agricultural encroachment during the refugee crisis.[12]

Between 12 and 17 Virunga mountain gorillas, or about 5 percent of the population, were killed as a direct result of military activity from 1992 to 2000, along with an unknown number of chimpanzees.[33] Although an end to the war has been declared, the forests of DRC continue to serve as hiding places and retreats for rebel forces, leading to continuing hunting.[46] The lowland areas of the parks are still largely inaccessible by park staff, and the status of the gorillas is unclear.[46]

Most bonobo populations are thought to have been largely unaffected by the war, but in those areas with greater human density and easier access, it is possible that populations have been greatly reduced.[59] In one northern part of the Lomako forest, for example, nest counts indicated a 75 percent decrease in bonobo density;[18] local reports suggest a combination of hunting losses and bonobo flight to less accessible areas.[3, 59] Hunting intensified in the late 1990s, in response to disruption of local agriculture and an increase in the number of commercial hunters entering the area.[3, 19] The trade was limited by the war, but is likely to return now that the conflict has subsided.

Protected areas less affected by war

Salonga NP, which has two sections covering a total area of 36 560 km^2, is the largest rain-forest reserve (and national park) in Africa. It was created in 1970 to protect endemic species, including the bonobo. It is one of only two protected areas supporting the species (the other being the Luo Reserve for Scientific Research, Wamba, at 358 km^2).[68] There is as yet no reported logging threat in Salonga NP, and its location in the center of the country has meant it has been less affected by the conflicts than other sites in DRC. The Mabali Scientific Reserve (19 km^2), like Luo, is a center for bonobo research, and receives local government protection.[60]

Other protected areas that host great apes include the Bili-Uere Hunting Zone (6 000 km^2), Bososandja Faunal Reserve (34 km^2), and Rubi-Tele Hunting Reserve (9 080 km^2). The proposed Lomako Forest Reserve (3 600 km^2) was first suggested by WWF–The Global Conservation Organization in the 1980s, being set aside by the logging company Siforzal that held the concession.[3, 59] The process of its approval for national designation was interrupted by the war.

Threats in peace time

The relative remoteness of prime timber areas and the country's poor transportation infrastructure mean that, until now, only low-volume, selective logging has been profitable, and then only in limited areas along large rivers. This has effectively protected much of the forest of DRC. It is estimated that 5 320 km^2 of forest (about 0.4 percent of the total) is cleared in the country each year.[21] A Forest Code was established in January 2003, under which the state owns all forests and defines legitimate uses for them. Legal mechanisms and a zoning system are to follow. Concerns have been raised that there

Staff at Kahuzi-Biega National Park with elephant and gorilla skulls.

Ian Redmond/UNESCO/Born Free Foundation

are no plans to recognize the resource rights of people traditionally resident in the forest, and that there has been little civil society involvement in the formulation of the code.[48] Forest exploitation taxes are low (US$0.06/ha), so large areas must be licensed to timber companies for government to obtain significant income from this source. The World Bank expects that 600 000 km^2 will be zoned as production forests.[5, 48]

Timber extraction will be made easier by planned new roads, supported by US$270 million of donor funds promised in April 2004.[59, 71] This is equivalent to the reconstruction of about 1 080 km of the country's 2 800 km of paved roads or 5 400 km of its unpaved roads.[71] As trade routes are revitalized, there is a danger that the bushmeat trade will follow the timber trade.[2]

Although not universal, hunting taboos have existed for all ape species in different areas of DRC. The loss of these taboos, and movement of people without them into great ape range areas, is a significant threat. For example, local people near Wamba have traditionally refrained from hunting bonobos for religious reasons.[8, 54] However, in the mid-1980s hunters began taking bonobos for meat.[61] Bonobo meat is avoided in many parts of DRC, and regarded as a delicacy in others.[59, 68] Historically, bonobos have also been hunted for their supposed medicinal and/or magical value, with charms made from body parts being available in some areas.[31] Relatively few people live in and around most of the range of the bonobo, however, so hunting pressure is low overall, though bonobos are very vulnerable to wire snares set to catch other species. This is a widespread practice, and one in which a very few people can set a large number of traps, all of which remain a threat for years even if they are abandoned or lost by the snare-setter.

A taboo against eating gorilla meat still exists in and around the mountain gorilla range area. Hunting of this species has become rare in the last 20 years due to dedicated protection and support from governments and conservation groups.[46] Historically, taboos in the east Congo Basin also protected many eastern lowland gorillas from being killed for food. Wire snares set to catch other mammals, such as antelopes, had represented the more serious threat. By 1998, at least one individual in each habituated eastern lowland gorilla group in the montane sector of Kahuzi-Biega NP had lost a hand to snares.[25, 75] As noted above, by 2001, most of these gorillas had been killed.

LEGISLATION AND CONSERVATION ACTION

National legislation

The Ministry of Environment, Nature Conservation, and Tourism is the government body responsible for nature conservation. The Nature Conservation Act of 1969 (Ordinance-Law 69.041) defines national parks. The 1982 hunting act (Law 82.002) defines faunal reserves and game reserves and lists animals for which hunting and trapping are prohibited.[14, 62] Unusually, since 1985, regional governments have had the right to set their own regulations on species protection and to specify hunting seasons, bypassing this law. In Zone Dekese (Province Kasai Occidental), bonobo hunting was permitted in 1997 with written permission for each incident. There were no such incidents, and their protected status was restored following a campaign by the Lukuru Wildlife Research Project team.[58, 59]

International agreements

DRC ratified or acceded to the Convention on Biological Diversity in 1994, the Convention on International Trade in Endangered Species of Wild Fauna and Flora (CITES) in 1976, the Convention on Migratory Species in 1990, the UN Convention to Combat Desertification in 1997, the African Convention on the Conservation of Nature and Natural Resources in 1976, and the World Heritage Convention in 1974. There are five World Heritage Sites: Garamba NP, Kahuzi-Biega NP, Salonga NP, Okapi Faunal Reserve, and Virunga NP. All are listed as World Heritage Sites in Danger due to human pressures. DRC also participates in UNESCO's Man and Biosphere (MAB) Programme.

Protected areas

There are four main categories of protected area in DRC: national parks (nine sites), game reserves (one site), forest reserves (seven sites), and faunal reserves (two sites).[14, 65] Other designations include areas set aside for scientific research, or as hunting zones and nature reserves. The existing protected area network covers an estimated 195 426 km^2 (8 percent of the land area).[65] Management of national parks, faunal reserves, and game reserves is delegated to the Institut Congolais pour la Conservation de la Nature (ICCN), which also manages scientific research. Effective control of many protected areas in the east of the country has been in the hands of rebel authorities in recent years.

Conservation and field projects

Most research into the populations, ecology, and social behavior of great apes in DRC has taken place in the east and north of the country. Research sites include Kahuzi-Biega NP,[9, 30] Salonga NP,[68] Okapi Faunal Reserve,[27] Wamba Scientific Reserve, the Luo Reserve for Scientific Research,[24] and Lui Kotal in Lomako forest.[29] Research was scaled down during the war, but persisted at several field sites.[60] In particular, the Lukuru Wildlife Research Project, Max Planck Institute, the Center of Research for Natural Sciences in DRC (CRSN), and the Wildlife Conservation Society (WCS) continued to work throughout the civil war. CRSN researchers continued a research project on sympatric gorillas and chimpanzees in Kahuzi-Biega NP, in cooperation with Kyoto University.[9, 76]

National organizations active in ape conservation in DRC include the Association de Femmes pour la Conservation et le Développement Durable (AFECOD), the Pole Pole Foundation (POPOF), and Programme d'Appui aux Initiatives de Développement Economique du Kivu (PAIDEK). Typically, these focus on community projects and alternative livelihoods. Active international nongovernmental organizations include the International Gorilla Conservation Programme (Virunga NP), WWF (Virunga and Garamba NPs), Lukuru Wildlife Research Project (Salonga NP), WCS (Salonga NP, Okapi Faunal Reserve, and the wider Ituri Forest Reserve), the Zoological Society of Milwaukee (Salonga NP), and the Dian Fossey Gorilla Fund (DFGF) (Mount Tshiaberimu and Virunga Conservation Center). For example, DFGF aims to establish a community conservation zone between Maiko NP and Kahuzi-Biega NP.[15] Unusually, the Lukuru Wildlife Research Project has purchased land rights to the Bososandja Faunal Reserve (which incorporates the range of the Bososandja bonobo community).

In 2004, bonobo conservation and research activities – underway and proposed by the Centre de Recherche en Ecologie et Forestrie, Lukuru Wildlife Research Project, Max Planck Institute, Vie Sauvage, Wamba Committee for Bonobo Research, WCS, and the Zoological Society of Milwaukee – included biomonitoring, building community relations, economic development of micro-projects outside the Salonga NP, employing more people in conservation and research, awareness-raising programs, infrastructure building and rehabilitation, inventory, surveying and mapping, training and support of poaching patrols, and research.[59] Other goals included establishing an education center, extending national park boundaries, formalizing the awareness-raising programs, demarcating protected area boundaries, building awareness among logging concessionaires, developing land management plans, providing appropriate equipment, training, and support to ICCN, and revitalizing the Lomami-Lualaba conservation site.

Ecotourism

Gorilla tourism brought 2 800 visitors to the Virunga NP between 1986 and 1990, who between them spent an estimated US$250 000. Meanwhile, between 1988 and 1991, 2 000 visitors to the Kahuzi-Biega NP generated an estimated US$200 000.[72] Ecotourism income has been minimal for the past decade, however, due to the conflicts; the death through poaching of habituated gorilla groups in the Kahuzi-Biega NP is a major setback.

Sanctuaries and rehabilitation

The Lola ya Bonobo sanctuary for orphan bonobos is located 25 km from Kinshasa, and run by Les

Jo Thompson/Lukuru Wildlife Research Project

Wild bonobo in Bososandja Community Forest.

Martha M. Robbins

The gorilla census team at Virunga National Park.

Amis des Bonobos du Congo. In May 2004, it had 38 bonobos, with numbers of infants arriving at the center increasing since the end of the war. Investigations for an appropriate site for eventual reintroductions are underway.[38] In November 2004, an ICCN meeting decided to create a sanctuary in the Goma region for confiscated infant eastern lowland gorillas.[43]

The Lwiro Primate Sanctuary, sited close to Kahuzi-Biega NP, cared for 13 orphaned chimpanzees and one bonobo at the end of 2003. The sanctuary is under the management of ICCN, and supported by the German overseas development agency GTZ, the Born Free Foundation, the International Primate Protection League (IPPL), the Jane Goodall Institute (JGI), and the Pan African Sanctuary Alliance (PASA).[4]

FUTURE CONSERVATION STRATEGIES

The National Biodiversity Strategy and Action Plan identifies the need for the conservation of gorillas and bonobos, which are highlighted within one of the priority projects – Project 3: Protection and rehabilitation plans for threatened species and ecosystems.[37]

A GRASP workshop was held in Kinshasa in 2002 to discuss conservation action for great apes in DRC, from which a draft action plan was produced.[40, 64] The most urgent recommendations from the working groups at the meeting were:

- undertake surveys of little-known areas to establish which apes survive and where, in particular in the Mayombe forest, Bas-Fleuve, which is contiguous with forests in Angola's Cabinda province;
- rehabilitate the neglected Maiko NP, which holds important populations of eastern lowland gorillas and chimpanzees;
- reclaim the parts of Kahuzi-Biega NP still outside the rangers' control and survey the area to establish how many apes have survived the onslaught of bushmeat hunters feeding the coltan miners;
- develop community conservation initiatives to create jobs in areas of rural poverty in locations important for great apes, such as around the village of Lomako;
- strengthen existing laws protecting great apes and improve awareness among law enforcement agencies and the courts;
- increase resources for sanctuaries to care for confiscated infant apes and develop the potential of sanctuaries for conservation education.

An action plan for bonobos was published in 1995[61] and endorsed by the IUCN Species Survival Commission. The plan calls for regional surveys to identify bonobo populations, conservation education throughout DRC, economic benefits for local people from conservation projects, and enforcement of antipoaching and habitat protection laws.

In 2003, an international workshop held in Japan sought to build on the 1995 plan.[60] It made a number of recommendations for conservation of the bonobo. These included field surveys (at Kokolopori, Lac Tumba, and Lomako), construction of research facilities (at Kokolopori, Lac Tumba, and Lui Kotal), completion of management plans (for Lomako forest and Salonga NP), community development activities (at Wamba and Lukuru/Bososandja), extending reserve boundaries (at Lui Kotal), and continuing or establishing research (at Wamba, Salonga, Lukuru/Bososandja, and Lui Kotal).

Various recommendations have been made for action in specific protected areas to conserve great apes. Prior to the coltan rush, it was suggested that conservation efforts for eastern lowland gorillas should be focused on the lowland sector of the Kahuzi-Biega NP and the adjacent Kasese region.[25] It had been considered essential to maintain the forest corridor between the lowland and montane populations to ensure gene flow between them.[51, 77] More recently, calls have been made for crisis management and damage assessment and limitation. The Itombwe forest has also been recommended as a focus for the conservation of eastern lowland gorillas,[25] and its population seems even more important given the loss of so many Kahuzi-Biega gorillas.

Overall priorities for bonobos include the purchase of forestry concessions in key areas for bonobos, the designation of these sites, as well as Bososandja Community Forest and the Lomami-Lualaba conservation site, as nationally protected areas, and the creation of the proposed Lomako Forest Reserve.[19, 60]

FURTHER READING

Hart, J.A., Hall, J.S. (1996) Status of Eastern Zaire's forest parks and reserves. *Conservation Biology* **10** (2): 316–324.

Hicks, T.H. (2005) *The Bondo Mystery Apes, Winter 2004 Field Data.* Unpublished report. http://www.karlammann.com/bondo-winter2004.html. Accessed June 8 2005.

Rainforest Foundation (2004) *New Threats to the Forests and Forest Peoples of the Democratic Republic of Congo. Briefing Paper February 2004.* The Rainforest Foundation, London. http://www.rainforestfoundationuk.org/files/RF%20UK%20Briefing%20Paper%20on%20DRC%20February%2004.pdf. Accessed June 2 2004.

Redmond, I. (2001) *Coltan Boom, Gorilla Bust.* Report for the Dian Fossey Gorilla Fund (Europe) and Born Free Foundation. http://www.bornfree.org.uk/coltan.

Thompson, J., Hohmann, G., Furuichi, T., eds (2003) *Bonobo Workshop: Behaviour, Ecology and Conservation of Wild Bonobos.* Inuyama, Japan.

UNSC (2002) *Final Report of the Panel of Experts on the Illegal Exploitation of Natural Resources and Other Forms of Wealth of the Democratic Republic of the Congo.* United Nations Security Council, New York. http://www.natural-resources.org/minerals/CD/docs/other/N0262179.pdf. Accessed June 18 2004.

MAP DATA SOURCES

Maps 16.7a and b Great apes data are based on the following source:

Butynski, T.M. (2001) Africa's great apes. In: Beck, B.B., Stoinski, T.S., Hutchins, M., Maple, T.L., Norton, B., Rowan, A., Stevens, E.F., Arluke, A., eds, *Great Apes and Humans: The Ethics of Coexistence.* Smithsonian Institution Press, Washington, DC. pp. 3–56.

With additional data by personal communication from Blake, S. (2005), Furuichi, T. (2003), Lanjouw, A. and Gray, M. (2003), Plumptre, A. (2004), Ron, T. (2003), Stokes, E. (2003), Thompson, J. (2003), and from the following sources:

Hall, J.S., Saltonstall, K. Inogwabini, B-I., Omari, I. (1998) Distribution, abundance and conservation status of Grauer's gorilla. *Oryx* **32**: 122–130.

Hohmann, G., Fruth, B. (2003) Culture in bonobos? Between-species and within-species variation in behaviour. *Cultural Anthropology* **44** (4): 563–571.

Hohmann, G., Fruth, B. (2003) Lui Kotal – a new site for field research on bonobos in the Salonga National Park. *Pan Africa News* **10** (2): 25–27.

Ilambu, O., Grossman, F., Mbenzo, P., Blake, S. (2005) *Monitoring of the Illegal Killing of Elephants in Central African Forests: Elephant and Ape Population Surveys in the Salonga National Park, Democratic Republic of Congo.* Report to CITES MIKE and the government of the Democratic Republic of the Congo.

Kalpers, J., Williamson, E.A., Robbins, M.M., McNeilage, A., Nzamurambaho, A., Lola, N., Mugiri, G. (2003) Gorillas in the crossfire: population dynamics of the Virunga mountain gorillas over the past three decades. *Oryx* **37**: 326–337.

Thompson, J. (2005) Field research at Lukuru, Democratic Republic of Congo. *Pan Africa News* **10** (2): 21–22.

Wamba Committee for Bonobo Research (2004) Latest News. http://www.pri.kyoto-u.ac.jp/shakai-seitai/shakai/BONOBOHP/English/News_e.html. Accessed May 1 2005.

For protected area and other data, see 'Using the maps'.

ACKNOWLEDGMENTS

Many thanks to reviewers Jo Thompson (Lukuru Wildlife Research Project), Annette Lanjouw (IGCP), and Juichi Yamagiwa (Kyoto University) for their valuable comments on the draft of this section.

AUTHOR

Nigel Varty, UNEP World Conservation Monitoring Centre

REPUBLIC OF EQUATORIAL GUINEA

BRIGID BARRY

BACKGROUND AND ECONOMY

The Republic of Equatorial Guinea is one of the smallest countries in Africa. Its two principal regions are Río Muni, bordering the Bight of Biafra between Gabon and Cameroon, and the island of Bioko. The total land area is 28 051 km^2, of which 26 017 km^2 is in the continental Río Muni. The climate is equatorial with a mean temperature of 26°C (range of 17–34°C). The average annual precipitation in Río Muni is less than 4 000 mm, while in southern Bioko it can reach 11 000 mm.[12, 24] The official languages are Spanish, French, and Fang, and the main ethnic groups are Fang, Bubi, Annobonese, Ndowe, Kombe, and Bujebas.[31]

Independence from Spain was ratified in 1968 and Equatorial Guinea remains the only Spanish-speaking country in sub-Saharan Africa. President Obiang Nguema Mbasogo has been in power since 1979. Under his rule, the country has experienced drastic demographic and economic changes. Equatorial Guinea is the third greatest producer of oil in sub-Saharan Africa, behind Angola and Nigeria. The country's gross domestic product (GDP) is US$2.2 billion (2002 estimate),[4] largely as a result of the expansion in the oil and gas sector.[4] GDP increased by 71.2 percent in 1997, 45.5 percent in 2001, and around 15 percent in 2003, which makes Equatorial Guinea the world's fastest growing economy.[4] The population is also rising dramatically, up from 358 000 in 1991 to 523 051 in 2004.[8] The population is growing at around 2.5 percent per year according to 2000 estimates.[4] Infrastructure, health, sanitation, and education remain underdeveloped.

With an estimated 64 percent of the population living in rural areas, most people still work in the agricultural sector.[35] The main food crops are malanga (*Xanthosoma* spp.), sweet potatoes, peanuts (cacahuete), cassava, sugar cane, maize, pineapples, bananas, and plantain.[12, 24, 34] There is evidence of repeated failed attempts to breed livestock and chickens. The lack of success is attributed to the tsetse fly (vector of trypanosomiasis) and the low prices of imported, frozen European chickens respectively. Fishing also remains relatively undeveloped in Río Muni, with European and Nigerian trawlers exploiting the waters.[24, 31] Limited domestic protein sources put pressure on Equatorial Guinea's wild animals, including the great apes, as a source of meat for the growing affluent population.

DISTRIBUTION OF GREAT APES

The western lowland gorilla (*Gorilla gorilla gorilla*) and the central chimpanzee (*Pan troglodytes troglodytes*) are both present in mainland Río Muni. The current numbers of apes are unclear. In the early 1980s, it was estimated that there were 1 000–2 000 gorillas and 600–1 500 chimpanzees, with 150–500 breeding female chimpanzees.[22, 25] A 1989–1990 census concluded that depending upon the calculation applied, there could be 990–2 450 individuals.[20] To date only four studies on the density of apes in Equatorial Guinea are known.

- Jones and Sabater Pí[26] spent 18 months in 1967 to 1968 at Monte Alén, Abuminzok-Aninzok, and Monte Okoro Biko (now called Monte Mitra). The conservation status of both species was considered critical due to human activities. Western lowland gorilla densities varied from 0.58 to 0.86/km^2 in stable groups of two to 12 individuals. Their home ranges averaged approximately 6.75 km^2 and they were found in open areas, with nests close to the ground in thick vegetation. Central chimpanzees were encountered at densities of 0.31–1.53/km^2 in fission-fusion groups of two to 23 individuals. Their home ranges averaged approximately 15 km^2. They were observed mostly in the upper strata of the forests, with nests in the canopy of small trees surrounded by taller primary trees.
- Gonzalez-Kirchner[20] studied western lowland gorilla populations throughout Río Muni for 18 months (1989 to 1990). Gorillas were present in 385 km of 783 km of transects. Based on this, a gorilla density of 0.45 of nesting gorillas per square kilometer was estimated for the surveyed areas. Site densities ranged from 0.12/km^2 in the Río Campo basin to 0.71/km^2 in the Nsork highlands region in the southeast of the country. Gorilla densities at Evinayong (0.26/km^2) and Niefang (0.59/km^2) were considerably lower than those estimated in the 1960s at Evinayong (0.58/km^2) and

Map 16.8 Great ape distribution in Equatorial Guinea

Data sources are provided at the end of this country profile

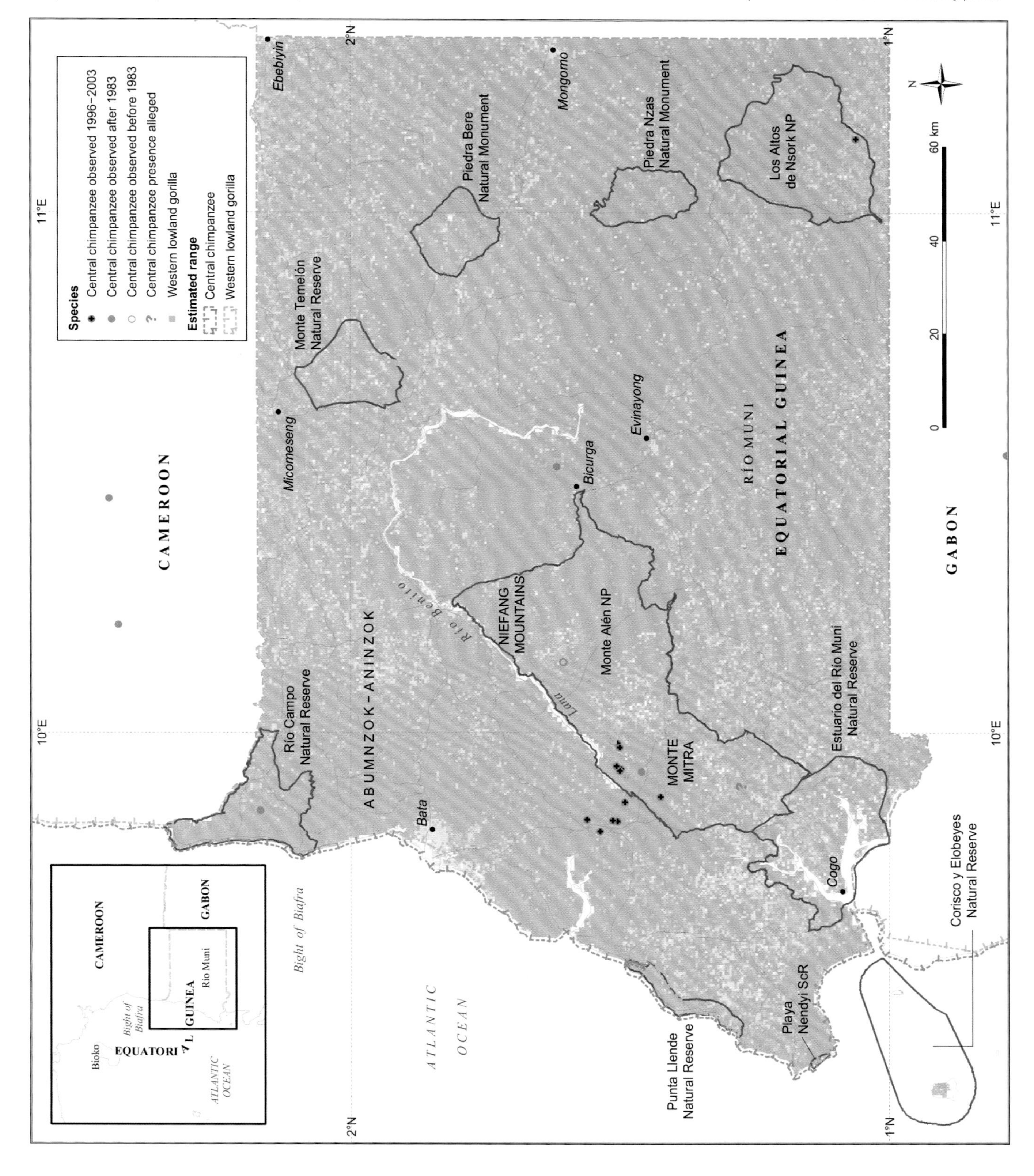

Noëlle Kümpel

A hunter with a dead mandrill, Equatorial Guinea.

Niefang (0.86/km^2).[26] A subsidiary study reported an abundant population of central chimpanzees in the Río Campo area at the same time.[21] Presence of gorillas in the Nsork highlands was confirmed by a 1998 survey, which also found chimpanzee nests in the same area.[30]

- Garcia and Mba[18] studied the distribution and population densities of apes for three months in 1994 as part of a longer primate study in Monte Alén National Park (NP), prior to its 1997 extension to include Monte Mitra in the south. They identified 208 gorilla nests. Gorillas were most abundant in secondary forests (5.15 nests per kilometer of the surveyed area), in mean group sizes of 3.3 individuals. Most were found in the montane areas crossed by the Niefang–Bicurga road, with fewer observations around the low-lying Lana River and the confluence of the Lana–Wele Rivers. Chimpanzees were found in all areas surveyed, with 333 nests detected. They were abundant in open primary forest (5.35 nests per kilometer of the surveyed area), with a mean group size of 2.3 individuals. This study found evidence of ape hunting, but concluded that they were not threatened within the park.
- Ghiurghi and Puit[19] spent four months (2003 to 2004) recording gorilla and chimpanzee nests along 68.3 km of transects in the southern extension of the Monte Alén NP. The nests of the two species together were at a low density, 0.82 nests per kilometer or 24.18 nests per square kilometer. There was very little evidence of a gorilla presence in the southern areas of the park, which may result from the absence of key food plants (*Afromomum* and Marantaceae species).

THREATS

Bushmeat hunting

The greatest current threat to apes in Río Muni is bushmeat hunting. The Fang, representing the largest ethnic group, regard meat as an essential component of their diet.[15] Fresh meat and fish is preferred over frozen alternatives. While domestic meat options (e.g. cattle imported from Cameroon) do exist, bushmeat is more popular with the Fang and slightly cheaper.[9] In combination with a scarcity of other fresh meat sources, this means that bushmeat is both a substantial part of the people's diet and a source of income for a large number of rural people. With the exception of coastal towns like Cogo, all the rural villages studied throughout Río Muni are active in bushmeat trapping and hunting.[2]

The regions north and south of the Wele River have experienced more extreme hunting pressure than others.[24] During a study period of 15 months, 80 people from Sendje, a village of 400 people on the edge of the Monte Alén NP, were recorded as actively shooting or trapping, and nearly 5 000 traps were counted both inside and outside the park[28] (see Box 13.4). In a previous study made over a period of 16 months, the village of Sendje alone harvested 11 376 kg of bushmeat. This was declared an unsustainable rate.[14, 16] Hunters, though increasing in abundance, are currently fewer than trappers because guns and cartridges are expensive and not readily available in Equatorial Guinea.[27]

The apes are able, often collectively, to escape from wire snares, although injury and/or limb loss is possible. However, hunters are increasingly killing apes for bushmeat, and as guns become more available, apes become easier targets. Among several of the Fang subtribes, apes, especially gorillas, were traditionally believed to cause infertility in women if eaten.[6] This meant that there was little demand and the cost of transporting the heavy corpses was not worth the market return. However, many previously taboo species are becoming marketable. A decade ago, bushmeat was mostly consumed locally,[12, 23] but the improving infrastructure allows meat to be distributed to the larger city markets.[2]

The two main bushmeat markets in Río Muni are both located in Bata: the Central and Mondoasi Markets. Tons of bushmeat are brought to the Bata markets annually from all over the country by traders and bush taxis, and openly traded. Although state forest guards are positioned at the main military barriers on the two roads entering Bata, little is done to monitor or prevent any bushmeat passing through. Bribes are extracted from bushmeat traders, and seizures of meat are rare.

Habitat destruction

Approximately 17 520 km^2 of forest was thought to remain in the whole of Equatorial Guinea in 2000, with an average annual decrease of forest cover between 1990 and 2000 of 110 km^2, or 0.6 percent.[17] Statistics for the area of plantations were not available. Prior to the 1960s, large areas of the forest of Río Muni were destroyed or degraded by commercial agriculture and logging. Following this, periods of political and social unrest led to the abandonment of commercial logging schemes, emigration, and concentration of the human population in the coastal regions. The political stability under President Obiang's rule has allowed an increase in forest exploitation and from the mid-1980s on there was a further estimated 20 percent reduction in forest cover.[12]

In 1999 alone, 788 000 m^3 of timber was logged in Río Muni for export. Logging concessions have been sold over most of the country, including large parts of the protected areas.[11, 31] In particular, forests in the Micomeseng–Ebebiyin–Mongomo region in the northeast have been severely damaged.

The pet trade

The growing economy has attracted non-African immigrants to the Río Muni region. Individual expatriates regularly buy live wild animals as pets, mainly African grey parrots (*Psittacus erithacus*). Although illegal, this trade is increasing. Baby apes are seen as 'cute' and some expatriates believe that purchasing them saves them from being eaten by locals. Hunters are reported to kill entire troops of apes to capture one orphan for sale.[6] They can earn more by selling one baby chimpanzee to an expatriate than selling 20 blue duikers (*Cephalophus monticola*) to the bushmeat market.[6, 27]

LEGISLATION AND CONSERVATION ACTION

Equatorial Guinea is a signatory to the Convention on International Trade in Endangered Species of Wild Fauna and Flora (CITES) (1992) and the Convention on Biological Diversity (1994). It is one of the few African countries that is not part of the 1968 African Convention on the Conservation of Nature and Natural Resources or the World Heritage Convention. No sites have been designated as Biosphere Reserves under UNESCO's Man and Biosphere (MAB) Programme.[24]

In 1999, Equatorial Guinea participated in the Yaoundé Forest Summit in Cameroon. The meeting established the Yaoundé Forest Declaration, which committed participants to create new protected areas and to have plans to combat illegal logging and poaching[38]. In 2000, the Law on Protected Areas of Equatorial Guinea was passed, which increased the number of protected areas to 13, covering a total area of 5 860 km^2 (approximately 20 percent of the country). Ten are found in Río Muni: two national parks (Los Altos de Nsork, Monte Alén); five natural reserves (Estuario del Río Muni, Río Campo, Monte Temelon, Punta Llende, Corisco y Elobeyes); one scientific reserve (Playa Nendyi); and two natural monuments (Piedra Bere, Piedra Nzas).[33, 37] Monte Alén NP covers an area of dense tropical forest, and is the largest and most effectively protected of these areas.[24] The park has been selectively logged in the past and so is largely made up of secondary forest. Hunting, logging, and agriculture are officially prohibited within the 2 000 km^2 park, but there are logging concessions running up to its boundaries.[29]

Conservation and field projects

Equatorial Guinea is part of the European Union program, Conservation and Rational Use of Forest Ecosystems in Central Africa (ECOFAC), which started in 1992. The role of ECOFAC is to support the management of the Monte Alén NP, and it has trained park rangers from villages surrounding the park to police illegal activities, monitor flagship species such as the apes, and to guide visiting researchers and ecotourists. ECOFAC has constructed a guesthouse in the park to encourage community-based conservation through tourism revenues.[10] Limited resources and a gap in funds, combined with lack of success in developing the tourism potential of the site, mean that hunting and trapping still continue within the park. The park's great apes are seldom killed this way.[29, 32]

Nationally, conservation issues are administered by the Instituto Nacional de Desarrollo Forestal y Gestión del Sistema de Areas Protegidas (INDEFOR), a conservation body within the Ministry of Infrastructure and Forests (Ministerio de Infraestructuras y Bosques). Most of the field staff are botanists, which places limits on the research and management programs set up for apes and other vertebrates.

In 2002, a team from the Zoological Society of London and Imperial College London started a project on the sustainability of bushmeat hunting in Río Muni, in conjunction with ECOFAC, INDEFOR, and Conservation International. Initial work focused on the incentives for hunting for villagers living close to Monte Alén NP, asking what socioeconomic circumstances drive the decision to hunt, prey selection, and impacts on the viability of prey populations.[27] Urban demand and the role of consumer preferences within the regional capital of Bata were also analyzed.[9] Further work will address reliance on bushmeat in terms of livelihoods and food security across Río Muni,[3] and the potential use of agricultural areas for sustainable bushmeat hunting.[36]

The Durrell Wildlife Conservation Trust, Jersey and the Wildlife Conservation Research Unit, Oxford carried out another bushmeat study in conjunction with INDEFOR for six months from October 2003 to March 2004. This followed up a bushmeat consumption study carried out by John Fa and colleagues in the 1990s. Household interviews, hunter interviews, and market data were gathered throughout the country with the aim of understanding the demand for and consumption of bushmeat from an economic and ecological perspective.[1]

Sanctuaries, education, and ecotourism

Despite the presence of orphan chimpanzees, there are no primate sanctuaries or rehabilitation centers in Equatorial Guinea.

A survey conducted in early 2004 in Bata found that environmental education, including conservation, was absent in all but one (a private institution) of 60 schools.[2]

Ecotourism could use Equatorial Guinea's natural beauty to provide an income for the local population. Unfortunately, in spite of existing infrastructure at Monte Alén and growing facilities in Bata, the large military presence has ensured that international tourism has been minimal. Tourists may find it difficult to obtain correct authorization to travel throughout the country and hold-ups at regular military blockades may prove tedious.

FUTURE CONSERVATION STRATEGIES

Research

A longer study on the ape populations of Río Muni is needed to reassess the densities and areas of high ape conservation priority. Research in the past has been sporadic and the information is now out of date. Possible research projects are outlined here.

- In a study using nests as a proxy for the relative abundance of the gorilla and chimpanzee populations, it was shown that there were apparent differences in the distributions of the two species between open and dense primary forest, and secondary forest. Further studies might elucidate whether these differences result from species-specific habitat selection or are related to past hunting pressures.
- Río Campo apparently had high chimpanzee densities in 1989–1990. The current status of chimpanzee densities needs to be assessed, and if confirmed to be high, the area should be included within protective management by ECOFAC.
- Gonzalez-Kirchner's study[20] found high densities of gorillas in the Nsork highlands. High elephant densities were also recorded in this area,[5] which indicates that hunting impacts had been limited, at least by the early 1990s. A thorough study of this district would help to establish it as an actively protected area.[13]

Hunting

In general, hunting should be controlled inside protected areas and managed to within sustainable yields. As hunting is thought to be the biggest threat to ape populations in Equatorial Guinea, the prohibiting legislation needs to be enforced and stricter controls and fines put in place for killing, transporting, and trading in great apes. Improved training and management would help forest guards and park rangers to carry out their part in this work. Conservation organizations and government agencies need to target the source of bushmeat by working with logging companies to provide alternative protein sources to workers and their families, and to prevent their involvement in the bushmeat trade. The development of the trade in Equatorial Guinea is largely due to the absence of other protein sources. Research and development in sustainable fisheries, domestic livestock, and other protein alternatives might reduce demand.[6]

National awareness

Environmental studies and conservation education need to be introduced into the national curriculum. A local nongovernmental organization, Tierra Viva (Living Earth), is currently seeking funds to extend its work to Río Muni and carry out conservation education in both urban and rural areas. It has been suggested that education programs be introduced to the state radio and television stations, focusing on sustainable use of forests and wildlife. Awareness programs are also required for the growing community of expatriates, who do not always realize that buying orphan apes fuels the market in wildlife. International conservation agencies also need to be made aware of the current state of conservation in Equatorial Guinea and brought to the country, thereby encouraging national interest in ape survival.

At all times, the local population must be integrated in the management of protected areas. All human activities must be kept to a minimum in core zones of the park while managing areas around the buffer zones.[15] Training and management of rangers in Monte Alén NP should resume and be extended to other protected areas around the country.

If the government were to sign and ratify the World Heritage Convention, Monte Alén could be proposed as a UNESCO World Heritage Site to inspire national interest and pride in the protected areas.[7]

Logging

Attention must be paid to ways of making timber extraction compatible with biodiversity conservation. Monitoring of natural regeneration is needed to determine sustainable felling cycles, extraction intensities, and harvest methods, and habitat restoration may be needed in buffer zones around the parks.

FURTHER READING

Fa, J. (1992) Conservation in Equatorial Guinea. *Oryx* **26** (2): 87–102.

MAP DATA SOURCES

Map 16.8 Great apes data are based on the following source:

Butynski, T.M. (2001) Africa's great apes. In: Beck, B.B., Stoinski, T.S., Hutchins, M., Maple, T.L., Norton, B., Rowan, A., Stevens, E.F., Arluke, A., eds, *Great Apes and Humans: The Ethics of Coexistence.* Smithsonian Institution Press, Washington, DC. pp. 3–56.

With additional data by personal communication from Kümpel, N. (2004) and from the following sources:

Garcia, J.E., Mba, J. (1997) Distribution, status and conservation of primates in Monte Alén National Park, Equatorial Guinea. *Oryx* **31** (1): 67–76.

Ghiurghi, A., Puit, M. (2004) *Inventaire des grands et moyens mammifères dans l'extension sud du Parc National de Monte Alén.* Rapport Technique. ECOFAC, AGRECO, SECA, CIRAD Fôret.

Gonzalez-Kircher, J.P. (1994) *Ecología y Conservación de los Primates de Guinea Ecuatorial.* Ceiba Ediciones, Cantabria.

Jones, C., Sabater Pí, J. (1971) Comparative Ecology of *Gorilla gorilla* (Savage and Wyman) and *Pan troglodytes* (Blumenbach) in Río Muni, West Africa. *Bibliotheca Primatologica* **13**: 1–95.

Larison, B., Smith, T.B., Girman, D., Stauffer, D., Mila, B., Drewes, R.C., Griswold, C.E., Vindum, J.V., Ubick, D., O'Keefe, K., Nguema, J., Henwood, L. (1999) *Biotic Surveys of Bioko and Rio Muni, Equatorial Guinea.* Center for Tropical Research, University of California, Los Angeles. Submitted to Biodiversity Support Programme. http://www.ioe.ucla.edu/CTR/reports/CARPE1.pdf. Accessed November 25 2004.

For protected area and other data, see 'Using the maps'.

ACKNOWLEDGMENTS

Many thanks to Allard Blom (WWF-US), Noëlle Kümpel (Imperial College London), Matthew Shirley (University of Florida), John Fa (Durrell Wildlife Conservation Trust), and Lise Albrechtsen (University of Oxford) for their valuable comments on the draft of this section. Thanks to Edmund McManus (UNEP-WCMC) for research into the literature.

AUTHOR

Brigid Barry, Tropical Biology Association

REPUBLIC OF GABON

AMBROSE KIRUI, LERA MILES, AND JULIAN CALDECOTT

BACKGROUND AND ECONOMY

The Republic of Gabon is located between the Congo Basin and the Gulf of Guinea, bordering Cameroon and Equatorial Guinea to the north, and Congo to the east and south. It has a total land area of 257 667 km^2, with an additional water area of 10 000 km^2.[4] The climate is humid equatorial with a mean temperature of 27°C and an annual rainfall ranging from 1 400 mm in the south to more than 3 200 mm in the north.[3, 16] In 2000, forests were estimated to cover 84 percent of the country with an approximate annual loss of 100 km^2 between 1990 and 2000.[9, 10]

In 2004, Gabon's human population was approximately 1.4 million (150 000 of whom were expatriates), growing at a rate of about 2.5 percent per year.[4] The official language is French, and the main ethnic groups are Fang, Bapounou, Nzebi, and Obamba. Gross domestic product (GDP) in 2002 was US$5 billion, with a relatively high gross national income (GNI) per person of US$3 110, six times higher than the sub-Saharan average.[4, 24]

Abundant natural resources, a small population, considerable foreign support, and a booming oil industry accounting for 50 percent of Gabon's GDP all help to explain the country's relative economic prosperity. Political stability has also played a role in the country's prosperity: following independence from France in 1960 there have been only two presidents, the current President El Hadj Omar Bongo having been in power since 1967. A multiparty system and a new constitution were introduced in the early 1990s, and there were local elections in 2002–2003 with a presidential poll scheduled for 2005. There is some political tension because of the long leadership tenure. Government priorities include lowering dependence on the oil industry, as the current petroleum fields are projected to run out in about 2015, and there are plans to develop other areas of natural resource exploitation such as forestry.[8]

DISTRIBUTION OF GREAT APES

Gabon has about 6 000 plant species, 446 bird species, and 190 mammal species.[5] Among them are the western lowland gorilla (*Gorilla gorilla gorilla*) and central chimpanzee (*Pan troglodytes troglodytes*). In the 1980s–1990s, Gabon was thought to hold approximately 40 percent of the world's gorillas (then perhaps 35 000 animals) and around 64 000 chimpanzees.[3, 13, 27] All Gabon's national parks contain at least one great ape species (Table 16.3). Although both species are recorded throughout the country, there are worrying signs that ape populations may have collapsed in recent years, which renders the impression given by Map 16.9 overoptimistic. Many populations may have been reduced to such a low density in areas affected by hunting and Ebola that their viability is threatened.

THREATS

The major threats to great ape populations in Gabon are illegal commercial hunting, rapid expansion of mechanized logging, and the Ebola virus, with the complicating pressure of human population growth. Up until the beginning of the 1980s it was thought that gorilla and chimpanzee populations were relatively stable,[11] but nest surveys carried out between 1983 and 2000 showed that ape populations had declined by half, largely due to bushmeat hunting and Ebola hemorrhagic fever outbreaks in 1994 and 1996.[28]

The Minkébé NP in northeast Gabon illustrates the impact of this viral disease. Pre-1994 data, the high frequency of encounters with gorillas by researchers in this forest during the 1980s, and 1990 estimates indicated healthy populations, but later surveys found few gorillas, despite the large area of suitable habitat and absence of obvious human disturbance.[14] The mean number of nests per nesting site declined from 6.0 in 1990 to 2.0 in 1998–2000, and the number of sites with only one nest increased from 20 percent to 60 percent in the same time period.[14] This population collapse has been attributed to Ebola, which has caused significant mortality in both gorillas and chimpanzees. The problem was originally restricted to the forest of Minkébé and adjacent parts of Congo, and it was hoped that the rivers bordering the area served as natural barriers to infected animals.[14] The apparent continued spread of Ebola in Congo in 2004, however, makes this view optimistic.[12]

Meanwhile, the timber industry is second only to the oil sector as an economic activity in Gabon. Initially, logging was concentrated along the coastal areas, but the opening up of the TransGabonais railway crossing the country from east to west has facilitated accessibility to formerly remote parts of central Gabon.[6, 17] The concession area for logging increased sevenfold between 1957 (16 000 km²) and 1999 (119 000 km²). Two thirds of Gabon's forests were logged during this period,[6] using selective logging methods that destroy 5–30 percent of the forest canopy.[17] Okoumé (*Aucoumea klaineana*, Burseraceae) is one of the most common tree species in Gabon and is widely exploited for commercial purposes, including the production of plywood.[17] Most forests where the species is found have been allocated as logging concessions. Though the government has expanded the protected area system, logging concessions were granted within reserves in the past.[6] The government now seems committed, however, to enforcing logging bans in the national parks.

Plans to log a concession that had been granted within the Lopé Reserve emerged in 1996, but were challenged, particularly by the European Union program, Conservation and Rational Use of Forest Ecosystems in Central Africa (ECOFAC), which was investing in conservation at Lopé. In 1997, the government of Gabon agreed to define a well protected core area of the reserve, and in 2000 it allocated a less biologically rich area to a logging company and added a smaller area of old growth forest originally slated for logging to the core.[6, 21] Lopé has since become a national park.

The industry has created two contributory factors for the development of the bushmeat trade in Gabon. First, its employees consume a large amount of bushmeat. An estimated 1 200 employees consumed up to 80 tons of bushmeat per year (67 kg per person per year) in a logging camp near Lopé in central Gabon.[6] In these circumstances, hunters have a large and regular clientele, and can make a transition from subsistence to profitable commercial hunting. Second, the improved infrastructure created by the logging companies has also made Gabon's urban markets more accessible. Hence, bushmeat consumption is a serious threat to ape populations in Gabon, against which the laws protecting both species have little weight, being scarcely enforced. Moreover, although some people in the southwest of the country are averse to eating ape meat, it is considered a delicacy elsewhere and is much sought after, especially in urban areas.[28]

Human encroachment and habitat fragmentation has also affected the social behavior of chimpanzee populations in Gabon. Research by White, carried out at Lopé before the increase in protection status, indicates that logging is asso-

Table 16.3 National parks of Gabon[18, 25, 26, 27, 31]

Park/reserve	Area (km²)	Percent terrestrial	Great apes present
Lopé	4 849	100.0	Gorillas and chimpanzees
Minkébé	7 567	100.0	Gorillas and chimpanzees
Pongara	930	84.3	Gorillas (?) and chimpanzees
Akanda	538	55.7	Chimpanzees
Waka	1 069	100.0	Gorillas and chimpanzees
Birougou	680	100.0	Gorillas and chimpanzees
Plateaux Batéké	2 049	100.0	Chimpanzees (orphan gorillas)
Ivindo (including the M'passa- (I'passa-) Makokou Biosphere Reserve)	3 003	100.0	Gorillas and chimpanzees
Monts de Cristal	1 196	100.0	Gorillas and chimpanzees
Mwagne	1 165	100.0	Gorillas and chimpanzees
Mayumba	972	6.5	Gorillas and chimpanzees
Gamba complex[a]	5 672	100.0	Gorillas and chimpanzees
Loango (comprising the former Iguela and Petit Loango Reserves)	1 152	100.0	Gorillas and chimpanzees
Moukalaba Doudou	4 496	100.0	Gorillas and chimpanzees
Wonga Wongué[b]	5 500	100.0	Gorillas and chimpanzees

a The Gamba complex is made up of the Loango NP and Moukalaba Doudou NP, with a reserve matrix between them.
b Presidential Reserve

Map 16.9 Great ape distribution in Gabon

Data sources are provided at the end of this country profile

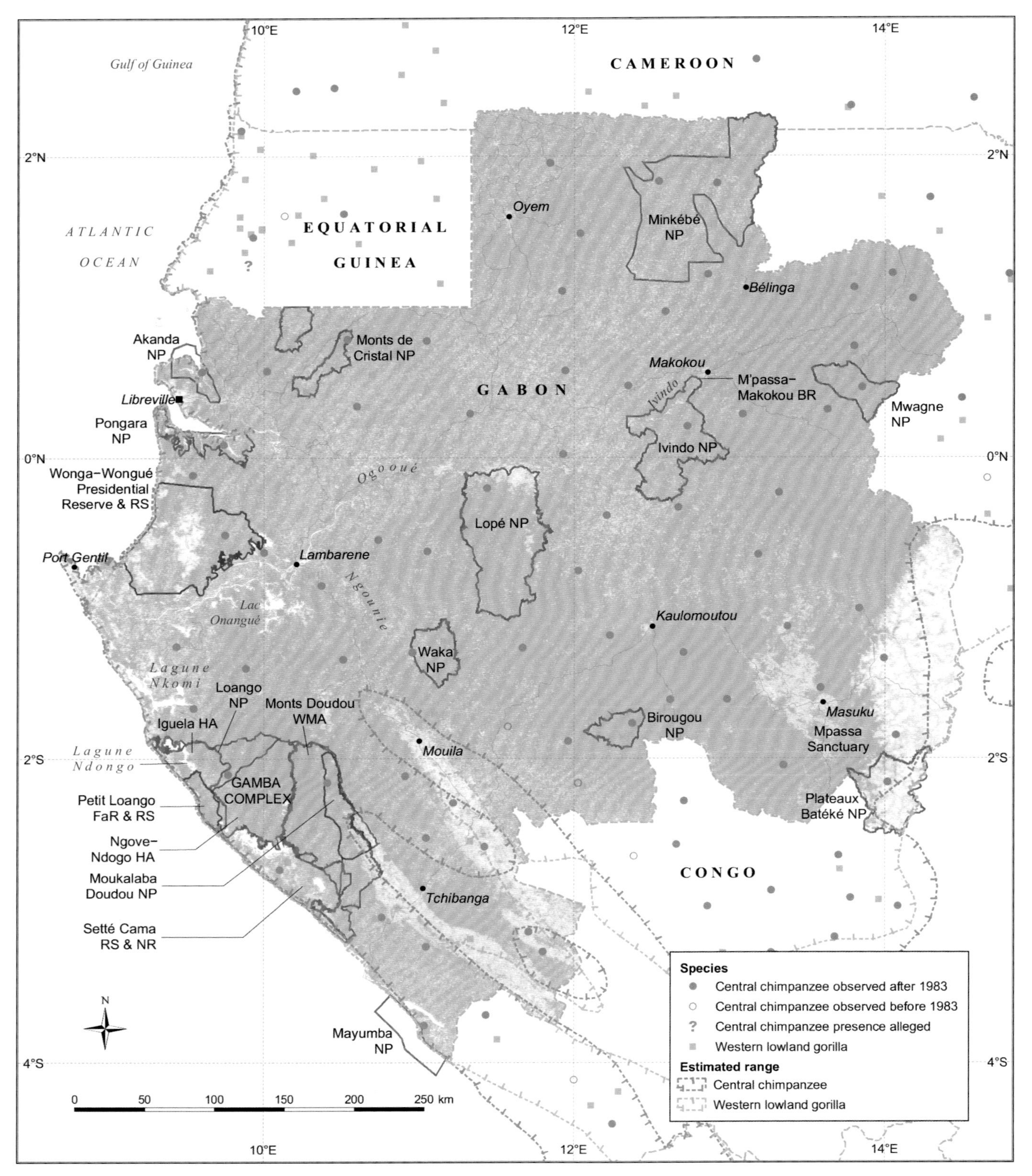

ciated with territorial conflicts among chimpanzees in which four out of every five chimpanzees may die.[17, 23]

LEGISLATION AND CONSERVATION ACTION

Legislation

Gabon has acceded to or ratified the Convention on Biological Diversity (1997), UN Framework Convention on Climate Change (1997), Convention on Wetlands of International Importance (Ramsar) (1987), and Convention on International Trade in Endangered Species of Wild Fauna and Flora (CITES) (1989), and is a party to both the 1983 International Tropical Timber Agreement and the 1994 International Tropical Timber Agreement.

The Gabon Wildlife and Forestry Policy (1992) demonstrates a political determination to undertake the planning and adequate ecological management of wildlife resources, the promotion of natural resources, and to play a leading role in biodiversity conservation. Recent efforts to ensure the sustainable management of natural resources have produced a protected area system totaling 40 000 km^2, and the designation of 80 000 km^2 of production forests and 100 000 km^2 of rural community forests,[5] as well as adherence to both the Brazzaville Process (Conference on Central African Moist Forest Ecosystems) and the Congo Basin Forest Partnership Convergence Plan.

Institutional and legal arrangements for protection and taxation in the forestry and wildlife management sectors have been reviewed. Under Law 1746/PR/MEFCR, Gabon has set up a wildlife management service and an antipoaching service. Gabon subscribes to subregional, regional, and international agreements including the Central Africa Protected Areas Network (RAPAC), the Brazzaville Declaration, and the Yaoundé Declaration.[5]

The former conservation law of 1982 no longer applies, having recently been replaced by a new Forestry Code. Gorillas and chimpanzees are now fully protected species under Gabonese law, having had temporary protected status since 1981 under the previous law.[15]

The Gabonese Ministry of Water and Forests is responsible for the management of natural resources in Gabon.

Protected areas

Gabon now has 13 national parks covering almost 11 percent of the country, and all containing at least one great ape species.[29, 30] The Setté Cama Nature Reserve (2 000 km^2) also contains chimpanzees.

International partnerships

The international conservation and sustainable development community is supporting a number of projects in partnership with the government of Gabon. Some of these are outlined here.

- ECOFAC has several ecotourism development projects, including a gorilla habituation program, and has supported the management and development of the Lopé NP since 1992.[17]
- WWF–The Global Conservation Organization has a Gabon branch, whose objectives in 2003 included policy implementation support to the Ministry of Water and Forests, supporting the management of Minkébé and Gamba areas, and involvement in a new EU–WWF network of partnerships on sustainable forestry.[32]
- WWF is also involved in extensive protection efforts with the government of Gabon. Since 1997, with funds from Netherlands Development Cooperation (DGIS) and the US Agency for International Development (USAID), WWF has been executing the Minkébé Conservation Project together with its main partner, the Gabonese Ministry of Water and Forests. They gazetted 5 665.5 km^2 of the Minkébé tropical moist forest as the main focus of a transborder complex of protected areas between Gabon, Congo, and Cameroon.[32]
- The integrated conservation, rational exploitation, and development program in the Gamba Protected Area Complex (Loango and Moukalaba Doudou NPs) has funding from US and German agencies.[19, 32]
- The long-term research center, Station d'Etudes des Gorilles et Chimpanzés (SEGC) in Lopé NP investigates various aspects of the forest ecosystem's ecology and dynamics as well as working on gorillas and chimpanzees.
- Global Forest Watch works with local organizations to collect and distribute information on forest developments and logging impacts.[2]
- The Wildlife Conservation Society (WCS) has been active in Gabon since 1985 and currently runs a major country program that includes activities in all the national parks, institutional support for the National Council for National Parks (CNPN) and the Ministry of Water and Forests, nationwide monitoring of the

bushmeat trade, new protected areas planning, and reduced impact logging. WCS also co-funds the SEGC long-term research center in Lopé and runs a training center in Lopé, used by both the forestry school and the national university.[31]

- The Institut de Recherche en Ecologie Tropicale (IRET), in the 100 km² Mpassa Biosphere Reserve near Makokou, was established in 1961 and is the oldest field research station in the country. It has been recently refurbished with funds from the EU and is equipped to receive students and scientists. The research station and Mpassa Reserve, now within the newly created Ivindo National Park, are an integral part of the regional and national conservation network.

Sanctuaries

The Projet Protection des Gorilles (PPG) in the Plateaux Batéké NP is supported by the John Aspinall Foundation (JAF), and aims to release orphan gorillas into the wild at the Mpassa Sanctuary on the Batéké Plateau.[7] It is currently holding about 20 gorillas for this purpose. The Petit Evengué program run by Operation Loango has six gorillas.[31] The Société d'Exploitation du Parc de la Lékédi (SODEPAL) has also provided sanctuary for chimpanzees and gorillas.[20]

FUTURE CONSERVATION STRATEGIES

Overview

The creation of 13 national parks in 2002 was in line with the priorities for Gabon of IUCN–The World Conservation Union, as recommended by its Tropical Forest Programme and Primate Specialist Group. The overall priority now is to develop professional management capacity in all of these national parks. Other priorities include legislative enforcement and the development of a national strategy for the use of wildlife resources. Priority actions to halt the decline of great apes in Gabon include the promotion of awareness on hunting and its implications, increased understanding and capacity to manage protected areas and the bushmeat trade, and further species population surveys.

Bushmeat trade and use

The capacity of wildlife departments needs to be strengthened, alternative sources of protein need to be explored, and laws governing the conduct of logging companies need to be applied. A bushmeat workshop was organized in 2002 by the Ministry of Water and Forests, and placed an emphasis on education, training, and awareness-raising, both for the general public via information campaigns, and for staff working in wildlife management.[22] Its specific recommendations included resource evaluation, the creation of an office for wildlife management, employment of provincial guards, creation of a joint action plan, and identification of funding sources.

Ebola virus

Since the Ebola virus poses such a threat to apes in Gabon, an effective plan to combat it is an urgent priority. Such a plan would need to address intensified research on reservoirs and hosts of the disease, including their ecologies, as well as on vaccines and their possible modes of deployment.[28]

FURTHER READING

Collomb, J.G., Mikissa, J.B., Minnemeyer, S., Mundunga, S., Nzao Nzao, H., Mapaga, J., Mikolo, C., Rabenkogo, N., Akagah, S., Bayani-Ngoye, E., Mofouma, A. (2000) *A First Look at Logging in Gabon.* Global Forest Watch, World Resources Institute, Washington, DC. http://www.globalforestwatch.org/common/gabon/english/report.pdf.

Tutin, C.E.G. (1999) Fragmented living: behavioural ecology of primates in a forest fragment in the Lope Reserve, Gabon. *Primates* **40** (1): 249–265.

Tutin, C.E.G., Fernandez, M. (1983) Gorilla and Chimpanzee census in Gabon. *IUCN/SSC Primate Specialist Group Newsletter* **3**: 22–23.

MAP DATA SOURCES

Map 16.9 Great apes data are based on the following source:

Butynski, T.M. (2001) Africa's great apes. In: Beck, B.B., Stoinski, T.S., Hutchins, M., Maple, T.L., Norton, B., Rowan, A., Stevens, E.F., Arluke, A., eds, *Great Apes and Humans: The Ethics of Coexistence.* Smithsonian Institution Press, Washington, DC. pp. 3–56.

With additional data by personal communication from Courage, A. (2004) and from the following sources:

Barnes, R.F.W., Jansen, K.L. (1987) *Forest Elephant Survey, Progress Report 1986–1987.* Report for Wildlife Conservation International (NYZS). Conservation International, New York.

Blom, A., Aler, P.T., Feistner, A.T.C., Barnes, R.F.W., Barnes, K.L. (1992) Primates in Gabon: current status and distribution. *Oryx* **26** (4): 223–234.

IUCN Conservation Monitoring Centre (1985) *The IUCN Directory of Afrotropical Protected Areas.* IUCN, Cambridge, UK and Gland, Switzerland.

Tutin, C.E.G., Fernandez, M. (1984) Nation-wide census of Gorilla (*Gorilla g. gorilla*) and Chimpanzee (*Pan t. troglodytes*) populations in Gabon. *American Journal of Primatology* **6**: 313–336.

White, R.J.T. (1994) Biomass of rain forest mammals in the Lopé Reserve, Gabon. *Journal of Animal Ecology* **57**: 345–367.

For protected area and other data, see 'Using the maps'.

ACKNOWLEDGMENTS

Many thanks to John Ady (UNEP-WCMC), Allard Blom (WWF-US), Sally Lahm (University of California, San Diego), and Lee White (Wildlife Conservation Society) for their valuable comments on the draft of this section.

AUTHORS

Ambrose Kirui, UNEP World Conservation Monitoring Centre
Lera Miles, UNEP World Conservation Monitoring Centre
Julian Caldecott, UNEP World Conservation Monitoring Centre

REPUBLIC OF GHANA

EDMUND MCMANUS

BACKGROUND AND ECONOMY

The Republic of Ghana is situated on the southern West African coastline, bordered by Togo to the east, Burkina Faso to the north, and Côte d'Ivoire to the west. It has a land area of 239 460 km^2, and is composed mostly of low plains, with a plateau in the south-central area. The climate is generally warm, with a strong moisture gradient. The southwestern part of the country has many moist forest fragments, which become progressively drier towards the north. This dry forest gives way to savanna bush in the more northern Brong-Ahafo region. In the southwestern areas where most chimpanzees are found (Map 16.10), two rainy seasons occur, from May to July and from September to October.[15]

Ghana has a culturally diverse population, estimated in 2003 to be about 20.3 million.[10] The area was once home to the powerful Ashanti kingdom, which resisted the expansion of the British colony of the Gold Coast through a series of wars in the 19th century. In 1901, the Ashanti kingdom and Northern Territories Protectorate were amalgamated into the Gold Coast, and neighboring German Togoland was placed under Gold Coast administration in 1919. In 1957, the Gold Coast became Ghana, the first African state to achieve full independence from the UK.

Ghana's first ruler, Kwame Nkrumah, was the pioneer of 'African socialism' and gradually developed close ties with the Soviet bloc.[10] He was overthrown by a military coup in 1966, the first of several that ended with the ascendancy of Flight Lieutenant Jerry Rawlings in the early 1980s. Rawlings seized power by force, assumed chairmanship of the ruling Provisional National Defense Council, abolished the constitution, dissolved parliament, and outlawed the opposition political parties. The ban on opposition parties was lifted by the mid-1980s, and a new constitution was drawn up and approved by referendum in 1992. The first two presidential elections, in 1992 and 1996, were endorsed by international observers, and returned President Rawlings to power. He stood down in 2000 after a constitutionally limited two terms of office, and was replaced through the election of opposition leader John Kufuor. In December 2004, Kufuor was re-elected for a second four year term of office.

The international community perceives Ghana as very much a success story, in terms of its pluralistic political system and smooth transitions of power.[7, 10] The Kufuor administration, however, inherited an economy in decline, and took strong measures that included the raising of fuel duties by

Map 16.10 Chimpanzee distribution in Ghana

Data sources are provided at the end of this country profile

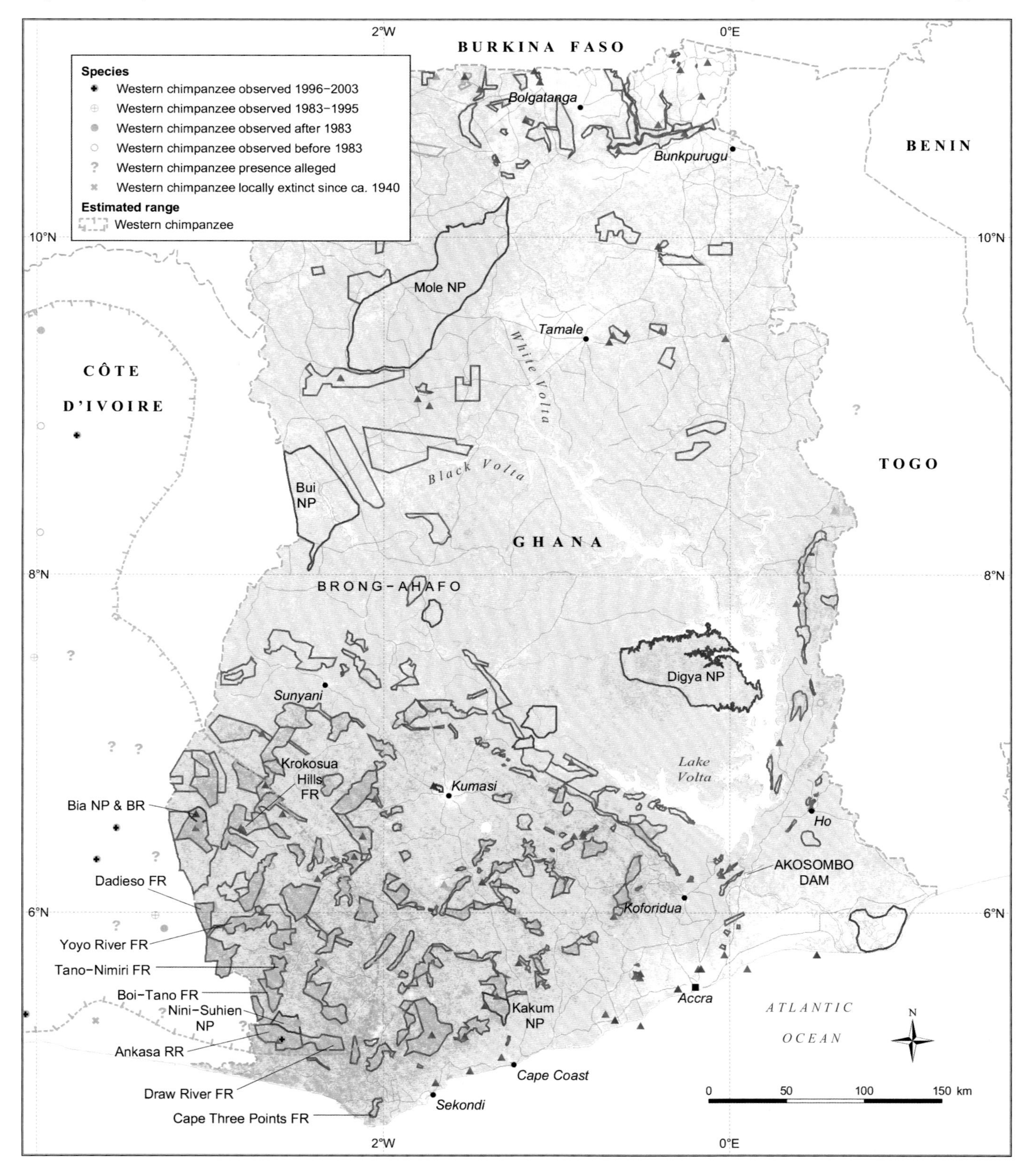

over 90 percent.[10] In 2001, a recovery in gold and cocoa prices helped Ghana attain macroeconomic stability, and in 2002, the country was granted 'heavily indebted poor country' (HIPC) status by the International Monetary Fund, making it eligible for certain forms of debt relief. Overall, poverty is declining, but unemployment remains high and food crop farmers remain vulnerable.[10] The government's economic and social policies are consistent with international development targets, but Ghana remains dependent on international financial and technical assistance.

About 57 percent of the country's land area is devoted to agriculture, particularly cocoa and oil palm plantations.[1] By 2004, subsistence agriculture accounted for 35 percent of gross domestic product (GDP) and employed 60 percent of the work force, mainly small landholders.[7] Some 300 000 people work as hunters.[1, 33] There is an estimated annual harvest of 385 000 tons of bushmeat (18.8 kg per person per year) worth about US$350 million, of which around 60 percent is sold in urban areas.[27] Gold, timber, and cocoa production are major sources of foreign exchange.

DISTRIBUTION OF GREAT APES

The only wild great apes found in Ghana are western chimpanzees (*Pan troglodytes verus*). They are limited to the southwest of the country, where they are probably found only in a few moist semi-deciduous and wet evergreen forest fragments. They may also occur in some moist evergreen and dry semideciduous forests, but this has not been confirmed.[22]

The most recent population estimate was made in 1995, when 1 500–2 200 chimpanzees were thought to be present in Ghana.[23] In 1979, there were thought to be fewer than 200;[17] by 1988, the estimate had risen to 300–500 chimpanzees.[31] These increases are a result of improved knowledge; the actual number of chimpanzees has almost certainly fallen over the period.

Extensive surveys were carried out in 1999 and 2001.[2, 20] Chimpanzees may still be present in the 78 km^2 Bia National Park (NP) and the 160 km^2 Nini-Suhien NP, but none were detected by these surveys in either location. Chimpanzee presence was confirmed in the Ankasa Resource Reserve (343 km^2),[22] Draw River Forest Reserve (235 km^2), and the Tano-Nimiri Forest Reserve (205 km^2). Chimpanzees are likely to occur in the Krokosua Hills (295 km^2) and Boi-Tano (128 km^2) Forest Reserves.[22] Chimpanzees may be locally extinct in several other forest areas including Kakum and Cape Three Points.[22] No chimpanzees were detected in the Yoyo River or Dadieso Forest Reserves, although anecdotal information from hunters suggests that they are present.[22]

THREATS

All things considered, it is possible that chimpanzees are nearly extinct in this country. Hunting is a significant threat to wildlife in Ghana,[20, 33] and it is estimated that 90 percent of Ghana's population eat bushmeat when they can.[26] Of the protected areas, Nini-Suhien NP, Bia NP, and Boi-Tano and Dadieso Forest Reserves have been subject to particularly heavy hunting pressure.[4, 22] Primates make up a small but significant fraction of the animals killed and traded: hunting pressure was a major factor behind the extinction of the Miss Waldron's red colobus monkey (*Procolobus badius waldroni*).[28] The pet trade is also thought to have affected the remaining wild populations of chimpanzees in Ghana.[24, 33]

In addition, Ghana lost much of its forested land in the 20th century.[8] Around 63 350 km^2 of forest was thought to remain by 2000, with an average annual loss of 1 200 km^2 (1.7 percent).[9] The major pressures on forest have been road construction, agriculture, and timber extraction, with mineral exploitation an emerging threat around the year 2000.[18, 19, 30] Demand for land and timber has led to major degradation of and encroachments on protected areas.[22]

LEGISLATION AND CONSERVATION ACTION

National laws and protected areas

Chimpanzees were wholly protected under the Wild Animals Preservation Act (Act 43 of 1961), and the Wildlife Conservation Regulations (1971) further strengthened the legal protection of chimpanzees. The Wildlife and Forest Policy, introduced in 1994, had the aim of promoting sustainable forest management.[25] Penalties for illegal hunting, logging, and other forest crimes were increased in 2002.[29]

Bia NP was gazetted in 1974 at 306 km^2, but was reduced to 78 km^2 in 1976, with the remainder of the park becoming a resource reserve with timber concessions.[3, 16] Bia was designated as a Biosphere Reserve in 1983.[16, 32] Nini-Suhien was designated as a national park in 1976.[16]

Most of the remaining closed forest is found in

the 18 000 km^2 of forest reserves,[1] which include production reserves where timber harvesting is permitted, and 4 500 km^2 of protective reserves where it is not.[12, 13] The protective reserves are intended to safeguard water bodies, areas of importance for biodiversity, and fragile ecosystems.[25] In addition, the new globally significant biodiversity area (GSBA) system is intended to prevent logging over much of the forest reserve area, mobilizing national and international resources in support of biodiversity conservation and alternative livelihoods. Local communities will receive Global Environment Facility (GEF) funds in exchange for not logging these areas.[11] Of those areas relevant to great ape conservation, the scheme covers Dadieso Forest Reserve, which has never been logged,[22] part of Yoyo River Forest Reserve (the remainder was mostly conceded for logging in 2001),[22] and parts of Draw River and Krokosua Hills Forest Reserves.

Sanctuaries and rehabilitation

A rehabilitation attempt concerning six chimpanzees in Bia NP failed in 1972.[5] A further attempt was initiated in 1994, but these animals are believed to have been killed by hunters.[22]

FUTURE CONSERVATION STRATEGIES

The *West African Chimpanzees: Status Survey and Conservation Action Plan* made the following recommendations for priorities in Ghana.[22]

- **Overall priorities.** Major efforts are needed to protect existing reserves, to improve forest management, and to collate basic ecological data.
- **Research and surveys.** Estimates are needed of chimpanzee numbers and distribution in the Ankasa Conservation Area (Ankasa Resource Reserve combined with the adjacent Nini-Suhien NP) and the Krokosua Hills Forest Reserve. Additional surveys in Yoyo River Forest Reserve, Bia NP, and Dadieso Forest Reserve would probably provide a sufficient baseline of chimpanzee status in Ghana. The results of these surveys could be used to develop a five year conservation program. All chimpanzee habitats should be mapped and recorded in a geographic information system (GIS) database to augment the development of management plans for chimpanzees and other wildlife.
- **Protection.** The Krokosua Hills Forest Reserve has been identified as the most likely site to support a sizeable chimpanzee population in Ghana (due to low human populations in and around the reserve), with chimpanzee sightings more frequent than in any other area in Ghana. This area should receive immediate capacity development in the form of training of park rangers and recruitment of local hunters as staff. In order to improve monitoring in protected areas there should also be a focus on training Wildlife Division staff in field identification of chimpanzees, and Ankasa Resource Reserve staff should complete their training.
- **Transfrontier conservation measures.** As the chimpanzee range in West Africa is continuous over many countries, joint and coordinated national cooperation and commitments to chimpanzee conservation will be needed. The Ghanaian government has stated that protocols for cooperating in the development of transboundary protected areas, including for Bia NP, are to be discussed.[14]

FURTHER READING

Grubb, P., Jones, T., Davies, A., Edberg, E., Starin, E., Hill, J. (1998) *Mammals of Ghana, Sierra Leone and The Gambia*. Trendrine Press, Zennor, UK.

Hall, J.B., Swaine, M.D. (1981) *Distribution and Ecology of Vascular Plants in a Tropical Rain Forest: Forest Vegetation in Ghana*. Dr W. Junk, The Hague.

Magnuson, L., Adu Nsiah, M., Kpelle, D. (2003) Ghana. In: Kormos, R., Boesch, C., Bakarr, M.I., Butynski, T.M., eds, *West African Chimpanzees: Status Survey and Conservation Action Plan*. IUCN/SSC Primate Specialist Group. IUCN, Gland, Switzerland. pp. 111–116.

Oates, J.F., Abedi-Lartey, M., McGraw, W.S., Struhsaker, T.T., Whitesides, G.H. (2000) Extinction of a West African red colobus monkey. *Conservation Biology* **14** (5): 1526–1532.

MAP DATA SOURCES

Map 16.10 Chimpanzee data are based on the following sources:

Butynski, T.M. (2003) The chimpanzee *Pan troglodytes*: taxonomy, distribution, abundance, and conservation status. In: Kormos, R., Boesch, C., Bakarr, M.I., Butynski, T.M., eds, *West African Chimpanzees: Status Survey and Conservation Action Plan.* IUCN/SSC Primate Specialist Group. IUCN, Gland, Switzerland. pp. 5–12.

Magnuson, L., Adu-Nsiah, M., Kpelle, D. (2003) Ghana. In: Kormos, R., Boesch, C., Bakarr, M.I., Butynski, T.M., eds, *West African Chimpanzees: Status Survey and Conservation Action Plan.* IUCN/SSC Primate Specialist Group. IUCN, Gland, Switzerland. pp. 111–116.

For protected area and other data, see 'Using the maps'.

ACKNOWLEDGMENTS

This country study draws extensively on the Ghana chapter from the IUCN/SSC *West African Chimpanzees: Status Survey and Conservation Action Plan.*[23] Many thanks to Lindsay Magnuson (College of the Redwoods) for her valuable comments on the draft of this section.

COMPILER

Edmund McManus, UNEP World Conservation Monitoring Centre

REPUBLIC OF GUINEA

MUHAMMAD AKHLAS

BACKGROUND AND ECONOMY

The Republic of Guinea is located on the West African coast, with Sierra Leone and Liberia to the south, Côte d'Ivoire and Mali to the east, and Senegal and Guinea-Bissau to the north, and has a land area of 245 857 km^2. Its climate is hot and humid but seasonal, with a rainy season from June to November and a dry season at other times, when the country is affected by dry, dusty harmattan winds off the Sahara. Guinea's terrain comprises a flat coastal plain with an undulating to mountainous interior. In 2000, total forest cover was estimated to be 69 290 km^2 or 28 percent of land area,[6] with about 17 percent (11 821 km^2) of it within 162 classified forests. In 2003, 3.6 percent of the terrestrial area was cultivated and 2.4 percent under permanent crops.[5] The remaining area is woodland savanna.

The population of Guinea was about 9.25 million people in 2004, growing at a rate of 2.4 percent annually.[5] The country's economy is based on a mixture of agriculture, which supports those people that live in rural areas, and mining. Guinea has more than 30 percent of the world's known bauxite (aluminum ore) and the mining sector accounts for about 75 percent of its exports by value.

The country achieved independence from France in 1958, and was ruled without elections by a military government until 1993, when the head of that government, General Lansana Conté, was elected president, with renewed mandates in 1998 and 2003. Unrest in Sierra Leone and Liberia has spilled over into Guinea on several occasions over the past decade, threatening stability and creating humanitarian emergencies. This has undermined investor confidence and the ability of the country to move out of poverty. The International Monetary Fund (IMF) and World Bank cut off most assistance in 2003, but economic activity was expected to strengthen in 2004 as the security situation started to improve.[5]

DISTRIBUTION OF GREAT APES

Guinea has about 15 species of primates, of which the western chimpanzee (*Pan troglodytes verus*) is the only great ape. There appears to be a reasonably healthy population that amounts to 36–51 percent of all western chimpanzees.[8, 13] It has therefore been suggested that the country may one day be the last stronghold of this subspecies,[14] but this will depend on the fate of the country's forests, and deforestation is an ongoing pressure in Guinea.

Chimpanzees are known to be widespread in Guinea. A questionnaire survey in the mid-1980s recorded their occurrence in 27–30 of the country's 34 prefectures.[20] A study published in 1998 confirmed their presence at 71 sites, and included a more detailed questionnaire survey that reported chimpanzees at 606 locations.[8] This led to an estimated national population of 8 113–29 011

Map 16.11 Chimpanzee distribution in Guinea

Data sources are provided at the end of this country profile

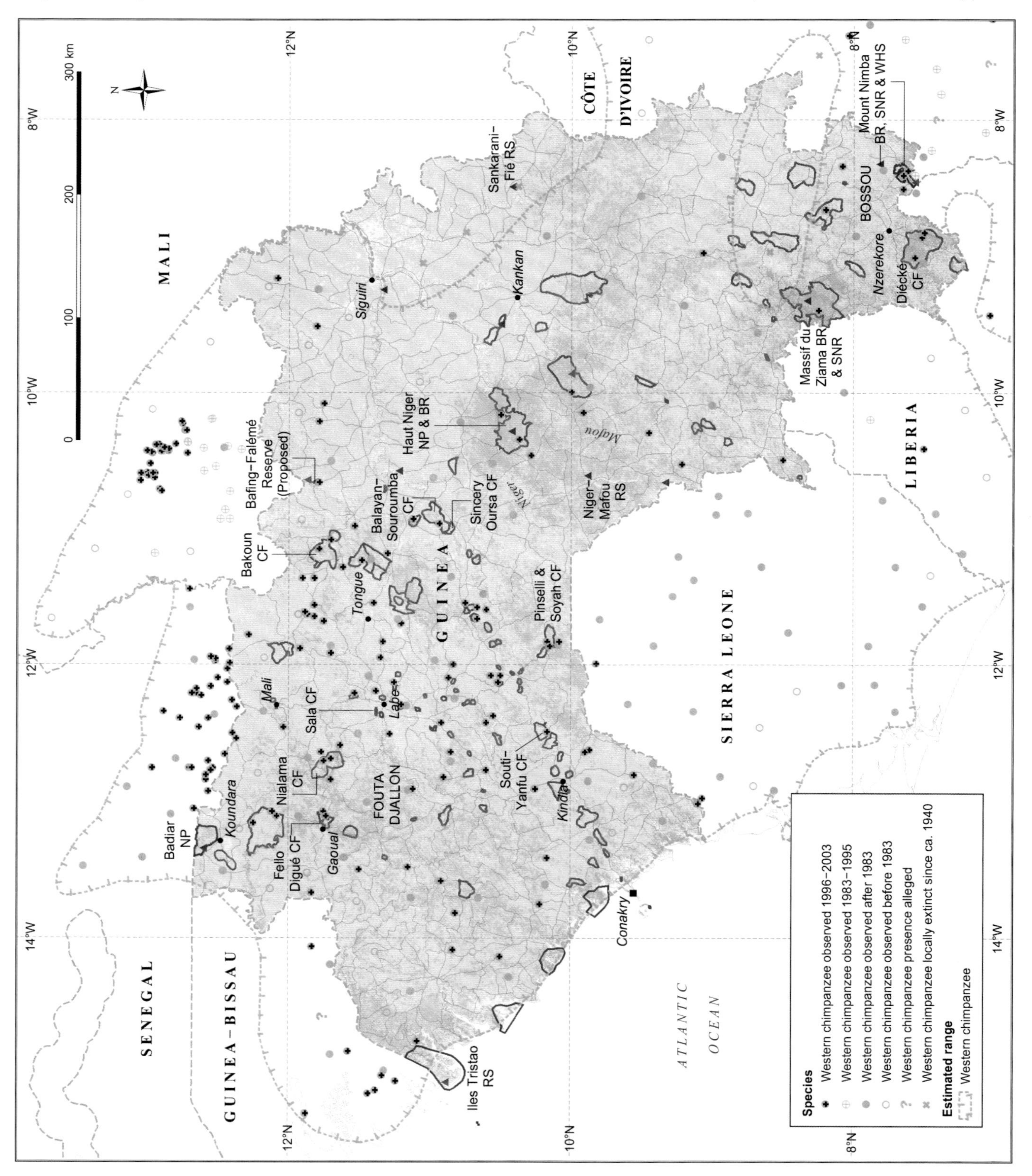

chimpanzees;[8] greater than the estimate of 1 420–6 625 individuals that arose from the 1980s survey. Earlier estimates had suggested that there were fewer than 12 500[22] and 8 000–10 000.[21] These figures were not generated using comparable techniques, and so it is not clear whether the chimpanzee population is stable.

Small populations are known to occur in forest areas throughout Guinea. Important sites include Mount Nimba Biosphere Reserve (including Bossou forest), Massif du Ziama Biosphere Reserve, Haut Niger National Park (NP), and the classified forests of Diécké, Nialama, Sala, Fello Digué, Balayan-Souroumba, Bakoun, and Souti Yanfou.[14] Population density can vary greatly at each site, for example at Haut Niger NP in 2001, where it ranged from 3.5/km^2 in gallery forests to 0.1/km^2 in wooded savannas.[7]

THREATS

Deforestation is being driven by population growth and agricultural expansion, aggravated by the use of fire to clear land in the dry season.[26] Selective logging is also a significant threat to chimpanzees, with a range of direct and indirect impacts on forest structure, connectivity, and species composition, and on chimpanzee ranging behavior, with logged forest being more likely to be cleared by settlers.[14, 16] Other pressures on chimpanzee habitats include mining and infrastructure development. The northern part of Mount Nimba Biosphere Reserve has been disturbed by iron-ore mining,[2] and the expansion of railway and road projects through protected areas has posed serious conservation problems.[23]

Hunting also affects Guinea's chimpanzee population. Commercial hunting for meat is widespread,[25] though not universal.[14] Farmers kill chimpanzees as crop pests, and they are sometimes hunted for the medicinal or magical properties of their body parts (for example, the blood is thought to cure epilepsy, and the meat is believed to strengthen young children).[8] Orphans are sold as pets. In the past, many chimpanzees were captured and shipped overseas for biomedical research.[14]

LEGISLATION AND CONSERVATION ACTION

Legislation

Guinea is party to the Convention on International Trade in Endangered Species of Wild Fauna and Flora (CITES). Guinea has also ratified the World Heritage Convention and the Convention for Co-operation in the Protection and Development of the Marine and Coastal Environment of the West and Central African Region (1981). Guinea has signed but not yet ratified the African Convention on the Conservation of Nature and Natural Resources.[1]

Chimpanzees are 'integrally protected' under Guinea's law governing the use of wildlife,[18] and may not be hunted, captured, detained, or exported without a scientific permit obtained from the government. Legislation typically has little impact, however, unless it is consistent with the prevailing culture, and only in some areas of Guinea, such as the Fouta Djallon area, are local people traditionally averse to the hunting of chimpanzees.[13, 14] These particular areas are relatively densely peopled, enabling a positive relationship between human and chimpanzee populations.[12]

Protected areas

There are three protected areas in Guinea containing chimpanzees. In the southeast, there are two strict nature reserves: Mount Nimba (130 km^2), designated as a Biosphere Reserve and a World Heritage Site in 1981, and the Massif du Ziama (1 123 km^2), a Biosphere Reserve since 1980. In central Guinea there is the Haut Niger National Park (NP) (6 000 km^2), a Biosphere Reserve since 2002. The core zone of Haut Niger NP, Mafou forest, makes up about 10 percent, with the remainder being buffer zones.[26] The Haut Niger NP has received significant funding and management support through the European Union (EU), but the other protected areas suffer from lack of management and resources.

Sacred sites

A number of areas in Guinea are protected for religious reasons; the best known of these is Bossou. This site consists of several small sacred hills, situated within a farming region of small villages and fields. Yukimaru Sugiyama and colleagues at the Primate Research Institute of Kyoto University have been researching chimpanzee socioecology, life history, demography, and tool use at Bossou since 1976.[24] The chimpanzees here are not completely isolated from others, for occasional visitors have been seen and maturing individuals have vanished. Bossou is one of the few sites at which there is evidence of male intercommunity transfer among chimpanzees.[11, 19] The nearest neighboring population is in the Nimba Mountains, 6–10 km away, and there are efforts to develop a forest corridor to link these areas (the Green Corridor Project).[9]

Education and sanctuaries

The Chimpanzee Protection Project (PCC) was funded by the EU in 1995–1999 and directed by Janis Carter. Major PCC objectives included the involvement of local people, including hunters, in chimpanzee population monitoring and village-scale environmental education, focusing on the areas around Bakoun and Nialama Classified Forests. Subsequently, Carter was able to continue certain education and long-term monitoring activities, supported by funds from the US Agency for International Development (USAID) and Friends of the Animals.

The educational activities conducted during the PCC and onwards included slide programs and the production of posters, brochures, and bumper stickers addressing the need to conserve and protect chimpanzees, the current legislation protecting them, and the impact of purchasing an orphaned infant. Radio programs were also produced in collaboration with Guinée Ecologie, a national nongovernmental organization involved in capacity building and awareness raising. A brochure entitled *Appel de Détresse* (Distress Call) was produced and distributed throughout Guinea, particularly in the prefecture of Pita where all schools and *sous-préfectures* were visited, educational materials distributed, meetings with elders held, and slide programs shown. In 2004, *Appel de Détresse* was being revised and translated into various local languages under a US Fish and Wildlife Services (USFWS) grant.

USAID and Winrock International later integrated the PCC education and monitoring components into their forest co-management program.[3, 4, 15] Chimpanzees are now monitored by local residents in the five co-managed forests of Bakoun, Nialama, Sincery Oursa, Balayan-Souroumba, and Souti Yanfou, and similar educational activities are ongoing in these areas.[4]

In the Bossou and Nimba region, environmental education activities coordinated by the Kyoto University Primate Research Institute have been ongoing since 1993. Since 2003, books aimed at raising awareness about chimpanzees and the environment have been donated to 16 schools. Environmental education sessions have been running in nine villages in the area of the Mount Nimba Biosphere Reserve, while pamphlets, badges, and videos have been distributed more widely. These activities have been financially supported by USFWS since 2003.[10]

In Fouta Djallon, the Wild Chimpanzee Foundation (WCF) has a program of education and awareness raising for chimpanzee conservation. Environmental education activities were scheduled to start in early 2005, and to include dramatic performances and distribution of newsletters.

Orphan chimpanzees are received for care and rehabilitation by the Centre de Conservation pour Chimpanzés (CCC), located near the village of Somoria within the Haut Niger NP, and supported by the US-based Project Primate. By November 2004, the sanctuary had taken in 37 chimpanzees.[17]

Tatyana Humle

Groundwork for the 'Green Corridor' linking Bossou to Mount Nimba.

FUTURE CONSERVATION STRATEGIES

In October 2004, a National Great Apes Survival Plan (NGASP) Workshop was held in the capital, Conakry. The NGASP is based on a previous workshop held by Conservation International in September 2002 and on the subsequent action plan of IUCN–The World Conservation Union for western chimpanzees.[12] This indicates the following priorities for conservation in Guinea.

- **Research and surveys.** There is a need for comprehensive information on the present status, number, distribution, threats, and conservation measures in Guinea, and on the nature of and potential for resolving competition between humans and chimpanzees over natural resources. Studies are needed to clarify aspects of chimpanzee behavior and social ecology, and their populations in various habitats; to identify practical ways to reconnect fragmented habitats through natural forest corridors; and to identify priorities for doing so.

- **Protected areas.** Priority areas include Fouta Djallon, Nimba Mountains, and Haut Niger. Comprehensive steps are needed to improve the standards of protection, planning, and management of existing protected areas, calling on the support of the international community and the involvement of government and local communities.
- **Peace and security.** International organizations may need to be involved to help resolve border issues and increase security in transboundary protected areas.
- **Capacity building.** There is a need to strengthen the capacity of protected area and wildlife management staff to implement national and international legislation.
- **Education and tourism.** Further education and awareness-raising programs regarding chimpanzee hunting, bushmeat, and the pet trade are needed among communities. Suggested educational targets in Guinea include schoolchildren and the military. Chimpanzee and wildlife-oriented tourism may be possible in some areas, but should be carefully regulated. Chimpanzee tourism should only be promoted if the study of individual populations leads to a positive assessment that they will not suffer negative impacts from habituation or undergo serious risks of disease transmission. Attention must be paid to the distribution of revenue, to ensure that chimpanzee conservation and local populations benefit. Finally, sanctuaries with secure financing are needed for the rehabilitation and educational use of orphaned and confiscated chimpanzees.
- **Development.** Environmental impact assessments are needed prior to the initiation of new mining or timber extractive activities in chimpanzee habitat, and guidelines to minimize impacts are required.

Matsuzawa Tetsuro

Chimpanzees at Bossou use a pair of stones as hammer and anvil to crack open oil palm nuts.

FURTHER READING

Barnett, A.A., Prangley, M.L. (1997) Mammalogy in the Republic of Guinea: an overview of research from 1946–1996, a preliminary check-list and a summary of research recommendations for the future. *Mammal Review* **27** (3): 115–164.

Humle, T., Matsuzawa, T. (2001) Behavioural diversity among the wild chimpanzee populations of Bossou and neighbouring areas, Guinea and Côte d'Ivoire, West Africa: a preliminary report. *Folia Primatologica* **72** (2): 57–68.

Kormos, R., Boesch, C., Bakarr, M.I., Butynski, T.M., eds (2003) *West African Chimpanzees: Status Survey and Conservation Action Plan.* IUCN/SSC Primate Specialist Group. IUCN, Gland, Switzerland.

Sugiyama, Y., Soumah, A. (1988) Preliminary survey of the distribution and population of chimpanzees in the Republic of Guinea. *Primates* **29**: 569–574.

Yamakoshi, G., Takemoto, H., Matsuzawa, T., Sugiyama, Y. (1999) Research history and conservation status of chimpanzees at Bossou, Guinea. *Primate Research* **15** (2): 101–114.

Ziegler, S., Nikolaus, G., Hutterer, R. (2002) High mammalian diversity in the newly established National Park of Upper Niger, Republic of Guinea. *Oryx* **36** (1): 73–80.

MAP DATA SOURCES

Map 16.11 Chimpanzee data are based on the following sources:

Butynski, T.M. (2001) Africa's great apes. In: Beck, B.B., Stoinski, T.S., Hutchins, M., Maple, T.L., Norton, B., Rowan, A., Stevens, E.F., Arluke, A., eds, *Great Apes and Humans: The Ethics of Coexistence.* Smithsonian Institution Press, Washington, DC. pp. 3–56.

Butynski, T.M. (2003) The chimpanzee *Pan troglodytes*: taxonomy, distribution, abundance, and conservation status. In: Kormos, R., Boesch, C., Bakarr, M.I., Butynski, T.M., eds, *West African Chimpanzees: Status Survey and Conservation Action Plan.* IUCN/SSC Primate Specialist Group. IUCN, Gland, Switzerland. pp. 5–12.

Kormos, R., Humle, T., Brugière, D., Fleury-Brugière, M-C., Matsuzawa, T., Sugiyama, Y., Carter, J., Diallo, M.S., Sagno, C., Tounkara, E.O. (2003) The Republic of Guinea. In: Kormos, R., Boesch, C., Bakarr, M.I., Butynski, T.M., eds, *West African Chimpanzees: Status Survey and Conservation Action Plan.* IUCN/SSC Primate Specialist Group. IUCN, Gland, Switzerland. pp. 63–76.

For protected area and other data, see 'Using the maps'.

ACKNOWLEDGMENTS

This country profile draws extensively on the Guinea chapter from the IUCN/SSC *West African Chimpanzees: Status Survey and Conservation Action Plan.*[12] Many thanks to four anonymous reviewers for their valuable comments on the draft of this section.

COMPILER

Muhammad Akhlas, University of East Anglia

REPUBLIC OF GUINEA-BISSAU

CLÁUDIA SOUSA, SPARTACO GIPPOLITI, AND MUHAMMAD AKHLAS

BACKGROUND AND ECONOMY

The Republic of Guinea-Bissau is one of the smallest countries on the Atlantic coast of West Africa, sandwiched between Senegal to the north and Guinea to the south and east. With an area of 36 125 km^2, it includes a number of small offshore islands – the Bijagos archipelago – that are separated from the mainland by wide intertidal mudflats. Guinea-Bissau's population, which includes about 20 ethnolinguistic groups, was approximately 2.4 million in 2004, and was growing at about 2 percent annually.[3]

After independence from Portugal in 1974, Guinea-Bissau established a one-party system and a centrally planned economy. A military coup in 1980 established a new system with a more pro-market stance, which won a mandate in the country's first elections in 1994, but there were repeated coup attempts through the 1980s and 1990s, one of which led to civil war in 1998. Intermittent fighting between Senegalese-backed government troops and a military junta destroyed much of the country's infrastructure and caused considerable damage to the economy. A brief return to democracy in 2000–2002 ended with another coup in September 2003 that installed the current government.

Like the other former Portuguese colonies that were abandoned to independence without much preparation in the mid-1970s (Angola, Mozambique, and East Timor) the country suffered terribly, and is now deeply impoverished. It depends mainly on fishing and farming. Cashew nut production is increasing, and most foreign exchange comes from the export of fish and seafood along with relatively small amounts of peanuts, palm kernels, and timber. Rice is the major crop and staple food. Offshore oil reserves could provide much-needed revenue in the long run, but are currently unexploited.[3]

Sixty percent (21 870 km^2) of Guinea-Bissau was forested in 2000, mainly with natural broadleaf humid or semidry forests. The country has the largest area of mangroves and coastal flats in Africa; originally 11 percent of the country was covered with mangroves.[9]

DISTRIBUTION OF GREAT APES

Guinea-Bissau has 11 species of wild primates, of which the western chimpanzee (*Pan troglodytes verus*) is the only great ape.[5, 13] Due to the very limited survey data, it is uncertain how many chimpanzees are found in Guinea-Bissau, but estimates range from 600 to 1 000 individuals.[7, 8] Until 1989, chimpanzees were thought to be extinct in the country,[11] but they were confirmed to be present after a comprehensive wildlife inventory was undertaken by Guinea-Bissau's Direção General das Florestas e Caça (DGFC) and the Canadian Centre

Map 16.12 Chimpanzee distribution in Guinea-Bissau

Data sources are provided at the end of this country profile

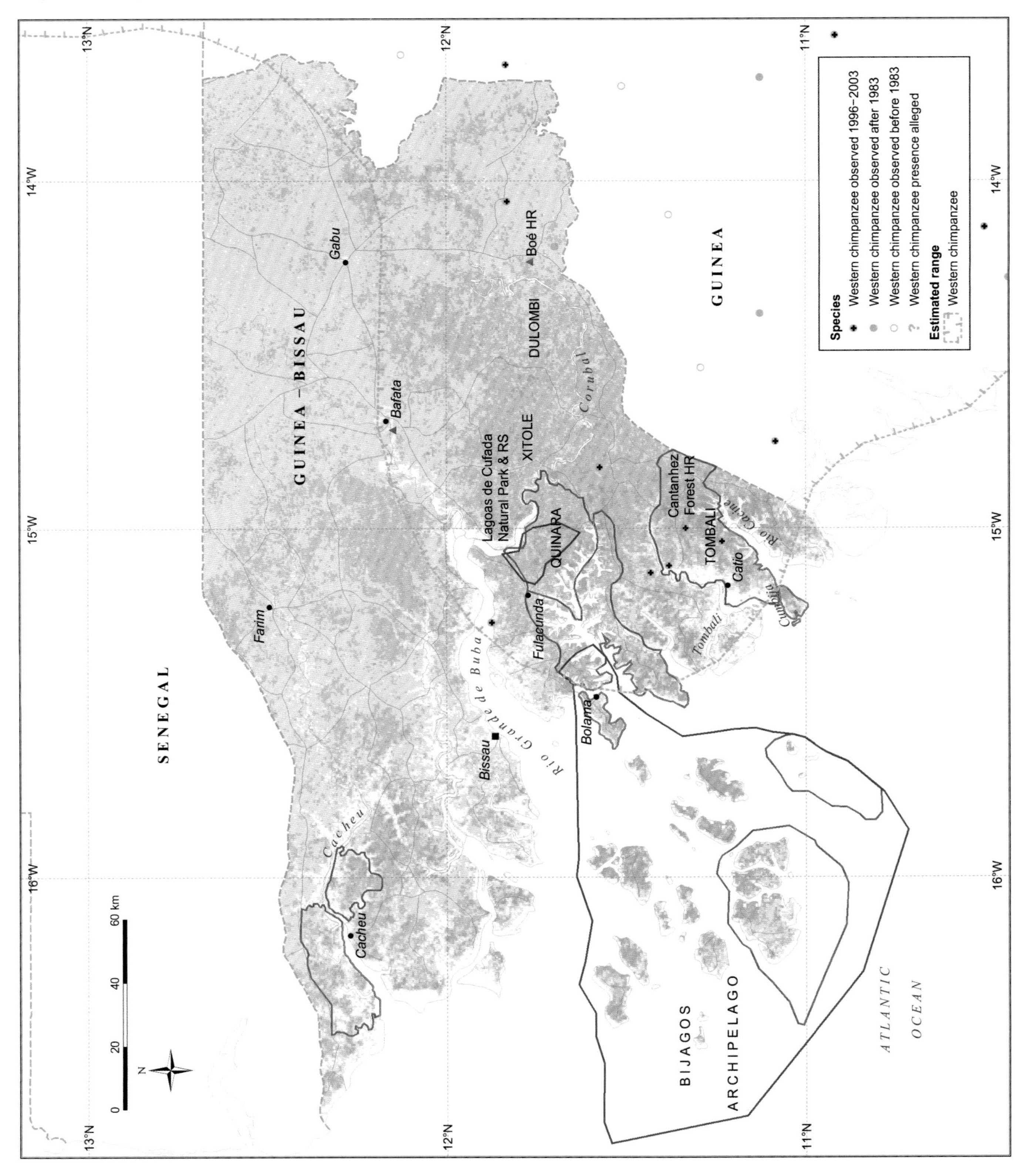

Cláudia Sousa

Cláudia Sousa

Left: Mangroves in Lagoas de Cufada Natural Park.
Right: Forest destroyed for cultivation.

for International Studies and Cooperation (CECI) with financial support from IUCN–The World Conservation Union.[12] More recent studies suggested their presence in the region of Xitole (an area once proposed as a national park to the north of the Corubal River), the Lagoas de Cufada Natural Park (a Ramsar Site), the North Bank of Rio Grande de Buba, Cantanhez Forest Hunting Reserve, and the Cacine Basin.[5, 6, 7, 8] The range is believed to extend through the country to the south of the Corubal River,[8] specifically in the Boé sector, between the Corubal River and the Guinea border, and in the southeastern regions of Quinara and Tombali.[7, 8]

THREATS

The major threat seems to be the destruction of chimpanzee habitat, especially primary forest.[8] Between 1990 and 2000, an estimated 220 km^2 of forest was lost each year.[4] Most land-use change within chimpanzee habitat is linked to local human population increase. Failure to take ecological constraints into account in the National Development Plan has also been identified as a leading factor.[9] Pressures include timber exploitation, bushfires, clearing for agriculture, fruit farming, and clearing of mangroves for rice cultivation. This is most common in the Tombali and Quinara regions. The Cantanhez Forest Hunting Reserve (Tombali region) is becoming seriously fragmented by banana, cashew, and other plantations. Consequently, crop-raiding by chimpanzees has also increased.

Chimpanzees are not generally eaten in Guinea-Bissau as they are considered too similar to humans. The young are sometimes taken for the local pet trade and chimpanzee skins are used in traditional medicine.[6] Accidental capture of chimpanzees in snares set for game animals such as duikers and other forest ungulates is also a threat.[6]

LEGISLATION AND CONSERVATION ACTION

Guinea-Bissau has signed the Convention on Biological Diversity (1995), the Convention on International Trade in Endangered Species of Wild Fauna and Flora (CITES) (1990), the Ramsar Convention (1991), and the Convention on Migratory Species (1995).

Chimpanzees are protected from hunting under Decree No. 21/1980. In addition, all hunting is prohibited in hunting reserves. Guinea-Bissau is still developing its protected area legislation.[7] One protected area falls within the range of the western chimpanzee: Lagoas de Cufada Natural Park, which was officially declared in 2000 and covers an area of 890 km^2.[2] Since 1990, about 44 percent of this park has been internationally recognized as a Ramsar Site.[2] Possibly of much greater significance for the short-term survival of chimpanzees in Guinea-Bissau is the traditional protection afforded to them in most parts of the country by their perceived close resemblance to people.[6, 8] In the Boé region they are thought to shelter the spirits of elders.[8]

FUTURE CONSERVATION STRATEGIES

Overview

There is ample scope for chimpanzee conservation measures in Guinea-Bissau. Three major actions needed are:[8] to obtain basic information on chimpanzee populations in the country; to undertake a feasibility study for establishing protected areas in the country and a transnational protected area along the border between Guinea-Bissau and Guinea; and to develop a national strategy for chimpanzee conservation.

Research and protected areas

Future research should aim at developing action plans for chimpanzee conservation through studies of their ecology, particularly in open woodlands, and through identifying appropriate protected areas for two viable populations at least: in Cantanhez Forest Hunting Reserve and in the Boé region.[6, 8] The basin of the Tombali, Cumbija, and Cacine Rivers, which includes Cantanhez, has long been recognized as a promising, high biodiversity area and recommended for protection.[6, 8, 9] The status of Cantanhez was in the process of being upgraded in 2001.[1]

Local participation

Guinea-Bissau's growing population depends heavily on forest resources. Biological conservation is therefore directly linked with economic growth and development. The participation of rural communities by assuring them the legal right to manage at least part of the natural resource base could well be the best long-term way to reconcile wildlife conservation and rural development. In addition, ecotourism programs could help emphasize the importance of the primates, while potentially boosting the local economy.[7, 8]

Capacity building

Long-term collaboration between government authorities, overseas governmental agencies, and nongovernmental organizations is required to support the creation of a national system of protected areas, and build national capacity to manage it effectively.

Education and tourism

General education, awareness-raising programs, and the involvement of local communities in conserving chimpanzees and managing their habitat can help greatly to relieve pressure on chimpanzees. There are currently no sanctuaries for orphaned chimpanzees. The establishment of a rehabilitation center and its use as an educational resource would support the development of future ape conservation in Guinea-Bissau. Alternatively, confiscated animals could be sent to sanctuaries in neighboring countries.[8]

FURTHER READING

Gippoliti, S., Dell'Omo, G. (1996) Primates of the Cantanhez forest and the Cacine basin, Guinea-Bissau. *Oryx* **30**: 74–80.

Jones, S. (1992) Guinea-Bissau. In: Sayer, J.A., Harcourt, C.S., Collins, N.M., eds, *The Conservation Atlas of Tropical Forests: Africa*. Macmillan, London. pp. 200–205.

MAP DATA SOURCES

Map 16.12 Chimpanzee data are based on the following sources:

Butynski, T.M. (2003) The chimpanzee *Pan troglodytes*: taxonomy, distribution, abundance, and conservation status. In: Kormos, R., Boesch, C., Bakarr, M.I., Butynski, T.M., eds, *West African Chimpanzees: Status Survey and Conservation Action Plan*. IUCN/SSC Primate Specialist Group. IUCN, Gland, Switzerland. pp. 5–12.

Gippoliti, S., Dell'Omo, G. (2003) Primates of Guinea-Bissau, West Africa: distribution and conservation status. *Primate Conservation* **19**: 73–77.

Gippoliti, S., Embalo, D.S., Sousa, C. (2003) Guinea-Bissau. In: Kormos, R., Boesch, C., Bakarr, M.I., Butynski, T.M., eds, *West African Chimpanzees: Status Survey and Conservation Action Plan*. IUCN/SSC Primate Specialist Group. IUCN, Gland, Switzerland. pp. 55–61.

For protected area and other data, see 'Using the maps'.

ACKNOWLEDGMENTS

This country study draws extensively on the Guinea-Bissau chapter from the IUCN/SSC *West African Chimpanzees: Status Survey and Conservation Action Plan*. Many thanks to Brigid Barry (Tropical Biology Association) for editorial assistance.

AUTHORS

Cláudia Sousa, New University of Lisbon
Spartaco Gippoliti, Conservation Unit, Pistoia Zoological Garden
Muhammad Akhlas, University of East Anglia

REPUBLIC OF LIBERIA

Gemma Smith

BACKGROUND AND ECONOMY

Situated in West Africa, Liberia is Africa's oldest republic, having been established in 1847 by freed American slaves (Americo-Liberians). Covering a total area of 111 370 km^2, it is bordered by Côte d'Ivoire to the east, Sierra Leone to the northwest, Guinea to the north, and the Atlantic Ocean to the south. The capital Monrovia is located on the coast and is the largest city in Liberia.

The country has flat coastal lowlands, inland rolling hills, plateaus, and tablelands, and mountains in the far north.[3, 15, 20] The lowlands, with riverine and coastal vegetation, mangrove swamps, lagoons, and alluvial sandbars, are about 579 km long and extend some 65 km inland. Most of the country's agricultural land is located in the hills behind the lowlands. Plateaus and tablelands (up to a height of 300 m) and mountain ranges (up to 610 m) occur beyond this area, mainly between the Lofa and Saint Paul Rivers in the northwest of the country. Highland areas, including the highest mountain in Liberia (Mount Wuteve, 1 380 m) occur in the north of the country, in Nimba and Lofa counties.[20] An estimated 34 810 km^2 of Liberia is currently still under forest: 31.3 percent of the total land area. This includes the largest remaining portion of upper Guinean rain forest. Legal and illegal logging are occurring rapidly, however, and forest cover was estimated by the Food and Agriculture Organization of the United Nations (FAO) to have declined by 7.6 percent between 1990 and 2000.[10] The Liberia Forest Reassessment project, which is a joint initiative of the government of Liberia, Fauna and Flora International (FFI), and Conservation International (CI), found total forest loss between 1987 and 2001 to be only 2.6 percent, giving an annual average forest loss rate of 0.2 percent.[16]

The country's people are mostly indigenous Africans, with Americo-Liberians and their descendants (colloquially referred to in Liberia as 'Congo' after their supposed geographical origins) comprising about 5 percent of the estimated population of 3.32 million. The annual population growth rate is about 1.7 percent.[6] The majority of the population lives in urban areas in central Liberia. An estimated 57.5 percent of the adult population (over the age of 15) is literate.[6]

Liberia has experienced intense and sustained political, social, and economic disruption since a military coup in 1980. There has been a series of conflicts since then, totaling 14 years of recurrent conflict. Until 1980, Liberia was relatively calm, but then Master Sergeant Samuel Doe overthrew President William Tolbert after food price riots. Doe being of the indigenous Krahn people, his coup marked the departure from power of the Americo-Liberians, who had tended to dominate since the country's establishment. It also heralded a period of instability as widespread human rights abuses followed, and it provoked tensions between the Krahn and other indigenous groups such as the Mandingo, Gio, and Mano.[11] The economy collapsed and all-out, ethnically based civil war began in 1989. Dissidents of Charles Taylor's National Patriotic Forces of Liberia (NPFL) overran much of the countryside, and an offshoot of the NPFL killed Doe in 1990.[11] These events prompted military intervention by the Economic Community of West African States (ECOWAS) to protect Monrovia. Liberian dissidents launched raids into Liberia from Sierra Leone, in retaliation for which the NPFL supported rebels in Sierra Leone, beginning Sierra Leone's own 10 year civil war (see Sierra Leone country profile).

The war in Liberia continued until 1996, when an ECOWAS-brokered peace agreement was signed, eventually leading to the election of Taylor as president. The conflict returned in 1999, however, and escalated thereafter.[5] Under intense pressure from ECOWAS and the International Contact Group on Liberia (comprising the European Union, USA, Nigeria, Morocco, UN, ECOWAS Secretariat, and Australia), the main factions came to sign a Comprehensive Peace Agreement in 2003, thereby exiling Taylor to Nigeria and creating a power-sharing National Transitional Government. A United Nations Mission in Liberia began to deploy in October 2003, and built up to a full strength of 15 000 peacekeeping troops by mid-2004. It has a broad and robust mandate, and in addition to peacekeeping, it addresses criminal justice, human rights, child protection, and public information, as well as the environment and forestry.

This troubled modern history has had devastating consequences for Liberia's economy, as infrastructure and social capital have been destroyed and there has been little investment. Few figures

Map 16.13 Chimpanzee distribution in Liberia

Data sources are provided at the end of this country profile

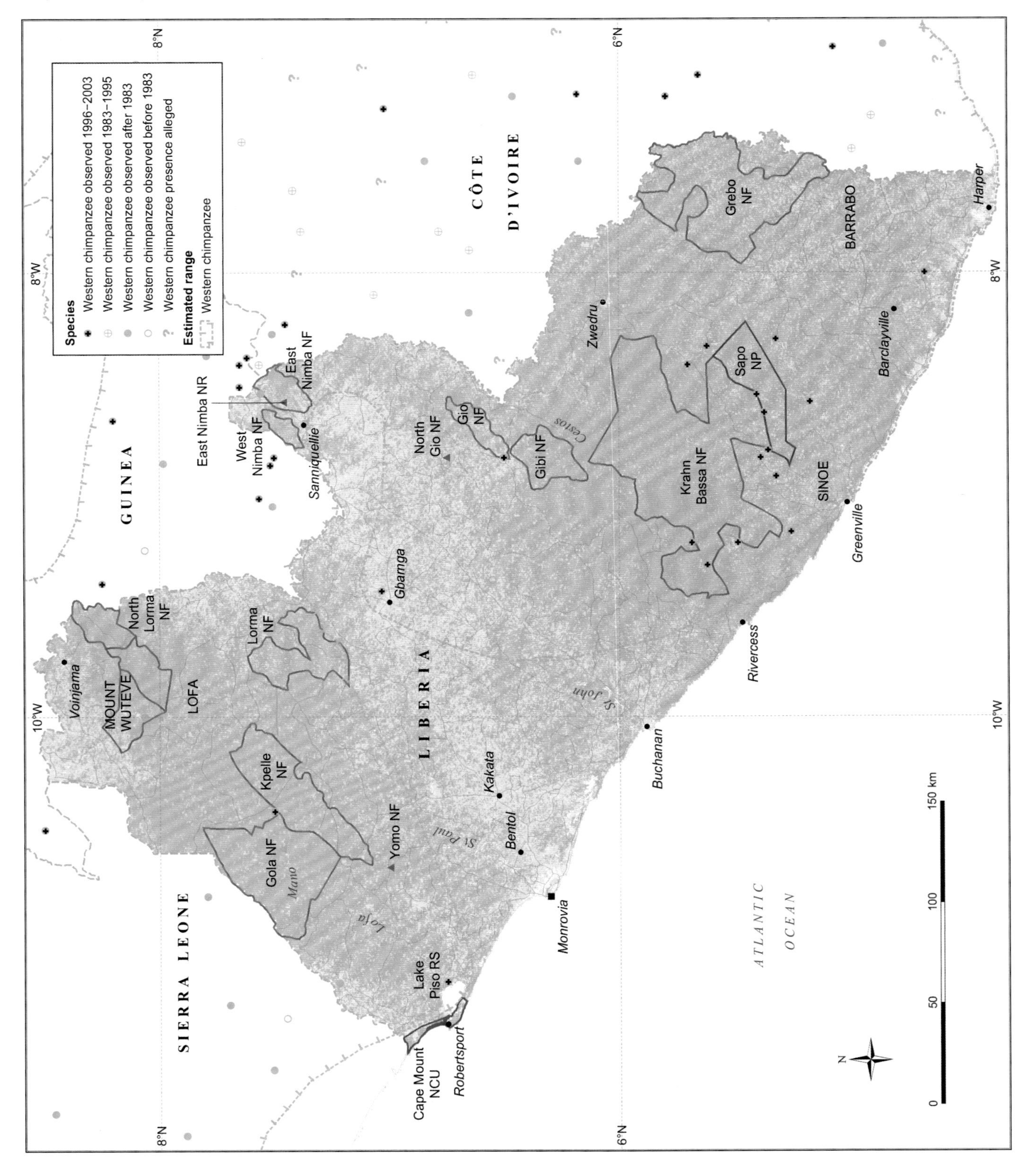

are available, but the most recent (2002) estimates by the International Monetary Fund (IMF) suggest a gross domestic product (GDP) of US$561.8 million and a gross national income (GNI) per person of US$169.20.[11] There was insufficient information available in 2003 or 2004 for the United Nations Development Programme (UNDP) to calculate a Human Development Index for Liberia, but the country is certainly among the world's poorest, with an average life expectancy of only 41.7 years.[19] International aid has increased with the signing of the Comprehensive Peace Agreement.[11]

DISTRIBUTION OF GREAT APES

The first national survey of Liberia's wildlife was undertaken by WWF–The Global Conservation Organization and the country's Forest Development Authority (FDA).[3] Twelve primate species, including the western chimpanzee (*Pan troglodytes verus*), live in Liberia. Chimpanzees were probably once present in all forested parts of Liberia,[17] and historically they have been widely recorded in southeastern Liberian forests in the Sapo, Krahn Bassa, Grebo, and Barrabo areas, as well as being noted in the northwestern and western forest blocks of the Gola and upper Lofa areas, and in upper Nimba county along the borders with Côte d'Ivoire and Guinea. Recent surveys have confirmed their survival in the East Nimba Nature Reserve, Grebo National Forest, Krahn-Bassa National Forest, and Sapo National Park (NP).[21] In the 1970s, it was estimated that there were 1 000–5 000 individuals. Surveys carried out between December 1989 and March 1990 indicated that most chimpanzees were found in high forest and older logged areas, where observations of nut-cracking sites and calls were relatively frequent.

THREATS

Large tracts of forest in the northwest and southeast of the country, in particular, are facing intense pressure from timber extraction and mining operations.[2, 9] Subsistence agriculture is spreading along roadsides, and around new laborer settlements in the forest.

Civil conflict has exacerbated threats to chimpanzee populations, both through military operations and the displacement of people. Refugees and internally displaced persons are a defining element of Liberia's post-conflict situation.[18] Nearly a million people have been displaced – close to a third of the country's population. The situation is particularly serious in northern Liberia, in the Mano River Union borders.[17] In the summer of 2003, factional fighting occurred in this area, and generated large numbers of refugees. Refugee populations are also known in the east and south near the borders with Côte d'Ivoire.

Chimpanzees have been protected by law in Liberia since 1964, but the hunting of chimpanzees for meat occurs throughout the country, and commercial hunting is seen as a particular threat.[17, 22] The rate of population decline, if any, is unknown.

Up to the 1960s, the Mano and Gio peoples of the Mount Nimba region hunted species such as chimpanzees to support their shifting cultivation lifestyle.[7] When mining operations began in this area, there was heavy exploitation of the local wildlife (including chimpanzees) to feed the mine-workers.[7] Rural communities in the southeast perceived chimpanzees as pests of their tree crops, and so they hunted them for bushmeat and for the pet trade. These pressures remain due to the continued migration of farmers and laborers from mining and logging industries, as well as military and non-military personnel, into rural areas. The growing local population increases subsistence hunting pressures, commercial hunting, and pet-trade demand. Liberian bushmeat is marketed in the Upper Guinea Forest subregion and may reach a global market. Internationally funded trade surveys were being undertaken in 2004, in an attempt to confirm this.

In some areas of Liberia, it is taboo to eat primate bushmeat, particularly within Islamic populations in the north of the country. In eastern Nimba and among some ethnic groups and clans in the southeast (for example the Sapo in Pynestown and Kpanyan districts of Sinoe county), it is taboo to eat chimpanzees. The Wehdjeh clan of the Sapo (near the northern border of Sapo NP) consider chimpanzees to be their relatives, from whom they are believed to have adapted some forest skills. Consequently, they are forbidden to kill chimpanzees.[17] There are reports, however, that along the Liberia–Sierra Leone border the species is hunted for body parts that are used for medicinal and magical purposes.

LEGISLATION AND CONSERVATION ACTION

Legislation

Liberia is party to the African Convention on the Conservation of Nature and Natural Resources, the Convention on International Trade in Endangered

Species of Wild Fauna and Flora (CITES), the Convention on Biological Diversity, the UN Convention to Combat Desertification, the World Heritage Convention, the UN Framework Convention on Climate Change, and the Convention on Wetlands of International Importance (Ramsar). Liberia finalized its National Biodiversity Strategy and Action Plan in mid-2004, and has submitted its first National Report to the UN Convention to Combat Desertification.

Existing legislation, and the associated institutional framework, should allow for the sustainable management of Liberia's environmental resources. There is an Environmental Protection Agency (EPA) Act, an Environmental Protection and Management Law, an Act Establishing a Protected Forest Areas Network, and a National Environmental Policy. Liberia's forests are managed by the Forest Development Authority which has issued logging concessions to 30 companies covering more than 50 000 km^2 in total.[4] The forest sector legislation has been evaluated by the Liberia Forest Reassessment project. While existing regulations appear appropriate, their enforcement has been weak.[4]

The Environmental Protection Agency was created by legislature on November 26 2002 and gazetted on April 30 2004. The EPA Act contains laws and policies covering the management of Liberia's environment.

Protected areas

Sapo NP was the first and, as of mid-2004, so far the only fully protected area in Liberia that had ever been managed for conservation.[17] Created in 1983, it has been the focus of much of the country's conservation effort. Originally 1 073 km^2 but expanded in October 2003 to 1 650 km^2, it comprises lowland rain forest, including swampy areas, and dryland and riparian forests.

Sapo NP contains what may be the most intact forest ecosystem in Liberia. It remains reasonably connected to several other forest areas to the north, west, and southeast, extending into Côte d'Ivoire. It is thus at the heart of the largest remaining forest block of the Upper Guinean forest ecosystem, providing habitat to species that need to range over large areas, such as forest elephants. A faunal monitoring program, established in 2001, found that the park harbors some of the richest and least disturbed wildlife in West Africa's rain forests. Chimpanzees have been surveyed,[17] but botanical surveys are less advanced: 353 plant species were collected in ten days in late 2002 in the western part of the park, including 78 that were endemic to the Upper Guinean forest and six that were new to science.[12] The area is far from fully secure, however, as illustrated by the looting of the park's infrastructure.[4]

The UK government's Darwin Initiative gave FFI a grant to restart active management of the park from April 2000 to September 2002. The funds were complemented by support from other donors including WWF-West Africa, the Whitley Foundation, and the Philadelphia Zoo. In 2002, the World Bank and the Global Environment Facility (GEF) supported preparation of a long-term management program via FFI, to be launched in early 2005. CI's Critical Ecosystem Partnership Fund has provided bridging funding between the end of the Darwin grant and the beginning of a GEF-supported program.

The Forest Development Authority and the Society for the Conservation of Nature of Liberia (SCNL) have carried out field activities. These resulted in provision of basic equipment and infrastructure, allowances, basic training, development of an 18 month operational plan, outreach to local communities (environmental awareness, provision of wells and latrines), and launching of a bio-monitoring program. A second Darwin Initiative grant was made to FFI to pilot communal forests around Sapo NP in 2004–2006. This is serving not only to secure the rights of rural communities to the forest resources that they traditionally depend upon for subsistence and small-scale commercial uses, but is also establishing a buffer zone around the park.

Conservation projects

Since 1997, a number of conservation projects have been funded and implemented by FFI and CI, the latter acting through its Critical Ecosystem Partnership Fund and Centre for Applied Biodiversity Science. Both have worked to support Liberian partners in restarting conservation and forest management, and to build the capacity of several Liberian organizations, including the Forest Development Authority, the SCNL, the National Environmental Commission of Liberia (NECOLIB) – now the Liberian Environmental Protection Agency, the Save My Future Foundation, and Green Advocates.

From 2001 to 2004, FFI and CI implemented the Liberia Forest Reassessment, a project that aims to establish the necessary information, tools, and policy environment for effective and sustainable forest and biodiversity management in Liberia.[12] It emphasizes correcting a historical imbalance that favored commercial use of forests over protecting representative samples of Liberia's biodiversity, and meeting the economic and cultural needs of rural Liberians. It is implemented in partnership with three national agencies (the Forest Development Authority, the EPA, and the Department of Statistics of the Ministry of Planning and Economic Affairs), with financial support from the European Commission and CI's Critical Ecosystem Partnership Fund.

The aims of the Liberia Forest Reassessment included an assessment of forest cover and the protection status and management of key forest areas, with a view to updating Liberia's system of legally protected forest areas. Such areas may be intended for forestry purposes, nature conservation, research, or low-impact human use (for example, as buffer zones and for recreation). It analyzed satellite imagery from the mid-1980s and 2001 to reassess the extent of, changes to, and quality of Liberia's forest cover, and established a geographic information system (GIS) database that has been continually expanded and updated with biophysical and socioeconomic information. Liberian forest policy was reviewed, recommendations for improvement provided, a new system of forest protected area categories agreed, and field surveys of significant forest blocks undertaken to obtain adequate socioeconomic and biological data with which to classify and manage Liberia's forests and biodiversity in the future.

Between 2003 and 2006, the Wild Chimpanzee Foundation (WCF), through CI's Critical Ecosystem Partnership Fund, is implementing a US$184 276 education and awareness project to improve the protection of wild chimpanzees in West Africa. Environmental education activities include dramatic performances and distribution of newsletters, and building capacity so as to generate support from local people. WCF is working in Sapo NP. The Liberian project partner is the Wildlife and National Parks Division of the Forest Development Authority.

From March to July 2004, a group called the Concerned Environmentalists for the Enhancement of Biodiversity, in partnership with Philadelphia Zoo and the Bushmeat Crisis Task Force, implemented a US$9 838 project to analyze the Liberian bushmeat trade. Surveys were conducted in Monrovia to obtain data on volumes and species traded, public opinion, and factors that affect the supply of bushmeat to the market (such as the price of gasoline and ammunition). Potential outlets through which bushmeat enters international markets were also studied.

Sanctuaries

There are no sanctuaries that accept newly orphaned chimpanzees in Liberia. Vilab II, of the Hepatitis Research Foundation, is a vaccine laboratory that is now slowly working towards the socialization and release of its ex-laboratory chimpanzees on to six small islands.[1, 8, 13, 14] Ninety chimpanzees had earlier been rehabilitated and released, but virtually all were shot or starved during the civil war.

FUTURE CONSERVATION STRATEGIES

Priorities for action

These include:

- comprehensive environmental assessment by government agencies supported by UN and nongovernmental scientific organizations;[20]
- assessment of the Liberian forestry sector and its impacts on chimpanzee populations;
- establishment and management of a protected area network representative of Liberia's biodiversity and protecting major populations of key species, including chimpanzees;
- adequate management of existing protected areas (Sapo, Nimba);
- biodiversity surveys and inventories, including identification of viable populations of chimpanzees; and
- a program of environmental education (public awareness).[18]

Improved environmental governance

The Forest Development Authority is the key government institution responsible for environmental administration (albeit with a focus on the forestry sector). However, it was looted during the conflict and is left with almost no implementation or enforcement capacity. It needs to be re-equipped and re-skilled.

The Environmental Protection Agency also requires strengthening and activation. Once this has occurred it would be able to coordinate, monitor, supervise, and consult on all activities in the protection of the environment and the sustainable use of natural resources.[4]

FURTHER READING

Agoramoorthy, G., Hsu, M.J. (1999) Rehabilitation and release of chimpanzees on a natural island – methods hold promise for other primates as well. *Journal of Wildlife Rehabilitation* 22 (1): 3–7.

FFI (n.d.) *Liberia – Conservation in a Post-Conflict Country.* http://www.fauna-flora.org/africa/feature_liberia1.html. Accessed June 30 2005.

UNEP (2004) *Desk Study on the Environment in Liberia.* United Nations Environment Programme, Geneva. http://postconflict.unep.ch/liberia/Liberia_DS_AGL.pdf. Accessed July 17 2004.

MAP DATA SOURCES

Map 16.13 Chimpanzee data are based on the following sources:

Butynski, T.M. (2003) The chimpanzee *Pan troglodytes*: taxonomy, distribution, abundance, and conservation status. In: Kormos, R., Boesch, C., Bakarr, M.I., Butynski, T.M., eds, *West African Chimpanzees: Status Survey and Conservation Action Plan.* IUCN/SSC Primate Specialist Group. IUCN, Gland, Switzerland. pp. 5–12.

Nisbett, R.A., Peal, A.L., Hoyt, R.A., Carter, J. (2003) Liberia. In: Kormos, R., Boesch, C., Bakarr, M.I., Butynski, T.M., eds, *West African Chimpanzees: Status Survey and Conservation Action Plan.* IUCN/SSC Primate Specialist Group. IUCN, Gland, Switzerland. pp. 89–98.

For protected area and other data, see 'Using the maps'.

ACKNOWLEDGMENTS

This country study draws extensively on the Liberia chapter from the IUCN/SSC *West African Chimpanzees: Status Survey and Conservation Action Plan.*[17] Many thanks to Chris Magin (Royal Society for the Protection of Birds) and Jamison Suter (Fauna and Flora International) for their valuable comments on the draft of this section.

COMPILER

Gemma Smith, UNEP World Conservation Monitoring Centre

REPUBLIC OF MALI

CHRIS DUVALL AND GEMMA SMITH

BACKGROUND AND ECONOMY

The Republic of Mali is a large, landlocked nation lying in central West Africa. It extends over 1 241 138 km², bordered by Algeria to the north, Niger to the east, Mauritania, Senegal, and Guinea to the west, and Côte d'Ivoire and Burkina Faso to the south.[11] Mali has eight administrative regions – Gao, Kayes, Kidal, Koulikoro, Mopti, Segou, Sikasso, and Tombouctou. Although its population of about 12 million[2] is relatively low for its large land area, the vast majority live in the densely populated southern third of the country. The population is growing at 2.8 percent per year.[2] Mali's people are predominantly Muslim but culturally diverse, with over 40 ethnic groups. The most numerous ethnic groups include the Mande or Manding (about 50 percent of Mali's total population, comprising Bamanan, Maninka, Soninke, and other groups), Peulh or Fulani (17 percent), Voltaic groups (12 percent, of which Bobo form the vast majority), Touareg and Moor (10 percent), and Songhai (6 percent).

Mali gained independence from French colonial rule in 1960 under the leadership of President Modibo Kéita. His single-party government was overthrown in 1968 in a military coup led by General Moussa Traoré. In 1991, following a popular uprising led by students in the capital Bamako, Lieutenant Colonel Amadou Toumani Touré overthrew the Traoré regime. Touré established a transitional government and instituted democratic reforms leading to elections in 1992. Alpha Oumar Konaré, representing the Alliance for Democracy in Mali, became the first president and was re-elected in 1997. Konaré pursued a series of profound political reforms to democratize and decentralize Mali's government and privatize parts of its economy, and Mali is now frequently cited as a model of successful, multi-party governance in Africa. In 2002, Touré

Map 16.14 Chimpanzee distribution in Mali

Data sources are provided at the end of this country profile

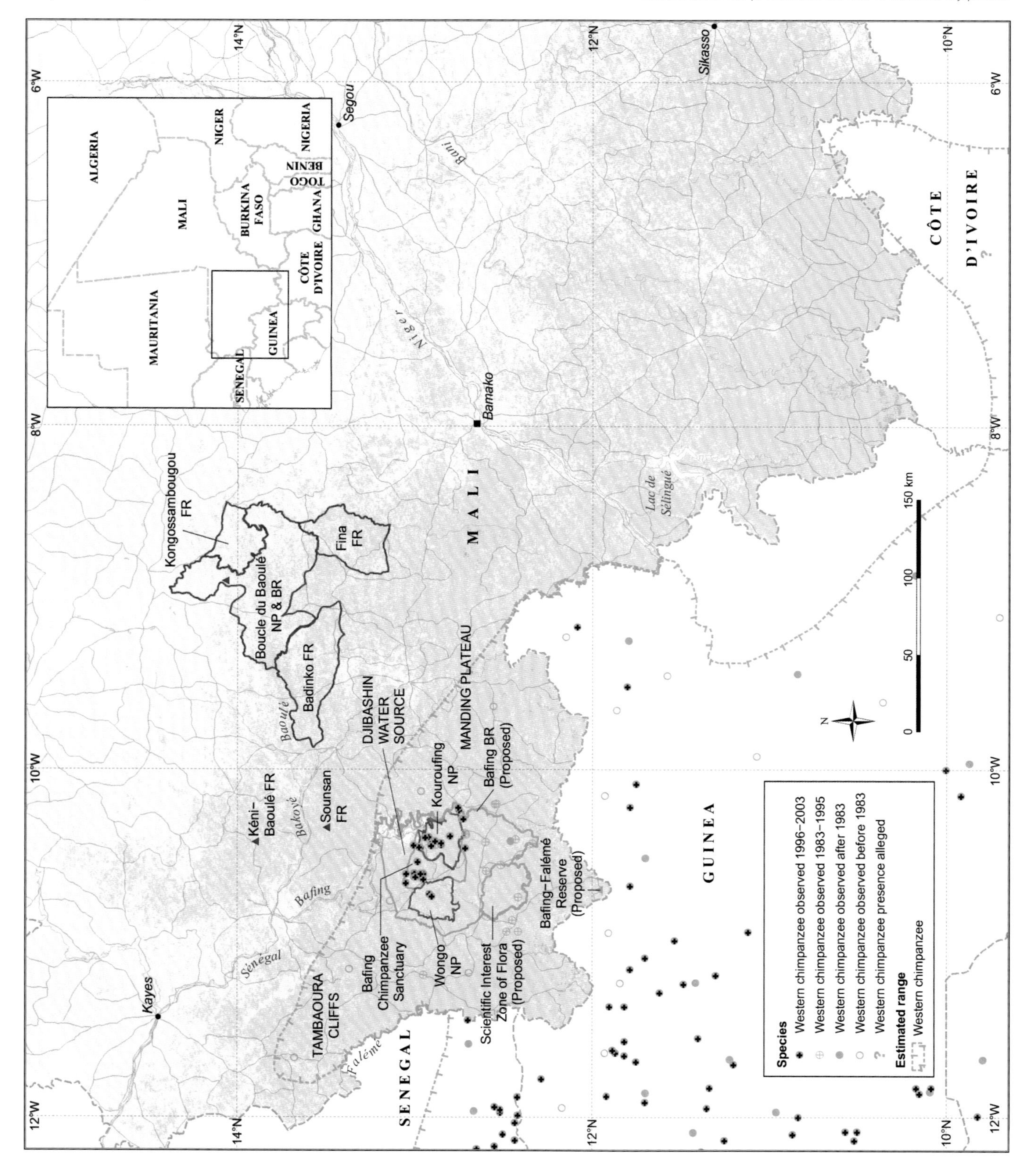

was elected as an independent candidate to replace Konaré.[11] Touré's domestic policies have focused on the struggling economy, Mali's unemployed and poorly educated youth, and the fight against corruption. Decentralization continues to increase the involvement of local civil society, nongovernmental organizations (NGOs), and other stakeholders in government and economic development, with a growing emphasis on involving women.

Mali remains one of the world's poorest nations. Gross domestic product (GDP) in 2004 was estimated at US$4.3 billion, representing a gross national income (GNI) of about US$290 per person.[25] Although large areas of the north are barren desert, the fertile Niger River basin and other parts of southern Mali support subsistence agriculture and animal husbandry, which have long dominated the economy. Manufacturing and industrial gold mining in southwest Mali have increased greatly in the past decade, and cotton, gold, and livestock are the nation's most valuable exports. Interannual variability in weather has a strong influence on agricultural production and therefore GDP. In 2002, foreign aid contributed about 14 percent of GDP,[25] and Mali has recently benefited from debt relief, including a 38 percent debt write-off by France in 2002, and US$765 million in debt relief from the International Monetary Fund (IMF) and the World Bank in 2003. Mali's debt burden remains high, and limits the government's ability to invest in sectors such as education, healthcare, natural resource conservation, and infrastructure development.

Mali's terrain is mostly flat and low-lying, with an average altitudinal range of 200–500 m.[23] The lowest elevation is 23 m above sea level (the Senegal River), and the highest is 1 155 m (Hombori Tondo rock outcrop). Most of southern Mali lies in the Niger River floodplain, with the only relief being small areas of undulating hills lying between the river's various tributaries. However, in western, central, and eastern Mali, sandstone massifs outcrop to form topographically complex highland areas that have a relatively high level of biological diversity due to the wide range of microhabitats supported by the varied topography.[7, 8, 14, 15] These areas include: the Dogon Cliffs in central Mali; the Adrar des Iforhas, located in the Sahara in the northeast; and in the southwest, the Manding Plateau, the only area where chimpanzees occur in Mali. Western Mali, including most of the Manding Plateau, is drained by the Senegal River and its main tributaries – the Bafing, Bakoyé, Baoulé, and Falémé Rivers – none of which have wide floodplains in Mali.[2]

More than two thirds of Mali is arid or semi-arid, with vegetation density increasing from north to south as annual rainfall increases from less than 100 mm in the Sahara to more than 1 500 mm in the south.[26] The country is divided latitudinally into three distinct zones of natural vegetation:[9, 17, 29] the Saharan zone (dominated by desert and semi-desert), the Sahelian zone (dominated by *Acacia* wooded grassland and deciduous shrubland), and the Sudano-Guinean woodland zone. The Sudano-Guinean woodland zone covers about 131 860 km^2 of Mali, 10.8 percent of its land area.[10] These woodlands, which have a relatively dense tree cover dominated by various species in the Combretaceae and Leguminosae, are prominent in the landscape in areas where chimpanzees occur. Mali's closed-canopy forests are mainly located in the southwest, and are primarily riparian gallery forests dominated by bamboo (*Oxytenanthera abyssinica*, Graminae) and raphia palm (*Raphia* sp., Palmae).[4] Additionally, small patches of non-riparian gallery forest occur on steep slopes and cliff edges in the Manding Plateau. The endemic tree *Gilletiodendron glandulosum* (Leguminosae) dominates these forests, and many species with edible fruits are also abundant.[7] Finally, several types of grassland and shrubland, with few trees and dominated by grasses, occur in large patches in areas with shallow or poor soil throughout southern Mali.

DISTRIBUTION OF GREAT APES

Western chimpanzees (*Pan troglodytes verus*) are the only species of great ape in Mali, as first reported by Sayer.[18, 19, 24] They are only known to occur in the southern part of the western region of the country, in the Manding Plateau near the borders of Guinea and Senegal.[6, 18] Most of southwest Mali has not been surveyed for chimpanzees, but it is estimated that the maximum current range of the species in Mali is 19 440 km^2.[6] Mali's chimpanzees are the northernmost natural population, occurring to about 13°10'N, including some of the driest environments in the animal's global range.[16, 20] In the recent past, chimpanzees were found farther north than at present;[24] this range decline is probably due to human threats, especially hunting.[6] It is not known if the Malian chimpanzee population is currently isolated from those in neighboring countries, but genetic analysis shows that they have not been isolated in the past.[12]

Chimpanzees inhabit areas of riparian and gallery forest in Mali, particularly non-riparian gallery forest patches in cliffs in the Manding Plateau.[7, 13, 19, 21] Chimpanzees also nest and feed in woodland situations, away from water sources, at least seasonally,[7, 13, 19] and evidence of chimpanzee activity (such as feeding remains) has been found in a wide range of habitats, including grasslands and bare sandstone with minimal shrub cover.[4]

There have been two efforts to estimate chimpanzee population density in Mali, both based on nest sampling. Pavy calculated a density of 0.25 chimpanzees per square kilometer in the northern portion of the proposed Bafing Biosphere Reserve,[21] while Granier and Martinez found evidence of 0.35–0.40 chimpanzees per square kilometer in three limited areas near the Guinea border, south, east, and west of the proposed reserve.[13] Both of these density estimates are among the highest recorded for chimpanzees in similar habitats,[13] and thus must be viewed cautiously as the likely upper limit of chimpanzee density in Mali. Recognizing that Granier and Martinez's estimate applies only to the limited areas where they collected data, Pavy's estimate is more likely to represent the maximum average population density for the animal's entire range in Mali. Thus, the maximum chimpanzee population size for Mali, based on the range area cited above, is 4 860 individuals, although the actual population size is certainly lower, since chimpanzees are patchily distributed.[5, 13]

Chimpanzees are apparently more abundant and widespread to the west of the Bafing River, south of the Manantali Dam,[6, 13] which is one of the least densely populated parts of West Africa south of the Sahara. On the west bank of the Bafing lie the only Malian protected areas with chimpanzees: Kouroufing and Wongo National Parks (NPs) and the Bafing Chimpanzee Sanctuary (a species-specific faunal reserve for naturally occurring populations), which together comprise the proposed Bafing Biosphere Reserve.[6] Nearly all research on chimpanzees in Mali has been undertaken in the Bafing area. An additional protected area has been proposed to the south of the proposed Bafing Biosphere Reserve; the transboundary Bafing–Falémé Reserve would link conservation efforts in Mali and Guinea in an isolated and biologically diverse area.[13]

THREATS

The populations of most large mammals in Mali have declined markedly in the last 50 years,[5, 21, 27] and chimpanzees are unlikely to be an exception to this pattern, though little is known for certain.[6] The main threats to chimpanzees in Mali are hunting and agricultural expansion, although Mali's chimpanzees appear to be relatively less vulnerable to these threats than most other natural populations.[6, 13]

Chimpanzees are rarely hunted in Mali relative to other large mammal species,[4, 13] but the animal's low population size in Mali means that even low absolute levels of hunting may pose a relatively large problem.[19] Malians most frequently hunt chimpanzees with guns to reduce losses of fruit from wild trees, although baboons and monkeys are the primary target species. In the parts of Mali where chimpanzees occur, indigenous religious beliefs that have no restrictions on eating primate meat predominate. Some people do eat chimpanzee meat, either as food or for medicinal reasons, but many consider chimpanzees to be too similar to humans to want to eat them.[4, 13] A less frequent reason for hunting chimpanzees is to supply the small, largely urban market for skins, body parts, and young chimpanzees. Many urban Malians consider chimpanzee skin and other parts to have medicinal properties or spiritual significance that people in rural areas do not recognize, whilst some wealthy Malians and European expatriates value chimpanzees as pets. Rural Malian hunters are aware of these urban markets, and many are willing to supply chimpanzees to them if an opportunity presents itself.[4] Further potential threat is posed by hunters from nations to the south, particularly Guinea, where chimpanzee meat is more widely consumed, who may extend their hunting activities northwards.

Habitat loss is probably the greatest medium- to long-term threat to chimpanzee survival in Mali,[6] and has four main causes: farming, livestock herding, logging, and mining. Although chimpanzees are most abundant in areas that are not farmed, such as cliffs and steep hills, agricultural expansion is a threat in some areas,[1, 13] particularly east of the Bafing River and south of the town of Manantali. Traditional swidden (slash-and-burn) farming requires large land areas, but in the long term is a sustainable approach in areas of low human population density. Population growth and increased market orientation are changing agricultural practices,[3] and permanent loss of woodland areas to agriculture is increasing around larger settlements and along all-season roads.[13] Since about 1989, following the construction of a bridge across the Bafing River at the Manantali Dam, livestock herders from northern Mali have entered the proposed Bafing reserve each dry season. Herders fell trees to provide forage for their animals, rather than relying solely on herbaceous vegetation. While the effects of their activities on wildlife have not been studied, local residents blame the herders for increased incidence of wildfires and a decrease in wildlife abundance, and the herders do visit areas and cut trees used by chimpanzees.[4] Small-scale logging has also reduced chimpanzee habitat since the filling of the Manantali Reservoir in 1988. Fishermen based along the Manantali Reservoir (including within the proposed reserve) illegally harvest large hardwood species for canoe-building, creating major canopy gaps in riparian forests.[4] Industrial mining operations near Kéniéba have cleared vegetation and soil from areas of known or potential chimpanzee habitat, produced much toxic and nontoxic waste, and triggered an influx of jobseekers.[13] Finally, proposals to build a major international highway just south of the proposed Bafing reserve will, if funded, probably increase all forms of habitat loss in the future. Habitat loss is a more difficult pressure to address than hunting, as it results from the economically important activities of farming, livestock herding, fishing, and mining.

LEGISLATION AND CONSERVATION ACTION

Mali has signed and ratified a number of international biodiversity and conservation conventions and agreements. These include the Convention on Biological Diversity, the Convention on Wetlands of International Importance (Ramsar), the Convention on Migratory Species, the Convention on International Trade in Endangered Species of Wild Fauna and Flora (CITES), the World Heritage Convention, the UN Convention to Combat Desertification, and the UN Framework Convention on Climate Change. As part of its obligations under one or more of these agreements, the Direction Nationale de la Conservation de la Nature (DNCN) has initiated several planning and management projects in the proposed Bafing Biosphere Reserve.

Protected areas are designated under Law No. 86-43/AN-RM, concerning hunting and the conservation of fauna and its habitat, and under Law No. 86-42/AN-RM (the Forest Code). Thirteen protected areas have been designated in Mali, covering about 5 percent of the country.[28] The newest of these are Kouroufing and Wongo NPs (both gazetted in 2001), and the Bafing Chimpanzee Sanctuary (gazetted in 2002). An extensive buffer zone and a *zone cynégétique* (a multipurpose area intended to protect defaunated woodland habitat whose exact status has not been determined) are proposed additions to the three gazetted areas. Altogether these protected areas form the proposed Bafing Biosphere Reserve, a total area of about 5 215 km^2.[6] A management plan for this reserve was drawn up in 2000–2001, but as yet only limited funds have been allocated for its implementation. First established as a faunal reserve in 1990, the Bafing Chimpanzee Sanctuary remains poorly developed. Since the early 1980s, chimpanzee protection has been one of the primary goals of conservation activities in the area. As for the proposed Bafing–Falémé protected area, the Direction Nationale de la Conservation de la Nature is actively undertaking natural resource assessments of the area and is likely to gazette it in the near future.

Ian Redmond

Chimpanzee habitat in the proposed Bafing Biosphere Reserve.

The management of wildlife in Mali is the responsibility of the Direction Nationale de la Conservation de la Nature,[6] which was established in 1998 to take over many functions of the former Direction Nationale des Eaux et Forêts. Chimpanzees are protected under Law No. 95-031, which governs wildlife management, and are listed in Appendix 1 of that law, which gives them complete protection against capture, habitat destruction, and hunting.[6] The minister responsible for the Direction Nationale de la Conservation de la Nature can grant exceptions to this rule for scientific collection and research, or the removal of dangerous animals. Other Malian laws, such as that governing forest management (Law No. 95-004), provide protection for chimpanzee habitat, particularly riparian and gallery forests, although the relevant parts of this law are rarely enforced.[6, 8]

All hunting and most woodcutting were illegal prior to the restoration of civilian rule in 1992, after which natural resource laws were modified to suit the democratic, decentralized principles of Konaré's reform program.[6] Current wildlife and forest management laws allow for some community-based resource control, although how this devolution of authority will be undertaken remains unclear. The state continues to exercise strong authority over natural resource management decisions in the Bafing area, including the decision made in 2001–2002 to forcibly remove several small settlements from within the protected area. Local residents did not support this decision, but from the standpoint of wildlife conservation, the move will probably reduce hunting and habitat loss. It may also increase human activities along the cliffs that are located predominantly outside of the currently gazetted protected areas, thus perhaps negatively affecting chimpanzees. In general, the Direction Nationale de la Conservation de la Nature has never had a large or highly active presence in the chimpanzee range areas. Many local residents and migrants from elsewhere in Mali hunt actively throughout the year, and some still farm in areas that have been designated off-limits.

TRADITIONS

Two ethnic groups, the Maninka and the Fulani dominate in the areas where chimpanzees are found.[6] Maninka are also numerous in parts of chimpanzee range in Senegal, Guinea, and Côte d'Ivoire, while Fulani also coexist with chimpanzees in Guinea.

Sedentary agriculturalists, the Maninka (or Malinké) are the main ethnic group in the Manding Plateau. Most rural Maninka are indifferent toward chimpanzees,[13] although attitudes toward the animal range from fear to respect, and from dislike (due to their consumption of wild foods valued by humans) to sympathy (due to their resemblance to humans). Some rural Maninka consider chimpanzees a pest of fruit trees and millet,[4, 13] and many consider the animal valuable, since consumption of its meat is believed to treat river blindness (onchocerciasis).[5] A traditional Maninka story known by some elders in the Bafing area tells that chimpanzees represent one of several human clans. In the distant past, a powerful sorcerer disfigured members of the chimpanzee clan and condemned them to eating raw food and remaining unclothed in the bush because the chimpanzees refused to abandon a settlement site coveted by the sorcerer.[4] This story suggests that on a cultural level Maninka recognize that humans and chimpanzees compete for natural resources, as recognized by many Maninka individuals[4] and by researchers.[7, 13, 22]

Within chimpanzee range in Mali, the Fulani (or Fula, Peulh, or Fulfulde) are mostly found in the southern part of the Manding Plateau, where they practice sedentary farming. Fulani to the north of the plateau practice pastoral livestock husbandry, and visit chimpanzee range only in the dry season. For the most part, Fulani and Maninka attitudes toward chimpanzees are similar, although Fulani people in southwest Mali are generally less likely to eat chimpanzee meat.[13]

Urban Malians often have different views and values concerning chimpanzees from rural people. First, many urban people do not know what chimpanzees are, and do not know the Maninka or Bamanan names for the animal. Second, few of those who do know of chimpanzees are indifferent to them – many urban Malians consider them to be ferocious and highly dangerous. Many people enjoy stories of the attacks or depredations of various wildlife species, including chimpanzees, although very few urban Malians have ever seen a chimpanzee, and even fewer have seen a noncaptive chimpanzee. Finally, many urban people value the skin of chimpanzees in the fabrication of amulets for the spiritual power it is believed to add to certain prayers or charms.[4]

FUTURE CONSERVATION STRATEGIES

Chimpanzee conservation in Mali is hindered by lack of resources and information on the best sites to focus conservation efforts. Two research projects conducted in 2003–2004 may improve the efficacy of future conservation actions regarding the protection of chimpanzees. Researchers working with the project Appui à la Gestion Intégrée des Ressources Naturelles (AGIR) assessed chimpanzee distribution and population density, and threats to chimpanzees at several sites in southwest Mali in 2003–2004.[13] Their findings provide baseline data for the evaluation of future conservation efforts, as well as improved understanding of the international significance of Mali's chimpanzees. Second, an assessment of chimpanzee habitat and threats posed by subsistence farming and hunting in the Bafing area was undertaken in 2003–2004 with the support of the Wildlife Conservation Society (WCS), the Great Ape Conservation Fund, Conservation International (CI), and the Milwaukee Zoological Society.[4] This research will help clarify ecological and spatial relationships between agriculture and human settlement, and chimpanzee distribution and abundance.

In *West African Chimpanzees: Status Survey and Conservation Action Plan*, Chris Duvall and colleagues recommended several conservation priorities for chimpanzees in Mali.[6] First, four focal areas for chimpanzee conservation and/or further research in Mali were identified: the proposed Bafing Biosphere Reserve, the area to the east of the Bafing River, the Tambaoura Cliffs (north of Kéniéba), and along the Falémé River on the Mali-Senegal border. Of these four areas, all but the Tambaoura Cliffs have been studied since the publication of the action plan. Second, it was recommended that financial support to the proposed Bafing reserve be increased with the aim of improving effective management. The same recommendation should be made for the proposed Bafing–Falémé transboundary protected area. Sustainable use of natural resources in these areas should be allowed as specified in the relevant legislation, which must be more effectively enforced to reduce poaching, livestock herding, logging, and farming in protected areas. Concerted efforts are needed to introduce and stimulate alternative development opportunities for local residents in these areas, as pioneered by the NGO Association Malienne pour la Conservation de la Faune et de l'Environnement (AMCFE). Finally, efforts are needed to increase public awareness of the international significance of Mali's chimpanzee population, and to dispel negative attitudes toward chimpanzees.[13] Environmental education activities such as those begun by the Wild Chimpanzee Foundation (WCF) must be continued and further expanded.

FURTHER READING

Duvall, C. (2003) Agriculture and chimpanzee survival in West Africa. In: Kormos, R., Boesch, C., Bakarr, M.I., Butynski, T.M., eds, *West African Chimpanzees: Status Survey and Conservation Action Plan.* IUCN, Gland, Switzerland. pp. 143–145.

Duvall, C. (2000) Important habitat for chimpanzees in Mali. *African Study Monographs* **21** (4): 173–203.

Moore, J. (1985) Chimpanzee survey in Mali, West Africa. *Primate Conservation* **6**: 59–63.

Touré, A.S., Maïga, A. (2001) *Deuxième rapport du Mali sur la diversité biologique.* http://www.biodiv.org/doc/world/ml/ml-nr-02-fr.doc. Accessed November 5 2004.

MAP DATA SOURCES

Map 16.14 Chimpanzee data are based on the following sources, with additional data by personal communication from Duvall, C. (2005):

Butynski, T.M. (2003) The chimpanzee *Pan troglodytes*: taxonomy, distribution, abundance, and conservation status. In: Kormos, R., Boesch, C., Bakarr, M.I., Butynski, T.M., eds, *West African Chimpanzees: Status Survey and Conservation Action Plan.* IUCN/SSC Primate Specialist Group. IUCN, Gland, Switzerland. pp. 5–12.

Duvall, C., Niagaté, B., Pavy, J-M. (2003) Mali. In: Kormos, R., Boesch, C., Bakarr, M.I., Butynski, T.M., eds, *West African Chimpanzees: Status Survey and Conservation Action Plan.* IUCN/SSC Primate Specialist Group. IUCN, Gland, Switzerland. pp. 41–50.

For protected area and other data, see 'Using the maps'.

ACKNOWLEDGMENTS

This country study draws extensively on the Mali chapter from the IUCN/SSC *West African Chimpanzees: Status Survey and Conservation Action Plan.* Many thanks to Ian Redmond (Ape Alliance/GRASP) for his valuable comments on the draft of this section, and to Jen Duvall for all her support.

AUTHORS

Chris Duvall, University of Wisconsin-Madison
Gemma Smith, UNEP World Conservation Monitoring Centre

FEDERAL REPUBLIC OF NIGERIA

EDMUND MCMANUS

BACKGROUND AND ECONOMY

The Federal Republic of Nigeria is a West African country bordered by Benin, Niger, Chad, and Cameroon, covering a land surface area of 923 768 km^2.[4] It is Africa's most populous country, with about 134 million people (estimates range from 128 to 137 million)[4,6] and an annual population growth rate of about 2.5 percent.[4] Nigeria became independent of Britain in 1960 as a federation of three regions (Northern, Western, and Eastern), which retained a large measure of self-government.[6] In 1963, Nigeria became a Federal Republic, with a fourth region (the Midwest), and a new constitution.

In January 1966, a group of mostly Igbo army officers overthrew the government and imposed a state of emergency. Anti-Igbo riots broke out in the north, and there was a counter-coup by northern troops in July. Thousands of Igbos were massacred in the north, and hundreds of thousands fled to their homelands in the southeast, where there were increasingly strong calls for secession. The new military leadership replaced the regions with 12 states in an attempt to increase local self-governance, but this gesture was rejected by the military government of the Eastern Region, under Colonel Odumegwu Ojukwu, who declared that region to be an independent state, the 'Republic of Biafra'. Civil war followed, and lasted until 1970, when the Biafran troops surrendered.

The 1970s, 1980s, and 1990s saw a succession of military coups, with long periods of military rule and intervals of civilian government.

The most recent dictatorship, that of General Sani Abacha from 1993 to 1998, dismantled all democratic structures. This regime oversaw the execution of Ken Saro Wiwa, the Ogoni activist and poet, and the systematic misappropriation of oil revenues. General Abacha alone is estimated to have stolen US$4.3 billion.[6] Nigeria was suspended from the Commonwealth from 1995 until after General Abacha's death in 1998. Abacha's successors repealed many military decrees and planned the restoration of a democratic, civilian government. Local elections were held in December 1998; the presidential election followed in February 1999, and was won by a former president, Olusegun Obasanjo.

President Obasanjo inherited an economically damaged country, which owed about US$28 billion to external creditors. Development has historically been hindered by pervasive corruption, which the Nigerian government is taking a number of steps to tackle, as well as working to improve relations with the International Monetary Fund (IMF), World Bank, and Paris Club of creditor nations. The economy has considerable potential to recover, as Nigeria is the leading sub-Saharan oil producer at 2.3 million barrels of oil per day.[6] Offshore oil production and investment in the gas sector are expected to continue growing rapidly.

The gross domestic product (GDP) in 2003 was US$35.1 billion,[6] or US$114.8 billion adjusted for purchasing power parity,[4] and was growing at 3.7 percent[6] per year. About 20 percent of GDP comes from oil sales.[4] Nevertheless, annual income per person in Nigeria is low, at US$314,[6] or US$900 adjusted for purchasing power parity.[4]

Nigeria now consists of 36 states, plus the Federal Capital Territory of Abuja. The climate is equatorial in the south, tropical in the center, and arid in the north. The terrain varies from the lowlands in the south to hills and plateaus in the central region (Jos Plateau), and from mountains in the southeast (extensions of the Cameroon Highlands) to plains in the north. Nigerian ecosystems range from tropical rainforest in the south, dry thorn scrub in the northeast, and montane communities on the Mambilla Plateau and Obudu highlands near the border with Cameroon, to the mangrove and freshwater swamp forests of the Niger Delta. These swamp forests were originally flanked to the east and west by lowland rain forest, much of which has now been replaced by cultivated land. However, the largely subsistence-based agricultural sector has not kept up with population growth: once a large food exporter, Nigeria now imports food. The largest remaining areas of closed-canopy rain forest are in the southeast, in Cross River state. These are contiguous with the forests of southwest Cameroon.

There are no great apes in the northernmost part of the country. In the regions inhabited by great apes, annual precipitation ranges from 1 500 to 4 000 mm, with a three to five month dry season from November to March.[13] About half of Nigeria's people live within the historical range of chimpanzees, across the southern part of the country.

Map 16.15 Great ape distribution in Nigeria

Data sources are provided at the end of this country profile

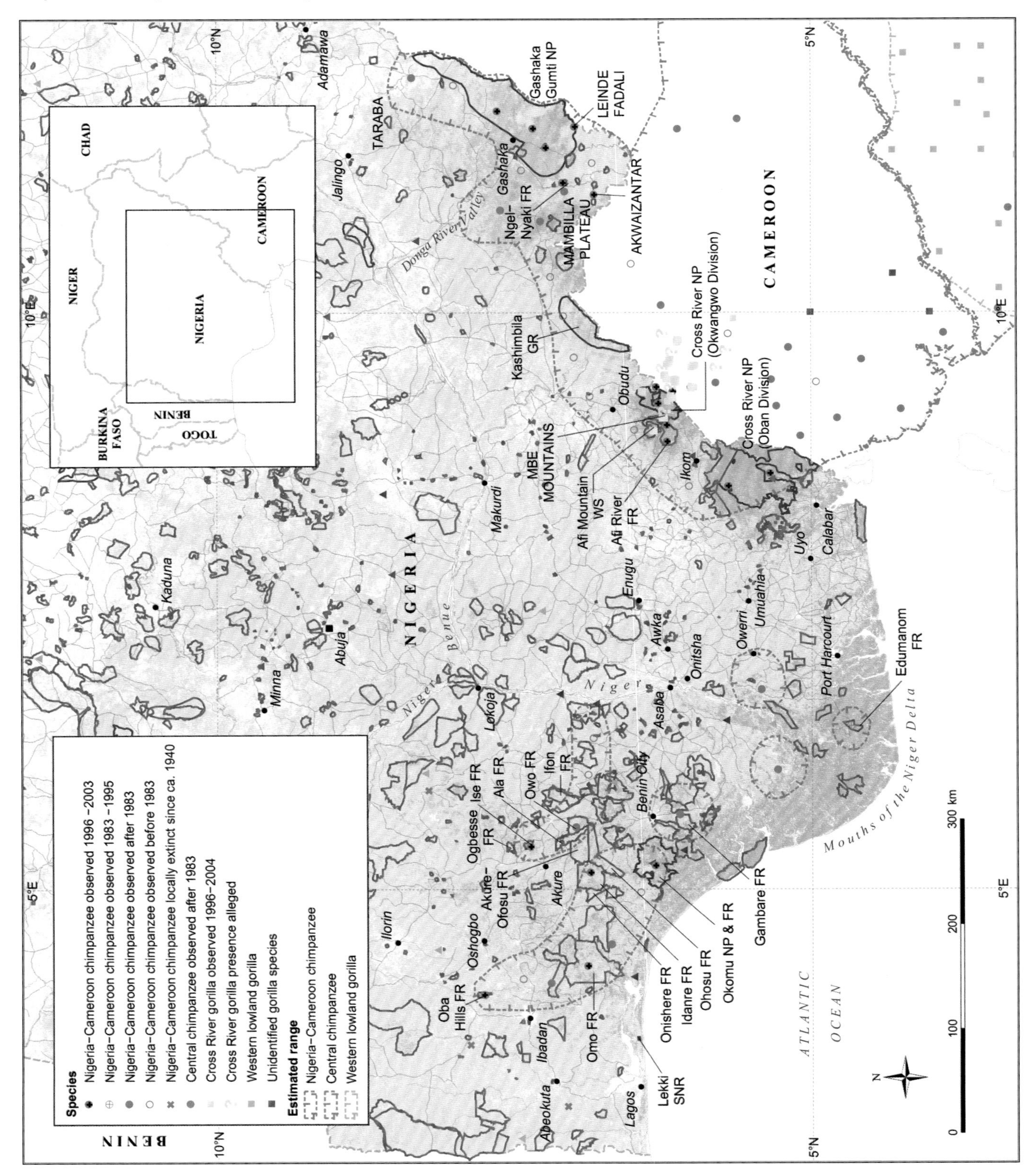

Tunde Morakinyo

Women from Old Ekuri, a village on the edge of Cross River National Park, collecting a forest vegetable called afang (*Gnetum africanum*).

DISTRIBUTION OF GREAT APES

By 1982, it was believed that gorillas had become extinct in Nigeria,[11] but later surveys demonstrated that they are still present. These Cross River gorillas are recorded in the Afi Mountain Wildlife Sanctuary of the Afi River Forest Reserve, in the Mbe Mountains community forest, and in the Okwangwo Division of the Cross River National Park (NP).[14] There are probably three distinct subpopulations, and a fourth shared with Cameroon. There are estimated to be approximately 80–100 individuals remaining in Nigeria.[13]

There is some uncertainty about the subspecific affinities of the chimpanzees found in Nigeria. Based on sequencing of mitochondrial DNA, chimpanzee populations in eastern Nigeria are related to populations in western Cameroon, but are not closely related to populations in the Upper Guinea region or to chimpanzees south of the Sanaga River in Cameroon. The chimpanzees in the Nigeria–Cameroon border region have accordingly been placed in the subspecies *Pan troglodytes vellerosus*.[8, 9] Depending on the analytical techniques used, chimpanzees in western Nigeria group either with *P. t. vellerosus* as shown on Map 16.15, or with *P. t. verus* of Upper Guinea. More research is needed, therefore, to clarify the affinities of chimpanzees in southwestern Nigeria.

A 2005 workshop in Brazzaville estimated that there are around 3 050 chimpanzees in Nigeria. The largest remaining population is probably that in Gashaka Gumti NP and its vicinity, with another large population occurring in the Cross River NP and its surroundings, including the Afi Mountain Wildlife Sanctuary.[13] The chimpanzee range can be divided into three main regions.

Southwest Nigeria and the Niger Delta

Surveys are few in western Nigeria and along the eastern edge of the Niger Delta. Chimpanzee populations here are small, highly fragmented, and severely threatened.[13] They have recently been confirmed to survive in the southeastern forests of the Niger Delta, and in Omo, Ise, Owo, and Okomu Forest Reserves. They are reported but not confirmed from the Oba Hills, Ala, Idanre, Ifon, and Ogbesse Forest Reserves,[1] and suspected in Okomu NP and the Akure-Ofusu, Onishere, and Ohusu Forest Reserves.

Gashaka–Mambilla

The largest chimpanzee population in the country is found in the Gashaka Gumti NP on the eastern border, in Taraba and Adamawa states. It hosts up to 1 500 chimpanzees. Chimpanzees also occur in neighboring areas, including the Mambilla Plateau (specifically in Ngel Nyaki Forest Reserve and Leinde Fadali and Akwaizantar forests) and the Donga River valley.[13]

Cross River state

There are thought to be around 400 chimpanzees living in the Okwangwo Division of Nigeria's Cross River NP, in the northern part of Cross River state. This population is contiguous with populations in the adjacent Takamanda Forest Reserve of southwest Cameroon and the Mbe Mountains Community Forest in Nigeria. The Mbe population is probably still tenuously linked to that in the Afi Mountain Wildlife Sanctuary and other areas of the Afi River Forest Reserve immediately to the west of Mbe, although the Ikom–Obudu highway runs between the Mbe Mountains and the Afi River Forest Reserve.[13] In the southern part of Cross River state, the Oban Division of Cross River NP and surrounding forest reserves and community forests are also important for chimpanzees.

THREATS

The major threats to gorilla and chimpanzee survival in Nigeria are hunting, forest degradation, and deforestation. In 2000, it was estimated that

135 170 km^2 of forest remained in Nigeria, with an average annual decrease of forest cover of 3 980 km^2 or 2.6 percent.[5] There are logging concessions in almost all forest reserves in Nigeria, although not all are being actively logged. Much illegal logging also occurs – by 1987, around 24 percent of Nigeria's protected land area had already been converted into farmland, plantations, and bush-fallow.[2] The expansion of agriculture, oil palm plantations, and road networks has led to the widespread degradation and fragmentation of great ape habitat.

Deepening poverty levels and an increased gap between rich and poor, accompanied by a growing awareness of wealthy lifestyles, combine to increase rural people's incentives to exploit forest habitats. The relationship with national park authorities is sometimes a difficult one, with resentment building when forest products such as plant material or bushmeat are confiscated.

Chimpanzee meat is not generally taboo in most of southern Nigeria, although local taboos may exist. There is less hunting of chimpanzees in Islamic areas in the northern parts of their Nigerian range (especially Gashaka Gumti NP), but hunting is nevertheless reported there.[16] It is unlikely that any populations are entirely free from hunting pressure, but it is more often opportunistic than planned. Infants are sometimes sold as pets, and the value of these infants may encourage hunters to target females with young.

Hunting has historically threatened the survival of Cross River gorillas. In 1989, it was suggested that twice as many were killed each year as were being born.[10] At that time a reasonable monthly wage was 150 naira per month, and a single gorilla carcass could fetch as much as 300 naira. About 15 communities hunted in the gorilla's range, and in 1986 just one of these was reported to have killed eight gorillas.[10] The hunting of gorillas is now much reduced, however. This is largely due to increased conservation education in Nigeria, beginning with a Nigerian Conservation Foundation (NCF) project that followed up on the 1989 survey, followed by the Okwangwo program of WWF–The Global Conservation Organization and, most recently, a Nigerian Conservation Foundation–Wildlife Conservation Society (WCS) program. There is an occasional report of a gorilla being killed by hunters in the Okwangwo Division of Cross River NP, but there is no direct evidence of any gorillas having been killed at Afi or Mbe in the last five years.[14]

Plans for road development in the vicinity of the Oban Division of Cross River NP are a current concern. Traffic may be increased on a road already running through the Oban Division from Calabar to the Cameroon border at Ekang, and there is also a plan to improve the transborder highway from Ikom to Mamfe.

LEGISLATION AND CONSERVATION ACTION

Legislation

Nigeria ratified the African Convention on the Conservation of Nature and Natural Resources in 1968, the Convention on International Trade in Endangered Species of Wild Fauna and Flora (CITES) in 1973, and the Convention on Biological Diversity in 1996. The Endangered Species Act of 1985 is the legal instrument through which the international treaties are enforceable. All wildlife in national parks is protected by law, but these laws are often not enforced. There is little control of hunting in forest reserves, but the designation at least means that the areas are intended to remain under forest.

Conservation and research projects

There are many conservation projects in Nigeria, mostly coordinated by local nongovernmental organizations. International organizations also work in the country, with many activities focused close to the border with Cameroon. A WWF program supported by the European Union was operating in the Okwangwo Division of the Cross River NP until 1998.[12] Rural development, education, and protection efforts were carried out in the park and adjacent Mbe Mountains, and a gorilla census was supported. WWF is not currently active in this area, but Fauna and Flora International (FFI) is assisting

Volker Sommer

Gashaka Gumti National Park (Gashaka sector) during the rainy season: chimpanzee habitat.

the Cross River State Forestry Commission to develop better management practices at Afi Mountain. To the northeast, there is a chimpanzee habituation and behavioral research project at Gashaka Gumti NP, involving researchers from University College London. WCS has a biodiversity research program in southeastern Nigeria, run jointly with the Nigerian Conservation Foundation. USAID and the US Fish and Wildlife Service fund its Cross River gorilla research and conservation component.

Three transboundary protected areas have been proposed and are now at various stages of discussion. One would unite the Okwangwo Division of Cross River NP with Cameroon's Takamanda Wildlife Reserve, one Gashaka Gumti NP with Cameroon's Tchabal-Mbabo NP, and a third, in the south, would unite the Oban Division of Cross River NP with Korup NP in Cameroon.[3, 15, 17]

Sanctuaries and rehabilitation

The Pandrillus Drill Rehabilitation and Breeding Centre provides a sanctuary to orphaned chimpanzees, but does not breed them.[18] Young chimpanzees are quarantined and prepared for forest life at the urban facility in Calabar, then move to a 0.02 km^2 enclosure at Afi Mountain, which contributes to conservation education in the area. There were 23 chimpanzees here in 2004. The center is involved in general conservation work in southeast Nigeria including creation of protected areas, community outreach, and education.

FUTURE CONSERVATION STRATEGIES

The following priorities for action are derived from Oates *et al.* (2003)[13] for chimpanzees and Sunderland-Groves and Oates (2003)[17] for gorillas. The latter are the recommendations from the Second International Workshop and Conference on the Conservation of the Cross River Gorillas held in Limbé Botanic Garden, Cameroon in 2003.

Research

Basic research into the ecology, distribution, and population biology of the Cross River gorillas should be expanded.

For chimpanzees, priority actions in the Gashaka–Mambilla area include supporting existing conservation and research activities in Gashaka Gumti NP, Nigeria. It has been recommended that basic surveys be conducted to assess chimpanzee distribution and numbers and evaluate existing and potential connectivity between forests and populations.

In southwest Nigeria and the Niger Delta, surveys are urgently needed to assess the distribution, abundance, and genetic affinities of chimpanzees. The most viable populations should receive effective protection, and at least one should be selected for a long-term research effort.

Protection

Protected areas and wildlife laws need to be made more effective, and the institutional capacity to enforce them enhanced. In particular, protection and law enforcement measures should be strengthened for all Cross River gorilla populations. There should be more initiatives to raise public awareness of conservation issues, including the impacts of the bushmeat trade.

For gorillas, three major initiatives are recommended: first, the establishment of a Cameroon–Nigeria transboundary protected area for the Takamanda–Okwangwo complex, in particular by upgrading the protection status of the Takamanda Forest Reserve; second, the development of land-use plans for the wider Takamanda–Mone–Mbulu area in Cameroon, which would include a network of protected areas and corridors; and third, a plan for the conservation of the Afi–Mbe–Okwangwo area in Nigeria, with a review of the management status for the Mbe Mountains and the maintenance of forested connections between gorilla habitats.

It has been recommended that formal protected area or reserve status be extended to all remaining chimpanzee habitats in Nigeria, including, in particular, the community forests in Bayelsa state (the proposed Edumanom Forest Reserve); the remnant forests of the Mambilla Plateau in Taraba state; and the Mbe Mountains in Cross River state. The Takamanda–Okwangwo transboundary protected area would benefit chimpanzees as well as gorillas.

Cross River gorilla management committees should be established in Cameroon and Nigeria. Capacity building for conservation and research is needed by the relevant institutions in Nigeria and Cameroon, including government departments, universities, and nongovernmental organizations. Finally, it is vital to incorporate local community needs into the development of management strategies, including the study of alternative livelihood options.

FURTHER READING

Gonder, M.K., Oates, J.F., Disotell, T.R., Forstner, M.R., Morales, J.C., Melnick, D.J. (1997) A new West African chimpanzee subspecies? *Nature* **388**: 337.

Ite, U.E., Adams, W.M. (1998) Forest conversion, conservation and forestry in Cross River state, Nigeria. *Applied Geography* **18** (4): 301–314.

Oates, J.F., McFarland, K.L., Groves, J.L., Bergl, R.A., Linder, J.M., Disotell, T.R. (2002) The Cross River gorilla: natural history and status of a neglected and critically endangered subspecies. In: Taylor, A.B., Goldsmith, M.L., eds, *Gorilla Biology: A Multidisciplinary Perspective*. Cambridge University Press, Cambridge, UK. pp. 472–497.

Osemeobo, G.J. (2001) Is traditional ecological knowledge relevant in environmental conservation in Nigeria? *International Journal of Sustainable Development and World Ecology* **8** (3): 203–210.

Sommer, V., Adanu, J., Faucher, I., Fowler, A. (2004) Nigerian chimpanzees (*Pan troglodytes vellerosus*) at Gashaka: two years of habituation efforts. *Folia Primatologica* **75** (5): 295–316.

Sunderland-Groves, J.L., Oates, J. (2003) Protection strategies for Cross River gorillas. *Gorilla Journal* **27**: 12–13. http://www.berggorilla.de/english/gjournal/texte/27crossr.html.

MAP DATA SOURCES

Map 16.15 Great apes data are based on the following sources:

Butynski, T.M. (2003) The chimpanzee *Pan troglodytes*: taxonomy, distribution, abundance, and conservation status. In: Kormos, R., Boesch, C., Bakarr, M.I., Butynski, T.M., eds, *West African Chimpanzees: Status Survey and Conservation Action Plan*. IUCN/SSC Primate Specialist Group. IUCN, Gland, Switzerland. pp. 5–12.

Oates, J., Gadsby, L., Jenkins, P., Gonder, K., Bocian, C., Adeleke, A. (2003) Nigeria. In: Kormos, R., Boesch, C., Bakarr, M.I., Butynski, T.M., eds, *West African Chimpanzees: Status Survey and Conservation Action Plan*. IUCN/SSC Primate Specialist Group. IUCN, Gland, Switzerland. pp.123–130.

With additional data by personal communication from Bergl, R. (2005) and from the following source:

Sunderland-Groves, J.L., Maisels, F., Ekinde, A. (2003) Surveys of the Cross River gorilla and chimpanzee populations in Takamanda Forest Reserve, Cameroon. In: Comiskey, J.A., Sunderland, T.C.H., Sunderland-Groves, J.L., eds, *Takamanda: The Biodiversity of an African Rainforest*. Smithsonian Institution, Washington, DC. pp. 129–140.

For protected area and other data, see 'Using the maps'.

ACKNOWLEDGMENTS

This country study draws extensively on the Nigeria chapter from the IUCN/SSC *West African Chimpanzees: Status Survey and Conservation Action Plan*, Oates *et al.* (2002),[14] and Sunderland-Groves and Oates (2003).[17] Many thanks to Richard Bergl (City University of New York), Alexander Harcourt (University of California, Davis), Bethan Morgan (Zoological Society of San Diego), John F. Oates (Hunter College, City University of New York), and Jason Sali (Development in Nigeria) for their valuable comments on the draft of this section, and to Brigid Barry (Tropical Biology Association) for editorial assistance.

COMPILER

Edmund McManus, UNEP World Conservation Monitoring Centre

REPUBLIC OF RWANDA

NIGEL VARTY

BACKGROUND AND ECONOMY

The Republic of Rwanda is a mountainous country 26 338 km^2 in area. It is located in the equatorial highlands of the Western or Albertine Rift Valley, bordered by the Democratic Republic of the Congo (DRC), Uganda, the United Republic of Tanzania, and Burundi. It has a temperate climate with two rainy seasons (March to May and October to December). The official languages are French, English, and Kinyarwanda, and Christianity is the majority religion. Rwandan society has traditionally been portrayed as consisting of three groups: Hutu (84 percent), Tutsi (15 percent), and Twa (1 percent), and historically the Tutsi provided the governing

Map 16.16 Great ape distribution in Rwanda

Data sources are provided at the end of this country profile

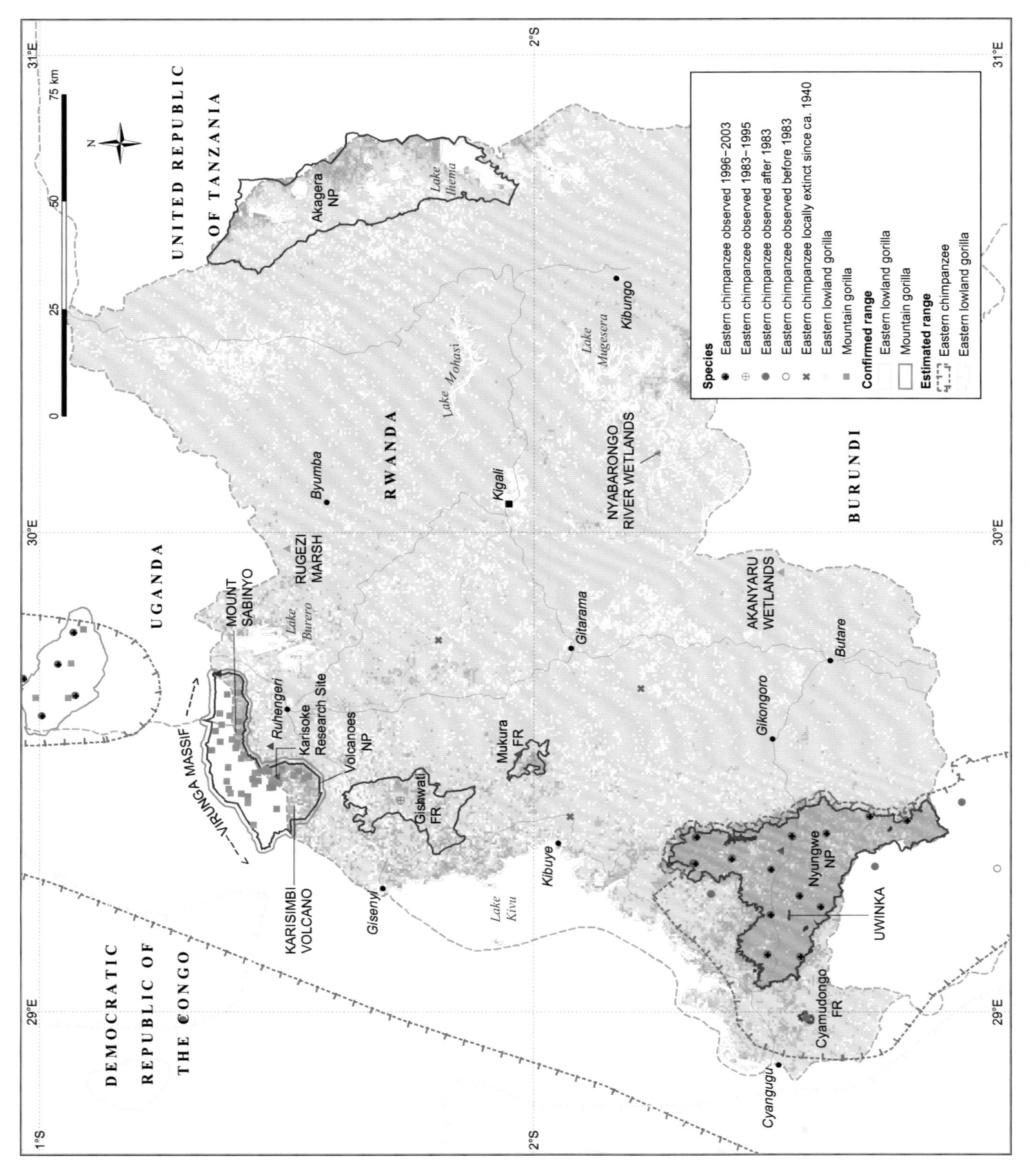

class. Rwanda gained independence from Belgium in 1962, and was subsequently ruled by Hutu-dominated governments.[10]

From 1972 to 1994, Rwanda was ruled by one party under President Juvénal Habyarimana, a Hutu. Following the fall of the Tutsi monarchy in 1961, the Tutsis were both excluded from power and subject to episodic massacres. Many fled the country, and some gathered in Uganda to form an armed movement (the Rwandan Patriotic Army, RPA) to resist Hutu dominance. The Rwandan Patriotic Army invaded Rwanda in 1990 but was fought to a standstill. A power-sharing agreement was brokered by countries in the region and signed in 1994, but immediately after signing it President Habyarimana was assassinated and a full-scale program of genocide against Tutsis and moderate Hutus began. Hutu peasants, incited to kill by ethnic propaganda, supported the extremists. At least 800 000 people were killed and over a million refugees escaped from the country.[19] The Rwandan Patriotic Army captured the capital, Kigali, in July 1994, and ended the genocide, forming the Government of National Unity to implement the earlier power-sharing agreement. France also intervened militarily at this time. Fighting nevertheless continued in and around Rwanda, and in 1996 spilled over into DRC (then Zaïre). The Rwandan Patriotic Army backed a rebellion there, which destroyed the refugee camps and resulted in the overthrow of President Mobutu in May 1997 (see DRC country profile). One result was the mass repatriation of refugees from DRC in late 1996.

By August 1998, continuing insecurity on Rwanda's western border led Rwanda, initially in coalition with Uganda, to intervene again in eastern DRC. The fighting involved DRC forces and their Zimbabwean, Chadian, Angolan, and Namibian allies on one side, and rebels supported by Uganda and Rwanda on the other. By the time that a ceasefire agreement was signed in autumn 1999, the rebels controlled large areas in the north and east. Further fighting and negotiations occurred, and a more complete accord was brokered by South Africa in 2002, allowing for Rwandan withdrawal of its troops from DRC, and disarmament, demobilization, and repatriation of Rwandan exiles.

In the aftermath of these events, the population was estimated at 7.8 million in 2003, with a growth rate of 1.8 percent.[5] In 2002, gross domestic product (GDP) was estimated to be US$1.7 billion, and gross national income (GNI) per person was only US$230.[48] The country's Human Development Index is ranked 159th out of 177 in the world.[42] Subsistence agriculture is by far Rwanda's most important sector, and in 2002 employed 90 percent of the workforce and contributed 45 percent of GDP.[5] About 42.6 percent of the country is occupied by permanent crops and arable land.[5] Natural forests cover about 462 km^2 (1.8 percent) of land area, according to FAO, although the area of protected forest suggests that this is an underestimate. There are a further 2 610 km^2 of forest plantations, largely of eucalyptus and pine.[8, 9] Most of Rwanda's natural forests lie in the afromontane region (1 600 m to 4 500 m, typically 2 000 m), and are characterized by clearings (thought to be the result of human disturbance) and dense understory typical of montane forests.[9] The country has had the highest rate of forest loss in the region.[8] Until recently, there were four main montane forests in Rwanda (Nyungwe, Gishwati, Mukura, and Volcanoes), all located in the west of the country. Nyungwe National Park (NP) and the adjacent Kibira NP in Burundi still form one of the largest remaining blocks of lower montane forest in Africa.[6, 44]

DISTRIBUTION OF GREAT APES

The mountain gorilla (*Gorilla beringei beringei*) and the eastern chimpanzee (*Pan troglodytes schweinfurthii*) occur in Rwanda. Mountain gorillas occur in the Volcanoes (Volcans) NP in northern Rwanda. The gorillas here comprise part of the larger single Virunga population, which includes those in the contiguous Mgahinga NP in Uganda and in the southern sector of Virunga NP in DRC. This mobile population cannot be precisely divided according to country, but has been censused repeatedly since the 1970s (see Chapter 8), making it one of the best monitored of all great ape populations. A partial census of the Virunga mountain gorillas was conducted in 2000,[18] followed by a full census in September and October 2003. The census was a joint effort of the International Gorilla Conservation Programme (IGCP), Wildlife Conservation Society (WCS), Dian Fossey Gorilla Fund International (DFGFI) and Europe (DFGFE) and Berggorilla & Regenwald Direkthilfe. It put the population at 380, which was a 17 percent increase (56 more gorillas) over the previous full census of 1989. Of these gorillas, about a third are estimated to be resident in Rwanda.[4, 18]

Eastern chimpanzees are restricted to the southwest of Rwanda, largely in Nyungwe NP, although they formerly occurred in other forest areas to the west[2, 41] Records also exist for a small area in the northwest, in the Gishwati Forest Reserve area,[2] but if they still exist here, they are very few.[30] There have been no recent national surveys, but in 1987–1989 there were estimated to be at least 500 chimpanzees remaining.[27]

THREATS

Rwanda has the highest human population density of any African nation, is overwhelmingly agrarian, and ranks among the 10 poorest nations in the world. These factors result in acute land and resource scarcity, and consequent pressure on natural habitats. Illegal clearing for subsistence crops, tree felling for timber and fuel (firewood is the country's main source of energy), fire (often set by honey gatherers), cannabis propagation, and mining all threaten Rwanda's forests.[6, 24] It is estimated that from 1990 to 2000, an average of 150 km² or 3.9 percent of the total was cleared each year. This includes natural and planted cover, but the figures of the Food and Agriculture Organization of the United Nations (FAO) indicate that there has been an increase in plantation cover over this period, implying that the loss is of natural forests.[7, 8] Nyungwe forest (now NP) covered 1 140 km² in 1960, but had been reduced to 945 km² by 1996, and Mukura forest covered 30 km² in 1960, but only 16 km² in 1996.[12] There has been little habitat loss in the Volcanoes NP in recent years,[33] but half the park was lost during the late 1960s to early 1970s.[30, 45] In 2004, there was a proposal to route electricity pylons through the park to the radio antennae on the summit of Karisimbi Volcano.[24]

In 1981, 110 mountain gorillas were thought to be living in the Volcanoes NP, with an additional 30–40 using the park some of the time.[16] By 2001, the estimated population was 129 animals.[4] This is consistent with an increased gorilla population in the Virungas as a whole over the last 20 years.[1, 38] While this is encouraging, the global mountain gorilla population is still very small and at risk.

Chimpanzees are thought to have declined in numbers in Rwanda over the last 50 years, though it is not clear whether this is more a reflection of a lack of observers in the field. There does seem to have been a marked reduction in range, however, with chimpanzees lost from the center-west region of Rwanda since 1940, and no records since 1983 in several other areas.[2]

Although the bushmeat trade increased following the genocide in 1994,[29] chimpanzees and gorillas are generally not hunted for food in Rwanda. Two nursing female gorillas were killed and a young male abducted from the Volcanoes NP in Rwanda on May 9 2002 by poachers hoping to sell the infant.[26] There is no established trade in baby gorillas, however, and a possible threat from poaching is inferred only from the fact that market prices are quite high.[21] That few young gorillas have been taken in recent years, even during periods of conflict, reflects the dedication of Volcanoes NP staff and international nongovernmental organizations in maintaining patrols of key gorilla areas. A much greater threat is the setting of wire snares for ungulates by people living adjacent to the Volcanoes NP in Rwanda, and by professional poachers. Two veterinary surgeons are permanently employed at the park to monitor the health of habituated groups of mountain gorillas, and to remove snares from the limbs of injured animals.[29, 33]

There is a high human population density around Nyungwe[24] and snares are common, with several thousand being collected annually by guards from the Office of Tourism and National Parks (ORTPN). Those that remain in the forest undoubtedly pose a threat to chimpanzees, but the numbers of deaths and injuries are unknown.[30] There is no sign that Nyungwe chimpanzees have been captured for the pet trade.[30] Mining and forest harvesting are ongoing in this park, and forest fires are frequent.[24]

Mountain gorillas can suffer from many of the same diseases as humans.[14] Close contact with tourist groups, researchers, and local people is believed to have caused a number of outbreaks of illness in the Virungas, which have impacts rivaling those of hunting.[47] In 1988, for example, measles or

Elizabeth A. Williamson

The influx of Rwandan refugees led to deforestation of large areas of Virunga National Park.

a related morbillivirus killed six habituated female mountain gorillas.[37, 47] In 1990, bronchopneumonia affected 26 of the 35 gorillas in a tourist group, killing two.[23] When the security situation allows, 70 percent of the Virunga gorillas are visited daily by more than 70 tourists and a similar number of guides, porters, rangers, and researchers.[3] This continued daily exposure to large numbers of people and their diseases is considered to be a major potential threat.[4] Groups of chimpanzees habituated for tourism and research are similarly vulnerable[4] and those in the Nyungwe NP could be at risk.

The war and genocide in Rwanda caused a massive increase in human traffic across the Virungas, followed by a sustained military presence. Parasites known to infect humans were subsequently found for the first time in mountain gorillas.[46] Between 1992 and 2000, 12–17 gorillas died from military gunfire or explosions in the Virungas,[18] representing between 4 and 5 percent of the 1989 population. The recent conflicts and subsequent resettlement of many tens of thousands of refugees and internally displaced people have also resulted in the loss and degradation of great ape habitat. The Gishwati Forest Reserve, for instance, was almost totally encroached during this period and only tiny degraded fragments remain.[6]

LEGISLATION AND CONSERVATION ACTION

Rwanda ratified or acceded to the Convention on Biological Diversity in 1996; the Convention on International Trade in Endangered Species of Wild Fauna and Flora (CITES) in 1980; the UN Convention to Combat Desertification in 1998; the African Convention on the Conservation of Nature and Natural Resources in 1979; the Convention on Migratory Species in 2005; and the World Heritage Convention in 2000. The country also participates in UNESCO's Man and Biosphere (MAB) Programme, under which a 151 km^2 Biosphere Reserve has been designated at the Volcanoes NP.

Chimpanzees and gorillas are protected by law, and national and international trade is restricted.[15] Ordinance 18/6/73, modified by Law-Decree 26/4/1973 and Law 34/2000, established the Office of Tourism and National Parks, and governs the creation and functioning of protected areas and hunting arrangements. Forestry matters are governed by Law 47/1988, which sets out the basic legislation concerning conservation and use of forests. Ministerial Decision 02/88 and Prime Ministerial Order 72/03 of October 14 2002 detail the organization of the Department of Forestry (DoF), which sits alongside the Department of Environment (DoE) within the Ministry of Lands, Environment, Forestry, Water, and Natural Resources (MINITERE). On the whole, the departments deal with policy development, and the agencies with implementation.[36]

Charles Mayhew/Tusk Trust

A juvenile mountain gorilla.

The Department of Environment has overall responsibility for biodiversity conservation. The management of forest reserves is the responsibility of the Department of Forestry. The Office of Tourism and National Parks, however, based within the Ministry of Commerce, Industry, Investment Promotion, Tourism, and Cooperatives, has responsibility for management of national parks and matters relating to ecotourism. The Volcanoes and Nyungwe NPs are therefore the responsibility of the Office of Tourism and National Parks, whereas the Mukura and Gishwati Forest Reserves are the responsibility of the Department of Forestry. The Rwanda Environmental Management Authority (REMA) has also been established, and is expected to take charge of a wide range of environmental projects.[36]

Protected areas

There are four main categories of protected area in Rwandan law: national parks, hunting reserves, special reserves, and forest reserves. There are currently three national parks, including Volcanoes NP and the recently designated Nyungwe NP, and three forest reserves, but no hunting reserves or special reserves.[20, 30, 43] Until 1997, about 15 percent of the country was included in national parks and reserves, but the Akagera NP was subsequently reduced to about 40 percent of its original size to allow resettlement of refugees. Today, the protected area system covers about 8 percent of the country.[20]

The Virunga NP on the borders of the present-day Rwanda, Uganda, and DRC was created in 1925 for the protection of the mountain gorilla, and was the first national park in Africa. The Rwanda portion was established as the separate Volcanoes NP in 1929. Its current area is 160 km^2. Of the hundreds of species found here, many are unique to the Western Rift region. CARE International and the International Gorilla Conservation Programme (IGCP) are working with local farmers to develop 'on-farm resources' to try to reduce pressures on the park.

A 1999 survey of Nyungwe forest (now NP) found chimpanzees to be widespread across the reserve, especially in the western section, and the reserve is thought to support a relatively large and viable population, living at one of the highest altitudes yet recorded for this species.[31] Some 100–200 chimpanzees may remain in the 980 km^2 park, but the most recent estimate dates from the 1980s;[40] more than 1 000 were thought to be present in the early 1980s.[39] Chimpanzees apparently still survive in the nearby 6 km^2 Cyamudongo Block to the southwest of Nyungwe, which has now been incorporated into the park.[20, 30] The park has been heavily disturbed in the past by mining (particularly for gold), road building, selective logging, reforestation with exotic species, and encroachment by local farmers.[22, 31] Chimpanzees were also previously known from the Gishwati Forest Reserve (near Ruhengeri), where at least 13 remained in 1985,[22] but after the civil conflict only 6 km^2 of this forest remains and it is doubtful that the species survives here.

Conservation action

Wildlife tourism, particularly visits to the Virunga gorillas, is seen as a key tool in protecting forest conservation areas in Rwanda. Tourism in the Volcanoes NP provided Rwanda with US$4–6 million in 1989, in direct and indirect revenues. Tourism at Nyungwe provided around US$0.5–1.0 million in 1990.

During the conflict, the headquarters of the Volcanoes NP was ransacked and destroyed. Some of the staff were killed and others temporarily evacuated. Tourism is now increasing again, but security concerns remain a major deterrent. In 2003, bookings for gorilla tourism were at about 60 percent capacity,[21] while visitors to Volcanoes NP were escorted by heavily armed guards. Fees were increased in 2004 to US$350 per tourist per hour of contact time with gorillas.[34] The income is spent on protecting the park and investing in local development projects.[17]

The Rwandan Office of Tourism and National Parks typically works with nongovernmental organizations to establish a common strategic framework and conservation goals, and a number of such organizations work in the Volcanoes NP area. Chief among them is IGCP, a joint initiative of the African Wildlife Foundation (AWF), Fauna and Flora International (FFI), and WWF–The Global Conservation Organization, which has been operating in Rwanda since 1978. IGCP continued to provide technical and material support to the park throughout the conflict in the 1990s, including emergency funding for basic park operations, salaries and rations for the patrols, equipment and medical supplies, rehabilitation of infrastructure, and clearing of mines. Other groups involved with gorilla conservation in Rwanda include the Dian Fossey Gorilla Fund International and the Dian Fossey Gorilla Fund Europe, which support research and community awareness, and work in Volcanoes NP, the Karisoke Research Center, and the Mountain Gorilla Veterinary Center (MGVC). Together these provide technical and financial assistance worth around US$1 million per year to the Office of Tourism and National Parks for management, research, and veterinary support at the park, and for local development activities around its borders.[30]

WCS has a long-standing Project for the Conservation of Nyungwe Forest (PCFN), which provides technical and financial assistance worth about US$200 000 per year to the Office of Tourism and National Parks for management, research, tourism development, rural public awareness, and staff training at Nyungwe NP. Ninety guards have so far been trained under the project.[30] The project established a research station and tourist facilities at Uwinka in the northwest sector of Nyungwe forest in the 1980s. These were looted during the genocide, but were saved from destruction by the French military intervention in Rwanda.[30] They have since been repaired and small streams of researchers and tourists have begun to return.

Following the deforestation and cultivation of a large area in the Mikeno sector of Virunga NP in DRC in 2004, the United Nations Environment Programme (UNEP), WWF, IGCP, the European Union, and Frankfurt Zoological Society released emergency funds to support the Congolese Institute for Nature Conservation (ICCN) with the construction of a dry stone wall to help restore the integrity of the park boundary.[11] By August 2004, over 7 km of the wall had been completed, and the work continues with a workforce of more than 2 000 people. Six Rwandan associations are building part of the wall, which runs along the international boundary. The construction of the wall clearly demarcates the boundary of the park and it is expected that it will, over time, come to be accepted as the limit of cultivation.

Much research has been conducted on the Virunga mountain gorillas.[13, 18, 33, 35, 45] Karisoke is the base for most research within Volcanoes NP, although research has been intermittent since the late 1980s due to political and civil unrest. It was established in 1967 and Dian Fossey ran the site until she was murdered there in 1985. In 1997, three Spanish medical workers in Ruhengeri were murdered, resulting in the departure of expatriate staff. Great insecurity followed and many people, including several Karisoke staff, were arrested and imprisoned.[30]

Organizations involved in the 2003 census of the Virunga gorillas included IGCP, the three park authorities (Rwanda's Office of Tourism and National Parks, the Uganda Wildlife Authority, and the DRC's ICCN), the Institute for Tropical Forest Conservation (ITFC) of Uganda's Mbarara University, the Dian Fossey Gorilla Fund, the Mountain Gorilla Veterinary Center, WCS, and the Max Planck Institute for Evolutionary Anthropology. Gorilla and habitat monitoring teams have been established in all the Virunga parks by IGCP and the park authorities.[21]

There are no great ape sanctuaries in Rwanda, but the Mountain Gorilla Veterinary Center, established in 1986 near the Volcanoes NP, looks after the health of wild gorillas and helped to care for baby gorillas confiscated by the authorities in 2002.[21] One infant was reintroduced to its natal group, and one was reared in captivity, but both died.

FUTURE CONSERVATION STRATEGIES

A workshop held in July 2003 led to the production of a national great ape survival plan (NGASP), including a list of potential conservation projects.[24] In addition, the National Biodiversity Strategy and

Ian Redmond/UNESCO

Guards at Volcanoes National Park receive new uniforms.

Action Plan (NBSAP) calls for improved conservation of protected areas. The measures proposed include development and implementation of land use and management plans for each protected area, continuing research at protected areas including inventory studies, and the sustainable use of biodiversity including development of ecotourism.[6] In November 2004, Rwanda became the first of the 23 great ape range states to officially endorse and distribute its national great ape survival plan. The chief recommendations of this plan are outlined here.

For gorillas:

- protect and conserve Volcanoes NP, at least within the current limits of the park, by diminishing poaching and other illegal activities;
- increase the benefits to local communities from conservation, and increase local support for conservation;
- increase awareness of conservation issues at the local and political levels;
- continue and increase research and monitoring for the gorillas and their habitat;
- improve human, gorilla, and livestock health standards; diminish disease among the gorillas through control of zoonoses and gorilla/livestock interaction; and
- increase tourism revenue through sustainable ecotourism.

For chimpanzees:

- reduce the area of degraded habitat and ensure no further loss of habitat;
- promote, implement, and enforce laws and policies regarding chimpanzees and their habitat;

- increase awareness of, and pride in, the importance of both chimpanzees and their habitat;
- improve community livelihoods around chimpanzee habitats through conservation;
- create a greater understanding of status, trends, and biology of chimpanzees; and
- improve protection and management of Nyungwe and Cyamudongo.

IGCP is developing a regional framework for conserving gorilla habitat, and there is a proposal to develop a tourism development plan for the Volcanoes NP and the Ruhengeri region. In addition, the park authorities of the three countries involved have accepted the creation of a transfrontier park, covering the whole range of the Virunga gorillas.[21] Further investigation into the possible effects of stress and transmission of disease from close human contact with habituated apes has been recommended[3, 4, 14] and recommendations and a 'best practice' code to reduce the risk of disease transmission from humans to apes has been suggested.[14, 47]

Continued close monitoring of the mountain gorillas, improved conservation awareness and revenue-sharing programs, and better information on the numbers and distribution of chimpanzees in Nyungwe NP have also been recommended.[28] A census of Nyungwe chimpanzees, funded by WCS and the University of Antioch, was ongoing in early 2004.

FURTHER READING

DoE (2003) *National Strategy and Action Plan for the Conservation of Biodiversity in Rwanda.* Department of the Environment, Ministry of Lands, Resettlement, and Environment, Kigali, Rwanda. http://www.biodiv.org/doc/world/rw/rw-nbsap-01-en.pdf. Accessed March 31 2003.

Kanyamibwa, S. (1998) Impact of war on conservation: Rwandan environment and wildlife in agony. *Biodiversity and Conservation* **7**: 1399–1406.

Monfort, A. (1992) Première liste commentée des mammifères du Rwanda. *Journal of African Zoology* **106**: 141–151.

Robbins, M.M., Sicotte, P., Stewart, K.J., eds (2001) *Mountain Gorillas: Three Decades of Research at Karisoke.* Cambridge University Press, Cambridge, UK.

MAP DATA SOURCES

Map 16.16 Great apes data are based on the following source:

Butynski, T.M. (2001) Africa's great apes. In: Beck, B.B., Stoinski, T.S., Hutchins, M., Maple, T.L., Norton, B., Rowan, A., Stevens, E.F., Arluke, A., eds, *Great Apes and Humans: The Ethics of Coexistence.* Smithsonian Institution Press, Washington, DC. pp. 3–56.

With additional data by personal communication from Plumptre, A. (2004) and from the following sources:

Kalpers, J., Williamson, E.A., Robbins, M.M., McNeilage, A., Nzamurambaho, A., Lola, N., Mugiri, G. (2003) Gorillas in the crossfire: population dynamics of the Virunga mountain gorillas over the past three decades. *Oryx* **37**: 326–337.

Lee, P.C., Thornback, J., Bennett, E.L. (1988) *Threatened Primates of Africa. The IUCN Red Data Book.* IUCN, Gland, Switzerland and Cambridge, UK.

MINITERE (2004) *Rwanda's National Great Apes Survival Plan 2004-2009.* Final draft. Ministry of Lands, Environment, Forestry, Water, and Natural Resources, Republic of Rwanda.

Plumptre, A.J., Masozera, M., Fashing, P.J., McNeilage, A., Ewango, C., Kaplin, B.A., Liengola, I. (2002) *Biodiversity surveys of the Nyungwe Forest Reserve in S.W. Rwanda.* WCS Working Paper 19. http://www.wcs.org/media/file/workingpaper19.pdf.

For protected area and other data, see 'Using the maps'.

ACKNOWLEDGMENTS

Many thanks to Alexander Harcourt (University of California, Davis), Annette Lanjouw (International Gorilla Conservation Programme), Andrew Plumptre (Wildlife Conservation Society), Elizabeth A. Williamson (University of Stirling), and an anonymous reviewer for their valuable comments on the draft of this section.

AUTHOR

Nigel Varty, UNEP World Conservation Monitoring Centre

REPUBLIC OF SENEGAL

EDMUND MCMANUS

BACKGROUND AND ECONOMY

The Republic of Senegal is the westernmost country of West Africa, bordering the Atlantic Ocean, with Guinea-Bissau and Guinea to the south, Mauritania to the north, and Mali inland to the east. The Gambia penetrates the central axis of the country along the Gambia River, and formed a brief political union with Senegal (under the name Senegambia) from 1982 to 1989.[5] Senegal itself became independent from France in 1960, its capital Dakar having until then been the administrative headquarters of French West Africa.[7] A succession of moderate, partly democratic governments held power during the 1960s and 1970s, and moderate, left-leaning governments won a series of regular elections through the 1980s and 1990s, with the opposition then winning the 2000 presidential and 2001 legislative elections.[7] There has been an armed separatist movement in the Casamance region of southern Senegal since 1982. This region is the richest agricultural area in Senegal, but as a result of conflict, it lacks basic infrastructure and good links to the rest of Senegal.

The total land area of Senegal is 196 190 km^2. Annual rainfall ranges from 200 mm in the northern regions to 1 600 mm in the south, with a mean daily temperature range of 18–35ºC. There is a dry season from November to April/May and a rainy season from May/June to October.[17] There are three main zones of natural vegetation, correlated with rainfall, from north to south. Northernmost is the Sahelian zone, with vegetation dominated by *Acacia* spp. and annual grasses. The central zone is Sudanian, with a range of vegetation types from wooded savannas to dry forests. Southernmost is the Guinean zone, characterized by less dry forests, with gallery forests along the river courses. Senegal replants forest trees on about 300 km^2 of land each year, though many young trees die due to lack of follow-up care.[9]

In mid-2004, there were an estimated 10.8 million people in Senegal, with a population growth rate of 2.6 percent.[5] There are several ethnic groups, including Wolof (43.3 percent), Pular (23.8 percent), Serer (14.7 percent), Diola (3.7 percent), Mandinka (3 percent), Soninke (1.1 percent), and others (5.4 percent), but the population is 95 percent Muslim.[7]

Even though many young people have migrated to urban areas, such as the capital Dakar, in search of employment, agriculture remains the chief industry.[21] The main crops are maize, rice, millet, sugarcane, alongside livestock, while cotton and peanuts are the chief export crops. The economy has benefited from a stable microeconomic policy since the late 1990s, but the country's gross domestic product (GDP) in 2002 was US$4.9 billion, or less than US$470 per person.[19] Senegal has many structural economic problems and poor social indicators (high unemployment, high illiteracy rates, and some poor health statistics, especially in rural areas). In 2001, Senegal was reclassified as a least developed country (LDC) by the UN and in 2004 was ranked 157th (out of 177) in the Human Development Index of the United Nations Development Programme (UNDP).[22]

DISTRIBUTION OF GREAT APES

The only great ape to be recorded in Senegal is the western chimpanzee (*Pan troglodytes verus*). It is restricted to the administrative region of Tambacounda, in the southeastern part of the country. The first systematic national survey of chimpanzee distribution was carried out in 2003–2004 and the results were at the analysis stage in 2004.[2]

Existing data suggest that approximately 10 percent of the country's chimpanzees are thought to live in the Niokolo-Koba National Park (NP),[4, 15] with the remainder of the population being distributed to the south and southeast of the park (Map 16.17). Of 10 areas around the park surveyed in 2000–2001, four were found to have relatively high concentrations of chimpanzee nests: Bandafassi, Fongolembi, Segou, and Tomboronkoto.[15] The Diarha River area serves as an important dry-season refuge for several families of chimpanzees. All populations are small.

THREATS

Although it is clear that Senegal's chimpanzee population has declined historically,[4] recent trends are unknown. In 1979 it was estimated at a maximum of 300,[10] but national estimates for other countries made at this time have since proved to be underestimates. The most recent population estimate, published in *West African Chimpanzees:*

Map 16.17 Chimpanzee distribution in Senegal

Data sources are provided at the end of this country profile

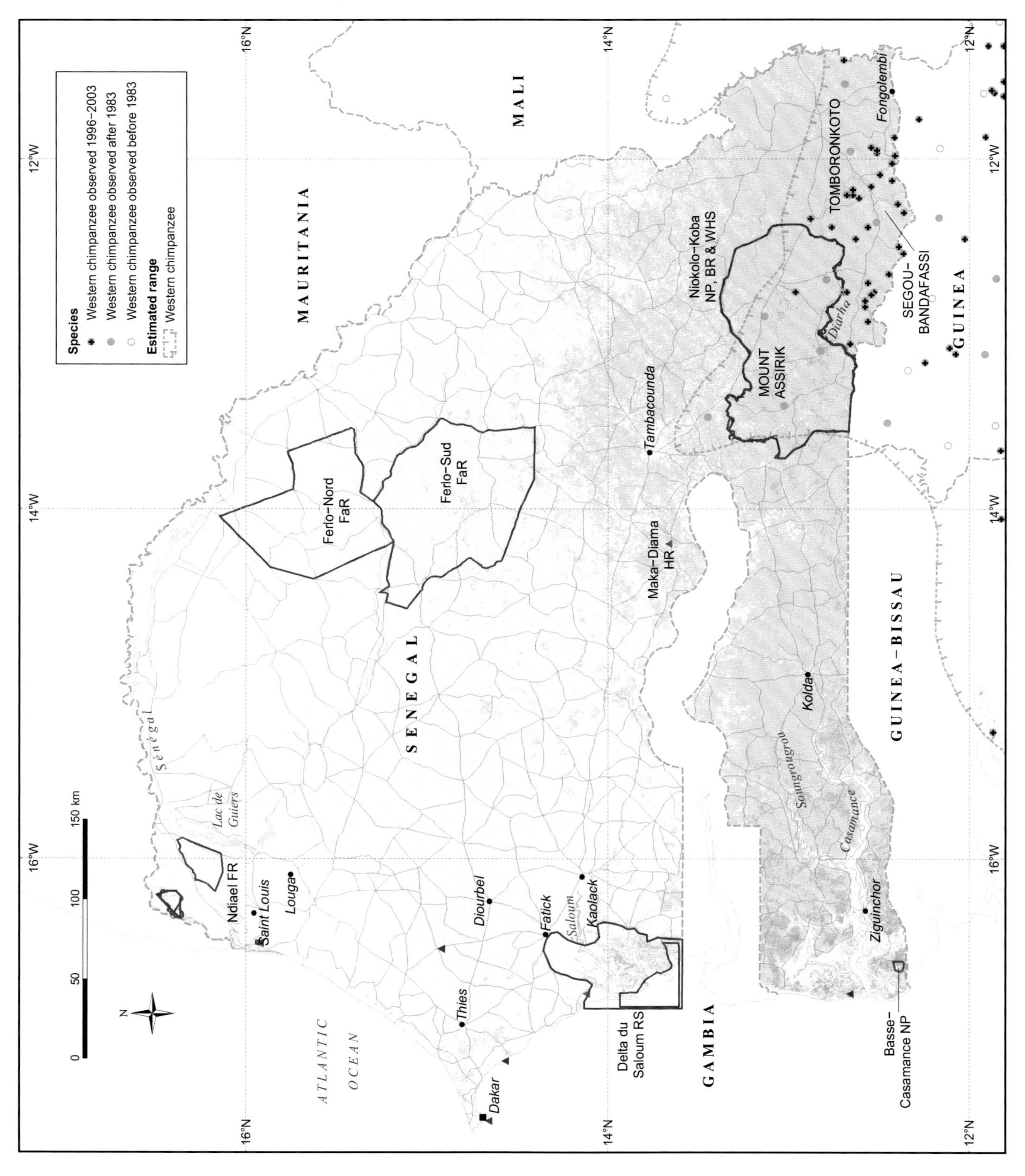

Status Survey and Conservation Action Plan, is that there are 200–400 chimpanzees living in the country.[4] This range is derived by extrapolating density figures taken from specific field sites within Senegal, estimates derived in other countries with similar habitat types, and responses to questionnaires and interviews.[1, 13, 15, 18] Widespread habitat destruction, fragmentation of remaining forest, and competition with humans over critical water and food sources are all threats to Senegal's chimpanzees, however, so the apparent population stability cannot be assumed to be real.

Habitat degradation and alteration are thought to be the most significant of these threats. It is clear that most of the country's small population of chimpanzees live outside the protected area network. They typically inhabit small areas of forest that are under increasing pressure, including the destruction of forest corridors that connect them to other forest blocks.[4] It is estimated that there are 62 050 km² of forest remaining, but most of this area is unsuitable for chimpanzees.[2] The deforestation rate is about 450 km² (0.7 percent) per year.[6]

Chimpanzees and humans compete for various wild foods, including honey and the fruit of the climber *Saba senegalensis* (Apocynaceae).[15] As the dry season progresses, many natural water sources dry up and in some areas chimpanzees and humans also compete for access to those few that remain.

Islam, the majority religion in Senegal, prohibits consumption of chimpanzee meat, and traditional beliefs also forbid the killing of chimpanzees on the grounds of their unique position among other primates.[14] Hunting does not appear to be a threat to chimpanzees in Senegal at present, although the home ranges of populations to the south of Niokolo-Koba NP are thought to extend across the border into Guinea, where hunting is more prevalent.

Since 1997, the Chimpanzee Rehabilitation Project in The Gambia has received five reports of captive baby chimpanzees in Senegal. Further investigation of the cause and scope of the problem has been recommended.[4]

LEGISLATION AND CONSERVATION ACTION

Legislation

Senegal is a party to the Convention on International Trade in Endangered Species of Wild Fauna and Flora (CITES), the Convention on Biological Diversity, the Convention on Migratory Species, the World Heritage Convention, and the African Convention on the Conservation of Nature and Natural Resources. The country provides chimpanzees with complete protection under the Code for Hunting and the Protection of Fauna.[4] Although it is theoretically possible to obtain a license to capture them for approved scientific purposes, no such licenses have been issued. Forest protection is provided for under the Forest Code. National plans relevant to the environment include: *Le Plan National d'Aménagement du Territoire*; *Le Plan National d'Action pour l'Environnement*; *La Stratégie Nationale et le Plan d'Action pour la Conservation de la Biodiversité*; and *Le Plan d'Action National de Lutte contre la Désertification*.

In 2003, a moratorium was placed on the granting of quarrying permits in forest reserves, in an attempt to reduce deforestation. Efforts to persuade companies involved in existing operations to move out were also underway.[9]

Protected areas

Senegal's national parks are managed by the Direction des Parcs Nationaux. The Niokolo-Koba NP, which is also a Biosphere Reserve and World Heritage Site, has an area of 9 130 km² and is the only protected area in Senegal to contain chimpanzees.[4] It is contiguous with the Badiar NP in Guinea, and there are moves to treat these two Biosphere Reserves as a Niokolo–Badiar transboundary protected area. The two management authorities already cooperate to some extent in scientific and technical matters.[23] The chimpanzee populations in the two parks appear to be isolated from one another.[8]

Niokolo-Koba NP was created as a hunting reserve in 1926, a forest reserve in 1951, a faunal reserve in 1953, and a national park in 1954, and was enlarged by a succession of decrees in 1962, 1965, 1968, and 1969. It rises from just above sea level to the summit of Mount Assirik at 311 m, and its ecosystems comprise about 55 percent grassland, 37 percent woodland, 5 percent bamboo, and 3 percent forest. The vegetation varies from a southern Sudanian type to Guinean, with savanna predominant, more luxuriant vegetation along the course of the rivers, and a varying cover of trees and bushes according to local topography and soils. The only people to live within the park are forest guards and workers at tourist camps.

Conservation and field projects

The Programme d'Education et de Recensement des Chimpanzés du Sénégal (PERCS) has carried

Officials at an illegal logging operation.

Georges Grépin

out chimpanzee surveys to obtain a complete understanding of their distribution, an estimate of their numbers, and more information about threats to their populations. Key populations were monitored over a year to identify migration patterns and dry-season refuge sites. Education activities were also carried out in critical areas, and a second phase is proposed to concentrate on educational activities and solutions to conflicts and competition over water.[3] PERCS works alongside the Direction des Parcs Nationaux, and receives financial support from Friends of Animals, a US nongovernmental organization.

The Niokolo-Koba NP is the site of the only long-term study of chimpanzee ecology in Senegal, with the first data having been collected by McGrew and colleagues in the 1970s.[11, 12] In May 2001, Pruetz established a research site in the Tomboronkoto region to study the ecology and behavior of 'savanna' chimpanzees, following up on work carried out by McGrew, Tutin, and colleagues at Mount Assirik.[15, 16, 20] The Tomboronkoto population lives close to humans and they are sometimes in conflict over resources, especially *Saba senegalensis*.[4] The Diarha River in the district of Salemata has several groups of chimpanzees and, following the PERCS surveys, has been selected as a second site for long-term monitoring to determine the number of chimpanzees using the area and to identify their migration routes.[4]

There are no known sanctuaries, rehabilitation centers, or reintroduction sites for chimpanzees in Senegal. The zoo in Dakar has a small captive collection.

FUTURE CONSERVATION STRATEGIES

The following recommendations were made in *West African Chimpanzees: Status Survey and Conservation Action Plan.*[4]

With only 200–400 chimpanzees remaining in Senegal, all locations that contain or that could support chimpanzee populations should be considered priority sites in need of special attention. Senegal should collaborate with Guinea on the protection of chimpanzee populations migrating across the Senegal–Guinea border. The relevant legislation should be amended to prohibit the capture and trade of chimpanzees for any reason other than for their conservation, thus excluding scientific purposes as a justification. Sustainable solutions to competition between humans and chimpanzees over water and the fruits of *Saba senegalensis* should be sought. More education work needs to be carried out throughout Senegal, emphasizing the role of habitat destruction as a threat to chimpanzee survival. Chimpanzee habituation for ecotourism may be detrimental to the already fragmented groups in the country, so a multidisciplinary board of scientists and conservationists should consider the issue on a case-by-case basis before such initiatives are approved.

FURTHER READING

McGrew, W.C., Baldwin, P.J., Tutin, C.E.G. (1980) Chimpanzees in a hot, dry and open habitat: Mt Assirik, Senegal, West Africa. *Journal of Human Evolution* **10**: 227–244.

Pruetz, J.D., Marchant, L.M., Arno, J., McGrew, W.C. (2002) Survey of savanna chimpanzees (*Pan troglodytes verus*) in Senegal. *American Journal of Primatology* **58**: 35–43.

UNESCO (2002) Biosphere Reserve Information. Senegal: Niokolo-Koba. http://www2.unesco.org/mab/br/brdir/directory/biores.asp?mode=all&code=SEN+03. Accessed November 17 2004.

MAP DATA SOURCES

Map 16.17 Chimpanzee data are based on the following sources:

Butynski, T.M. (2003) The chimpanzee *Pan troglodytes*: taxonomy, distribution, abundance, and conservation status. In: Kormos, R., Boesch, C., Bakarr, M.I., Butynski, T.M., eds, *West African Chimpanzees: Status Survey and Conservation Action Plan.* IUCN/SSC Primate Specialist Group. IUCN, Gland, Switzerland. pp. 5–12.

Carter, J., Ndiaye, S., Pruetz, J., McGrew, W.C. (2003) Senegal. In: Kormos, R., Boesch, C., Bakarr, M.I., Butynski, T.M., eds, *West African Chimpanzees: Status Survey and Conservation Action Plan.* IUCN/SSC Primate Specialist Group. IUCN, Gland, Switzerland. pp. 31–39.

For protected area and other data, see 'Using the maps'.

ACKNOWLEDGMENTS
This country study draws extensively on the Senegal chapter from the IUCN/SSC *West African Chimpanzees: Status Survey and Conservation Action Plan.* Many thanks to Georges Grépin and an anonymous reviewer for their valuable comments on the draft of this section.

COMPILER
Edmund McManus, UNEP World Conservation Monitoring Centre

REPUBLIC OF SIERRA LEONE

Edmund McManus

BACKGROUND AND ECONOMY
The Republic of Sierra Leone is situated between Guinea and Liberia and is one of the smallest countries in West Africa, with a land area of about 71 620 km^2 divided into the Eastern, Northern, and Southern Provinces and the Western Area. It has wooded hill country, upland plateaus, mountains in the east, and a coastal belt mostly covered by mangrove swamps in the west. Sierra Leone has a seasonal tropical climate, with a rainy season from May to December and a dry season from December to April, when dry dust-laden harmattan winds blow in from the Sahara. Mean annual rainfall ranges from 1 830 mm in the northern savannas to 5 230 mm on the coast, making it one of the wettest places in West Africa, and the annual temperature range is 21–36°C.[6] Sierra Leone became independent from the UK in 1961.

The population was about 5.9 million in 2004, with a growth rate of 2.3 percent.[2] In 2002, gross domestic product (GDP) was US$782.9 million, with a gross national income (GNI) of less than US$140 per person.[15] English is the official language (although limited to the literate minority), but English-based Krio (Creole) is understood by 95 percent of the population. The main ethnic groups are Temne (who are dominant in the north) and Mende (who are dominant in the south), at 30 percent of the population each, with Islam as the main religion (60 percent of population); sizeable minorities practice traditional beliefs (about 30 percent) or some form of Christianity.[2] Agriculture plays a major role in the economy with about two thirds of the working-age population engaging in subsistence farming.[2, 6] Diamonds, rutile (titanium oxide), bauxite, cocoa, coffee, and fish have contributed most of Sierra Leone's exports historically, but diamonds have become the major source of income.

Much trade in 'conflict' diamonds was used to sustain armed factions and their mercenaries in a 1991–2002 civil war between the government and the Revolutionary United Front (RUF). The civil war resulted in tens of thousands of deaths and the displacement of more than 2 million people (well over one third of the population), many of whom are now refugees in neighboring countries. With the support of a UN peacekeeping force and contributions from the international community, the demobilization and disarmament of the Revolutionary United Front and Civil Defense Force combatants have been completed. National elections were held in May 2002 and the government is continuing slowly to re-establish its authority. The gradual withdrawal of most of the UN peacekeeping mission in Sierra Leone in 2004 and the security situation in neighboring Liberia may present challenges to the continuation of Sierra Leone's stability. Fighting among disparate rebel groups, warlords, and youth gangs in Guinea, Liberia, and Sierra Leone has created insurgences, street violence, looting, arms trafficking, ethnic conflicts, and refugees in border areas,[2] and all these factors contribute to seriously hamper economic development. Nevertheless, in May 2004, Sierra Leone held its first local elections in 32 years, and the country continues to recover

Map 16.18 Chimpanzee distribution in Sierra Leone

Data sources are provided at the end of this country profile

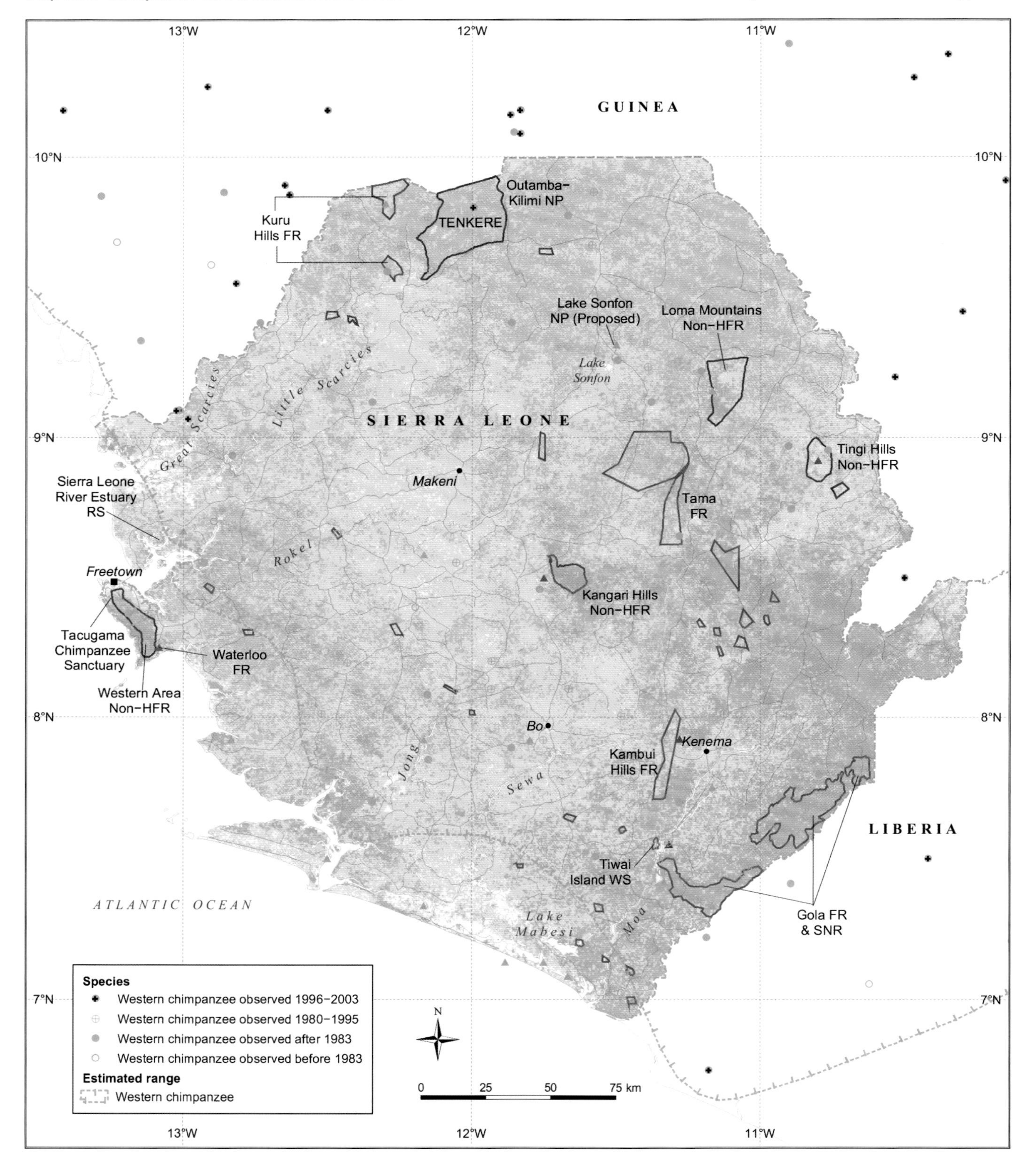

socially and economically from the civil conflict. The latter devastated its infrastructure, however, and Sierra Leone remains at the bottom rank (out of 177 countries) in terms of the UN's 2004 Human Development Index.[16]

DISTRIBUTION OF GREAT APES

The only species of great ape found in Sierra Leone is the western chimpanzee (*Pan troglodytes verus*). It has a widespread but sparse distribution, occupying habitats ranging from young secondary to primary forest, riverine, gallery, and savanna woodlands, and logged forests.[1, 7] Chimpanzees can be found in lowland rain forests in the east and south, in montane vegetation of the Loma and Tingi, and in the woodland-savanna ecosystem of Outamba in the north of the country. It is thought, however, that few sites now have viable populations or offer suitable habitats.[6]

The recent history of conflict has limited access to researchers, and so recent information on the distribution and numbers of chimpanzees in Sierra Leone is scarce. In the early 1990s, chimpanzee numbers were estimated at about 2 000 individuals.[13] It is believed that the population has decreased in size since then. Surveys and anecdotal data show that historically, chimpanzees occurred in Outamba-Kilimi National Park (NP), the Western Area Non-Hunting Forest Reserve, Tiwai Island Wildlife Sanctuary, the Gola Forest Reserves, the proposed Lake Sonfon NP, Loma Mountains Non-Hunting Forest Reserve, Kambui Hills Forest Reserve, and Tingi Hills Non-Hunting Forest Reserve, all of which add up to a total area of 2 835 km^2.[6] A study in the early 1980s estimated 49–60 individuals to be living in the Kilimi section of the Outamba-Kilimi NP,[7] and the continued presence of chimpanzees in this area has recently been confirmed.[17] A group of 27 individuals was sighted in Tenkere, Outamba,[6] with a total population estimate for Outamba being 200–300. It is thought that the park could support a potential population of 600–700 individuals.[6]

The three Gola reserves (Gola East, Gola West, and the largest, Gola North) lie in the Eastern Province. They were designated for timber exploitation in the 1920s, but now comprise the country's largest single area of lowland tropical rain forest.[4, 8] Over 80 percent of Gola North remains unlogged, but satellite images reveal the existence of a logging road running from Liberia into the reserve. The images showed no sign of major logging activity, however, and it is possible that with the recent change in government in Liberia the road may not be in use (see the Liberia country profile).

There have been multiple sightings of solitary chimpanzees in the Tiwai Island Wildlife Sanctuary; for example, 22 individuals in young forest and 38 in old forest in the mid-1980s.[3] At that time, observations in the Gola reserves were minimal, and the population was estimated to be sparse. In the 1980s, the Loma Mountains were proposed as an important area for chimpanzees. Plans for a biomedical research facility there were scrapped after international protest. Later surveys confirmed that hunting and agriculture in the area may have significantly reduced the population and little sign of its existence was found.[6]

THREATS

All the records of chimpanzee presence mentioned above are in protected areas. The remaining populations outside these areas are becoming isolated,[6] and declining due to inadequate protection from tree cutting, hunting, and the pet trade.

Rapid population growth has led to increasing pressure on the environment. Overharvesting of timber, the expansion of cattle grazing, and slash-and-burn agriculture have resulted in deforestation and soil exhaustion. It is estimated that 10 550 km^2 of forest remains, and that there is an average annual loss of 360 km^2 or 2.9 percent.[5] Plantations of cash crops are replacing the remaining areas of forest and otherwise reducing the area of habitat that may be suitable for chimpanzees. In the Eastern Province, diamond mining has resulted in large-scale modification of terrestrial habitat, including forest clearance.

Although the overall trend in Sierra Leone is one of forest loss, there have been small areas of forest regrowth due to reduced habitation during the war; around the internal refugee camps there was heavy resource exploitation. The total area of forest estimated by the Darwin-funded Habitat Audit Project is about 5 000 km^2 lower than the estimate of the Food and Agriculture Organization of the United Nations (FAO), but the estimated rate of forest loss between 1985–1986 and 2000–2003 is almost identical, at 382 km^2/year.[17] The Habitat Audit Project identified no significant areas of cash crops and concluded that the few large plantations in Sierra Leone had been abandoned for many years, although some rehabilitation is going on. Satellite images of the diamond mining areas

Glyn Davies

A recently caught chimpanzee being taken to the bushmeat market.

indicate a dynamic landscape. In the north, there are areas where the bush-fallow system is breaking down, because the crop-forest rotation is too short to be sustainable, but in the south and east the system appears to be stable. On the Freetown Peninsula there is uncontrolled development along the coast, and a new road is being built that will further increase it. The forest on the hills above the new developments is being cleared very rapidly.

Chimpanzees have been hunted and trapped in Sierra Leone for at least 300 years. The demand for live chimpanzees for medical research overseas led to a boom in the export of chimpanzees in the 1970s. Between 1973 and 1978, two wild animal exporters from Sierra Leone are reported to have shipped 1 582 live chimpanzees to countries overseas,[14] in particular to Japan, the USA, and the Biomedical Primate Research Centre in Rijswijk, the Netherlands. The last export to Europe happened in 1984.[12]

During the recent civil conflict, protected areas lacked any form of management, which led to the uncontrolled poaching of bushmeat.[9] The Kenema bushmeat market in the southeast is among those known to sell primate meat. The quantity of chimpanzee meat in the bushmeat trade remains unknown. In areas where firearms or their cartridges are unaffordable to hunters, nets are used to trap chimpanzees for bushmeat or for the pet trade. In some rural areas, chimpanzee bones are used in traditional medicine in the belief that they increase strength and vitality.

LEGISLATION AND CONSERVATION ACTION

Sierra Leone is a signatory to the Convention on International Trade in Endangered Species of Wild Fauna and Flora (CITES), the Convention on Biological Diversity, and the International Tropical Timber Agreement. Ratification for all conventions is pending.[6] The national law protecting chimpanzees is the Third Schedule of the Wildlife Conservation Act of 1972. Chimpanzees are listed as 'protected animals' and the young of the genus *Pan* are listed as 'specifically protected'. Under the Fourth Schedule of the same act, however, an individual is allowed two kills of chimpanzees.[6] These laws were under revision at the time of writing.

The long-running civil war and prevailing insecurity that followed have inhibited engagement by the international conservation community in Sierra Leone. Earlier projects included surveys by Teleki and Baldwin (1981), Hardy (1984), Davies (1987), and Hanson-Alp (1989), but there are no chimpanzee-oriented conservation projects currently underway.[6]

There is one sanctuary affiliated with the Pan African Sanctuary Alliance (PASA) in Sierra Leone. This is the Tacugama Chimpanzee Sanctuary, which evolved as a response to the number of chimpanzees kept as pets in the country; by June 2003 it held 62 rescued chimpanzees.[6] An acute viral infection in 2004 killed five infant chimpanzees at the sanctuary, prompting an emergency veterinary intervention by PASA with the support of the Zoological Society of London.[10] The sanctuary was later awarded a grant by the US government with which to install electric fencing around its 15 km^2 enclosure.[11]

FUTURE CONSERVATION STRATEGIES

The following recommendations follow the *West African Chimpanzees: Status Survey and Conservation Action Plan.*[6]

Research

Basic data collection should be undertaken in priority sites such as the Gola Forest Reserves, Outamba-Kilimi NP, Kuru Hills Forest Reserve, Tiwai Island, Loma Mountains, and Western Area Non-Hunting Forest Reserve. A nationwide population census of chimpanzees should also be initiated, partly to investigate sightings of chimpanzees near human settlements in various parts of the country.

Protection

The Outamba-Kilimi NP is believed to be the stronghold for chimpanzee conservation in Sierra Leone, so it should be a target of conservation activities through antipoaching activities and promotion of tourism and conservation research, possibly through community programs. Foot and vehicular patrols should be undertaken periodically in protected areas with viable populations of chimpanzees. The Wildlife Conservation Act of 1972 should also be reviewed.

Education

Public awareness-raising programs should be initiated, with at least one such program targeting people in the rural parts of chimpanzee range areas. Chimpanzee suppliers should be a focus of such initiatives. International support should be provided to the Tacugama Chimpanzee Sanctuary as it houses confiscated apes, protects wild chimpanzee habitat, and provides environmental education.

Capacity building

Staff of the Wildlife Conservation Branch should be strengthened in numbers and better trained. The capacity of local nongovernmental organizations should also be enhanced.

Population management

Establishing and maintaining corridors between fragmented habitats that hold isolated populations of chimpanzees will be crucial to long-term survival of Sierra Leone's chimpanzees. Patches of natural vegetation, riparian forests, and mature bush-fallows should all be considered as critical habitats when designing landscapes to accommodate the conservation needs of chimpanzees.

FURTHER READING

Alp, R. (1997) "Stepping-sticks" and "seat-sticks": new types of tools used by wild chimpanzees (*Pan troglodytes*) in Sierra Leone. *American Journal of Primatology* **41** (1): 45–52.

Fimbel, C. (1994) The relative use of abandoned farm clearings and old forest habitats by primates and a forest antelope at Tiwai, Sierra Leone, West Africa. *Biological Conservation* **70** (3): 277–286.

Sept, J.M., Brooks, G.E. (1994) Reports of chimpanzee natural history, including tool use, in 16th-century and 17th-century Sierra Leone. *International Journal of Primatology* **15** (6): 867–878.

Squire, C.B. (2001) *Sierra Leone's Biodiversity and the Civil War.* Biodiversity Support Program, Washington, DC. http://www.worldwildlife.org/bsp/publications/africa/176/titlepage.htm. Accessed June 12 2005.

MAP DATA SOURCES

Map 16.18 Chimpanzee data are based on the following sources with additional data by personal communication from Davies, G. (2005):

Butynski, T.M. (2003) The chimpanzee *Pan troglodytes*: taxonomy, distribution, abundance, and conservation status. In: Kormos, R., Boesch, C., Bakarr, M.I., Butynski, T.M., eds, *West African Chimpanzees: Status Survey and Conservation Action Plan.* IUCN/SSC Primate Specialist Group. IUCN, Gland, Switzerland. pp. 5–12.

Grubb, P., Jones, T., Davies, A., Edberg, E., Starin, E., Hill, J. (1998) *Mammals of Ghana, Sierra Leone and The Gambia.* Trendrine Press, Zennor, UK.

Hanson-Alp, R., Bakarr, M.I., Lebbie, A., Bangura, K.I. (2003) Sierra Leone. In: Kormos, R., Boesch, C., Bakarr, M.I., Butynski, T.M., eds, *West African Chimpanzees: Status Survey and Conservation Action Plan.* IUCN/SSC Primate Specialist Group. IUCN, Gland, Switzerland. pp.77–87.

For protected area and other data, see 'Using the maps'.

ACKNOWLEDGMENTS

This country study draws extensively on the Sierra Leone chapter from the IUCN/SSC *West African Chimpanzees: Status Survey and Conservation Action Plan.* Many thanks to Glyn Davies (Zoological Society of London), Hassan Mohamed (Ministry of Lands, Country Planning, Forestry, and the Environment, Sierra Leone), and Richard Wadsworth (Centre for Ecology and Hydrology, UK) for their valuable comments on the draft of this section

COMPILER

Edmund McManus, UNEP World Conservation Monitoring Centre

REPUBLIC OF SUDAN

NIGEL VARTY

BACKGROUND AND ECONOMY

Sudan, with an area of 2 505 813 km^2, is the largest country in Africa. It is dominated by the Nile and its tributaries, and has borders with Egypt, Libya, Chad, the Central African Republic (CAR), the Democratic Republic of the Congo (DRC), Uganda, Kenya, Ethiopia, and Eritrea. It has over 800 km of coastline on the Red Sea along its northeastern border, and is generally flat with mountains in the east and west. Sudan obtained independence from joint Anglo-Egyptian rule in a complex process that began with the 1952 Revolution in Egypt and ended with self-rule in 1956.[10] The first coup d'état and military government followed within two years, and since then there have been only two periods of truly civilian rule, in 1965–1969 and 1986–1989, both terminated by military coups; the current form of government remains that of military dictatorship.[10]

Throughout this history, southern Sudan has suffered from conflict between government forces and rebels. A peace agreement was signed in 1972, but its terms gradually unraveled during the 1970s, and by 1983, when the government imposed Islamic Sharia law throughout the country, southern resistance forces were remobilizing into what became the Sudan People's Liberation Army (SPLA). The conflicts resulted in some 2 million deaths, hundreds of thousands of refugees and internally displaced people, and widespread famine. It has been a complex struggle, which is sometimes oversimplified as a clash between southern, non-Muslim peoples seeking independence, and northern, Muslim peoples seeking to maintain national integrity. In fact, much of the fighting has been factional and both northern and southern opinion is divided. By 2002, war-weariness and international pressure finally brought the government of Sudan and the SPLA's political arm, the Sudan People's Liberation Movement (SPLM), to the negotiating table. Agreements have since been reached on a wide range of issues that provide the political framework for a comprehensive peace agreement.

The population of Sudan, which consists of about 500 ethnic groups, was estimated to be 38.11 million in 2003, with an annual growth rate of 2.7 percent.[6] In 2002, gross domestic product (GDP) was US$13.5 billion annually, or less than US$360 per person.[26] Agriculture is Sudan's most important sector, employing 80 percent of the workforce and contributing 43 percent of GDP.[6]

Sudan has a tropical south and arid desert in the north. Woodland savannas and forests cover 616 270 km^2 of the country,[9] although this includes 6 410 km^2 of plantations. Approximately 68 percent of the country's forest biomass resources are found in the war-affected south of the country.[13] Arable land and permanent crops account for only 7.1 percent of the territory.[6] Tropical moist forest is confined to a few small, scattered localities near the CAR, DRC, and Uganda borders, representing the very northernmost Congo Basin forests. They include small areas on the Aloma Plateau and in the Yambio area; the Azza forest in the Meridi district; the Talanga, Lotti, and Laboni forests at the base of the Imatong Mountains; the Kinyeti Valley; and parts of the Boma Plateau.[11] Closed forest also occurs as extensive galleries along rivers and in depressions in the southwest of Sudan.

DISTRIBUTION OF GREAT APES

The eastern subspecies of the chimpanzee, *Pan troglodytes schweinfurthii*, occurs in Sudan. Relict populations of eastern chimpanzees are thought to survive in the southwest of Sudan (west of the White Nile) and in forest areas on the border with DRC and CAR, such as the Aloma Plateau near Yei.[2, 7, 15] There is only one recent (post-1983) record, found in the border region in the vicinity of Bengangai Game Reserve.[17] This area of Sudan has been a stronghold for southern forces during the civil war, which has prevented safe access for surveys.

There has been no national census of chimpanzees or recent studies of the species in Sudan, and its current status is unknown.[3, 21, 23] The population is estimated to be 200–400 animals.[3] The species was already considered to be 'highly endangered' in Sudan by 1988.[18]

THREATS

Sudan's first civil war ended in 1972, and peace allowed considerable wildlife work to be done up to 1983, when it was brought to an end by the beginning of another, much longer civil war. Recent surveys of wildlife resources of certain areas of

Map 16.19 Chimpanzee distribution in Sudan

Data sources are provided at the end of this country profile

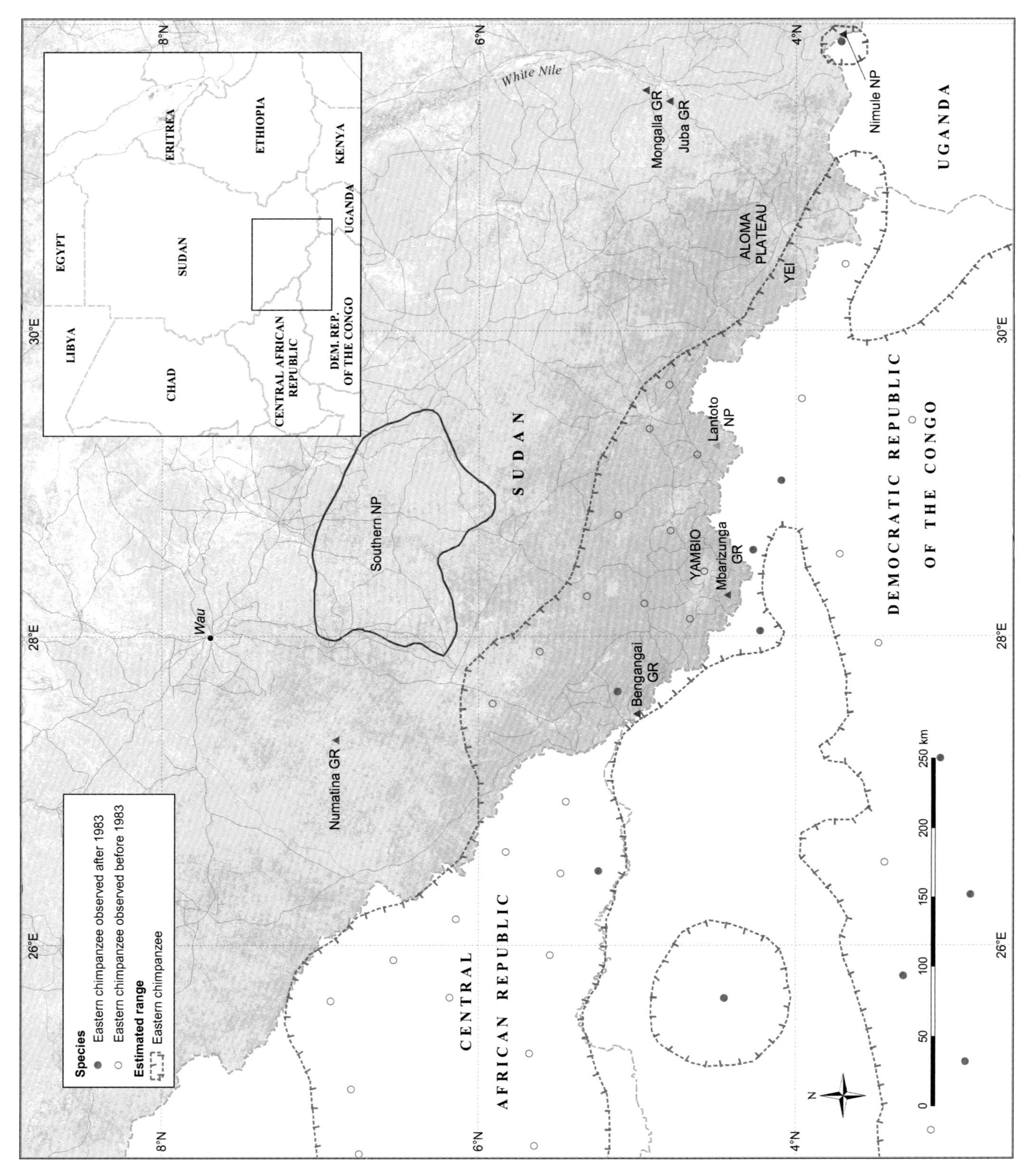

southern Sudan have begun only recently as part of rebuilding efforts in anticipation of lasting peace,[5] but there is no information on the rates of decline of chimpanzee populations. Poaching is known to be rife, however, in areas where chimpanzees have been recorded, and there appears to have been a marked contraction in range over the last 50 years.[3]

Hunting

Hunting for bushmeat is considered a major threat to wildlife in southern Sudan, and certainly threatens any chimpanzees remaining in the country. Hunting greatly expanded during the conflict because of the proliferation of firearms, and the consumption and sale of bushmeat is now commonplace.[4] Many former fighters and others now keep firearms, mainly AK-47s, ostensibly to protect their livestock. The continuing lack of security has made it difficult to restock to previous livestock levels, and wild animals are viewed as an alternative source of meat. The Sudan People's Liberation Movement has attempted to control poaching of major species, and has directed its soldiers not to kill endangered species.[5]

Capture of young chimpanzees for the pet trade has been identified as a threat in Sudan.[7] In October 2003, Sudanese officials intercepted a shipment of 10 young chimpanzees of unknown origin (thought likely to have been captured in DRC) at Yambio in southern Sudan.[1] Four of the chimpanzees were taken to the Sweetwaters Chimpanzee Sanctuary in Kenya, which had received four other animals from Sudan earlier in 2003. If Sudan was not the source of these chimpanzees, it is certainly being used as a conduit to illegal markets.

Habitat loss

The woodland and forest cover of Sudan was estimated at 34 percent of the total land area at the time of independence in 1956, but had declined to 17 percent by 2000.[4] It is estimated that 9 590 km^2 of forest or 1.4 percent of the total is cleared in Sudan each year,[9] although some authors give lower figures.[8] The rate of deforestation in the southern areas thought to support chimpanzees is not known, but these areas were strongholds of the Sudan People's Liberation Army during the conflict and both sides cleared forest around large settlements and military areas as a defensive measure. The civil war drove many millions of people from rural areas, however, and killed many more, which has resulted in the abandonment and regeneration of bush and savanna forest in some places.[5]

Disease

There was an outbreak of Ebola hemorrhagic fever in Yambio, southern Sudan in May–June 2004.[25] Transmission from animals was not implicated in this outbreak, and there is no evidence that the disease has affected chimpanzees.

Cessation of hostilities

The end of the civil war in the south is likely to be followed by large-scale deforestation as a result of returning residents clearing land for agriculture, cutting trees for timber to build houses, and for fuel to meet domestic energy needs.[4] Given the insecurity of the last four decades there has been little forestry activity in the south, and existing plantations would probably be unable to meet demands. Hunting for game is also likely to increase significantly as farmers and refugees seek food while waiting for crops to grow. This could put any remaining chimpanzees at risk.

LEGISLATION AND CONSERVATION ACTION

Although national laws have been passed governing wildlife conservation, most refer to the situation in the north of the country and so have not been applicable in the south, the regional government of which has developed its own set of legislation.

Chimpanzees are protected under the 1975 Wildlife Conservation and Parks Act of the Southern Regional Government,[7] but in 1986, this was repealed (in the north) by the national Wildlife Conservation and National Park Ordinance. Forest reserves and other protected forests are covered under the national Forests Act 1989 (No. 14 of 1989), which allows for the creation of a number of forest-reserve categories, and the 2002 Forestry and Renewable Natural Resources Law.

The Wildlife Conservation General Administration (WCGA) is the national government agency responsible for wildlife conservation and management and the creation and management of protected areas, although it only operates in the north of the country. The management of forests, including the creation of forest reserves, comes under the National Forestry Corporation, within the Ministry of Agriculture and Forestry, but its activities have also been almost exclusively confined to the north in recent years due to the civil war. The Higher Council for Environment and Natural

Resources (HCENR) coordinates environment-related projects and programs, and was the lead agency in the development of the National Biodiversity Strategy and Action Plan.[12]

As part of the peace process, the existing governmental organizations of the Sudan People's Liberation Movement will continue overseeing sector-specific areas in southern Sudan, which include both forestry and wildlife departments. This joint arrangement is expected to last for several years during the transition period before the south decides whether to separate from the north or to rejoin it within a single state.

International agreements

Sudan ratified the Convention on Biological Diversity in 1995, the Convention on International Trade in Endangered Species of Wild Fauna and Flora (CITES) in 1982, the UN Convention to Combat Desertification in 1995, the African Convention on Conservation of Nature and Natural Resources in 1973, and the World Heritage Convention in 1974. The country also participates in UNESCO's Man and Biosphere (MAB) Programme, under which two Biosphere Reserves (the Dinder NP and the Radom NP) have been designated.

Protected areas

There are three main protected-area categories in Sudan: national parks (nine terrestrial and two marine); game (faunal) reserves (22); and sanctuaries (including three wildlife sanctuaries), as well as several other types of protected area.[4, 20, 24] According to the Ministry of Agriculture and Forests, protected areas in Sudan cover an estimated 14 percent of the country.[19] There are also many forest reserves in Sudan, particularly in the south, but data on many of these reserves are lacking because records were lost or destroyed during the civil war. Most parks and protected areas are inadequately staffed and financed.

Chimpanzees are thought to occur in the Mbarizunga and Bire Kpatuos Game Reserves,[7] and the Bengangai Game Reserve.[7, 14, 16, 20] There is no recent information on the occurrence of chimpanzees at these sites. Plotting of their recent historical range in Sudan suggests eastern chimpanzees may also occur in the Lantoto NP.

Due to the long-running civil war in southern Sudan there have been no conservation projects for the eastern chimpanzee. Nor have there been any detailed chimpanzee research projects within the last 20 years. The US Agency for International Development (USAID) has been supporting work to determine the current status of protected areas, wildlife, and biodiversity conservation in Sudan through its Sudan Transitional Assistance for Rehabilitation (STAR) program,[4] and aims to re-create institutional capacity for conservation and wise management of natural resources.[5] USAID has also been funding the New Sudan Wildlife Society (a nongovernmental organization) to carry out reconnaissance surveys in a number of national parks, including Southern NP, considered the most likely protected area still to support chimpanzees, due to its size and remoteness.[5]

There are no ape sanctuaries in Sudan, but a coalition of organizations, including Born Free, New Sudan Conservation Society, Kenya Wildlife Service, Sudan Conservation Authority, International Fund for Animal Welfare, and the Sweetwaters Chimpanzee Sanctuary, has been helping with the rescue and ongoing care of orphaned chimpanzees from southern Sudan.

FUTURE CONSERVATION STRATEGIES

There is no specific conservation action plan for chimpanzees in Sudan, although there are some published recommendations dealing with the areas in which chimpanzees are thought still to occur. Although there is reported to be considerable support among the southern Sudanese and their government to re-establish protected areas and re-constitute them,[5] in practice, most protected areas are left open to human settlement, cultivation, and livestock grazing. There are also few land-use plans, so surveying wildlife areas and preparing wildlife-management plans merit priority attention.[4, 18]

There have been calls for the Sudan People's Liberation Movement to consider making "unequivocal and very public pronouncements of absolute prohibitions against the hunting of elephant, rhino, and chimpanzee".[4] Incentives have been proposed to encourage armed groups to decommission their firearms. There is also the need to develop a strategy for the conservation and sustainable management of the country's forests and woodlands, particularly for high forest areas in the far south, and capacity building and institutional strengthening in natural resource planning and strategy.[4]

Given that Sudan has just been through more than 30 years of civil war, with millions killed or displaced, and with very few financial resources, wildlife conservation is unlikely to be a high priority for either the northern or the southern government, and outside agencies and nongovernmental organizations are expected to provide most of its funding for the foreseeable future. The over-arching priority is to continue to survey present conditions as peace emerges, determining the status of wildlife populations and developing a prioritized plan for their conservation and management, taking careful account of the immense human needs in the country, particularly in the south. Given the international conservation interest in chimpanzees, it is not unreasonable to expect donor support for any areas that are found to harbor remnant populations of these great apes.

FURTHER READING

Catterson, T. (2003) *Environmental Threats and Opportunities Assessment. USAID Integrated Strategic Plan in the Sudan 2003–2005.* USAID/REDSO/NPC and the USAID Sudan Task Force, Washington, DC. http://www.usaid.gov/locations/sub-saharan_africa/sudan/sudan_isp_a4.pdf. Accessed June 12 2005.

El Moghraby, A.I. (2001) State of the environment in Sudan. In: *UNEP EIA Training Resource Manual.* UNEP. pp. 27–36. www.unep.ch/etu/publications/11)%2027%20to%2036.pdf. Accessed June 12 2005.

HCENR (2000) *The Sudan's National Biodiversity Strategy and Action Plan.* Higher Council for Environment and Natural Resources, Ministry of Environment and Tourism, and IUCN with support from UNDP, Khartoum. http://www.biodiv.org/doc/world/sd/sd-nbsap-01-p1-en.pdf to http://www.biodiv.org/doc/world/sd/sd-nbsap-01-p7-en.pdf. Accessed June 12 2005.

MAP DATA SOURCES

Map 16.19 Chimpanzee data are based on the following source:

Butynski, T.M. (2001) Africa's great apes. In: Beck, B.B., Stoinski, T.S., Hutchins, M., Maple, T.L., Norton, B., Rowan, A., Stevens, E.F., Arluke, A., eds, *Great Apes and Humans: The Ethics of Coexistence.* Smithsonian Institution Press, Washington, DC. pp. 3–56.

For protected area and other data, see 'Using the maps'.

ACKNOWLEDGMENTS

Many thanks to Thomas Catterson (USAID Sudan International Consultant, Forestry, Natural Resources & Environmental Management) for his valuable comments on the draft of this section.

AUTHOR

Nigel Varty, UNEP World Conservation Monitoring Centre

REPUBLIC OF UGANDA

Nigel Varty

BACKGROUND AND ECONOMY

The Republic of Uganda lies on the Equator in East-Central Africa, mostly between the northern and western shores of Lake Victoria and the Rift Valley. Its territory includes nearly half of Lake Victoria and substantial parts of Lakes Edward and Albert. Hence land occupies only about 85 percent (199 710 km^2) of its total area of 236 040 km^2. Large swamps amount to another 5 percent of total area, while forests cover 41 900 km^2 of Uganda, representing about 21 percent of the land area. Most forest is natural woodland, but 430 km^2 of plantations are recorded.[14] The main forest areas are in the west of the country.

Uganda became independent from the UK in 1962, with an elected prime minister (Milton Obote), who was overthrown in a coup d'état in 1971.[15] The new dictatorship, led by Idi Amin, persecuted and expelled members of the Ugandan Asian community and the country's intellectuals. The United Republic of Tanzania invaded Uganda in 1979 with the support of exiled Ugandan groups. Amin was overthrown, and elections in 1980 returned Obote to power. The validity of the elections was challenged by

Yoweri Museveni's National Resistance Army (NRA), which initiated and eventually won a civil war, installing Museveni as president in 1986. He took charge of a country in which 1.5 million people had been killed or maimed in war, 2 million were refugees, and the economy was in ruins. Considerable progress has since been made in restoring peace and in rebuilding infrastructure and civil institutions. Museveni was elected to office in the elections of 1996 and 2001.

The southern half of Uganda is now relatively peaceful, but large areas of the north and east are still troubled by violent insurgency.[15] The main rebel group is the spiritualist Lord's Resistance Army (LRA), which, despite its stated aim of wishing to govern the country according to the Ten Commandments, has been committing widespread atrocities since 1986. It is believed to have abducted some 20 000 children over the years, using them as expendable fighters, and the number of displaced people living in refugee camps in the north is estimated at about 1.6 million.[15] The LRA had been supported by Sudan, which suspected Uganda of supporting its own rebel group in the south (see also the Sudan country profile), but in 1999, both countries reached an agreement not to support the respective insurgents. A succession of broken ceasefires and failed attempts at negotiations between the government and the LRA mean that the conflict continues.

The 2002 census estimated Uganda's population at 24.7 million, with a growth rate of nearly 3 percent per year.[10, 15] In 2003, the gross domestic product (GDP) was US$6 billion, or about US$270 per person,[15] and the country's Human Development Index rank was 146th out of 177.[47] Permanent crops and arable land account for 34.1 percent of the territory and agriculture is the most important sector of the economy, employing more than 80 percent of the work force and accounting for 43 percent of GDP.[10]

DISTRIBUTION OF GREAT APES

The mountain gorilla (*Gorilla beringei beringei*) and the eastern chimpanzee (*Pan troglodytes schweinfurthii*) occur in Uganda. Mountain gorillas occur in only two localities: in the Mgahinga Gorilla NP in the Virunga mountains at the extreme southwest of the country bordering Rwanda and DRC, and in Bwindi Impenetrable NP, about 25 km away. Only 12 mountain gorillas were thought to be present in the Mgahinga Gorilla NP by 1998,[6] but these are part of a larger Virunga population. This was estimated to number 380 individuals in 2003, distributed among the contiguous Volcanoes NP in Rwanda and the southern section of the Virunga NP in DRC.[21] There are about 320 gorillas in Bwindi Impenetrable NP. Hence, this park contains just under half the world's mountain gorillas, although some primatologists consider the Bwindi gorillas to be a separate subspecies.[39]

The eastern chimpanzee occurs across the forests and woodlands of western Uganda,[4] having been recorded in 21 different forest blocks since 1994.[35] Population estimates have slowly been revised upwards, although populations have been declining. In 1979, it was thought that there were fewer than 3 000 individuals, of which 750 were breeding females.[45] In 1989, it was estimated that there were about 4 000 chimpanzees in Uganda.[44] This was revised down to 3 300 in a chimpanzee population and habitat viability assessment (PHVA) in 1997.[12] However, the latest estimate, based on data gathered between 1994 and 2002, gives a figure of about 4 950 (range 4 000–5 700).[35]

THREATS

The mountain gorilla populations appear to be stable (Bwindi)[31] or increasing slightly (Virunga population),[21] but they and their habitat are still under threat from a variety of causes. Chimpanzee numbers have declined drastically in recent decades,[26] probably as a result of habitat loss.

Habitat loss

It is estimated that 910 km^2 of forest is cleared in Uganda annually, representing about 2 percent of the total each year during the period 1990–2000.[14] Tropical high forest covered about 13.7 percent of the land area in 1900, but had been reduced to about 3.6 percent by 2000.[2] Illegal logging, generally involving pit sawing, is widespread in the forest reserves of Uganda, particularly for valuable timber species such as the mahoganies *Khaya* and *Entandrophragma*.[35] Charcoal burning is also having an impact on some forest reserves, such as Kasyoha-Kitomi and Kalinzu. Indeed, 90 percent of the nation's energy source is derived from fuel wood.[46]

Encroachment of forest reserves by surrounding farming communities is reducing the area of critical chimpanzee habitat, but the greatest conversion of forest to agriculture is currently taking place outside forest reserves, with large-

Map 16.20 Great ape distribution in Uganda

Data sources are provided at the end of this country profile

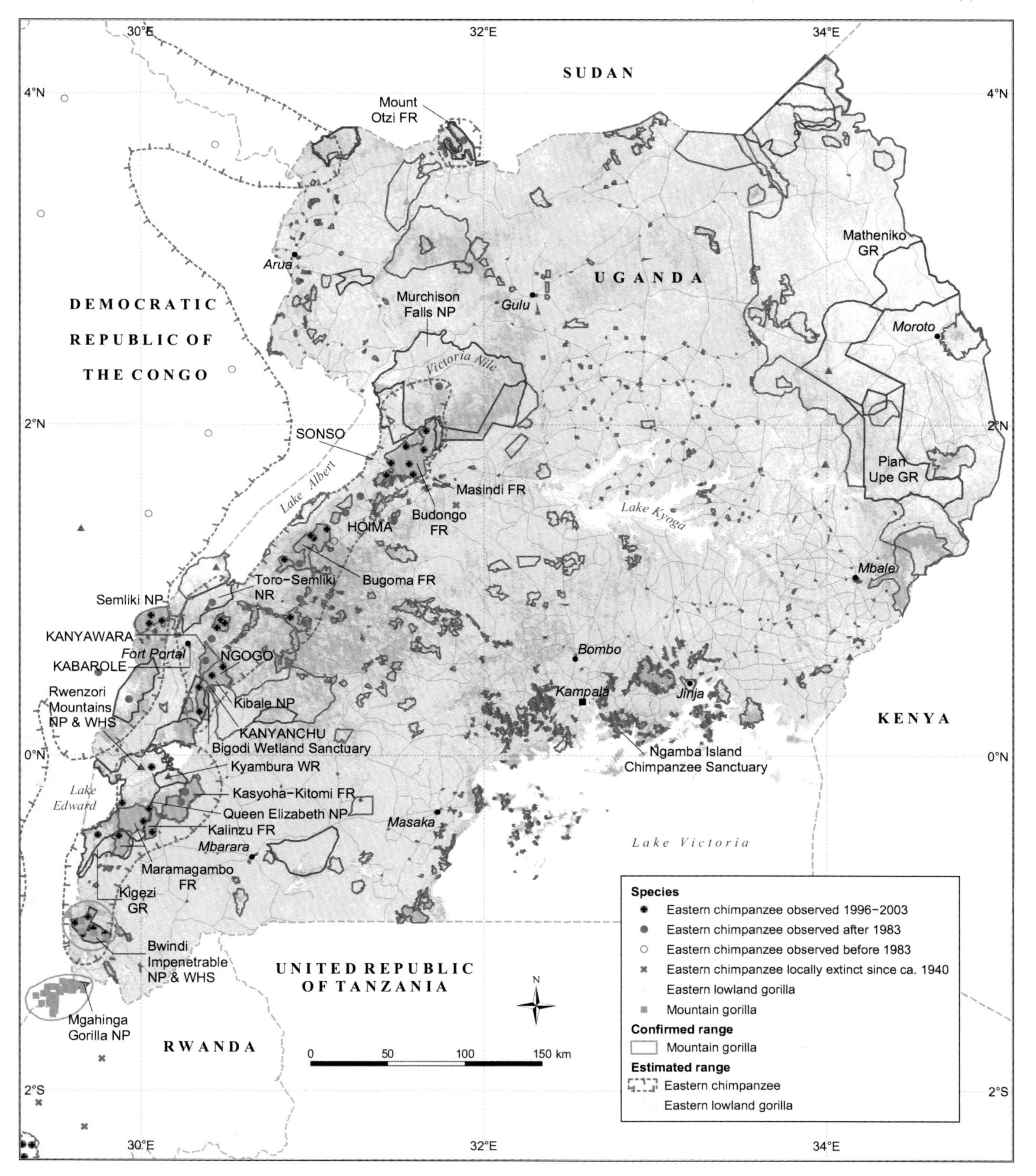

scale plantings of tobacco and cocoa in Hoima district, sugar cane and tobacco in Masindi district, and tea around forests in Kabarole district. This reduces forest connectivity and dispersal opportunities for chimpanzees, hence preventing gene flow. Comparison of satellite imagery between the mid-1980s and 2000–2001 showed that about 800 km^2 of chimpanzee habitat forest was lost in this period.[36]

Other threats include mining, particularly in the Kasyoha-Kitomi forest, most recently for coltan, which is used in the manufacture of semiconductors for mobile phones and other electronic goods.[35]

Hunting

Hunters in Uganda still sometimes kill mountain gorillas,[33] but regular monitoring and patrolling has reduced the incidence in recent years. Ugandans, for the most part, do not eat chimpanzees. Although signs of hunting were recorded in all the forests surveyed by Plumptre and colleagues, the Ruwenzori Mountains NP is the only one where chimpanzees are regularly hunted for meat.[35] There is increasing concern, however, that refugees from DRC, who do not share the same cultural taboos, could have a serious impact on remaining populations. In addition, gorillas and chimpanzees are affected by snares set for animals such as duikers and bush pigs. Snared hands or feet can be seriously mutilated or even lost, and the resulting infections can be fatal.[48] More than half the population (nine adults) in the Kalinzu forest, and up to 25 percent of chimpanzees in Kibale and Budongo forests have snare-related injuries.[12, 22, 48, 49] In Budongo's 'Sonso' habituated group, at least two deaths are suspected to be a result of snaring.

Chimpanzees have been hunted and killed by local people if crops around forest reserves and national parks have been raided, particularly if cash crops are involved. There are reports of spears and bows-and-arrows being used to kill animals at Bugoma for taking cacao, and at Budongo for raiding sugar cane fields.[35] Evidence of hunting of chimpanzees for body parts for ritual use has also been found at Kalinzu Forest Reserve.[35]

The collection of young apes to be sold as pets or for entertainment is less common in Uganda than in many other countries, but many young chimpanzees are smuggled over the border from DRC. There has been a recent increase in the number of chimpanzees confiscated (three to four a year), possibly because of the withdrawal of Uganda's military presence from DRC.[35]

Disease

A number of diseases have been reported in gorilla and chimpanzee groups in Uganda, some of which appear to have been contracted from close contact with humans or their livestock. Habituated gorillas in Bwindi Impenetrable NP have been reported suffering from scabies (*Sarcoptes scabiei*). One infant is known to have died from the disease, with five juveniles badly affected.[19, 30] The source of the infections is thought to have been humans or domestic animals living near this forest.[30] High rates of infection with *Cryptosporidium parvum* have also been reported in the Bwindi gorillas.[18] The human–ape disease risk is particularly worrying as increasing numbers of gorilla and chimpanzee groups are being habituated for tourism in Uganda, due to tourist demand and the potential financial benefits.[4, 6, 50] The enforcement of regulations put in place to reduce the risk of disease transmission is still fairly weak.

LEGISLATION AND CONSERVATION ACTION

National and international law

Uganda has ratified or acceded to the Convention on Biological Diversity (1993), the Convention on International Trade in Endangered Species of Wild Fauna and Flora (CITES) (1991), the UN Convention to Combat Desertification (1997), the Convention on Migratory Species (2000), the African Convention on Conservation of Nature and Natural Resources (1977), and the World Heritage Convention (1987). Two World Heritage Sites have been inscribed – the Bwindi Impenetrable NP and Ruwenzori Mountains NP. The country also participates in UNESCO's Man and Biosphere (MAB) Programme, and Queen Elizabeth NP has been designated as a Biosphere Reserve.

Chimpanzees are protected by law, and national and international trade is regulated or restricted.[26] There are two main statutes that most concern protection of great apes in Uganda – the Forests Act (1964) and the Uganda Wildlife Statute (1996), respectively executed through the National Forestry Authority (NFA) (formerly the Forest Department) and the Uganda Wildlife Authority (UWA), in the Ministry of Tourism, Trade, and Industry. The Uganda Wildlife Statute provides for the establishment of wildlife conservation areas, which fall under two categories: wildlife protected areas (national parks or wildlife reserves) and wildlife management areas (wildlife sanctuaries and community wildlife areas).

In 2004, Uganda was criticized for negotiating the transfer of three chimpanzees to Changsha Zoological Garden in China as a gift of state.[13] An injunction was issued at the request of the Greenwatch nongovernmental organization (NGO), preventing the transfer until a hearing in May 2004.

Protected areas

Altogether Uganda's protected areas cover about 33 000 km^2, or 14 percent of the land area of the country.[23] There are three main categories of protected area in Uganda: national parks (10 sites), wildlife reserves (10 sites), and forest reserves (710 sites). Forest reserves represent close to 50 percent of the protected land cover.[23] While most are less than 10 km^2 in area, some 2 600 km^2 are represented by just seven 'central' forest reserves: Budongo, Bugoma, Kalinzu, Kasyoha-Kitomi, Mabira, Maramagambo, and Sango Bay Forest Reserves.[23, 32] Some forest reserves include plantations of exotic species such as eucalyptus.[32]

Mgahinga Gorilla NP is Uganda's smallest national park, at only 34 km^2,[7] and harbors part of the Virunga population of mountain gorillas. The major problem facing the conservation of the park has been habitat loss or modification due to human population growth. The open woodland, which was once a favored gorilla habitat, was completely settled and cultivated prior to the park being gazetted in 1991. The settlers were evicted and the ecosystem is now regenerating. Sucker, the warden who had put in place the measures that began to restore the park, was murdered in the area in 1994.[24, 25] Pressures include hunting, grazing of livestock, beekeeping, and collection of bamboo, honey, water, and firewood.[1, 6, 8]

Gordon Miller/IRF

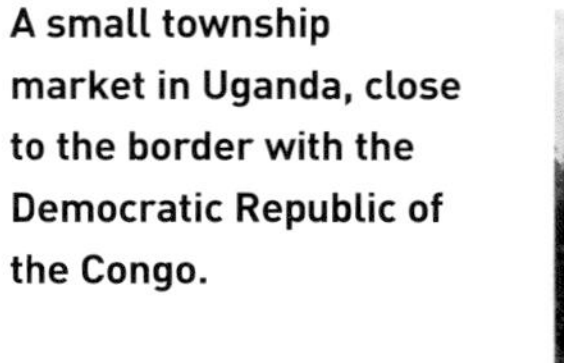

A small township market in Uganda, close to the border with the Democratic Republic of the Congo.

Bwindi Impenetrable NP (321 km^2) is the largest remaining fragment of natural forest in southwest Uganda. The terrain is extremely rugged,[31] with steep hills and narrow valleys, and is one of the few large expanses of forest in East Africa where lowland moist forests blend in an undisturbed continuum into montane forests. Among other differences, there is much less bamboo in Bwindi than in the Virunga region.[39] Bwindi is surrounded by one of the densest human populations in Africa (100–450/km^2), and surrounding lands are intensely cultivated. Gorilla-based tourism started in April 1993 and has been very successful. Since the genocide in Rwanda, the Bwindi gorillas have become the main gorilla population visited by tourists. To sustain the park in the long term, the Bwindi Impenetrable Forest Conservation Trust and a program to share tourism revenues with local communities have been established.[34] There have, however, been conflicts over revenue-sharing arrangements.[17]

Kibale NP (780 km^2) is the most important protected area for chimpanzees in Uganda. It supports a population of around 1 430–1 530, about 25 percent of Uganda's chimpanzees.[9, 35, 37] The density of chimpanzees recorded here (2.32/km^2) is the highest of any forest surveyed in Africa.[35] Three chimpanzee communities are habituated: Kanyawara and Ngogo for behavioral research purposes, and Kanyanchu for tourism activities (see Box 4.6). Forest covers nearly 60 percent of the park, with degraded forest, which is largely secondary forest recovering from agricultural use in the southern sector, covering another 9 percent.[9] Its designation as a national park in 1993 has reduced illegal logging and agricultural encroachment, but crop raiding by animals, including chimpanzees, and the laying of snares in the forest by villagers, still contribute to strained relations between park managers and local communities.[35] Near the ecotourism area at Kanyanchu, chimpanzees have killed several human infants, which has also led to conflict between the park and the local people. These killings were not carried out by habituated chimpanzees.[51]

Chimpanzees occur in many other protected

areas in Uganda, including Semliki NP, Rwenzori Mountains NP, Semliki and Kyambura Wildlife Reserves, and at least 11 forest reserves.[35] Several of these areas combined are estimated to hold over 80 percent (4 058) of the total population. These are Budongo Forest Reserve (639 animals), Bugoma Forest Reserve (628), Ruwenzori Mountains NP (500), Kibale NP (1 429), Kasyoha-Kitomi Forest Reserve (406), Kalinzu Forest Reserve (234), and Maramagambo forest inside Queen Elizabeth NP (222).[35] Chimpanzees have also been recorded in the Mount Otzi Forest Reserve on Uganda's northern border with Sudan.[11] The remaining chimpanzees are found in small numbers in isolated forest pockets between these larger forest blocks, particularly in the Hoima, Kibale, and Masindi districts.[37] Many of these isolated chimpanzee communities are not in protected areas.

Research and field conservation

Uganda has a history of research on chimpanzees dating to 1962 in Budongo,[38] with a particularly long program of research in Kibale forest.[16, 42, 43, 52] Research on mountain gorillas has been ongoing since the 1950s.[40] Following increased security in Uganda in the 1990s, several other studies were started in Budongo Forest Reserve, Semliki Wildlife Reserve, Bwindi Impenetrable NP, and Kalinzu Forest Reserve.[35] Comparative studies have been ongoing between the gorillas of Bwindi and Mgahinga, and of the dietary differences between gorillas and chimpanzees at Bwindi Impenetrable NP.[41] As elsewhere, the commitment of particular individuals has led to the development of these long-term study sites. In Uganda these include Struhsaker, Isabirye-Basuta, and Wrangham in Kanyawara; Mitani and Watts in Ngogo; Reynolds in Sonso and Budongo; Hunt in Semliki; Stanford in Bwindi; and Hashimoto and Furuichi in Kalinzu. National conservation research facilities include the Institute of Tropical Forest Conservation (ITFC), which belongs to Mbarara University of Science and Technology, and is based at Ruhija, and Makerere University's field station at Kanyawara, north of Kibale NP, which is the base for some of the longest ever field studies of primates.

In addition to conservation work promoted by international researchers, a number of international NGOs have been active in great ape conservation in Uganda. A partnership of the Jane Goodall Institutes (JGI) and the Uganda Wildlife Authority has been carrying out snare-removal programs in

Gordon Miller/IRF

Bwindi Impenetrable National Park.

Kibale NP and Budongo forest, in collaboration with the Kibale Chimpanzee Project (KCP) and Budongo Forest Project (BFP) respectively. These efforts have also had the additional effect of reducing forest degradation.[3] IUCN–The World Conservation Union is also involved in several conservation and development projects around major conservation areas important for chimpanzees, including Kibale NP and Semliki NP. The Jane Goodall Institutes and Wildlife Conservation Society (WCS) have worked with the Uganda Wildlife Authority and Forest Department (now the National Forest Authority) to develop a national species action plan for chimpanzee conservation, following their joint work on the national chimpanzee survey.[37]

The Jane Goodall Institute-Uganda works with local communities to resolve human–chimpanzee conflict issues through collaboration on snare-removal programs, alternative income-generating schemes, and sustainable development issues. It also facilitates environmental education programs for primary-school children, particularly those living near chimpanzee habitats, and plans to develop the same schemes for secondary schools and adults. The Jane Goodall Institute is also developing 'Roots & Shoots' environmental action groups for local communities.

Other conservation and development projects include activities of WWF–The Global Conservation Organization around Ruwenzori Mountains NP, and CARE International's activities around Bwindi Impenetrable NP and Mgahinga Gorilla NP. WWF and WCS also support the Institute of Tropical Forest Conservation, and WWF, together with the African Wildlife Foundation (AWF) and Fauna and Flora International (FFI), supports the International Gorilla Conservation Programme (IGCP), which runs and supports projects at Mgahinga Gorilla NP and Bwindi Impenetrable NP.

Staff at Kibali National Park.

Julia Lloyd

Tourism

Nature-oriented tourism, including gorilla and chimpanzee watching, has been developed at several sites in Uganda, including Bwindi Impenetrable NP (gorillas), Mgahinga Gorilla NP (gorillas), Kibale NP (chimpanzees), Budongo Forest Reserve (chimpanzees), Semliki Wildlife Reserve (chimpanzees), and Kyambura Wildlife Reserve (chimpanzees). An economic evaluation of gorilla tourism in Uganda commissioned by IGCP estimated that in 1994–1999, gorilla tourism attracted net foreign exchange earnings of about US$7.7 million, generated US$15.4 million in sales, contributed US$4.77 million in government taxes, and US$6.93 million to the national economy.[20] Gorilla watching at Bwindi Impenetrable NP generated up to 50 percent of the income of Uganda's national park system before an incident in March 1999, when Rwandan rebels traveling from DRC murdered eight tourists and a park warden. However, visits are recovering and the Uganda Wildlife Authority opened a new gorilla group for tourism at Bwindi in July 2002. Uganda is considered one of the best countries to see chimpanzees in the wild,[27] and Kibale is a particularly successful site for this. Questions have been raised, however, about the net benefits of great ape tourism for great ape conservation;[4, 6] great ape tourism can have potentially damaging impacts on the animals themselves[50] as well as on their habitats.[35] One issue, for example, is whether the development of extensive trail systems does excessive damage to the forest understory and improves access for hunters. Research to clarify the significance of such impacts for mountain gorillas and chimpanzees is underway.[28, 29] Meanwhile, it can be noted that trail systems that are frequented by tourists, researchers, and guards are likely to deter poachers, potentially offsetting any negative impacts of trail cutting.

Sanctuaries

The Uganda Wildlife Education Centre is the official holding facility for all confiscated animals in Uganda, and is also home to a community of eight chimpanzees as well as a variety of other animals.

Ngamba Island Chimpanzee Sanctuary was established in 1998 to care for confiscated orphan chimpanzees in Uganda. It is managed by the Chimpanzee Sanctuary & Wildlife Conservation Trust (CSWCT), which is a collaboration of six international trustees: the Born Free Foundation, the International Fund for Animal Welfare, the Jane Goodall Institutes, the Environmental Conservation Trust of Uganda, the Uganda Wildlife Education Centre, and the Uganda Wildlife Society. Originally, 19 chimpanzees were relocated from the Uganda Wildlife Education Centre in Entebbe and Isinga Island of Queen Elizabeth NP, but, following an influx of orphans, the sanctuary now cares for over 35 chimpanzees. Ngamba Island is 0.45 km^2 in area and is located in Lake Victoria, 23 km southeast of Entebbe. The island provides a semi-natural environment for the chimpanzees, but as the forest is not large enough to sustain this population, the animals are provisioned. The sanctuary is run as an education center and is a popular visitor destination for day and overnight visits, attracting more than 300 visitors per month. The Chimpanzee Sanctuary & Wildlife Conservation Trust works with and benefits neighboring island communities through education, health, sanitation, and 'sustainable living on Lake Victoria' programs, micro-finance and loan schemes, and provides casual labor and permanent employment. The trust also runs training programs to build the capacity of Ugandan nationals in captive chimpanzee management.

FUTURE CONSERVATION STRATEGIES

In 1999, the Jane Goodall Institutes and the Wildlife Conservation Society commenced a four year program, in collaboration with the Uganda Wildlife Authority and the Uganda Forest Department, to evaluate the current status of chimpanzees.[35] This program is built on a population and habitat viability assessment undertaken in 1997 by the Conservation Breeding Specialist Group of IUCN.[12] The results have been used to develop a five year national action plan,[37] which is currently being reviewed by the Minister of Tourism, Trade, and Industry.

The five year goal of the national chimpanzee action plan[37] is: "To strengthen the protection of chimpanzees and enhance the viability of populations in major forest blocks by establishing corridors." In order to achieve this goal, six objectives were defined:

- reduce fragmentation and loss of key chimpanzee habitats;
- reduce conflict between local communities and chimpanzees;
- promote awareness of chimpanzee values;
- reduce levels of human-caused deaths and injuries, and the pet trade;
- enhance corporate social responsibility where it affects chimpanzees;
- minimize the risk of disease transmission between people and chimpanzees.

Several projects and activities were identified for each objective, including:

- develop a monitoring system that will allow the Forest Department (now the National Forest Authority) to detect illegal activities quickly and to bring them under control;
- develop and fund snare-removal projects in Bugoma, Budongo, and Kasyoha-Kitomi;
- raise the awareness of law enforcement agencies (customs, police, etc.), international organizations (UN, diplomatic corps, airline companies, Interpol), for example by developing posters and leaflets for distribution, and ensure that relevant knowledge is incorporated in law enforcement training;
- work with private landowners to identify and maintain habitat corridors (particularly around Bugoma and Budongo forests);
- investigate what attributes corridors must have if chimpanzees are to be able to move freely along them;
- design chimpanzee-proof beehives for local communities around forest blocks;
- develop and fund community conservation, education, and development projects with people living around Budongo, Bugoma, and Kasyoha-Kitomi forests;
- implement a national environmental awareness campaign, with a focus on great apes, especially for primary schoolchildren, but also for secondary schoolchildren, and incorporate within the national curriculum;
- improve environmental awareness among corporations, which will lead to the development and implementation of environmental policies to promote chimpanzee conservation;
- develop a health monitoring program for chimpanzees;
- fund improvements in public health facilities available to communities in areas near tourism and research sites;
- evaluate and develop standardized chimpanzee research and tourism guidelines;
- fund and develop a project to establish health guidelines for tourists visiting chimpanzees;
- improve chimpanzee tourism experience in Uganda and enhance its product marketing (e.g. improve facilities, invest in habituation, improve guiding experience);
- evaluate the impact of ecotourism on chimpanzees, local people, and the environment.

The IUCN/SSC Primate Specialist Group's *African Primates: Status Survey and Conservation Action Plan*[34] additionally recommends (among other things) continued monitoring and protection of the Virunga and Bwindi gorillas, studies to assess the most appropriate tourism strategy and arrangements for the two areas, and improved programs for conservation awareness.

Research and education

Research is needed into the effects of human disturbance on gorillas and other wildlife at Bwindi Impenetrable NP. Areas of concern include tourism and the multiple-use zones that have been established around the edges of the park for beekeeping, collection of non-timber forest products, etc.[31]

The development of conservation–education programs has been identified as a particular priority for schools in the Biiso, Budongo, Kijunjuba, and Pakanyi (south) areas around Budongo Forest Reserve, and in the Mabale area around Bugoma Forest Reserve.[3]

Julia Lloyd

Tourism at Kibali National Park.

FURTHER READING

Edroma, E., Rosen, N., Miller, P., eds (1997) *Conserving the Chimpanzees of Uganda: Population and Habitat Viability Assessment for* Pan troglodytes schweinfurthii. IUCN/SSC Conservation Breeding Specialist Group, Apple Valley, Minnesota.

Howard, P.C., Davenport, T.R.B., Kigenyi, F.W., Viskanic, P., Baltzer, M.C., Dickinson, C.J., Lwanga, J., Matthews, R.A., Mupada, E. (2000) Protected area planning in the tropics: Uganda's national system of forest nature reserves. *Conservation Biology* **14** (3): 858–875.

McNeilage, A., Plumptre, A.J., Brock-Doyle, A., Weber, A. (2001) Bwindi Impenetrable National Park: gorilla census 1997. *Oryx* **35** (1): 39–47.

Plumptre, A.J., Arnold, M., Nkuuta, D. (2003) *Conservation Action Plan for Uganda's Chimpanzees 2003–2008*. Wildlife Conservation Society/Jane Goodall Institute.

Plumptre, A.J., Cox, D., Mugume, S. (2003) *The Status of Chimpanzees in Uganda*. Albertine Rift Technical Report Series **2**. Wildlife Conservation Society, New York.

Reynolds, V. (2005) *The Chimpanzees of the Budongo Forest*. Oxford University Press, Oxford.

Struhsaker, T.T. (1997) *Ecology of an African Rain Forest: Logging in Kibale and the Conflict between Conservation and Exploitation*. University Press of Florida, Gainesville.

MAP DATA SOURCES

Map 16.20 Great apes data are based on the following source:

Butynski, T.M. (2001) Africa's great apes. In: Beck, B.B., Stoinski, T.S., Hutchins, M., Maple, T.L., Norton, B., Rowan, A., Stevens, E.F., Arluke, A., eds, *Great Apes and Humans: The Ethics of Coexistence*. Smithsonian Institution Press, Washington, DC. pp. 3–56.

With additional data by personal communication from Plumptre, A. (2004) and from the following sources:

Kalpers, J., Williamson, E.A., Robbins, M.M., McNeilage, A., Nzamurambaho, A., Lola, N., Mugiri, G. (2003) Gorillas in the crossfire: population dynamics of the Virunga mountain gorillas over the past three decades. *Oryx* **37**: 326–337.

Plumptre, A.J., Cox, D., Mugume, S. (2003) *The Status of Chimpanzees in Uganda*. Albertine Rift Technical Report Series **2**. Wildlife Conservation Society, New York.

For protected area and other data, see 'Using the maps'.

ACKNOWLEDGMENTS

Many thanks to Maria Arnold (Australian National University), Thomas Butynski (Conservation International), Alexander Harcourt (University of California, Davis), Joanna Lambert (University of Oregon), Julia Lloyd (Jane Goodall Institute-Uganda), Cherie Montgomery-Lianda (Chimpanzee Sanctuary & Wildlife Conservation Trust), Andrew Plumptre (Wildlife Conservation Society), Vernon Reynolds (Budongo Forest Project), Craig Stanford (University of Southern California), and Richard Wrangham (Harvard University) for their valuable comments on the draft of this section.

AUTHOR

Nigel Varty, UNEP World Conservation Monitoring Centre

UNITED REPUBLIC OF TANZANIA

JARED BAKUSA AND EDMUND MCMANUS

BACKGROUND AND ECONOMY

The United Republic of Tanzania is so called because it was forged from the union in 1964 of Tanganyika and Zanzibar, shortly after their independence from UK-administered UN trusteeship in 1961, and from the UK in 1963, respectively. Situated in East Africa, bordering the Indian Ocean, Tanzania's neighboring countries are: Burundi, Rwanda, Uganda, Kenya, Mozambique, Malawi, Zambia, and the Democratic Republic of the Congo (DRC). It has a population of approximately 36 million people, with an annual growth rate of 1.95 percent.[3] The land area is 886 037 km^2 and agriculture accounts for half of the US$10.1 billion gross domestic product (GDP)[24], or about US$280 per person per year,[6, 24] provides 85 percent of exports, and employs 80 percent of the work force. Topography and climatic conditions, however, limit cultivated crops to only 4 percent of the land area. Tanzania is one of the poorest countries in the world, and in 2004 was ranked 162nd out of 177 by the Human Development Index.[22] Gold, oil, and gas exploitation, however, along with macroeconomic reforms, have helped support GDP growth, which is expected to be more than 5.2 percent in 2004.[3] Tanzania is peaceful and stable with few tribal or regional divisions.[6]

DISTRIBUTION OF GREAT APES

Only the eastern chimpanzee (*Pan troglodytes schweinfurthii*) is naturally present in Tanzania. Chimpanzees generally prefer evergreen or semi-deciduous forest, which in Tanzania occurs most often at lower altitudes and along river and stream valleys.[14, 19] They are found in the regions of Kigoma and Rukwa in the western part of the country, along the shores of Lake Tanganyika, and further inland to the west of the Ugalla River, including the Tongwe East Forest Reserve.[11]

A small population of chimpanzees, probably of the western subspecies, (*P. t. verus*), was introduced to the forested Rubondo Island in Lake Victoria in 1966–1969. These 17 animals were originally caught in Guinea, Sierra Leone, and Côte d'Ivoire and kept in captivity in Europe before being released into the area that is now the Rubondo National Park (NP) (457 km^2). The population is estimated to have at least doubled since the animals were introduced.[8, 12, 13]

THREATS

Thanks to local traditions opposed to eating primates, chimpanzees have historically faced little threat from hunting in Tanzania. Recent decades, however, have seen an influx of refugees from countries such as DRC that do have a culture of primate eating. Tanzanians sometimes kill chimpanzees to obtain body parts for traditional medicine.[7, 21] About 388 110 km^2 of forest remains, and is declining at an average rate of 910 km^2 (0.2 percent) each year,[5] but these figures may not accurately reflect serious local environmental problems of deforestation, desertification, and soil degradation. Using 1991 and 2003 satellite images, for example, forest loss outside Gombe NP has been estimated at 4 percent per year[20] (see Box 13.2). This can be explained in part by human population pressure, since the annual population growth rate in the Kigoma region increased from 2.8 percent (1978–1988 census) to 4.8 percent (1988–2003 census). As a result, Gombe NP (35 km^2) is now almost isolated as a forest island, where once it was a small sample of a much larger forested landscape. Subsistence farming in marginal areas such as Ugalla is a growing cause of forest destruction.

In 1979, it was estimated that there were fewer than 2 000 eastern chimpanzees in Tanzania, with fewer than 480 breeding females.[9] By 2000, the estimate was essentially unchanged, at 1 500–2 500 individuals,[2] but some decline must be assumed given the drastic habitat loss in the areas along the Tanganyika Lake shore and north of Malagarasi River. More than 80 percent of the Luiche, Mlele, and Mkuti Forest Reserves, for example, has been converted to farmland or burned for charcoal.[20] Population estimates exist for Ugalla, Lwazi,[10] and Gombe NP,[2] where there are fewer than 90 individuals,[21] and Mahale Mountains NP, where there were believed to be about 700 individuals in 1967.[17] The precise numbers of chimpanzees elsewhere in Tanzania are not known, and it is uncertain whether the population has actually been stable over this period.

LEGISLATION AND CONSERVATION ACTION

The government of Tanzania is a signatory to the Convention on Biological Diversity (ratified 1996), and the UN Convention to Combat Desertification (ratified April 1997). Chimpanzees and their habitats are legally protected under the National Forestry Policy (1994), National Parks Ordinance (1959), and the Wildlife Policy of Tanzania (1998).

The Jane Goodall Institute (JGI)

Chimpanzees in Gombe National Park.

Map 16.21 Chimpanzee distribution in the United Republic of Tanzania

Data sources are provided at the end of this country profile

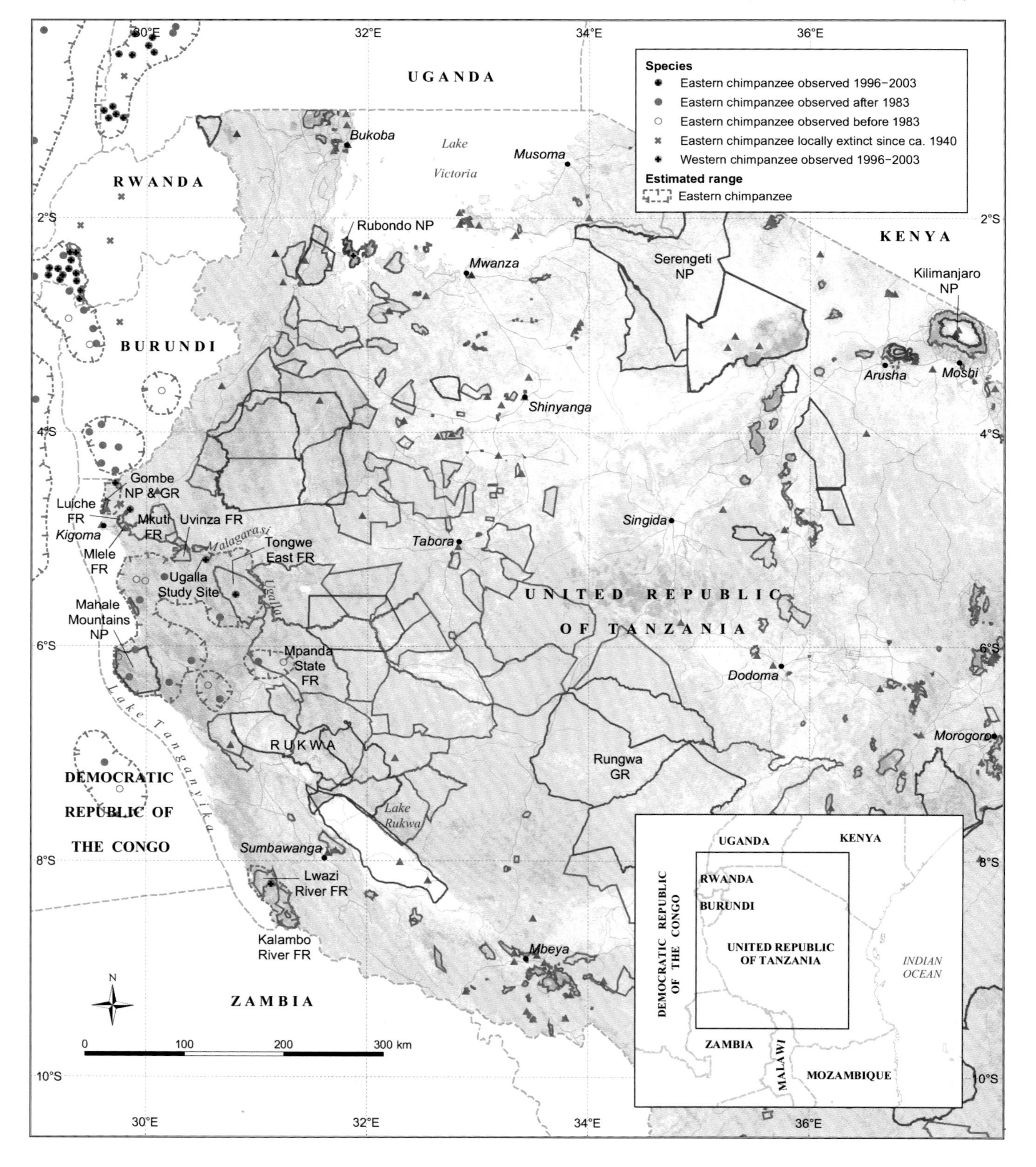

Gombe, at the northern end of Lake Tanganyika, was declared a game reserve in 1943 and a national park in 1968. The chimpanzees of Gombe NP have been studied continuously for many years and are well protected.[19] The Jane Goodall Institute (JGI), founded in 1977 by Jane Goodall, provides support for field behavioral research of chimpanzees in the park. To help conserve chimpanzees in western Tanzania, the Jane Goodall Institute founded the Lake Tanganyika Catchment Reforestation and Education (TACARE) project, which seeks to involve local people in the restoration of forests. It works with 33 villages and aims to promote the preservation of primate habitat, community-centered conservation, education, and youth engagement, enhanced roles for women, and control of the bushmeat trade. Activities include the planting of trees, prevention of soil erosion, and the promotion of family planning and AIDS awareness. Habitat loss is driven in part by a need for fuel for cooking, and therefore alternative sources of energy are being developed as part of the TACARE project.

A recent assessment of the TACARE project by the Jane Goodall Institute and USAID, using a threat reduction assessment (TRA) method and remote sensing and geographic information system (GIS) data, identified five major direct threats to forests outside Gombe NP.[1] These were conversion of forests to subsistence farming, conversion to cash crops such as oil palm, local-scale logging, firewood extraction, and burning. The assessment concluded that to reduce these threats there is a need for more strategic conservation-planning approaches and more spatially and temporally focused conservation actions at the scale of specific forest patches. As a result, the Jane Goodall Institute is now adopting the 'Conservation by Design' planning process and tools that were developed by The Nature Conservancy (TNC), in order to improve the effectiveness of its conservation actions.

The Mahale Mountains area was designated as a national park in 1985. Toshisada Nishida and his colleagues have conducted research on Mahale chimpanzees since 1965.[15] The Frankfurt Zoological Society has been providing support to park operations in the Mahale Mountains NP since 1985. This support has been expanded through the Mahale Ecosystem Management Project (MEMP), funded by the European Union, which is scheduled to run from 2003 to 2008. The goal of the project is to conserve the Mahale ecosystem's outstanding forest and aquatic biodiversity while strengthening the livelihood and environmental security of park-adjacent communities. To achieve this the Mahale Ecosystem Management Project is working to develop conservation-compatible, income-generating activities, build community-based institutions for sustainable development, enhance dialog between the park and the community, identify ecosystem management priorities and threats, develop general management plans for the Mahale Mountains NP and wider ecosystem, and support park administration, resource protection, and ecotourism.

General management plans for Rubondo, Mahale, and Gombe NPs are currently in preparation through collaboration between Tanzania National Parks (TANAPA), the Frankfurt Zoological Society, and the Jane Goodall Institutes.

Other field research programs include one in the Ugalla area that is currently focused on chimpanzee ecology and distribution, and another on Rubondo Island to look at the ecology of the island, the social and ecological adaptation of the chimpanzees to their new habitat, and their relationship with other wildlife (indigenous and introduced) on the island.

Since the mid-1990s, the Jane Goodall Institutes have helped establish chimpanzee sanctuaries throughout Africa.

Michael Huffman

Kansyana Valley, Mahale National Park, western Tanzania. Located along the shores of Lake Tanganyika, this is the central habitat of the Mahale chimpanzees studied since 1965.

FUTURE CONSERVATION STRATEGIES

Tanzania is free of the overwhelming systematic problems that determine conservation agendas elsewhere in Africa, such as a rampant bushmeat trade, widespread and accelerating logging, and warfare with consequent refugee and displacement problems. Accordingly, conservation needs in Tanzania are mainly 'tactical' and site-based, revolving around the need to stabilize land use in and around particular protected areas, and to expand or restore them as necessary. The chimpanzees of Gombe NP, for example, are threatened since the park is too small to sustain them in the long term, now that it has become a forest island surrounded by bare hills.[21] In all likelihood, the viability of this population depends on an increase, not merely a stabilization, of chimpanzee habitat outside the current park boundaries.[4] Mahale NP also faces habitat degradation threats, despite its much greater size.[18] Both Gombe and Mahale chimpanzees also suffer serious threats from disease, much of which may be introduced by proximity to people and human settlements.[16, 21, 23] To address these challenges, long-lasting programs are needed that should be conceived and discussed with local experts in the surrounding areas, and should take into account the dual objectives of arresting the rapid degradation of lands and forests, and improving the standard of living of the villagers.

FURTHER READING

Goodall, J. (1990) *Through a Window: My Thirty Years with the Chimpanzees of Gombe.* Houghton Mifflin Company, Boston.

Massawe, E.T. (1992) Assessment of the status of chimpanzee populations in western Tanzania. *African Studies Monographs* **13** (1): 35–55.

Nishida, T., ed. (1990) *The Chimpanzees of the Mahale Mountains.* University of Tokyo Press, Tokyo.

Tutin, C.E.G., White, L.J.T., Mackanga-Missandzou, A. (1997) The use by rain forest mammals of natural forest fragments in an Equatorial African savanna. *Conservation Biology* **11** (5): 1190–1203.

MAP DATA SOURCES

Map 16.21 Chimpanzee data are based on the following source:

Butynski, T.M. (2001) Africa's great apes. In: Beck, B.B., Stoinski, T.S., Hutchins, M., Maple, T.L., Norton, B., Rowan, A., Stevens, E.F., Arluke, A., eds, *Great Apes and Humans: The Ethics of Coexistence.* Smithsonian Institution Press, Washington, DC. pp. 3–56.

With additional data by personal communication from Idani, G. and Ogawa, H. (2003), Pintea, L. (2004), and from the following sources:

Moscovice, L.R., Huffman, M.A. (2002) The chimpanzees of Rubondo Island, Kakakuona. *Tanzanian Wildlife* **27**: 56–60.

Ogawa, H. (1997) The discovery of chimpanzees in the Lwazi River Area, Tanzania: a new southern distribution limit. *Pan Africa News* **4** (1). http://mahale.web.infoseek.co.jp/PAN/4_1/4(1)-01.html. Accessed June 12 2005.

Schoeninger, M.J., Moore, J., Sept, J.M. (1999) Subsistence strategies of two "savanna" chimpanzee populations: the stable isotope evidence. *American Journal of Primatology* **49** (4): 297–314.

For protected area and other data, see 'Using the maps'.

ACKNOWLEDGMENTS

Many thanks to Michael Huffman (Kyoto University), Lilian Pintea (The Nature Conservancy), Janette Wallis (University of Oklahoma), and Michael Wilson (Gombe Stream Research Center) for their valuable comments on the draft of this section.

AUTHORS

Jared Bakusa, University of Dar Es Salaam

Edmund McManus, UNEP World Conservation Monitoring Centre

CHAPTER 17

Asia

REPUBLIC OF INDONESIA

KIM MCCONKEY, JULIAN CALDECOTT, AND EDMUND MCMANUS

BACKGROUND AND ECONOMY

The Republic of Indonesia declared its independence from the Netherlands in 1945 and, after a Dutch 'police action', this was recognized by all in 1949. The independence agreement excluded West New Guinea (Papua/Irian Jaya), which remained under Dutch control until 1962, when it was absorbed into Indonesia.[8] Initially the country had a federal structure but this was dissolved in 1950; it became a unitary republic under its founding president, Sukarno. The first election, held in 1955, had an inconclusive outcome; this led to the introduction of martial law in 1957 and a period of communist-influenced 'guided democracy' in 1959–1965. A coup and counter-coup in 1965 led to the suppression of the Communist Party of Indonesia (PKI), the massacre of hundreds of thousands of its supporters, and the installation of President Suharto; he ruled with US backing in an authoritarian manner until his fall from power in the midst of the Asian economic crisis of 1997–1998. Since then, Indonesian institutions and society have been changing rapidly under the successive presidencies of B.J. Habibie, Abdurrahman Wahid, and Megawati Sukarnoputri. Following the first direct presidential election, Susilo Bambang Yudhoyono and Jusuf Kalla were inaugurated as president and vice president on October 20 2004.

Indonesia is made up of about 17 000 islands with an aggregate land area of nearly 2 million km^2. These islands have some 50 000–80 000 km of coastline between them. They are set in 3 million km^2 of territorial sea that extends for 5 100 km between mainland Asia and Australia, linking the Indian Ocean to the Pacific. Although traditionally regarded as a Southeast Asian nation, and indeed a founder member of the Association of Southeast Asian Nations (ASEAN), much of Indonesia's territory lies in areas that are not Asiatic in either a biogeographic or a cultural sense. Great apes are restricted to the islands of Sumatra (475 000 km^2) and Borneo (740 000 km^2, of which 536 000 km^2 is in Indonesian Kalimantan) in Western Indonesia, which rise from the Asian (Sunda) continental shelf alongside Java (133 000 km^2).

Indonesia had a total population of about 235 million people in 2003[3] with an annual growth rate of 1.5 percent. This low rate reflects long-standing government sponsorship of a nationwide *dua cukup* (two's enough) family-planning strategy. Nearly two thirds of all Indonesians live on Java and on the nearby islands of Madura and Bali. This is because of their long history of advanced civilization, supported by irrigated farming on their fertile volcanic soils. The Indonesian people are culturally diverse, with several hundred distinct ethnolinguistic groups – many of them in West Papua, part of what is culturally the richest island in the world. The Indonesian language is used for formal purposes throughout the country. It arose from Malay, an Austronesian trading language used throughout the Malay Archipelago (which includes Malaysia and the Philippines).

In 1988, nearly 10 percent of Indonesian land area was classified as arable and 7.2 percent was under permanent crops.[3] Subsistence farming and fishing is important to local people throughout the country, and many areas have been converted to agricultural plantations, often of oil palms (*Elaeis guineensis*). Between 1967 and 2000, the total area under oil palm plantations in Indonesia grew from less than 2 000 km^2 to over 30 000 km^2.[27] The overall economy has long been dominated by the extractive industries, however, with oil, gas, and hard-rock minerals providing much of the nation's foreign exchange. The timber industry is also important,

MAP 17.1 Orangutan distribution in Indonesia

Data sources are provided at the end of this country profile

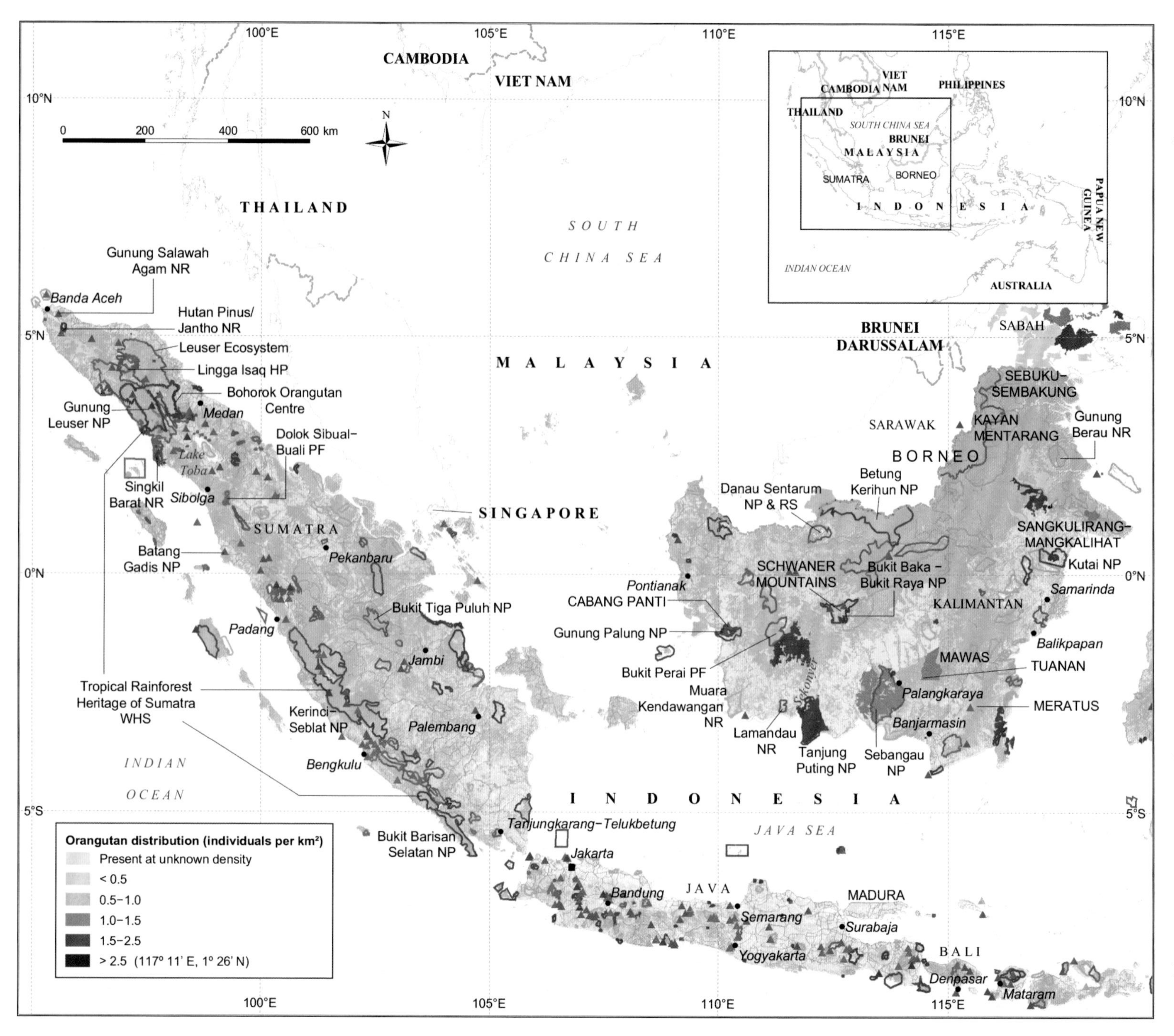

having begun in the early 1960s in Kalimantan; by 1988, concessions had been awarded covering about 434 000 km^2 or three quarters of the total Indonesian forest estate.[4] Following a government decision in the 1970s to phase out exports of logs, Indonesia exported just 1.4 million m^3 as logs and 8.2 million m^3 as plywood in 1991.[7] The total log harvest, meanwhile, rose from about 16 million m^3 in 1981 to about 26 million m^3 in 1987;[9] the annual target harvest for 1995–2000 was over 37 million m^3. The Asian financial crisis, however, intervened in 1997, and the fall of the Suharto regime in 1998 seriously reduced the ability of the government to regulate the timber industry. It is estimated that in 2002 the demand for timber from Indonesia's wood-processing industries was 63 million m^3, while the annual allowable cut, set by the government, was 12 million m^3.[14] The shortfall is being made up by illegal logging, which produces 50.7 million m^3 of logs annually, resulting in state financial losses of at least Rupiah 30.42 trillion (US$3.18 billion) in lost taxes,[1] and putting overwhelming pressure on Indonesia's remaining protected forest estate.[5, 12] In December 2004, a tsunami wave destroyed settlements along the coast of Aceh, killing thousands of people. It is thought that orangutans were little affected, but the resultant demand for timber for rebuilding poses a new threat to the forests of northern Sumatra (see Box 11.2).

DISTRIBUTION OF GREAT APES

Indonesia is home to both orangutan species: *Pongo abelii* in the northern part of Sumatra and all three subspecies of *Pongo pygmaeus* in Kalimantan (Indonesian Borneo): *P. p. morio* in the province of East Kalimantan, *P. p. pygmaeus* in West Kalimantan, and *P. p. wurmbii* in Central Kalimantan. (See also Map 17.1.) *P. pygmaeus* also occurs in Sarawak (Malaysian Borneo), but 80 percent of the population, and most suitable habitat, occurs in the Indonesian sector of the island.[20] The distribution of the apes has declined in parallel with the reduction in the distribution of forest in Indonesia.

Orangutan conservationists use 'habitat blocks' as their basic planning unit, each being an area of connected habitat that is separated from all others by normally impassable barriers, such as major rivers or wide swaths of cultivation. A habitat block therefore corresponds to a separate population, one that is not easily colonized by individuals from other populations. Once habitat blocks are identified, conservation planning then requires knowledge of the population size and rate of change within each.[22]

In Sumatra, 13 habitat blocks have been identified, representing together almost 9 000 km^2. In 2003, there were thought to be about 7 000 orangutans inhabiting these areas,[17] with three populations each containing over 1 000 individuals. Orangutans appear to have become locally extinct to the south of Lake Toba, with the exception of two small populations. The Leuser Ecosystem (Box 11.2) contains four habitat blocks: West Leuser (2 508 individuals), East Leuser (1 052 individuals), the Trumon-Singkil Swamps (1 500 individuals), and the Tripa Swamps (280 individuals). A priority is to reconnect these four units to make a single population of about 5 340 orangutans. The Tripa Swamps are already very fragmented and would be difficult to reconnect to West Leuser, but a degraded forest corridor still exists between West Leuser and the Trumon-Singkil Swamps that is being rehabilitated by the Leuser Development Programme (supported by the European Commission).

In West Kalimantan, the 2002 estimate for the total available habitat for *P. pygmaeus* amounted to about 85 000 km^2,[17] divided between 306 habitat blocks and inhabited by only 2 000–2 500 individuals.[17] There is a relatively large *P. p. wurmbii* population in Central Kalimantan, with a current estimate of over 32 000. This includes the areas of Tanjung Puting, Gunung Palung, and Sebangau National Parks (NPs),[10] and the Kapuas-Barito floodplain (Mawas).[24] An important new area, surveyed for the first time in 2003, is Arut-Belantikan in the foothills of the Schwaner Mountains.[17, 22] It has a total area of 5 600 km^2 of dipterocarp forest, with an estimated population of 6 000–6 500 *P. p. wurmbii*. The main population of *P. p. morio* in East Kalimantan is in the Berau area (including Gunung Gajah), where an estimated 1 558 orangutans survive.[16, 22] Several smaller populations exist, adding approximately 1 500 individuals to this subspecies population.

THREATS

Consistent with the nature of the Indonesian land-development process, which has been based largely on logging and the expansion of plantations, both orangutan species are threatened mainly by significant and ongoing habitat loss and forest fragmentation,[20] aggravated by hunting and persecution as agricultural pests.[12] These continue, despite the fact that orangutans have been protected in Indonesia since 1924. Conservation efforts have failed to slow the decline of orangutan habitat. About 55 000 km^2 of breeding habitat for Bornean orangutans was lost between 1993 and 2002, and much of the remaining forest is affected by logging and forest fires.[22] The probability of forest fires is increased by logging, which opens and fragments the forest and allows normally moist forest areas to dry out and become more prone to ignition. Very little forest below 1 000 m is expected to survive past the year 2010 in either Sumatra or Kalimantan.[12, 26] In November 2003, a landslide caused a flash flood in Gunung Leuser NP which killed over 140 people and destroyed a tourist village at Bukit Lawang in the regency of Langkat. Other floods in the region have been linked to illegal logging.

Infrastructural development also threatens orangutan habitat, as road building has long been seen as one of the most tangible expressions of government investment in development. The road-building lobby is powerful, conservation interests are relatively weak, and numerous protected areas throughout the country have had roads built through them. That this danger persists is shown by the Ladia Galaska road project in Sumatra, which will, if continued, bisect one of the largest remaining orangutan populations, in the Leuser Ecosystem. It would also open up a very large new area for commercial exploitation, whether legal or not, and is hence backed by powerful lobbies within local government, business, and the armed forces. As elsewhere, roads bring with them greater access for people who may hunt, and for settlers who clear forest and may kill orangutans that raid their newly established crops. Kalimantan's peat-swamp forests also suffer from canals (see Box 10.3) built to float out illegal logs, which are draining and killing the forests.

The net result is that both orangutan species are seriously endangered in Indonesia, making it urgently necessary to identify locations where there is a realistic chance of protecting viable populations, and to direct appropriate investments to them.

LEGISLATION AND CONSERVATION ACTION

National legislation

Conservation in Indonesia is based on Act No. 5 of 1990, Concerning Conservation of Living Resources and their Ecosystems. This lists species (including orangutans) covered by the Convention on International Trade in Endangered Species of Wild Fauna and Flora (CITES) and defines a variety of protected areas based on two main categories: sanctuary reserves (Article 14, comprising strict nature reserves and wildlife sanctuaries) and nature conservation areas (Article 29, comprising national parks, grand forest parks, and natural recreation parks).

The same law further allows for management zoning (Articles 32 and 34), the constitution of Biosphere Reserves (Article 18), protection of endangered and rare species (Articles 20–25), and refers to buffer zones for protected areas (Articles 16 and 29). Management of the conservation area system and wildlife protection are the responsibility of the Ministry of Forestry's Directorate General of Forest Protection and Nature Conservation (PHKA), in collaboration with local government and the police. The legal and institutional basis for enforcement of conservation law has never been completely clear in Indonesia, however, and there are many presidential, ministerial, and local government decisions (*keputusan*) that have the force of law. For example, in 1995, Law UU 280/kpts II/1995 made it illegal to release orangutans in areas where wild orangutans still persist (leading to the closure of Bohorok in Sumatra as a rehabilitation and release center).

Protected areas

In Indonesia, conservation planning went hand-in-hand with forest and land-use planning. The development of the timber industry was based on a consensus classification of forest function (*Tata Guna Hutan Kesepakatan*, TGHK). The TGHK arose in 1970–1985, based on discussions among various government agencies to produce maps showing agreed allocations of forest land to various categories of permanent use.[6] The five main categories and their forestry uses were:

- nature reserve (no timber extraction);
- protection forest (no timber extraction);
- limited production forest (for non-industrial selection felling);
- regular production forest (for industrial selection or clear felling, according to forest type); and
- conversion forest (for clear felling and conversion to other uses).

This classification system took little notice of any traditional land claims by local communities, who were not consulted.

The categorization program was also flawed

due to insufficient information to support spatial and forestry planning,[6] but other mapping operations were undertaken to correct this. The most comprehensive of these was by the Ministry of Transmigration in the late 1980s; this mapped land use and land capability for the whole country outside Java and Bali, with a view to finding suitable places to receive officially sponsored settlers.[19] The Ministry of Public Works later integrated these Regional Physical Planning Program for Transmigration (RePPProT) maps with those from the TGHK as well as district and provincial planning maps. These were then updated using field observations and new remote imagery to show actual forest cover, protected forests, nature conservation areas, sanctuary reserves, the alignments of existing as well as proposed roads, and other development projects. Act No. 24 of 1992 provided a comprehensive legislative context for a national system of spatial planning, although cases of conflict between planned and actual uses of land continued to occur.

Planning for a national system of protected areas in Indonesia was done in parallel with TGHK, and later with RePPProT. By 1990, Indonesia had gazetted 303 terrestrial nature reserves of various kinds totaling 160 000 km^2 or 8.2 percent of land area, and another 20 000 km^2 at 175 sites had been proposed as such reserves.[18] These areas were selected to include viable and representative samples of most ecosystems, and populations of most native species.[15] More than 300 000 km^2 had meanwhile been designated through TGHK and RePPProT as protection forest. The main role of these designations was the safeguarding of both water catchments and steep slopes. Conservation efforts during the 1990s were guided by a national biodiversity action plan[2] and policy analyses such as the *Indonesian Country Study on Biological Diversity.*[13] These were being supported by official donors to the extent of about US$12 million per year. The development of an effective system of protected areas therefore seemed possible, giving good coverage for most components of Indonesia's biodiversity.

Many reserves had little effective management, however, reflecting the limited resources of the responsible department within the Ministry of Forestry, the Directorate General of Forest Protection and Nature Conservation (then the PHPA, now the PHKA). At that time, rates of expenditure by the PHPA and its partners averaged about US$75 per km^2 per year for the reserve system as a whole.[13] This can be compared with the 300–500 percent higher levels of expenditure on priority nature reserves in Thailand and China, as well as with the recommended minimum figure of US$300–400 per km^2 per year for the management of national parks, suggested by IUCN.[11] At least US$130 million per year would have been required to have brought average expenditure rates in Indonesia into the same general range as in Thailand and China.

SOCP

Orangutans in transit to their release site in Jambi province, Sumatra.

The Asian financial crisis in 1997 meant that efforts to increase investment were derailed. The fall of the Suharto regime in 1998, after more than 30 years in power, caused grave disruption to the ability of central government to impose an orderly system for forest management or conservation on the outer islands, such as Sumatra and Kalimantan. Indonesia is organized into provinces, *kabupaten* (regencies or districts, each headed by a *bupati*, regent or resident); *kecamatan* (subdistricts, each headed by a *camat*); and various kinds of community. As a consequence of the political events of the late 1990s, a significant transfer of authority

SOCP

A young orangutan at the Sumatran Orangutan Conservation Programme's quarantine site.

Rondang Siregar

Bornean orangutans at the Wanariset Rehabilitation Centre, Indonesia.

from the center to the regions was agreed, with considerable authority granted to the *kabupatens* to make their own decisions regarding land and forest use. The result in many areas was a rapid increase in logging and forest clearance, both legal and illegal.[12]

The *kabupaten* is a political unit too large to be easily accountable to its inhabitants, but small enough to be influenced by private companies and rich or well connected individuals. There are signs, however, that local people are becoming more environmentally assertive, and in response to local public demand, some *kabupatens* have proposed new protected areas to Jakarta. Batang Gadis NP in northern Sumatra is the most recent example, and reflects long-suppressed wishes, expressed as early as 1928, by traditional leaders to protect the Leuser Ecosystem. Similarly, local people in the Kayan Mentarang area in East Kalimantan had lobbied for a national park for decades, supported by WWF–The Global Conservation Organization. Meanwhile, central government has been attempting to bring illegal logging under control. In 2004, it proposed a law that would punish convictions for illegal logging or the starting of fires with a minimum jail sentence of 12 years, or by death in exceptional cases.

In July 2004, an area of 25 950 km^2 jointly covered by the Gunung Leuser, Kerinci-Seblat, and Bukit Barisan Selatan NPs was designated as the Tropical Rainforest Heritage of Sumatra World Heritage Site.[23] Gunung Leuser NP is part of the Leuser Ecosystem area and supports orangutans, while Kerinci-Seblat NP is the site of reported 'orang pendek' sightings (see Chapter 1).

Conservation projects

Both national and international nongovernmental organizations (NGOs) found it hard to work in Indonesia under the Suharto regime, as they tended to be viewed with suspicion by the hard-line militarists who held most power at that time. They received some political protection, however, from the Ministry of State for Population and Environment (KLH) and its long-time Minister Emil Salim. Some became established by working carefully with central government (e.g. the WWF-Indonesia program), by gaining the strong support of provincial governments (e.g. the operation run by Biruté Galdikas in Central Kalimantan), or through making alliances at both levels. Conservation, research, and activities relating to sanctuary rehabilitation or reintroduction often had common roots during this period, although they have since become somewhat differentiated. They can now be classified roughly into:

- quarantine activities, in which formerly captive orangutans are cared for, such as those of the Sumatran Orangutan Conservation Programme (SOCP), the Orangutan Foundation International and UK, and the Borneo Orangutan Survival Foundation (BOSF);
- rehabilitation and reintroduction activities, such as those at Meratus, Tanjung Puting, and Lemandau in Kalimantan; as well as at Bukit Tiga Puluh NP in Jambi province, Sumatra;
- field research on wild orangutans, at, for example, Cabang Panti, Tanjung Puting, Kutai, Tuanan, or Mawas in Kalimantan; and Ketambe, Bohorok, and Dolok Sibual-Buali in Sumatra); and
- habitat-conservation projects, such as the Leuser Development Programme, and projects at Mawas, Tanjung Puting NP, and Gunung Palung NP.

In Sumatra, the SOCP has established a release program in Bukit Tiga Puluh NP in Jambi province. It also conducts most of the survey and monitoring work concerning the status and distribution of wild Sumatran orangutans. The Research, Monitoring, and Information Division of the Leuser Management

Unit manages research activities within the Leuser Ecosystem.

In Kalimantan, the Orangutan Foundation International and UK fund patrols in Tanjung Puting NP, rehabilitate and release orphan orangutans in Lamandau Nature Reserve, and support research into conservation and forest restoration. The BOSF rehabilitates and releases orphans in the Balikpapan area and in other parts of Kalimantan, and is involved in proposals to protect the Mawas area, 5 000 km^2 of peat-swamp forest inhabited by orangutans. Several partner NGOs in other countries include Balikpapan Orangutan Society–USA (BOS-USA), who provide support for the work of the BOSF.

The Orangutan Conservation Forum (OCF) is a group of orangutan-focused conservation organizations with educational programs. The OCF was originally planned at a meeting in Palangkaraya, Central Kalimantan, in 2002; it aims to act as a centralized body for communication and facilitation of the sharing of information between all groups and individuals involved in orangutan conservation, as well as other environmental education. It will also play a key role in advising upon the Indonesian national great ape survival plan (NGASP).

FUTURE CONSERVATION STRATEGIES

The forces pressing towards the complete destruction of lowland forests in Sumatra and Kalimantan are very powerful, and the outlook for species that inhabit those forests is worrying. There are signs of hope, in that people in some areas are starting to demand protected areas and local government leaders are starting to grant them with the endorsement of the national authorities. Public awareness of the plight of orangutans is rapidly increasing in Indonesia, and a political willingness to act against illegal logging is becoming established. The urgency for action varies among habitat units and depends upon the current rate of logging and size of the orangutan population. For some habitat units, the need for action is immediate if orangutans are to survive. If habitat loss can be controlled, then actions to reduce fragmentation will become more relevant and valuable. Some of the major priorities for action presented in the 2004 *Orangutan Population and Habitat Viability Assessment Final Report*[22] are listed below.

- **Studies.** Systematic surveys throughout the orangutan range are needed to ensure that all priority sites are identified and research protocols are standardized. Collaborative field efforts need to be initiated, with umbrella projects for both islands. In particular, research at Ketambe in the Leuser Ecosystem should be continued and expanded. Forest loss should be monitored and a sustainable source of funding for long-term orangutan research *in situ* should be sought. Participation by local NGOs should be encouraged.
- **Protection.** Surveillance and enforcement teams require training in the techniques needed to identify and protect populations, keystone resources, corridors, and essential habitats outside current protected areas. Helicopter surveillance should be considered. Conservation policy should be integrated into governmental policy.
- **Education.** The development of an increased awareness of preservation needs is required via education programs in both schools and national institutions. International media coverage of the current status of orangutan populations should be sought.
- **Regional actions in Sumatra.** Restore the connections between habitat blocks in the Leuser Ecosystem and surrounding areas (i.e. connect West and East Leuser, and Trumon-Singkil and West Leuser habitat blocks), and ensure that the Ladia Galaska road scheme does not pass through Gunung Leuser NP.
- **Regional actions in Central Kalimantan.** Extend the northern boundary of the Tanjung Puting NP to include the north shore of the Sekoyner River and to establish a corridor to the eastern forest. Fill in the canals that have been cut into peat swamps, to float out illegally cut logs in the Sebangau catchment.
- **Regional actions in East Kalimantan.** Establish nature reserves at Sangkulirang-Mangkalihat and Sebuku-Sembakung.
- **Regional actions in West Kalimantan.** Build capacity at Gunung Palung NP. The BOSF is encouraged to continue developing and refining innovative models for the sustainable conservation of the Mawas orangutan population.
- **Development.** Improve habitat quality in degraded areas, possibly through enrichment planting. Increase sustainable economic alternatives for communities surrounding critical orangutan habitats.
- **Rehabilitation and translocation.** Priority should be given to conserving the wild

population (efforts *in situ*), as opposed to ex-captive care (conservation efforts *ex situ*). Rehabilitation centers should be licensed and monitored. Record keeping should be improved and made freely available.

- **Coordination**. An Orangutan Scientific Commission should be established, and the plans for the Orangutan Conservation Forum realized using funds committed by NGOs at the 2002 meeting in Palangkaraya to initiate this process.
- **International policy**. The World Heritage Species concept was endorsed; the governments of Indonesia, Malaysia, Sabah, and Sarawak are encouraged to promote orangutans as one of the world's first such species.

FURTHER READING

Curran, L.M., Trigg, S., McDonald, A., Astiani, D., Hardiono, Y., Siregar, P., Caniago, I., Kasischke, E. (2004) Lowland forest loss in protected areas of Indonesian Borneo. *Science* **303**: 1000–1003.

Jepson, P., Jarvie, J., MacKinnon, K., Monk, K.A. (2001) The end of Indonesia's lowland forests. *Science* **292**: 859–861.

Robertson, J.M.Y., van Schaik, C.P. (2001) Causal factors underlying the dramatic decline of the Sumatran orang-utan. *Oryx* **35**: 26-38.

UNESCO (2004) *Tropical Rainforest Heritage of Sumatra.* World Heritage. http://whc.unesco.org/pg.cfm?cid=31&id_site=1167. Accessed December 5 2004.

van Nieuwstadt, M.G.L., Sheil, D., Kartawinata, K. (2001) The ecological consequences of logging in the burned forests of East Kalimantan, Indonesia. *Conservation Biology* **15** (4): 1183–1186.

Also see Global Forest Watch resources on Indonesian deforestation: http://www.globalforestwatch.org/english/indonesia/maps.htm.

MAP DATA SOURCES

Map 17.1 Orangutan data are based on the following sources, with additional information by personal communication from Meijaard, E. (2005) and Singleton, I. (2005):

Dadi, R.A., Riswan (2004) Orangutan distribution polygons: developed at the Leuser Management Unit as part of the Leuser Development Programme, funded by the European Commission and the government of Indonesia. Leuser Management Unit, Sumatra, Indonesia. Based on technical criteria set by Singleton, I. Main sources of field data: van Schaik, C., Idrusman, Singleton, I., Wich, S. Additional information from Dadi, R., Griffiths, M., Priatna, D., Rijksen, H., Riswan, Robertson, Y., Universities of Bristol and Bogor Expedition to Sumatra (Burton, J., Bloxam, C., Kuswandono, Long, B., McPherson, J.), and members of the Leuser Management Unit's Antipoaching Unit.

Meijaard, E., Dennis, R., Singleton, I. (2004) Borneo Orangutan PHVA Habitat Units: Composite dataset developed by Meijaard & Dennis (2003) and amended by delegates at the Orangutan PHVA Workshop, Jakarta, January 15–18 2004.

Singleton, I., Wich, S., Husson, S., Stephens, S., Utami Atmoko, S., Leighton, M., Rosen, N., Traylor-Holzer, K., Lacy, R., Byers, O., eds (2004) *Orangutan Population and Habitat Viability Assessment: Final Report.* IUCN/SSC Conservation Breeding Specialist Group, Apple Valley, Minnesota.

Please also see full acknowledgments for the Sumatra data in Chapter 11.

For protected area and other data, see 'Using the maps'.

ACKNOWLEDGMENTS

Many thanks to Ashley Leiman (Orangutan Foundation) and Ian Singleton (Sumatran Orangutan Conservation Programme) for their valuable comments on the draft of this section, and to Mike Griffiths (Leuser International Foundation) for information on the Leuser Ecosystem.

AUTHORS

Kim McConkey, UNEP World Conservation Monitoring Centre
Julian Caldecott, UNEP World Conservation Monitoring Centre
Edmund McManus, UNEP World Conservation Monitoring Centre

MALAYSIA

KIM McCONKEY, JULIAN CALDECOTT, AND EDMUND McMANUS

BACKGROUND AND ECONOMY

Malaysia comprises most of the Malay Peninsula (West Malaysia) and the northern and northwestern parts of Borneo (East Malaysia), with the southern South China Sea between them. It was formed in 1963 through a federation of the former British colonies of Malaya (which had become independent from the UK in 1957, and has increasingly come to be known as Peninsular Malaysia), Sarawak (which had been independent since the 1840s under the Raj of James Brooke and his successors, and was a colony only briefly after the Second World War), North Borneo (renamed Sabah in 1963), and Singapore (which then seceded from Malaysia in 1965). The federal constitution allocates roles, rights, and responsibilities among the state and federal governments and, in particular, gives the states control over most aspects of land and forest use. Wildlife and forest management, therefore, are primarily the responsibility of the state governments.

Malaysia is a middle-income country, the economy of which has diversified into high-technology manufacturing, tourism, and other services, from a strong base in plantation agriculture, production forestry, oil, and gas. The country's total land area is about 328 550 km^2, and its population was about 23 million in 2003, growing at nearly 1.9 percent per year.[3] The western part of the peninsula contains a series of large towns, cities, and industrial zones, while the interior and eastern parts remain largely agrarian or else under large plantations (mainly of oil palm) or natural forest (e.g. the Main Range Mountains, Taman Negara National Park (NP), and the Krau Wildlife Reserve). East Malaysia is relatively sparsely populated. Although it has urban areas (e.g. Kota Kinabalu and Sandakan in Sabah and Kuching, Sibu, and Miri in Sarawak) it is largely under shifting cultivation managed by indigenous Dayak peoples (e.g. Kadazan-Dusun in Sabah, Iban, and Bidayuh, and several 'Orang Ulu' peoples in Sarawak), or large, recent, oil palm and other plantations, or natural forests mostly managed for timber production, plus a number of important protected areas. Malaysia's economy grew by 4.9 percent in 2003, despite a difficult first half when investor confidence was shaken by the sudden acute respiratory syndrome (SARS) epidemic and by the war in Iraq.

Sarawak (12 500 km^2) is Malaysia's largest state and occupies 17 percent of the island of Borneo. Of its total land area, 37 percent is classified as permanent forest estate, 3.3 percent is gazetted under state law as protected areas (national parks, wildlife sanctuaries, and nature reserves), while another 6 percent has been proposed for protection.[9] Sarawak has been a major exporter of timber since the 1960s, when logging began in the peat-swamp forests of the coastal plain; in the 1970s this began to move into the hilly interior which, by the late 1980s, had come to dominate log production. Total production increased steadily from about 5 million m^3 per year in the late 1970s to about 14 million m^3 per year in the late 1980s. Since the early 1990s, over 2 000 km^2 of forest in Sarawak has been logged each year.

Sabah (73 371 km^2) occupies 10 percent of the island of Borneo,[5] and about half remains under natural forests,[6] although oil palm and pulpwood plantations continue to expand. Eastern Sabah was almost completely uninhabited until about 1960; now only 25 percent of land area remains under lowland forest, and much of this is logged. From the late 1960s into the 1990s, Sabah's forests were managed in ways that resulted in severe depletion of the state's timber reserves,[10] and they are now virtually exhausted.[7] In response to growing environmental concern and awareness, Sabahan institutions have steadily developed policies, plans, and laws to promote the survival of Sabah's biological resources, contributing to the development of a more sustainable approach to forest and biodiversity management. This process was matched by the growth in nature-oriented tourism within Sabah, leading to the realization that wildlife – especially orangutans – and forests could prove very valuable if used in new and nondestructive ways.

DISTRIBUTION OF GREAT APES

The northwest Bornean orangutan (*Pongo pygmaeus pygmaeus*) occurs in Sarawak, and the northeast Bornean orangutan (*P. p. morio*) occurs in Sabah. (See also Map 17.2.)

In Sarawak, significant populations of orangutans occur only in the south-central interior, in and around the Lanjak-Entimau Wildlife Sanctuary on the border with Indonesian West Kalimantan

MAP 17.2 Orangutan distribution in Malaysia

Data sources are provided at the end of this country profile

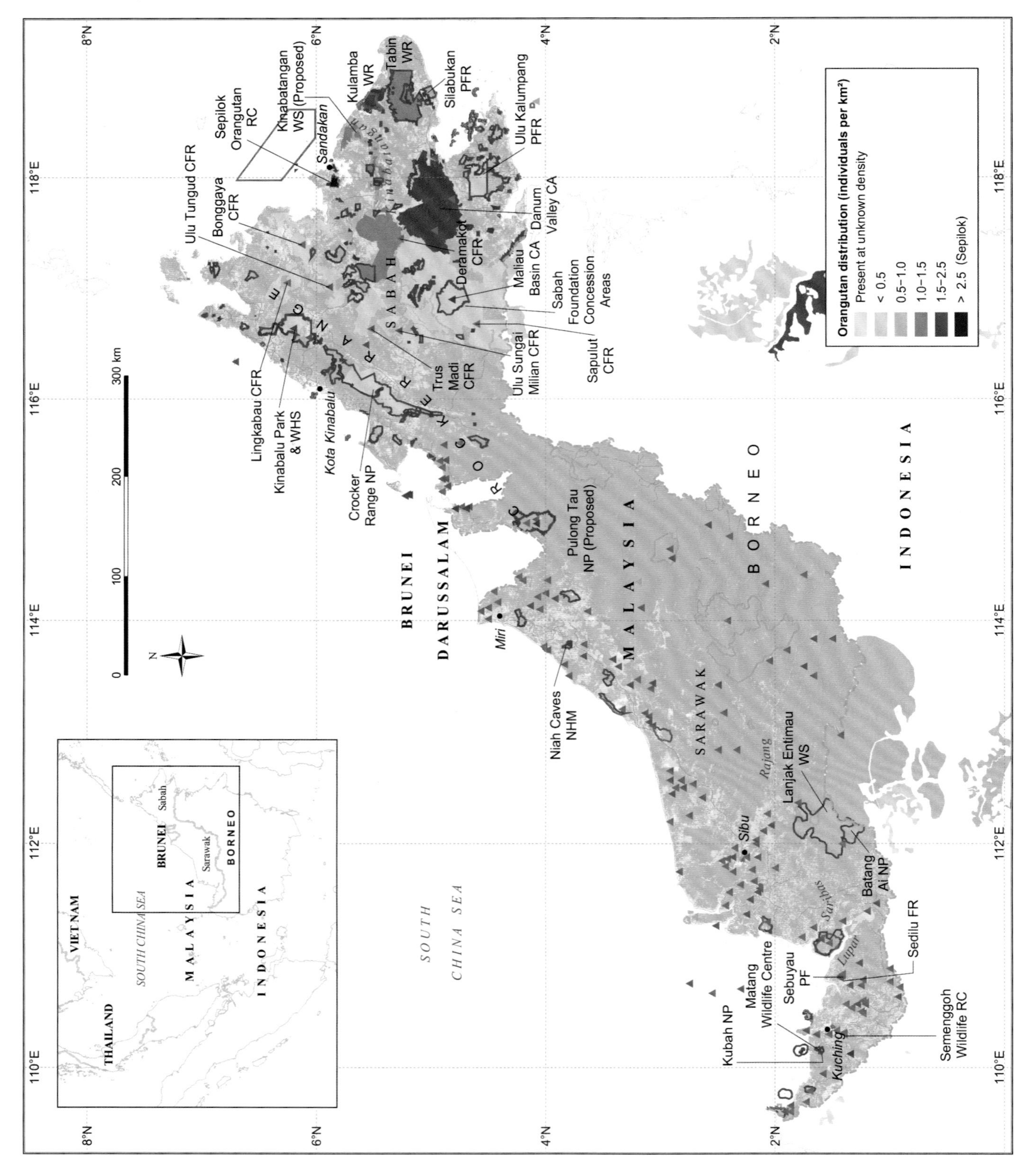

(which has a contiguous population in the Betung Kerihun NP).[3] Small orangutan populations have been documented at Sedilu (two animals), Sebuyau (over 30 animals), and possibly in pockets of swamp forest in the coastal region and mouths of the Lupar and Saribas Rivers. Wandering individuals are occasionally encountered as far north as the Tamabu Range (Pulong Tau) and Brunei Darussalam. The locations of the source populations, for these wanderers and therefore for the subspecies (*P. p. pygmaeus* or *P. p. morio*), in northern interior Sarawak and Brunei Darussalam are not known. The virtual absence of orangutans north of the Rajang River may be a consequence of prehistoric hunting or of insufficient fruit supplies in these dipterocarp-rich forests to sustain breeding populations, or a combination of both factors. Certainly, orangutan bones are common in the deposits at Niah Caves National Historic Monument in northern Sarawak, indicating a hunting history spanning several thousand years. Modern hunting, especially by Iban people who have extensively settled southern Sarawak, has greatly reduced orangutan populations; much of this area has also been converted to a mosaic of shifting cultivation, with little intact forest in many places. Elsewhere, logging has also had a negative effect on orangutan population numbers.

In Sabah, orangutans are patchily distributed over their range. The greatest abundance of orangutan nests is recorded in lowland forests below 500 m above sea level. Recent aerial and ground censuses estimated that about 11 000 orangutans (95 percent confidence intervals: 8 000–18 000) were present in 16 major populations (i.e. populations with more than 50 individuals, their degree of true isolation from one another being largely unknown), mainly in the eastern and central parts of Sabah.[1]

In the northern and western parts of Sabah, only two significant (though small and isolated) populations are now found: in Crocker Range NP, with about 180 orangutans, and in Kinabalu Park and World Heritage Site, with about 50. The Sepilok Orangutan Rehabilitation Centre in eastern Sabah also has about 150 ex-captive rehabilitants. The main protected populations include those in the Tabin Wildlife Reserve (with about 1 285 individuals), Lower Kinabatangan Wildlife Sanctuary (1 125 individuals),[2] Kulamba Wildlife Reserve (730 individuals), and Danum Valley Conservation Area (730 individuals).

More than 60 percent of Sabah's orangutans live outside protected areas, in production forests that have been through several rounds of timber extraction and that are still exploited for timber.[1] The two largest orangutan populations are found in logged production forests in the eastern parts of the Sabah Foundation (Yayasan Sabah) forest concession (with 3 300–11 900 individuals) and on the north side of the Upper Kinabatangan River (approximately 2 300 orangutans, 1 000 of them in the Deramakot Forest Reserve).[1] In addition, the Trus Madi, Ulu Sungai Milian, and Sapulut forests together host perhaps 520 orangutans; and the Ulu Tungud, Lingkabau, Bonggaya, Ulu Kalumpang, and Silabukan Forest Reserves together have about 500 orangutans in total.

Orangutan abundance in production forests is directly correlated with the type of forest management. The highest orangutan densities were identified in those forests implementing reduced-impact logging, showing that uncontrolled logging activities have a negative impact on orangutan abundance.

THREATS

Hunting in Malaysian Borneo is largely of only historical (or prehistorical) significance in helping to explain the modern distribution of orangutans, as almost all the surviving Sarawak population is now within secure protected areas, and there appears never to have been much hunting in central and eastern Sabah. Other, very serious, threats are at work in Sabah; these are interactive, cumulative, and potentially devastating to orangutans. One factor is the conversion of forest to plantations, particularly of oil palms (*Elaeis guineensis*) and fast-growing pulpwood trees such as *Acacia mangium*; this has occurred or is proposed over enormous areas of fertile soil in eastern Sabah. Forest conversion has contributed to the loss of at least 35 percent of orangutan habitat since the mid-1980s.[6, 9] An equally severe threat is logging, especially multiple re-entry logging of dipterocarp-rich forests. A third severe threat is fire, to which both logged forests and pulpwood plantations are dangerously vulnerable in Sabah's seasonal climate, especially when droughts are prolonged during El Niño climatic events.

The current network of protected areas in Sabah harbors about 4 500 orangutans, representing about 34 percent of the total number found in the state.[1] These populations are very fragmented;

KOCP

A Bornean orangutan in a *Ficus* tree, Kinabatangan.

in the long term, most of these are vulnerable to extinction through the effects of inbreeding, drought, fire, disease, or localized hunting.[9]

LEGISLATION AND CONSERVATION ACTION

Legislation

Orangutans are protected in Peninsular Malaysia under the Protection of Wildlife Act (1972), in Sabah by the Sabah Wildlife Conservation Enactment (1997), and in Sarawak by the Wild Life Protection Ordinance (1998). Their habitats may also be protected in various ways under forest law. In Sabah, parks are gazetted under the Parks Enactment (1984); wildlife sanctuaries are gazetted under the Wildlife Conservation Enactment (1997). In Sarawak, parks, wildlife sanctuaries, or nature reserves are gazetted under the Wild Life Protection Ordinance (1998) or the National Parks and Nature Reserves Ordinance (1998).

Protected areas

Sarawak has two protected areas that are significant for orangutan conservation: the Lanjak-Entimau Wildlife Sanctuary (1 870 km^2), and the Batang Ai NP (250 km^2); between them, these contain an estimated 1 400 orangutans. Hunting and illegal logging are only minor problems in these areas, but could become serious if not monitored, especially because the areas are contiguous with Indonesia, where illegal logging is rampant.

Close to 8 percent of Sabah's land area is included in the system of national parks and other categories of existing reserves.[9] Several of the areas protected under state law contain orangutan populations in the 500–1 500 range, and are therefore particularly significant for the conservation of orangutans.[1, 12] These are the Tabin Wildlife Reserve (1 225 km^2); the Lower Kinabatangan Wildlife Sanctuary (260 km^2), which is adjacent to 257 km^2 of unprotected and managed forest; and the Kulamba Wildlife Reserve (204 km^2). The Kinabatangan Orangutan Conservation Project (KOCP) was established in 1998 to secure the population in the Kinabatangan floodplain of eastern Sabah, especially in and around the Lower Kinabatangan Wildlife Sanctuary (see Box 10.2). Finally, the Danum Valley Conservation Area (438 km^2) is set aside as a protected area within the Sabah Foundation timber concession and officially designated as a class I protection forest reserve.[11] The Danum Valley orangutan population is connected with those inhabiting the broader Sabah Foundation concession production forests.

Other protected areas containing orangutans are the Crocker Range NP (2 400 km^2), Kinabalu Park (750 km^2), and the Sepilok Orangutan Rehabilitation Centre (43 km^2).

Most of Sabah's orangutans do not occur in official protected areas. They are instead found in various production forest landscapes in the eastern Sabah Foundation concession (an area of over 4 000 km^2); to the north of the Upper (Ulu) Kinabatangan River (some 2 000 km^2); in the Trus Madi, Ulu Sungai Milian, and Sapulut forests (which cover about 3 300 km^2 in total); and in the Ulu Tungud, Lingkabau, Bonggaya, Ulu Kalumpang, and Silabukan Forest Reserves (covering 2 621 km^2 in total).

Some populations in managed forests (Tabin, Trus Madi, and Sabah Foundation) are thought currently to be below the habitat's carrying capacity, and hence capable of increasing in future. Others, however (Kulamba and Lower Kinabatangan), are thought to be too dense for the habitat to sustain them in the long term. This is a result of the movement of refugee orangutans from surrounding areas, where habitat has been recently destroyed.[12]

Sanctuaries and rehabilitation

The Matang Wildlife Centre in Kubah NP, the Semenggoh Wildlife Rehabilitation Centre in Sarawak, and the Sepilok Orangutan Rehabilitation Centre in Sabah are all involved in the care and rehabilitation of confiscated young orangutans.

FUTURE CONSERVATION STRATEGIES

The Sabah government has recognized the recommendations of an International Workshop on Orangutan Conservation in Sabah, in August 2003, including those outlined below.[12]

- **Forest management.** Sabah's forests should be managed for orangutan conservation, by reviewing current and future plans for forest management in light of a state wildlife strategy formulated by the Sabah Wildlife Department, by enhancing collaboration among relevant management authorities, and by issuing practical guidelines to foresters, especially in the forest management units which harbor over 60 percent of Sabah's orangutans.
- **Agriculture.** Agricultural practices should incorporate the needs of orangutans through sensitive protection measures for small-scale agriculture and the strict control of land development for oil palm plantations in orangutan habitat regions, including the enforcement of Section 38 of the Wildlife Conservation Enactment.
- **Tourism industry.** Policies should be adopted for the enhancement and development of sustainable and responsible orangutan tourism in Sabah, both to minimize its impact on the environment and to enhance the conservation of orangutan populations.
- **Off-site conservation (*ex situ*).** Current conservation activities *ex situ* should continue to be enhanced to complement on-site conservation measures (*in situ*).
- **Research.** Current research on Sabah's orangutans should continue to be promoted and enhanced, especially through activities carried out at local universities, institutions, and departments.
- **Public awareness.** Awareness of orangutan needs and the legal framework for their protection must be heightened, especially among policy makers, forestry and plantation managers, and workers.

In the context of these overall goals, priority actions would include those listed below.

- **Research.** There is a need for in-depth field studies to investigate further the impacts of logging and associated human activities (such as illegal killing) on orangutan ecology and survival in unprotected forests; to assess the true role of these habitats for ape conservation; and to design forest-management strategies that could allow the long-term survival of orangutans outside protected areas, while providing opportunities for the sustainable use of natural resources.
- **Protection.** Training should be provided for surveillance and enforcement teams. Populations, keystone resources, corridors, and essential habitats outside current protected areas should be identified and protected.
- **Education.** Raise awareness of preservation needs, by education programs in schools, media contacts, and through public and governmental lectures.
- **Development.** Improve habitat quality in degraded areas through enrichment planting. Conservation policy should be integrated into governmental policy.
- **Rehabilitation and translocation.** In allocating scarce resources, priority should be given to conserving wild populations *in situ*, as opposed to the care of former captives *ex situ*. Rehabilitation centers should be licensed and monitored, and all pertinent regulations should be followed. Rehabilitation methods should be appropriately assessed, and the most effective methods should be used. Record keeping should be improved and records made freely available to the conservation community.

During a Population and Habitat Viability Assessment Workshop held in January 2004, it was recognized that several orangutan populations require special attention and these were designated as 'Orangutan High Priority Areas'. Furthermore, site-specific measures were agreed to protect orangutan populations in commercial forest reserves (e.g. Sabah Foundation, North Kinabatangan, and Trus Madi), where the priorities are:

Ian Redmond

A palm oil plantation in Malaysia.

- to keep under natural forest those areas in which the largest orangutan populations occur;
- to use reduced-impact logging systems, as developed in Deramakot Forest Reserve in the Upper Kinabatangan;
- to initiate studies on the long-term impacts of forest exploitation in commercial forests on orangutan ecology and survival;
- to monitor orangutan population trends through regular aerial surveys of their nests;
- to develop and implement forest-management plans with all relevant stakeholders, with special attention to orangutans;
- to enhance awareness of orangutan conservation through education campaigns conducted with the workers, contractors, managers, and all relevant stakeholders.

For orangutan populations in protected areas (e.g. Lower Kinabatangan Wildlife Sanctuary, Tabin Wildlife Reserve, Kulamba Wildlife Reserve), the priorities are:

- to enhance protection against illegal logging or any other threats to the habitat;
- to reduce conflicts with agriculture by identifying ways to deal with problem orangutans;
- to connect currently isolated protected areas by creating forest corridors between them;
- to monitor orangutan populations through ground and aerial surveys;
- to promote research activities in those protected areas;
- to develop orangutan-based ecotourism that will provide economic opportunities to local communities.

FURTHER READING

Ancrenaz, M., Gimenez, O., Ambu, L., Ancrenaz, K., Andau, P., Goossens, B., Payne, J., Tuuga, A., Lackman-Ancrenaz, I. (2005) Aerial surveys give new estimates for orang-utans in Sabah, Malaysia. *PloS Biology* **3** (1): e3. http://dx.doi.org/10.1371/journal.pbio.0030003. Accessed December 8 2004.

MacKinnon, J.R. (1971) The orang-utan in Sabah today. A study of a wild population in the Ulu Segama Reserve. *Oryx* **11**: 141–191.

McMorrow, J., Talip, M.A. (2001) Decline of forest area in Sabah, Malaysia: relationship to state policies, land code and land capability. *Global Environmental Change* **11**: 217–230.

Rijksen, H.D., Meijaard, E. (1999) *Our Vanishing Relative: The Status of Wild Orang-utans at the Close of the Twentieth Century.* Kluwer Academic Publishers, Dordrecht.

MAP DATA SOURCES

Map 17.2 Orangutan data are based on the following sources:

Ancrenaz, M., Lackman-Ancrenaz, I. (2004) *Orang-utan Status in Sabah: Distribution and Population Size.* Kinabatangan Orang-utan Conservation Project, Sandakan, Malaysia.

Meijaard, E., Dennis, R., Singleton, I. (2004) Borneo Orangutan PHVA Habitat Units: Composite dataset developed by Meijaard & Dennis (2003) and amended by delegates at the Orangutan PHVA Workshop, Jakarta, January 15–18 2004.

Singleton, I., Wich, S., Husson, S., Stephens, S., Utami Atmoko, S., Leighton, M., Rosen, N., Traylor-Holzer, K., Lacy, R., Byers, O., eds (2004) *Orangutan Population and Habitat Viability Assessment: Final Report.* IUCN/SSC Conservation Breeding Specialist Group, Apple Valley, Minnesota.

For protected area and other data, see 'Using the maps'.

ACKNOWLEDGMENTS

Many thanks to Marc Ancrenaz (Hutan), Melvin Gumal (Wildlife Conservation Society), Geoffrey Davison (National Parks Board, Singapore), and Lee Shan Khee (UNEP-WCMC) for their valuable comments on the draft of this section.

AUTHORS

Kim McConkey, UNEP World Conservation Monitoring Centre
Julian Caldecott, UNEP World Conservation Monitoring Centre
Edmund McManus, UNEP World Conservation Monitoring Centre

Afterword

Russell A. Mittermeier

Great apes, in many ways, are among the most interesting creatures on our planet. First of all, they are our closest living relatives, with chimpanzees differing from us by a mere 1.24 percent of their genetic makeup. Given this close relationship, studies of wild populations of great apes provide a window into our evolutionary history, as well as breakthroughs into the understanding of how our minds work, how we learn, and how we live together in societies. What is more, great apes may provide clues for cures to many human illnesses, especially given that they often share the same diseases with humans without necessarily showing the same symptoms. In addition, great apes play a key ecological role in the tropical forest systems in which they live, acting as major seed dispersers in many ecosystems and often even modifying the architecture of the forest with their nest-building and feeding activities.

Great apes are also without a doubt among the most charismatic and best known animals in the world. They have played important roles in many different parts of the world, both in the countries in which they live and many others that have learned about them through books and films – both fanciful and scientific – and, increasingly through tourist visits to the often remote areas in which they still occur. They have been particularly important to the African and Southeast Asian cultures that live with them, with many stories and legends that compare them to humans and encourage their protection by placing taboos on killing or eating them, or sometimes regard them with great fear and awe. Their resemblance to humans is striking to all who observe them. They use tools, they show empathy, and some, like the chimpanzees, even hunt in groups, in ways that must be very similar to what we ourselves did at an early stage in our history.

Great apes are also a more diverse group than is generally realized. The public usually thinks of three great apes (chimpanzees, gorillas, orangutans), or maybe four (the bonobo) if they are paying attention. But recent studies have indicated that there are at least six species in three genera, and no less than 13 taxa. These include two species and four to five taxa of gorillas, four taxa of chimpanzees, two species and at least four different kinds of orangutans, and the bonobo. As we delve further into remote parts of the forest and learn more about these animals through genetic studies, it is likely that this number will increase even more.

Unfortunately, as we learn more about them, we are also seeing their rapid decline almost everywhere that they occur. According to the Global Mammal Assessment, an international effort to assess the conservation status of all mammal species carried out under the auspices of the Species Survival Commission of IUCN–The World Conservation Union, at least three great ape taxa are Critically Endangered and close to extinction. These are the Cross River gorilla from the Nigeria–Cameroon border, the eastern lowland gorilla from the eastern Democratic Republic of the Congo, and the Sumatran orangutan from northern Sumatra: all three recently featured in a list of the Top 25 Most Endangered Primates on Earth. The *Red List* assessments for all the great ape species and subspecies are still underway, but it is virtually certain that all will continue to be classified as Endangered or Critically Endangered for the foreseeable future.

In other words, our closest living relatives are in big trouble. As is so clearly indicated in this atlas, their habitats have been destroyed or dramatically modified in many of the places in which they occur, and several species hang on in the tiniest of fragments. Those that still have a reasonable amount of habitat available – as in the Congo forest of Central Africa – are under heavy pressure from logging and the associated bushmeat trade (not subsistence hunting but a commercial trade to serve a luxury market). What is more, bushmeat consumption of this kind is not just bad for great apes, it is clearly an enormous human health hazard as well, with strong linkages having been established between consumption of great ape meat and outbreaks of the deadly Ebola virus in human populations.

What can we do? Well, some of the solutions are outlined in this atlas. First and foremost, we need more protected areas for great apes, more parks and reserves that are well managed and protect as many of the remaining populations as possible. To do this, we also need to demonstrate that these protected areas and the great apes and other creatures living in them provide benefits to local populations who share the broader environment. That said, recent Ebola outbreaks in West Africa have demonstrated that protected areas alone are not enough. We need to keep close track of these populations, and monitor their health and viability on an ongoing basis. And we need to do everything possible to put an end to the highly destructive bushmeat trade, not only for the great apes themselves but for those human populations that suffer so much from the transfer of diseases like Ebola through consumption of great ape meat.

To be sure, there are many challenges on the horizon and much that needs to be done, and finding long-lasting solutions will not be easy. Nonetheless, we who have been working with primates for so many years are optimistic that solutions can be found and that we should be able to maintain viable populations of all great apes, and indeed all nonhuman primates, in their natural habitats. Indeed, the rapid increase in the number of people interested in great apes and the increasingly strong organizations working on them bode well for the future.

This atlas, by bringing together such an enormous body of data on the great apes and presenting it in such an attractive and useable format, makes an enormous contribution to conservation efforts on behalf of the great apes. I would like to offer my congratulations to those who have worked so hard over the past few years to bring it to fruition.

Russell A. Mittermeier
Chair, IUCN/SSC Primate Specialist Group
President, Conservation International

ANNEX

Great Apes Survival Project GRASP and Partners

The Great Apes Survival Project (GRASP) is an innovative and ambitious project of the United Nations Environment Programme (UNEP) and the United Nations Educational, Scientific and Cultural Organization (UNESCO) with an immediate challenge – to lift the threat of imminent extinction faced by gorillas, chimpanzees, bonobos, and orangutans. GRASP's mission is "to halt the decline in great ape populations by ensuring that all those who have something to contribute have the opportunity to do so."

Despite the dedicated efforts of many individuals and organizations, the great apes are on the very edge of extinction. In response to the current crisis, Klaus Toepfer, the Executive Director of UNEP, launched GRASP – a new approach to save the great apes and their habitat. Through high-level technical visits, field projects, and National Great Ape Survival Plan (NGASP) policy-making workshops in African and Southeast Asian great ape range states, as well as political lobbying and awareness raising in donor countries, GRASP has made a strong case for the value it adds to great ape conservation efforts.

GRASP, a World Summit on Sustainable Development Type II Partnership, is a dynamic alliance bringing together UN agencies, governments, nongovernmental organizations (NGOs), foundations, and private-sector interests. By using close links with governments through the UN, GRASP can promote its message at the highest political levels. As such, it is uniquely placed to inform policy makers, to mobilize and pool resources for effective action, to ensure maximum efficiency, and to provide a communication platform in order to bring the decline of great ape populations to a halt.

Since its inception, GRASP activities have helped define the strategies GRASP might adopt to address this crisis, given its unique position as a truly international alliance among a diversity of stakeholders. GRASP Patrons have provided their world-renowned expertise and reputations to bring further attention to the plight of the great apes. Technical missions to the range states have catalyzed government action to respond to the crisis. NGASP workshops have helped great ape range countries develop conservation strategies. GRASP funding of NGO partner projects has involved local communities and brought about immediate successes in the field. Intergovernmental conferences, meetings with key GRASP Partners, and other forms of policy implementation have consolidated the GRASP Partnership and linked it to relevant biodiversity mechanisms and multilateral agreements. Information and awareness activities through such media as television and newspaper articles, publications, documentary films, and side events have raised the profile of the GRASP Partnership at the global level.

GRASP aims:

- to lift the threat of immediate extinction;
- to raise funds for great ape conservation;
- to develop a global strategy to coordinate efforts to halt the decline of great ape populations and ensure the long-term survival of their natural habitat;

- to educate local people and encourage sustainable community activities;
- to provide alternative income opportunities to hunting, logging, and mining, such as sustainable agriculture, ecotourism, etc.;
- to improve the infrastructure of protected areas;
- to improve the capacity of government wildlife agencies;
- to exemplify the added value of a UN-facilitated global partnership, with range state governments and NGOs assuming increasing control over the process.

The endangered great apes share their habitat with many millions of people in West, Central, and East Africa and in Southeast Asia. The majority of these people live below the poverty line. The need to link the welfare of humans and wildlife is a central objective of the GRASP Partnership.

Further progress will depend critically on raising new and additional resources – from country donors, from foundations and the private sector, and from existing mechanisms and national grant schemes. To ensure the long-term conservation of viable populations of the wild great apes and their habitat, the international community in the widest sense has to provide effective and coherent support to assist the efforts being made by the great ape range states.

Contact

Great Apes Survival Project Secretariat
United Nations Environment Programme
PO Box 30552
Nairobi
Kenya
Tel: (254 20) 624163/621234
Fax: (254 20) 623926
E-mail: grasp@unep.org
http://www.unep.org/grasp
http://www.unesco.org/mab/grasp

GRASP PARTNERS

African Wildlife Foundation (AWF), founded in 1961, is a conservation organization focused solely on the African continent. AWF's programs and conservation strategies are based on sound science and designed to protect both the wildlife and wild lands of Africa and ensure a more sustainable future for Africa's people. Since its inception, AWF has protected endangered species and land, promoted small business growth for African communities as a means to improve livelihoods, and trained hundreds of Africans in conservation. AWF has worked to protect mountain gorillas in Rwanda, Uganda, and the Democratic Republic of the Congo for 27 years, and is a co-founder and funder of the International Gorilla Conservation Programme described below. **http://www.awf.org**

Ape Alliance (ApAl) provides a forum for the discussion of issues relating to apes. An international coalition of organizations and individuals working for the conservation and welfare of apes, it undertakes collaborative action to tackle the problems they face, both in the wild and in captivity, through specialist working groups. It brings together about 70 organizations and hundreds of individuals, all working for apes. The website links to all member organizations and includes an interactive noticeboard for anyone wishing to do something for apes. **http://www.4apes.com**

Australian Orangutan Project (AOP) works to ensure the survival of both Sumatran and Bornean orangutan species in their natural habitat and promote the welfare of all orangutans. AOP raises awareness of the need to preserve orangutan populations in their natural habitat and the intrinsic value of individual orangutans, and raises funds to assist *in situ* orangutan projects in their conservation and welfare work. AOP supports many orangutan conservation organizations, as well as its own projects. AOP is a nonprofit organization staffed by volunteers. No salaries are paid to AOP volunteers and most services are donated. Therefore a very high percentage of donations go straight to the active welfare of orangutans and on habitat protection. **http://www.orangutan.org.au**

Balikpapan Orangutan Society-USA (BOS-USA) is dedicated to the conservation of orangutans and their habitats in Malaysia and Indonesia on the islands of Sumatra and Borneo. BOS-USA also raises awareness on the orangutans' plight and funds conservation projects. **http://www.orangutan.com/**

Berggorilla & Regenwald Direkthilfe (BRD) focuses on the eastern gorillas by supporting projects contributing to their conservation, for example by providing necessary equipment to rangers and park managers. BRD also supports projects for the conservation of certain populations of western gorillas that are particularly at risk. In addition, BRD supports public awareness activities, population censuses, and ecological studies. **http://www.berggorilla.de/**

Bonobo Conservation Initiative (BCI) has a mission to promote conservation of the bonobo and its tropical forest habitat in the Democratic Republic of the Congo. BCI uses a multi-sectoral approach emphasizing stakeholder involvement and addressing the needs of local populations, as well as building capacity of Congolese institutions and NGOs. BCI has established over a million acres under accords for community-based reserves, is working to convert logging concessions for conservation, and is developing sustainable development projects in bonobo habitat. It is conducting bonobo and ecological surveys and information exchange programs in critical bonobo habitat, motivating local involvement and leadership in conservation, and promoting bonobo awareness. BCI is also dedicated to supporting bonobo research and raising international awareness about this important species of great ape and its habitat. **http://www.bonobo.org/**

Born Free Foundation (BFF) campaigns for the protection and conservation of animals in their natural habitat and against the keeping of animals in zoos and circuses and as exotic pets. BFF works with sanctuaries for orphaned great apes in the Democratic Republic of the Congo, Cameroon, and Uganda, and helped to establish the Pan African Sanctuaries Alliance. It supports the rangers who patrol the Kahuzi-Biega National Park in the Democratic Republic of the Congo and monitors the habituated eastern lowland gorillas. It has provided the core of the GRASP Technical Support Team since 2001. **http://www.bornfree.org.uk**

Bristol Zoo Gardens (BZG) supports conservation projects in the UK and overseas in partnership with governmental and nongovernmental organizations. It has been working in Cameroon with the Ministry of the Environment and Forests (MINEF) and the Living Earth Foundation on community engagement and support in relation to apes living in and around the Dja Biosphere Reserve, and with MINEF and the Cameroon Wildlife Aid Fund to care for the orphans of the bushmeat trade at sanctuaries at the Mvog-betsi Zoo and the Mefou National Park, and to provide conservation education programs to schools and visitors to the sanctuaries. **http://www.bristolzoo.org.uk**

Budongo Forest Project (BFP) undertakes research on and conservation of chimpanzees and other wildlife in Budongo Forest Reserve, Uganda, as well as undertaking background studies of the forest itself and the surrounding population. It works closely with Makerere University Faculty of Forestry and Nature Conservation, the Royal Zoological Society of Scotland, and St Andrews University. **http://www.budongo.org**

Bushmeat Crisis Task Force (BCTF) is a consortium of organizations and individuals dedicated to the conservation of wildlife threatened by unsustainable exploitation for meat, in Africa and around the world. We help members identify and implement effective and appropriate solutions to this 'bushmeat crisis', by managing scientific information to support education and training, engagement with key decision makers in government and private industry, and raising public awareness. **http://www.bushmeat.org**

Care for the Wild International (CFTWI) promotes the conservation and welfare of wild animals in need throughout the world. It runs an adoption program, which supports orphaned orangutans in a rescue center in Central Kalimantan, Indonesia, and funds habitat protection and antipoaching initiatives to safeguard orangutans in the wild.
http://www.careforthewild.com

Conservation International (CI) has as its mission to conserve the Earth's living natural heritage and global biodiversity, and to demonstrate that human societies are able to live harmoniously with nature. CI's work, through its regional programs and funding support to partners, contributes to the protection of great apes and other species found in ape habitat. **http://www.conservation.org**

Dian Fossey Gorilla Fund Europe (DFGFE) works to save gorillas from extinction and ensure local people genuinely benefit from their natural heritage. Based in the UK, the fund currently manages more than 20 projects designed to integrate traditional conservation and research with economic development and education. **http://www.dianfossey.org**

The Dian Fossey Gorilla Fund

Dian Fossey Gorilla Fund International (DFGFI) carries out extensive gorilla protection and science programs in Rwanda (Karisoke Research Center) and in the Democratic Republic of the Congo (Kabara Research Center, Tayna Center for Conservation Biology). It is the lead partner in a 38 700 km^2 landscape in eastern Congo, protecting eastern lowland gorillas by supporting community-based reserves and national parks. The DFGFI supports 400 field staff, antipoaching patrols, and education, community, health, and economic development programs. **http://www.gorillafund.org**

Discovery Initiatives (DI) promotes ecotourism, working in cooperation with the conservation community with the intention of enhancing the work of those involved in local conservation projects by linking into the demand and monies from nature travel. It arranges quality escorted small group or tailor-made tourist travel itineraries to great ape habitat in Rwanda, Uganda, Congo, Gabon, Cameroon, Central African Republic, Malyasia (Borneo), and Indonesia. **http://www.discoveryinitiatives.com**

EARTHWATCH INSTITUTE

Earthwatch Institute engages people worldwide in scientific field research and education to promote the understanding and action necessary for a sustainable environment. It has supported several years of fieldwork on great apes. **http://www.earthwatch.org**

Fauna and Flora International (FFI) acts to conserve threatened species and ecosystems worldwide. Together with AWF and WWF, FFI is responsible for the International Gorilla Conservation Programme (see below), which is working towards the conservation of mountain gorillas. **http://www.fauna-flora.org**

Filmmakers for Conservation (FFC) promotes global conservation through the making, broadcasting, and distribution of films, and helps conservation organizations and filmmakers worldwide to make more, better-informed, and effective conservation films. It aims to educate, motivate, and inspire new audiences to actively participate and support conservation. **http://www.filmmakersforconservation.org**

THE GREAT APE WORLD HERITAGE SPECIES PROJECT

Great Ape World Heritage Species Project (GAWHSP) works to secure the passage of an international declaration and convention designating the four great apes as world heritage species. It has been established in recognition of the outstanding universal value of each of the great apes, and in response to the unprecedented and imminent threats to their survival. **http://www.4greatapes.com**

Institute for Tropical Forest Conservation (ITFC) is the lead organization undertaking conservation-oriented research and training in the Albertine Rift montane forests of southwest Uganda, in particular Bwindi Impenetrable National Park, Mgahinga Gorilla National Park, and Echuya Forest Reserve. The institute was established in 1991 as a semi-autonomous branch of Mbarara University of Science and Technology. **http://www.must.ac.ug/faculties/tropical_forest.htm**

IFAW.org

International Fund for Animal Welfare (IFAW) works to improve the welfare of wild and domestic animals throughout the world, prevent cruelty to animals, and promote animal welfare and conservation policies that advance the wellbeing of both animals and people. IFAW and other partner organizations are working to find practical solutions to the bushmeat crisis. **http://www.ifaw.org**

International Gorilla Conservation Programme (IGCP) acts to ensure the conservation of mountain gorillas and their regional afromontane forest habitat in Rwanda, Uganda, and the Democratic Republic of the Congo, with activities including research and transboundary collaboration. IGCP is a coalition of AWF, FFI, and WWF, working in partnership with the protected area authorities and other local stakeholders in the region. **http://www.mountaingorillas.org**

International Ranger Federation (IRF). The IRF is an international network of national, state, and community ranger associations dedicated to raising the professional standards of rangers worldwide and providing a voice for those working at the 'grass roots' of conservation and protected areas. Members include rangers from a number of great ape range states. **http://www.int-ranger.net**

Jane Goodall Institute (JGI) has as its mission to inspire and empower people to take informed, compassionate action to make the world a better place for people, animals, and the environment. JGI does this through world class primate care and research; community centered conservation; and environmental and humanitarian education. **http://www.janegoodall.org**

Kinabatangan Orangutan Conservation Project (KOCP) was set up by the French NGO Hutan in 1998 in collaboration with the Sabah Wildlife Department. It is located in the village of Sukau, close to the Kinabatangan Wildlife Sanctuary in Sabah, Malaysia (Borneo). The primary goal of this project is to achieve the long-term viability of orangutan populations in Sabah. The project now comprises a highly motivated team of 35 trained staff from the Kinabatangan community. **http://www.boh.com.my/pl/pubdoc/43191**

KOCP

Living Earth Foundation is an international nongovernmental organization that encourages people to learn and work together to resolve the environmental issues which concern them. It carries out environmental education and capacity development in partnership with corporations, communities, and governments. It is involved in the conservation of great apes and their habitat through its Africa Programmes. **http://www.livingearth.org.uk**

Orangutan Foundation (OF) conserves the orangutan and its rain-forest habitat while conducting long-term research on the ecology of orangutans and other rain-forest fauna and flora within their habitat. OF objectives are to support conservation work in Indonesia and Malaysia and to raise funds and awareness in the UK and worldwide. In Indonesia, OF actively supports the protection of Tanjung Puting National Park, and other protected and nonprotected areas of critical orangutan habitat. It also operates a rehabilitation program that returns orangutans to a life in the wild. **http://www.orangutan.org.uk**

Pan African Sanctuary Alliance (PASA) is an alliance of 18 primate sanctuaries from all over Africa, designed to bring the sanctuaries together for long-term planning, and to improve collaboration between sanctuaries and primate experts. PASA also organizes workshops on topics such as the veterinary care of orphaned primates. **http://www.panafricanprimates.org**

PanEco Foundation for Sustainable Development and Intercultural Exchange & the Sumatran Orangutan Conservation Programme (PanEco-SOCP) works on all aspects of the conservation of Sumatran orangutans. In 1999, the PanEco Foundation and the Indonesian government Department of Forest Protection and Nature Conservation (PHKA) formally established the Sumatran Orangutan Conservation Programme (SOCP). SOCP also involves the Frankfurt Zoological Society and the Indonesian Foundation for a Sustainable Ecosystem (YEL). Based in Sumatra, activities include the confiscation of illegal captive orangutans, their reintroduction to form a new population in the wild, research into orangutan behavior and ecology, surveys and monitoring, public education, and habitat protection. PanEco itself is also engaged in a diverse array of other projects focusing on sustainable development and raising environmental awareness in both Switzerland and Indonesia. **http://www.paneco.ch; http://www.sumatranorangutan.org**

Tayna Centre for Conservation Biology (TCCB) aims to protect gorillas and chimpanzees in the Tayna Gorilla Reserve. It was formed under the umbrella of the Union of Associations for Gorilla Conservation and Community Development in East Democratic Republic of Congo (UGADEC), and receives support from DFGFI. **http://www.gorillafund.org/002_site_ind_frmset.html**

Tusk works to support wildlife and habitat conservation and promotes sustainable rural community development across Africa. Tusk supports the Chimfunshi Chimpanzee Sanctuary in Zambia, as well as an orphaned chimpanzee release program in Conkouati National Park, Congo. Tusk also funds different conservation and development projects in and around Virunga National Park and Walikale in the Democratic Republic of the Congo as a key to conserving both mountain and eastern lowland gorillas and their habitat. **http://www.tusk.org**

UNEP World Conservation Monitoring Centre (UNEP-WCMC) provides objective, scientifically rigorous products and services that include ecosystem assessments, support for implementation of environmental agreements, regional and global biodiversity information, research on threats and impacts, and development of future scenarios for the living world. UNEP-WCMC has produced this atlas in support of GRASP. **http://www.unep-wcmc.org**

Volcanoes Safaris runs gorilla safaris in the gorilla habitats of Mgahinga and Bwindi National Parks in Uganda and Volcanoes National Park in Rwanda. It works closely with local communities and conservation organizations to help the development of tourism, enhance private sector capacity, and help communities around Volcanoes National Park and Nyungwe in Rwanda. **http://www.volcanoessafaris.com**

Wild Chimpanzee Foundation (WCF) works to preserve the remaining wild chimpanzee populations and their natural habitat throughout their range in Africa. WCF is initially concentrating its efforts in different West African countries while also starting some activities in Central Africa. **http://www.wildchimps.org**

Wildlife Conservation Society (WCS) works to save wildlife and wild lands through careful science, international conservation, education, and the management of the world's largest system of urban wildlife parks. WCS has been working to protect all four subspecies of gorilla as well as working to protect chimpanzees and orangutans in their native habitats. **http://www.wcs.org**

WWF–The Global Conservation Organization acts to conserve the natural environment and ecological processes worldwide. WWF is involved in the conservation of the great apes in part through its support to the International Gorilla Conservation Programme and through its African Great Apes Programme. **http://www.panda.org**

ZSL
LIVING CONSERVATION

Zoological Society of London (ZSL) aims to achieve and promote the worldwide conservation of animals and their habitats. ZSL's Bushmeat and Forests Conservation Programme is centered on equatorial Africa, and focuses on bushmeat research in West Africa, and national park development in Gabon and the Democratic Republic of the Congo. **http://www.zsl.org**

ENVIRONMENTAL CONVENTIONS

Convention on Biological Diversity (CBD) aims at conservation and sustainable use of biological diversity, and the fair and equitable sharing of benefits arising from the use of genetic resources. The CBD's program of work on protected areas, which seeks to establish effective networks of protected areas, and its expanded program of work on forest biological diversity, which promotes sustainable harvesting of timber and non-timber forest resources and forest law enforcement, are of particular relevance to the conservation of great apes. All individual ape range states are Parties to the CBD. **http://www.biodiv.org**

Convention on International Trade in Endangered Species of Wild Fauna and Flora (CITES) aims to ensure that no species of wild fauna or flora becomes or remains subject to unsustainable exploitation because of international trade. All great apes are listed in Appendix I of CITES. Trade in specimens of these species is permitted only in exceptional circumstances, and for non-commercial purposes. **http://www.cites.org**

Convention on Migratory Species (CMS) aims to conserve terrestrial, aquatic, and avian migratory species throughout their range. CMS is interested in the conservation of the mountain gorilla, which crosses the mountainous border areas between Uganda, Rwanda, and the Democratic Republic of the Congo. **http://www.cms.int**

World Heritage Convention (WHC) is based on the premise that certain places on Earth are of outstanding universal value and should therefore form part of the common heritage of humankind. Countries which have ratified this agreement are committed to identifying and safeguarding the world's most outstanding natural and cultural heritage. Countries nominate sites in their territories which they believe are of World Heritage quality, and these are inscribed on the World Heritage List if they meet the criteria. A number of these sites are critical for the survival of great apes. **http://www.unesco.org/**

DONOR GOVERNMENTS

These are governments that offer funding for the conservation of great apes.

GREAT APE RANGE STATES

The range states consist of 21 states in Africa and two in Southeast Asia:

Angola
Burundi
Cameroon
Central African Republic
Congo
Côte d'Ivoire
Democratic Republic of the Congo
Equatorial Guinea
Gabon
Ghana
Guinea
Guinea-Bissau
Indonesia
Liberia
Mali
Malaysia
Nigeria
Rwanda
Senegal
Sierra Leone
Sudan
Uganda
United Republic of Tanzania

PRIVATE SECTOR

GRASP acknowledges the importance of the private sector and seeks to encourage private sector investment and involvement in the conservation of great apes.

INDEX

Note: Page references in **bold** refer to illustrations and maps; those in *italic* refer to boxes and tables

D

G

H

I

M

N

Q

R

U

V

W

X

Y

Z